ACS SYMPOSIUM SERIES **502**

Isotope Effects in Gas-Phase Chemistry

Jack A. Kaye, EDITOR
National Aeronautics and Space Administration

Developed from a symposium sponsored
by the Division of Physical Chemistry
at the 201st National Meeting
of the American Chemical Society,
Atlanta, Georgia,
April 14–19, 1991

American Chemical Society, Washington, DC 1992

Library of Congress Cataloging-in-Publication Data

Isotope effects in gas-phase chemistry / Jack A. Kaye, editor.

p. cm.—(ACS symposium series, ISSN 0097–6156; 502)

"Developed from a symposium sponsored by the Division of Physical Chemistry at the 201st National Meeting of the American Chemical Society, Atlanta, Georgia, April 14–19, 1991."

Includes bibliographical references and indexes.

ISBN 0–8412–2471–4

1. Chemical reaction, Conditions and laws of—Congresses. 2. Isotopes—Congresses.

I. Kaye, Jack A., 1954– . II. American Chemical Society. Division of Physical Chemistry. III. Title: Gas-phase chemistry. IV. Series.

QD501.I813 1992
541.3'88—dc20   92–23134
                                                                    CIP

The paper used in this publication meets the minimum requirements of American National Standard for Information Sciences—Permanence of Paper for Printed Library Materials, ANSI Z39.48–1984. ∞

Copyright © 1992

American Chemical Society

All Rights Reserved. The appearance of the code at the bottom of the first page of each chapter in this volume indicates the copyright owner's consent that reprographic copies of the chapter may be made for personal or internal use or for the personal or internal use of specific clients. This consent is given on the condition, however, that the copier pay the stated per-copy fee through the Copyright Clearance Center, Inc., 27 Congress Street, Salem, MA 01970, for copying beyond that permitted by Sections 107 or 108 of the U.S. Copyright Law. This consent does not extend to copying or transmission by any means—graphic or electronic—for any other purpose, such as for general distribution, for advertising or promotional purposes, for creating a new collective work, for resale, or for information storage and retrieval systems. The copying fee for each chapter is indicated in the code at the bottom of the first page of the chapter.

The citation of trade names and/or names of manufacturers in this publication is not to be construed as an endorsement or as approval by ACS of the commercial products or services referenced herein; nor should the mere reference herein to any drawing, specification, chemical process, or other data be regarded as a license or as a conveyance of any right or permission to the holder, reader, or any other person or corporation, to manufacture, reproduce, use, or sell any patented invention or copyrighted work that may in any way be related thereto. Registered names, trademarks, etc., used in this publication, even without specific indication thereof, are not to be considered unprotected by law.

PRINTED IN THE UNITED STATES OF AMERICA

1992 Advisory Board

ACS Symposium Series
M. Joan Comstock, *Series Editor*

V. Dean Adams
Tennessee Technological
  University

Mark Arnold
University of Iowa

David Baker
University of Tennessee

Alexis T. Bell
University of California—Berkeley

Arindam Bose
Pfizer Central Research

Robert F. Brady, Jr.
Naval Research Laboratory

Margaret A. Cavanaugh
National Science Foundation

Dennis W. Hess
Lehigh University

Hiroshi Ito
IBM Almaden Research Center

Madeleine M. Joullie
University of Pennsylvania

Mary A. Kaiser
E. I. du Pont de Nemours and
  Company

Gretchen S. Kohl
Dow-Corning Corporation

Bonnie Lawlor
Institute for Scientific Information

John L. Massingill
Dow Chemical Company

Robert McGorrin
Kraft General Foods

Julius J. Menn
Plant Sciences Institute,
  U.S. Department of Agriculture

Vincent Pecoraro
University of Michigan

Marshall Phillips
Delmont Laboratories

A. Truman Schwartz
Macalaster College

John R. Shapley
University of Illinois
  at Urbana–Champaign

Stephen A. Szabo
Conoco Inc.

Robert A. Weiss
University of Connecticut

Peter Willett
University of Sheffield (England)

Foreword

THE ACS SYMPOSIUM SERIES was first published in 1974 to provide a mechanism for publishing symposia quickly in book form. The purpose of this series is to publish comprehensive books developed from symposia, which are usually "snapshots in time" of the current research being done on a topic, plus some review material on the topic. For this reason, it is necessary that the papers be published as quickly as possible.

Before a symposium-based book is put under contract, the proposed table of contents is reviewed for appropriateness to the topic and for comprehensiveness of the collection. Some papers are excluded at this point, and others are added to round out the scope of the volume. In addition, a draft of each paper is peer-reviewed prior to final acceptance or rejection. This anonymous review process is supervised by the organizer(s) of the symposium, who become the editor(s) of the book. The authors then revise their papers according the the recommendations of both the reviewers and the editors, prepare camera-ready copy, and submit the final papers to the editors, who check that all necessary revisions have been made.

As a rule, only original research papers and original review papers are included in the volumes. Verbatim reproductions of previously published papers are not accepted.

M. Joan Comstock
Series Editor

Contents

Preface ix

1. Isotope Effects in Gas-Phase Chemical Reactions and
   Photodissociation Processes: Overview.. 1
   Jack A. Kaye

   THEORETICAL STUDIES OF ISOTOPE EFFECTS
   IN CHEMICAL REACTIONS

2. Variational Transition-State Theory with Multidimensional,
   Semiclassical, Ground-State Transmission Coefficients:
   Applications to Secondary Deuterium Kinetic Isotope Effects
   in Reactions Involving Methane and Chloromethane 16
   Donald G. Truhlar, Da-hong Lu, Susan C. Tucker,
   Xin Gui Zhao, Angels Gonzalez-Lafont, Thanh N. Truong,
   David Maurice, Yi-Ping Liu, and Gillian C. Lynch

3. HN_2 and DN_2 Resonance Spectra: Scattering and Stabilization
   Calculations.. 37
   Hiroyasu Koizumi, George C. Schatz, and Joel M. Bowman

4. Isotope Effects in Addition Reactions of Importance
   in Combustion: Theoretical Studies of the Reactions
   $CH + H_2 \rightleftarrows CH_3^* \rightleftarrows CH_2 + H$.. 48
   Albert F. Wagner and Lawrence B. Harding

   EXPERIMENTAL STUDIES OF ISOTOPE EFFECTS
   IN REACTIONS OF NEUTRAL SYSTEMS

5. Hydrogen Isotope Effects in Chemical Reactions and
   Photodissociations... 66
   B. Katz and R. Bersohn

6. Isotope Effects at High Temperatures Studied by the Flash
   or Laser Photolysis–Shock Tube Technique 80
   J. V. Michael

7. **Deuterium Substitution Used as a Tool for Investigating Mechanisms of Gas-Phase Free-Radical Reactions** 94
   P. H. Wine, A. J. Hynes, and J. M. Nicovich

8. **Kinetic Isotope Effects in Gas-Phase Muonium Reactions** 111
   Susan Baer, Donald Fleming, Donald Arseneau, Masayoshi Senba, and Alicia Gonzalez

9. **Mass-Independent Isotopic Fractionations and Their Applications** 138
   M. H. Thiemens

10. **Heavy Ozone Anomaly: Evidence for a Mysterious Mechanism** 155
    S. M. Anderson, K. Mauersberger, J. Morton, and B. Schueler

11. **Isotopic Study of the Mechanism of Ozone Formation** 167
    N. Wessel Larsen, T. Pedersen, and J. Sehested

12. **Negative-Ion Formation by Rydberg Electron Transfer: Isotope-Dependent Rate Constants** 181
    Howard S. Carman, Jr., Cornelius E. Klots, and Robert N. Compton

EXPERIMENTAL STUDIES OF ISOTOPE EFFECTS
IN REACTIONS OF IONIC SYSTEMS

13. **Isotope Effects in the Reactions of Atomic Ions with H_2, D_2, and HD** 194
    Peter B. Armentrout

14. **Non-Mass-Dependent Isotope Effects in the Formation of O_4^+: Evidence for a Symmetry Restriction** 210
    K. S. Griffith and Gregory I. Gellene

15. **Temperature, Kinetic Energy, and Rotational Temperature Effects in Four Reactions Involving Isotopes** 225
    A. A. Viggiano, Robert A. Morris, Jane M. van Doren, John F. Paulson, H. H. Michels, R. H. Hobbs, and Christopher E. Dateo

16. **Isotope-Exchange Reactions Within Gas-Phase Protonated Cluster Ions** 246
    Susan T. Graul, Mark D. Brickhouse, and Robert R. Squires

ISOTOPE EFFECTS IN PHOTODISSOCIATION PROCESSES

17. Mechanisms for Isotopic Enrichment in Photochemical Reactions 264
    Moshe Shapiro

18. State-Selected Dissociation of *trans*-HONO($\tilde{A}$) and DONO($\tilde{A}$): Effect of Intramolecular Vibrational Dynamics on Fragmentation Rate, Stereochemistry, and Product Energy Distribution 279
    R. Vasudev, S. J. Wategaonkar, S. W. Novicki, and J. H. Shan

19. Isotopic Dependence of the Methyl-Radical Rydberg 3s Predissociation Dynamics 297
    S. G. Westre, P. B. Kelly, Y. P. Zhang, and L. D. Ziegler

20. Laser-Stimulated Selective Reactions and Synthesis of Isotopomers: New Strategies from Diatomic to Organometallic Molecules 310
    J. E. Combariza, C. Daniel, B. Just, E. Kades, E. Kolba, J. Manz, W. Malisch, G. K. Paramonov, and B. Warmuth

21. Multiphoton Ionization Dynamics Within $(CH_3OH)_n Cr(CO)_6$ van der Waals Clusters: Isotope Effects in Intracluster Energy Transfer 335
    William R. Peifer and James F. Garvey

APPLICATIONS OF ISOTOPE EFFECTS

22. Deuterium Fractionation in Interstellar Space 358
    E. Herbst

23. Deuterium in the Solar System 369
    Yuk L. Yung and Richard W. Dissly

24. Kinetic Isotope Effects and Their Use in Studying Atmospheric Trace Species: Case Study, $CH_4 + OH$ 390
    Stanley C. Tyler

25. Key Sulfur-Containing Compounds in the Atmosphere and Ocean: Determination by Gas Chromatography–Mass Spectrometry and Isotopically Labeled Internal Standards 409
    Alan R. Bandy, Donald C. Thornton, Robert G. Ridgeway, Jr., and Byron W. Blomquist

INDEXES

Author Index **425**

Affiliation Index **426**

Subject Index **426**

Preface

CHEMISTS DO EXPERIMENTS and make calculations of isotope effects in chemical reactions and in photodissociation processes. Other scientists are using knowledge of isotope effects to better understand complex physical and chemical systems, such as the earth's atmosphere, planetary atmospheres, and interstellar space. This book brings them together and showcases some of the recent developments in the area of isotope effects. Some of these developments have been pursued in related disciplines; results have been published in journals, such as the *Journal of Geophysical Research* and *Geophysical Research Letters*, that are not necessarily standard reading material for the more traditional chemistry community. The material in this book crosses a number of boundaries—experiment–theory, neutral–ionic systems, very small–intermediate-size systems, and basic science–applications. The only restriction is that consideration is limited to gas-phase processes.

Several interesting issues are described both in the overview chapter and in the individual chapters in this volume, but it is worth pointing out some that receive particular emphasis.

1. The observation that three-body recombination of O and O_2 in the laboratory preferentially forms isotopically heavy O_3 in a mass-independent manner that appears to be related to molecular symmetry and is undoubtedly related to the observation of enhanced heavy O_3 (mass 50) in the earth's atmosphere. The enhancement in the laboratory is surprisingly large (10–15%) for a heavy-atom isotope effect. Other large heavy-atom isotope effects have since been reported and are discussed.

2. The interest in accurate knowledge of the isotope effect in the reaction of OH with CH_4 (both for ^{13}C and D substitution) for the range of temperature appropriate to the earth's atmosphere.

3. The ability to calculate accurate primary and secondary isotope effects for simple chemical reactions by using variational transition-state theory.

4. The ability to measure directly and to calculate the branching between D- and H-atom production in the photolysis of molecules containing both.

Scientists from the United States, Canada, Israel, Denmark, and Germany contributed to this book. Among them, Susan Graul received the Division of Physical Chemistry's Nobel Laureate Signature Award for Graduate Studies in Chemistry.

Acknowledgments

The symposium on which this book is based benefited from the support of the ACS Division of Physical Chemistry and its Subdivision of Theoretical Chemistry. Joel Bowman, John Tully, and Fritz Schaefer played key roles in approving the symposium and encouraging the organizer throughout the process. Support for the participation of the non-North American speakers was provided by a Type SE grant from ACS's Petroleum Research Fund (PRF). Division secretary-treasurer Edward Eyring provided valuable assistance in obtaining the PRF grant and in handling reimbursement to the speakers. This book could not have been published without the interest and efforts of several staff members of ACS Books, notably A. Maureen Rouhi, Cheryl Shanks, Barbara Tansill, and Betsy Kulamer.

Although he was unable to submit a chapter to this book, I would like to acknowledge the contribution to the field of isotope chemistry made by Professor Aron Kuppermann of the California Institute of Technology. Besides serving as the thesis advisor for several of the contributors to this book (Bowman, Garvey, Kaye, Schatz, and Truhlar), he has made numerous important contributions to the field of chemical dynamics that have helped in better understanding isotope effects in chemical reactions.

JACK A. KAYE
National Aeronautics and Space Administration
Washington, DC 20546

May 14, 1992

Chapter 1

Isotope Effects in Gas-Phase Chemical Reactions and Photodissociation Processes
Overview

Jack A. Kaye

Earth Science and Applications Division, National Aeronautics and Space Administration Headquarters, Washington, DC 20546

The origins of isotope effects in equilibrium and non-equilibrium chemical processes are reviewed. In non-equilibrium processes, attention is given to isotope effects in simple bimolecular reactions, termolecular and complex forming bimolecular reactions, symmetry-related reactions, and photodissociation processes. Recent examples of isotope effects in these areas are reviewed. Some indication of other scientific areas for which measurements and/or calculations of isotope effects are used is also given. Examples presented focus on neutral molecule chemistry and in many cases complement examples considered in greater detail in the other chapters of this volume.

The study of isotope effects on the rates and products of chemical processes has long been useful in increasing chemists' understanding of the detailed nature of chemical reactions. For example, by substituting deuterium (D) for hydrogen (H) in a simple bimolecular abstraction reaction, information about the contribution of quantum mechanical tunneling to the total reaction rate becomes available. In reactions which may occur either by simple abstraction or by a more complicated mechanism, such as complex formation followed by dissociation, isotopic substitution may help discriminate between pathways.

While chemists have been studying isotope effects, other scientists have been using these measurements to aid in their interpretation of observations of isotopic composition of natural systems. By combining observed isotopic compositions with knowledge of the rates and isotope effects of each of the processes producing and removing a given molecule in a system, one should be able to validate one's knowledge of that substance's chemistry in the system. Alternatively, if the isotopic composition and either the isotope-specific production or loss rates are known, the other may be inferred from the available observations using a simple model for that molecule's chemistry. Such analysis has been used by scientists as diverse as geochemists, atmospheric scientists, and astronomers. Examples from several of these areas will be shown below.

A number of different types of isotope effects have applications in the understanding of detailed chemical mechanisms and of the chemical processes which can account for the observed isotopic composition of some substance.

This chapter not subject to U.S. copyright
Published 1992 American Chemical Society

These may be broken down into several categories. The first of these are the equilibrium isotope effects, in which a difference in isotopic composition in reaction reagents and products arises from the free energy difference between them. This can occur entirely in one chemical phase or can involve two phases. For example, the vapor pressure difference between normal and isotopically substituted forms of water (H_2O, HDO, $H_2^{18}O$) leads to observed differences in the isotopic distribution of liquid water on the Earth's surface and water vapor in the atmosphere (*1*). For equilibrium isotope effects to hold, the reagents and products must remain in close proximity over time scales significantly greater than those of the reactions connecting them. Equilibrium isotope effects can usually be understood using the techniques of statistical mechanics, as will be summarized below.

Non-equilibrium isotope effects arise when the reagents and products of a chemical reaction do not co-exist over time scales long relative to the time scales of the reaction. This can typically happen when one of the reaction products is itself reactive so that the reaction is essentially irreversible. In a laboratory reaction conditions may be established so that only one chemical reaction is under study. Many non-equilibrium isotope effects are studied within the framework of transition state theory (*2*). In this theory, statistical mechanical techniques are used to relate the reaction rate to the free energies of the reagents and an intermediate ("transition") state connecting reagens and products. It is then assumed that there is sufficiently rapid exchange between the reagents and the transition state that equilibrium statistical mechanics may be used to describe their interaction. Such models need to account for more complex dynamical effects, however, including quantum mechanical tunneling and the effects of long-lived metastable states which may give rise to quantum mechanical dynamical resonances. Simple and improved transition state theories have been successfully used to study isotope effects in non-complex forming bimolecular reactions.

For complex forming bimolecular reactions, more complicated theories must be used to study isotope effects. These usually are based on the Rice-Rampsberger-Kassel-Marcus (RRKM) theory, a variant of transition-state theory which relates the reaction rate to the properties of reagents and products and some critical configuration between them (*3*). This theory allows for the determination of both pressure and temperature dependence of the reaction rates and, with inclusion of appropriate intermolecular force information, the dependence of the reaction rate on the collision partner (third body). Most such studies have been carried out for perdeuteration of radical-radical recombination reactions, although attention is starting to be paid to reactions involving substitution for only one atom. Some important atmospheric reactions involve reaction of a radical and a stable molecule which can form a long-lived complex and must thus be studied with an RRKM-type theory.

One type of isotope effect which has recently become known is a symmetry-related one, which is not simply understood in terms of transition state or RRKM-type theories. This was first observed in the gas phase formation of ozone (O_3) in the recombination of atomic oxygen (O) and molecular oxygen (O_2) (*4*). In this case, the rate of formation of partially isotopically substituted ozone (for example, mass 50, formed by the combination of two ^{16}O and one ^{18}O atoms) was found to be faster than that of unsubstituted (mass 48) ozone. Similar but weaker isotope effects have been formed in the recombination reactions of O with carbon monoxide (CO) to form carbon dioxide (CO_2) and of sulfur pentafluoride (SF_5) to form disulfur decafluoride (S_2F_{10}) (*4*). The origin of these isotope effects is still not clearly understood, and there is no accurate quantitative theory which explains the observations. The enhanced formation of partially substituted ozone is apparently intimately associated with the observation of enhanced amounts of monosubstituted ozone ($^{49}O_3$, $^{50}O_3$) in the Earth's stratosphere (5).

There are also isotope effects in the photodissociation of small molecules. These can arise from several different mechanisms. First, the shift in energies of vibrational and rotational states of both the ground and electronically excited states of a molecule means that there will be a shift in the energy of transitions connecting specific ground and excited state levels on isotopic substitution. If isotopic substitution results in a loss of symmetry, certain transitions not allowed in the unsubstituted molecule may become so in the substituted form, meaning that the substituted form may have additional absorptions not found in the normal molecule. This is the case for atmospheric photolysis of $^{34}O_2$ (=$^{16}O^{18}O$), which unlike $^{32}O_2$(=$^{16}O^{16}O$) is not restricted to only odd rotational states in the ground state, and thus has twice the number of transitions in the Schumann-Runge band region (6). Symmetry considerations may also affect product state distributions in photolysis of polyatomics, such as that of ozone, in which there are more rotational states available to $^{34}O_2$ than $^{32}O_2$ in the photolysis of the asymmetric isomer of $^{50}O_3$(=$^{16}O^{16}O^{18}O$) (7,8). Finally, there are other dynamical isotope effects, in which isotopic substitution seems to affect the energy flow within the photolyzing molecule (e. g. HONO vs. DONO (9), perdeuteration of methanol ligands in methanol clusters of chromium hexacarbonyl (10)).

While much of the interest here is on the magnitude of the isotope effect on the rate of a chemical reaction, the branching among competing channels for reactions involving isotopic substitution is also important. For example, in attempting to understand the ^{18}O distribution of atmospheric CO (11), it is important to understand not only the relative rates of its reaction with hydroxyl (OH) for normal CO(=$^{12}C^{16}O$) and ^{18}O-substituted CO (12,13), but whether or not O-atom exchange can occur in otherwise non-reactive collisions (14).

The recent increase in attention paid to isotope effects, especially non-equilibrium ones, has arisen from several factors. First, improvements in both experimental and computational techniques have meant that many questions about the effects of isotopic substitution on the rates and products of chemical reactions are now answerable for the first time. Second, increased observations of the isotopic composition of constituents in natural systems has meant that there is now increased need for accurate knowledge of these isotope effects. This is especially true for information on heavy-atom isotope effects (^{12}C-^{13}C, ^{14}N-^{15}N, ^{16}O-^{18}O, ^{32}S-^{34}S). Third, the identification of large (several to greater than 10 percent) symmetry-related isotope effects has opened up a new area of research for chemical scientists.

In the remainder of this overview, these different isotope effects in gas phase processes are surveyed, with applications, primarily in the area of atmospheric chemistry, to be addressed later in this volume being highlighted. Much of the basic material on this subject has been discussed in a previous review article (15), and is only briefly outlined here. Most of the examples cited here will be recent data not available during the preparation of the previous article. Examples are given for neutral systems; isotope effects in ion-molecule collisions are explored in three chapters of this book (16-18), and their applications to chemistry in interstellar space are explored in another (19).

Equilibrium Isotope Effects

For chemical reactions involving exchange of isotopes between molecular species, the equilibrium isotope effect is given by the reaction's equilibrium constant. Using the schematic representation

$$AX + BY \rightarrow AY + BX \tag{1}$$

one has in the usual notation (for pressure or concentration)

$$K_{eq} = [AY][BX]/([AX][BY]) \tag{2}$$

and in statistical mechanical notation

$$K_{eq} = [Q_{AY}Q_{BX}/(Q_{AX}Q_{BY})]\exp(-\Delta V/k_BT) \tag{3}$$

where Q_i is the partition coefficient of species i with the zero of energy at its classical equilibrium potential and ΔV being the change in classical equilibrium potential energy in the reaction, which has the form

$$\Delta V = V_{e,AY} + V_{e,BX} - V_{e,AX} - V_{e,BY} \tag{3a}$$

The calculation of the partition functions based on molecular information (bond lengths, bond angles, vibrational frequencies, etc.) is straightforward using standard techniques. The major quantum mechanical contribution to the equilibrium constant comes from the vibrational component of the partition functions, which under the harmonic approximation has a form $\exp(-h\Sigma v_{ij}/2k_BT)$ for each species (where k_B is used for Boltzmann's constant), where v_{ij} is the ith vibrational frequency of molecule j and the sum runs over all vibrational frequencies in molecule j. This term corresponds to the zero-point energy in each vibration. The overall vibrational contribution to the equilibrium constant will thus be of the form

$$K_{eq}(vib) = \exp\{-hc[\Sigma(v_{i,AY} - v_{i,BX}) - \Sigma(v_{i',AX} - v_{i',BY})]/2k_BT\} \tag{4}$$

where the summation is over the vibrational frequencies n_{ij} or $n_{i'j}$ for modes i or i' of species j. The indices i and i' are shown as being different in order to emphasize that the number of vibrational degrees of freedom in molecules AY and AX need not be the same. From this equation one can see that the position of the equilibrium will depend on the relative vibrational frequencies of the reagent and product molecules. Further, the equilibrium constant will tend to be more non-statistical at low temperatures.

Applications of equilibrium statistical isotope effects have been made in various types of applied problems. For example in understanding the D/H distribution in methane-hydrogen mixtures, one must consider the isotope fractionation between the two molecules $CH_4 + HD \rightarrow CH_3D + H_2$ (5)

Accurate calculations of the equilibrium constant for this reaction show that large deviations from the statistically expected value are found at low temperatures, but that at higher temperatures the deviation becomes quite small (20). The fractionation expected for this system in the deep atmospheres of the outer planets has been extensively studied. The equilibrium constant for this reaction, commonly known as the fractionation factor, is important if one is to infer the relative abundance of H and D in the planet's atmosphere from observation of CH_3D and CH_4 rather than the far more abundant HD and H_2. Calculations based on thermochemical equilibrium assumptions suggest a ratio of approximately 1.37 at the high temperatures of the deep Jovian and Saturnian tropospheres where isotope exchange between CH_4 and H_2 may be sufficiently rapid that equilibrium holds (*21*).

Equilibrium isotope effects are also expected in phase change processes, such as the evaporation of water, and their magnitudes are related to the free energy difference between the liquid and gas phases (*22*). This is very much a function of the zero-point energy difference associated with isotopic substitution, which explains why the vapor pressure isotope effect for HDO is some 8 times that of $H_2^{18}O$, even though the mass difference from unsubstituted water is larger in the latter. The effects of the differences are observed in both tropospheric (*23*) and stratospheric (*24,25*) water vapor as well as in precipitation (*26*).

Non-Equilibrium Isotope Effects

Simple Bimolecular Reactions. The basic transition state theory approach for the rate of non-complex forming bimolecular chemical reactions relates the reaction rate (k_{rate}) to the partition function of the reactants (A,B) and transition state (AB*) of a reaction

$$A + B \rightarrow AB^* \rightarrow Products \qquad (6)$$

by the expression

$$k_{rate} = \kappa(k_BT/h)[Q^{\neq}_{AB}/(Q_A\Phi_B)]\exp(-\Delta V^{\neq}/k_BT) \qquad (7)$$

where $Q^{\neq}_{AB}$ and Q_A are partition functions for the transition state and species A, respectively (with zeros of energy being their classical equilibrium energies), Φ_B is the partition function per unit volume of species B with its zero of energy at its classical equilibrium energy, and $\Delta V^{\neq}$ is the change in classical equilibrium energy in proceeding from the reactants to the transition state (*27,28*). In eq. (7), κ is a dynamical correction factor needed to account for the neglect of tunneling and the assumption that all reactions which "pass over the transition state" lead to chemical reaction (*2*).

Isotopic substitution affects many of the terms in eq. (7). First, it can have an effect on the $\Delta V^{\neq}$ term, especially since the effect of isotopic substitution is usually less (per vibrational degree of freedom) in the transition state than it is in the reactant molecule. There will also be an effect on the partition coefficients (translational, rotational, and non-zero-point vibrational parts). Finally, the dynamical correction factor κ will also be isotope-dependent. This will be particularly true for light atom substitution (e. g. D, T, μ, for H) when it is the light atom which is being transferred between groups in a reaction of the type

$$X + HY \rightarrow HX + Y \qquad (8)$$

In that case, there can be appreciable differences in the tunneling probabilities for H, D, T, and μ. There can also be changes in the kinematics of the reaction due to the mass change; for example in the H-atom exchange reaction

$$H + H_2 \rightarrow H_2 + H \tag{9}$$

substitution of μ or T for the central H makes the reaction almost a prototypical heavy-light-heavy and light-heavy-light reaction. This changes the dependence of the reaction probability with translational energy (which is the fundamental quantity behind κ); if there are low-energy dynamical resonances in a reaction, isotopic substitution can have a major affect on their structure as well (29).

Most of the studies of isotope effects in bimolecular reactions have focused on D-H substitution. Two recent examples of these relevant to atmospheric chemistry include the reactions of H_2 with O and OH:

$$O + H_2 \rightarrow OH + H \tag{10}$$
$$OH + H_2 \rightarrow H_2O + H \tag{11}$$

The former is nearly thermoneutral and has a large (approximately 10 kcal/mole) barrier to reaction, while the latter is strongly exothermic and has a much smaller barrier to reaction. Isotope effects for reaction (10) have been measured by Gordon and co-workers (30-32), while that of reaction (11) has been measured by room temperature by Ehhalt et al. (33); reaction rates of OH with D_2 have been measured by various groups in the past as well (34).

In both cases, reaction with H_2 is faster than HD; the fractionation factor (ratio of rates with H_2 to HD) is approximately 1.5 for reaction (10) (30) and 1.65 for reaction (11) at the measured temperatures. Intramolecular kinetic isotope effects were also measured for reaction (10), showing that reaction to form OH is favored over that forming OD by an amount which increases from 2.4±0.3 at 500K to 16.0±2.2 at 339K (31). At high (400-500K) temperatures, there is good agreement between the observed branching ratios and those calculated using theoretical techniques including tunneling (35,36), but at low temperatures (339K), the observed HD/DH branching ratio exceeds the calculated one by a factor of two.

Two other simple reactions for which isotope effects have been measured are

$$H + HBr \rightarrow Br + H_2 \tag{12}$$
$$H + Br_2 \rightarrow HBr + Br \tag{13}$$

For the former, the room temperature isotope effects (37) could be well reproduced by transition state calculations, but the activation energies could not (the observations found an activation energy for the D + DBr reaction to be twice that for H + HBr, which could not be obtained with the calculations). In the latter reaction, the isotope effects for the H- and D-substituted reactions could be reproduced by the calculations (38), but the results of experiments involving muonium substitution, especially the observation of a negative temperature dependence, could not (39,40).

There have been far fewer measurements of heavy atom isotope effects in simple chemical reactions. Perhaps the best studied such reaction is that of OH with CH_4

$$OH + CH_4 \rightarrow H_2O + CH_3 \tag{14}$$

for which the ratio of the rates of reaction with $^{12}CH_4$ to $^{13}CH_4$ was found to be 1.0054±0.0009 over the temperature range from 273-353K (41,42). A detailed theoretical treatment of the isotope effect of this reaction has recently appeared, also suggesting little or no temperature dependence for the ratio of reaction rates (43). The observed isotope effects are important in understanding the D/H content of atmospheric methane (42). It is worth noting that the ^{12}C-^{13}C isotope effect in this system is probably better understood than the larger D-H isotope effect (44).

Complex Forming Bimolecular and Termolecular Reactions. Most of the early experimental work on termolecular reactions involved perdeuteration of one or both active reactants (15), while there have been some measurements of isotope effects in singly-substituted complex forming reactions. These have included experimental studies of D/H substitution in the reaction (45)

$$C(^1D) + H_2 \rightarrow CH + H \tag{15}$$

and theoretical studies for the reaction (46)

$$O(^1D) + H_2 \rightarrow OH + H \tag{16}$$

and for $^{12}C/^{13}C$ and $^{16}O/^{18}O$ substitution in the reaction (12,13)

$$CO + OH \rightarrow CO_2 + H \tag{17}$$

There has been a greatly enhanced consideration of D/H substitution in complex-forming reactions in recent years. One of the more complete studies (47) has been of the reaction $D + CH_3 \leftrightarrow CH_3D^* \rightarrow CH_3D, H + CH_2D$ (18) and the comparison of its rate with that of the unsubstituted one

$$H + CH_3 \leftrightarrow CH_4^* \rightarrow CH_4 \tag{19}$$

The D-substituted system is more complex than the unsubstituted one because of the possibility of a slightly exothermic H/D atom exchange, forming the $CH_2D + H$ product channel in addition to the 3-body recombination process forming CH_3D. The dynamics of the two reactions differ because of the existence of this additional channel, which means that (18) will be in its high pressure limit at low pressures where (19) is still in its falloff region. The ratio (approximately 0.4) of the rate of disappearance of D in (18) at room temperature (1.88±0.11x10^{-10}) cm^3 molec^{-1} sec^{-1} at 200 torr) to that obtained for H loss in (19) by extrapolating to the high pressure limit (4.7x10^{-10} cm^3 molec^{-1} sec^{-1}) is significantly smaller than suggested by both transition state (approximately 0.7) and microcanonical RRKM theories (0.7-0.8) (47). The problem is not likely due to tunneling and could be due to differences in vibrational coupling between CH_4 and CH_3D.

One termolecular reaction for which there have been several recent studies of the effects of isotopic substitution is

$$OH + NO_2 + M \rightarrow HNO_3 + M \tag{20}$$

for which studies involving both D- (45,46) and ^{18}OH-substitution (14,47) have been carried out. No significant D/H isotope effect was found for reaction (20); Smith and Williams found the ratio of the rates with OH to that with OD to be 0.99±0.17 at 298K and 18 torr Ar (45), while Bossard et al. found the ratio at room temperature to be unity within experimental error over a range of pressures throughout the falloff region in He, N_2, and SF_6 (46).

By use of ^{18}O-substituted OH in reaction (20), an additional product channel (O-atom exchange) becomes available, analogous to the D-atom exchange channel in reaction (18). If one assumes that O-atom exchange will take place in 2/3 of the complex-forming collisions between ^{18}OH and NO_2, the loss rate for ^{18}OH in this system will correspond to 2/3 of the high pressure limiting rate for the recombination reaction (14). Comparison of the rate thus inferred agreed with that obtained by extrapolation of the temperature dependence by several groups but not with the value inferred from vibrational deactivation of OH(v=1) by NO_2 (14,45). In the latter technique, it is assumed that vibrational deactivation is very rapid in complex-forming reactions, so that the rates of the two processes are then simply related.

In addition to these studies of isotopic substitution in complex-forming radical-radical reactions, there have also been several recent studies of isotopic substitution in radical-molecule reactions. For example, in the reaction

$$OH + HNO_3 \rightarrow H_2O + NO_3 \qquad (21)$$

perdeuteration caused there to be a minimum in the reaction rate near 323K which is not found in the unsubstituted reaction (48). At room temperature the isotope effect (ratio of the rates of the unsubstituted and perdeuterated reactions) is approximately 11. The existence of a minimum in the recombination rate for the perdeuterated version of (21) suggests that there are two competing mechanisms - a complex-forming/dissociation process at low temperatures, and a direct abstraction reaction at high temperatures. The existence of an apparent abstraction pathway was not clear from the observations of the rate of (21). The origin of the large room temperature isotope effect and the quantitative origins of the differing temperature dependences of these reactions are not yet understood.

A second radical-molecule reaction for which the effect of D/H substitution has been studies over a range of temperature is

$$CH + H_2 \leftrightarrow CH_3^* \rightarrow CH_3, CH_2 + H \qquad (22)$$

where, in particular, early work on the unsubstituted reaction was been supplemented by recent work on the reaction of both CH and CD with D_2 over a broad temperature range (49). There has also been a set of theoretical calculations of the various D/H isotope effects in this system (50). In the CH + D_2 reaction, there is a rapid isotope exchange channel, forming CD + HD, which leads to removal of CH. An approximately linear Arrhenius plot is found for the temperature range from 298-1260K. For the CD + D_2 reaction at 100 torr, on the other hand, a minimum is found near 500K, again demonstrating two competing mechanisms - an endothermic addition-elimination channel and a complex stabilization which dominates at the low and moderate temperatures typically studied in laboratory experiments.

Effects of single and multiple deuteration have also been investigated in the reaction of CH with HCN (51)

$$CH + HCN \rightarrow H_2CCN^* \rightarrow H + HCCN \qquad (23)$$

All four possible H-D isotopic combinations of (23) have been studied (CH + HCN, CH + DCN, CD + HCN, CD + DCN), with the rate constant varying within 20 percent of the average of the rates for the four reactions. CH + HCN is the slowest, while CD + HCN is fastest, with CH + DCN being a close second. The ratio of the rate of the perdeuterated reaction to that of the unsubstituted one is similar to that in the CH + H_2 system. The mixed isotope reactions are believed to be fastest because of the existence of a D/H exchange reaction, especially at temperatures above 475K. At lower temperatures, the dominant pathway is formation of H + HCCN.

Several other radical-molecule reactions for which the effects of D-H substitution are recently studied are reviewed in the article by Wine et al. in this volume (*52*); in addition, the effect of isotopic substitution on the reaction of CH_3 with H_2 is discussed in (*18*).

Symmetry-Related Isotope Effects. A new isotope effect recently discovered is that which is believed to be related to issues of molecular symmetry in that reactions in which one of two or more like atoms undergoes isotopic substitution, the rate becomes enhanced. If an isotope effect is truly symmetry-related, its magnitude will be mass-independent. Thus, for atoms such as oxygen, for which there are three stable isotopes (^{16}O, ^{17}O, ^{18}O), the magnitude of a symmetry-related isotope effect will be the same for both ^{17}O and ^{18}O substitution. This is in contrast to usual mass-dependent isotope effects, such as discussed in the above sections, for which one might expect the isotope effect for ^{17}O to be roughly half that for ^{18}O.

The reaction for which the largest symmetry-related isotope effect has been observed is formation of O_3 from O and O_2:

$$O + O_2 + M \rightarrow O_3 + M \qquad (24)$$

The symmetry relationship of the isotope effect was first noted by Heidenreich and Thiemens (*53*), who found that the deviation of the amounts of ^{17}O and ^{18}O in ozone formed in a discharge from that expected based on normal abundance was the same, rather than the 2:1 ratio for $^{18}O:^{17}O$ expected for normal isotopic fractionation. Much of the early interest in this work stemmed from the demonstration that chemical processes could produce mass-independent isotopic fractionation in oxygen. Previously mass independent fractionations were observed in meteoritic samples and were attributed to nucleosynthesis because no known chemical processes could produce them (*54*). The relationship between the mass-independent fractionation and the isotopic anomaly in meteorites has not been firmly established, however.

Following this discovery, there have been numerous laboratory experiments aimed at characterizing the isotope effect and its dependence on method of preparation of O, temperature, and pressure. The groups of Thiemens and Mauersberger have been especially active in these areas. Contributions of these groups to this issue are included in this volume (*4,5*). It is worth noting that the isotope effect can be quite large - in excess of 10 percent under appropriate conditions. The symmetry-relatedness has been clearly demonstrated in several experiments, including the use of isotopic labeling to show that formation of mass-51 ozone (^{16}O-^{17}O-^{18}O and its isomers), which consists only of asymmetric molecules, is the fastest of all ozone isotopomers (*55*), and the use of laser spectroscopy to show that the bulk of the enhancement of production of mass-50 ozone is in the asymmetric isomer (*56*). These experiments are discussed more fully elsewhere in this volume (*4,5*).

Additional reactions which have been shown to have "not-strictly-mass-dependent" isotope effects are

$$O + CO + M \rightarrow CO_2 + M \qquad (25)$$

$$SF_5 + SF_5 + M \rightarrow S_2F_{10} + M \qquad (26)$$

The magnitude of the isotope effect decreases from (24) through (26), which goes along with the inverse of the density of states of the polyatomic complex first formed in the recombination reactions. Some of the qualitative ideas of the origins of the isotope effect suggest that the magnitude of the isotope effect should be inversely correlated with the density of states (4,5). In reactions (25) and (26), the isotope effects are not equal for the different mass atoms involved (^{17}O-^{18}O in (25), ^{33}S-^{34}S in (26)), leading to mass-dependences which are neither strictly mass dependent (having the usual 2:1 ratio) or mass independent (having equal ratios).

It should be emphasized that there is no quantitative theory for the origin of these isotope effects. There is, in fact, considerable controversy about their origins and magnitudes. For example, on the basis of a theory for the isotope effect in reactions (24) and (25) Bates (57) recently questioned many of the observed isotope effects, interpreting them as artifacts of the experimental procedures. A clear understanding of the origins of the isotope effects in these systems is needed.

Two other systems for which symmetry related isotope effects have been recently reported are the formation of O_4^+ in the collision of O_2 and O_2^+ (58) and of CS_2^- in collisions of Rydberg atoms and CS_2 (59). Quantitative explanations of these phenomena are not yet available, either.

Photodissociation Processes. There are several different ways in which isotopic substitution can influence a molecule's photodissociation. First, isotopic substitution can lead to a shift in the absorption spectrum of a molecule. Early work in this area, especially on the effects of perdeuteration on ultraviolet absorption cross-sections, has been reviewed previously (15). More recently there have been studies on the wavelength shift in going from $^{48}O_3$ to $^{54}O_3$ (60,61). In the recent study (61) of Anderson et al. on the effect of isotopic substitution on the position of the Wulf bands in ozone, shifts of +30 to -50 13 cm^{-1} were observed (for a transition centered at approximately 9990 cm^{-1}). Second, isotopic substitution can lead to the removal of symmetry restrictions on molecular wave functions, allowing for the occurrence of transitions which cannot occur in the unsubstituted molecule. An example would be the allowance of odd rotational levels in the ground electronic state of $^{16}O^{18}O$ not permitted in $^{16}O^{16}O$ by the Pauli Principle (6). This can affect not only the ultraviolet absorption of O_2, but it can affect branching in the photodissociation of O_3 because O_2 is a product of O_3 photolysis (7,8). Third, isotopic substitution can affect line intensities in individual absorption spectra. In the case of $^{16}O^{18}O$, transitions corresponding to those in $^{16}O^{16}O$ may differ in intensity by as much as 40% in the Schumann-Runge band region (62).

Substantial recent attention has gone into the study of isotope effects on branching ratios in photodissociation processes where there are two or more product channels. For example, in photodissociation of simple iodides, the iodine atom can be formed in either its ground ($^2P_{3/2}$, abbreviated I) or excited ($^2P_{1/2}$, abbreviated I*) states. It is of interest how the branching ratio (I/I*) is affected by isotopic substitution in the rest of the photolyzing molecule. Such substitution can provide an important test of models of photodissociation because in the Born-Oppenheimer approximation the potentials governing the photolysis should be independent of isotopic substitution (*63*). Thus, a model (including dynamical model and potential surface) which simulates the branching in the photodissociation of one iodide should also work in that of the substituted one.

For example, in the photolysis of HI and DI, calculations (*64*) showed that in the frequency range from 35000 to 40000 cm^{-1}, the I*/I branching ratio in HI photolysis exceeds that for DI, while at higher energies (> 40000 cm^{-1}), the branching ratio is higher in DI photolysis. These calculations agree with several experimental results. A more detailed explanation of the origin of these effects is given in the article by Shapiro in this volume (*63*).

The situation is significantly more complex in the photolysis of CH$_3$I and CD$_3$I, however. Calculations suggest that the relative I*/I ratios for the two molecules may vary with wavelength - the I* yield from CD$_3$I was found to be smaller than that of CH$_3$I at 248 nm but to be larger at 266 nm.(*65*). Comparison of calculated results to experiments is complicated by the range of experimental values for the I*/I branching ratio , especially for CH$_3$I at 248 nm. In general it appears that experiments show more formation of I* in the photolysis of CD$_3$I at this wavelength. It has recently been suggested (*66*) that models for I*/I branching in CH$_3$I photolysis must include bending modes. These have not been included in most models, which treat CH$_3$I photodissociation as a collinear, quasi triatomic process (H$_3$-C-I).

Another type of branching which has been extensively looked at in recent years is D-H branching in photodissociation of molecules containing both atoms. Experiments have been done on both ground vibrational states and vibrationally excited molecules. Experiments on ground vibrational states are summarized in the chapter in this volume by Katz and Bersohn (*67*). Molecules studied to date include HDO, CHDO, HCCD, and various D-containing isotopomers of CH$_4$. All of the molecules studied to date have preferential formation of H on photolysis (after accounting for the number of D and H atoms in the parent molecule). The reason for a given isotope effect differs from molecule to molecule, however. Among the factors which affect D/H branching are the nature of the initially excited state (bound or repulsive), the difference in zero-point energy between the two isotopomers, and the relative rates of radiative decay, internal conversion, and intersystem crossing (*67*). Theoretical pictures have been developed to facilitate studies of branching in photodissociation (*63,68,69*), and, in the case of photolysis of HDO at 157 nm, there is excellent agreement between the calculated (*69*) and measured (*70*) D/H branching ratios as well as the absorption cross section for HDO, D$_2$O, and H$_2$O.

The theories also examine the effects of reagent excitation and suggest that isotope-specific photodissociation can be performed for a variety of molecules (*68-70*) if the reagent molecules are appropriately excited. In an early demonstration of this effect, Vander Wal et al. (*71*) excited HDO to the v_{OH}=4 state and photolyzed the resulting molecule at 266.0, 239.5, and 218.5 nm. They obtained preferential formation of OD at 266 and 239.5 nm, with the OD/OH ratio being greater than 15. At 218.5 nm, on the other hand, the OD/OH ratio was unity within experimental error. More recently, preferential formation of OD in the photodissociation of HDO excited to the v_{OH}=1 state was also observed (*72*). In this experiment, the OD/OH ratio for photolysis at 193 nm was greater than or equal to three. Calculations for photodissociation of HDO excited to several different vibrational levels are summarized by Shapiro (*63*) in this volume. Also in this volume are calculations reviewing strategies by which internal excitation can be used to selectively photolyze a variety of isotopically-substituted molecules (73).

Such processes may be useful on a commercial scale. For example, Zittel and Wang (74) recently carried out two-step laser photodissociation of OCS over a range of temperatures (296-150 K) to enrich sulfur and oxygen isotopes. In these experiments, the $2\nu_2$ state of OCS was first populated prior to photodissociation at 249 nm. The enhancement factor for ^{33}S and ^{36}S may be sufficient that the photolytic process could compete with currently used enrichment processes on the basis of economics, but numerous cycles would be needed in order to enrich oxygen or carbon isotopes. It is possible that the use of more highly excited intermediate levels would improve the efficiency of the process, however.

Summary and Conclusions

The study of isotope effects in gas phase chemical reactions and photodissociation processes can play an important role in improving our understanding of how they occur on a molecular scale. They can also provide quantitative information which can be used to test our understanding of various chemical and physical systems, especially those in the Earth's atmosphere and in various extraterrestrial environments (comets, planetary atmospheres, interstellar space). This is true not only for D/H substitution, as has been carried out most extensively, but also for heavy atom substitution as well. Improvements in measurement capability for both isotope effects in reaction rates and for the isotopic content of real systems have made the quantitative study of isotope effects more important than previously.

As a result of these studies, some interesting surprises have arisen. For example, the enhancement in heavy ozone in the stratosphere appears to be related to the enhanced formation of heavy ozone in the laboratory, although the mechanism of both processes is not well understood. These effects were completely unexpected, especially the presence of large (> 10%) heavy atom isotope effects for ozone formation. There is also the possibility that bond-selective photodissociation in isotopically substituted molecules may provide new methods for isotope enrichment, which could have commercial applications.

Although much of this overview has emphasized the areas in which there is disagreement between theory and experiment, the progress in both experiments reactions and photodissociation processes has been substantial in the recent past. Continued advances in experimental and theoretical techniques may provide for further improvements in our ability to address these issues and to help verify models of complex physical and chemical systems.

Acknowledgments. I thank two anonymous reviewers for helpful comments on this manuscript and Rose Brown for her assistance in the preparation of camera-ready copy.

Literature Cited

1. Dansgaard, W. *Tellus* **1953**, *5*, 461.
2. Johnston, H. S. *Gas Phase Reaction Rate Theory*; Ronald: New York, NY, 1966.
3. Robinson, P. J.; Holbrook, K. A. *Unimolecular Reactions*; Wiley-Interscience: New York, NY, 1972.
4. Thiemens, M. H., *this volume*.
5. Anderson, S.; Mauersberger, K.; Morton, J.; Scheuler, B., *this volume*.

6. Cicerone, R. J.; McCrumb, J. L. *Geophys. Res. Lett.* **1980**, *7*, 251.
7. Valentini, J. J.; Gerrity, D. P.; Phillips, D. L.; Nieh, J.-C.; Tabor, K. D. *J. Chem. Phys.* **1987**, *86*, 6745.
8. Valentini, J. J. *J. Chem. Phys.* **1987**, *86*, 6757.
9. Vasudev, R. et al., *this volume*.
10. Peifer, W. R.; Garvey, J. F.; *this volume*.
11. Stevens, C. M.; Krout, L.; Walling, D.; Venters, A.; Engelkmeir, R.; Ross, L. E. *Earth Planet. Sci. Lett.* **1972**, *16*, 147.
12. Stevens, C. M.; Kaplan, L.; Gorse, R.; Durkee, S.; Compton, N.; Cohen, S.; Bielling, *Int. J. Chem. Kinet.* **1980**, *12*, 935.
13. Smith, H. G. J.; Volz, A.; Ehhalt, D. H.; Kneppe, H. *Anal. Chem. Symp. Ser.* **1982**, *11*, 147.
14. Greenblatt, G. D.; Howard, C. J. *J. Phys. Chem.* **1989**, *93*, 1035.
15. Kaye, J. A. *Rev. Geophys.* **1987**, *25*, 1609.
16. Armentrout, P. B., *this volume*.
17. Viggiano, A. A. et al., *this volume*.
18. Truhlar, D. G.: Lu, D.-h.; Tucker, S. C.; Zhao, X. G.; Gonzalez-Lafont, A.; Truong, T. N.; Maurice, D.; Liu, Y.-P.: Lynch, G. C. *this volume*.
19. Herbst, E., *this volume*.
20. Bottinga, Y. *Geochim. Cosmochim. Acta* **1969**, *33*, 49.
21. Beer, R.: Taylor, F. W. *Astrophys. J.* **1978**, *219*, 763.
22. Bigeleisen, J.; Stern, M. J.; van Hook, W. A. *J. Chem. Phys.* **1963**, *38*, 497.
23. Ehhalt, D. H. *Vertical Profiles of HTO, HDO and H_2O in the Troposphere, Rep. NCAR-TN/STR-100*, Nat'l. Cent. for Atmos. Res.: Boulder, CO, 1974.
24. Rinsland, C. P.; Gunson, M. P.; Foster, J. C.; Toth, R. A.; Farmer, C. B.; Zander, R. *J. Geophys. Res.* **1991**, *96*, 1057.
25. Dinelli, B. M.; Carli, B.; Carlotti, M. *J. Geophys. Res.* **1991**, *96*, 7509.
26. Dansgaard, W. *Tellus* **1964**, *16*, 436.
27. Kreevoy, M. M.; Truhlar, D. G. In *Investigations of Rates and Mechanisms of Reactions, Vol. 6, Part 1*, Editor D. Bernasconi; John Wiley, New York, NY, 1986, pp. 13-95.
28. Tucker, S. C.; Truhlar, D. G. In *New Theoretical Concepts for Understanding Organic Reactions*, Editors J. Bertran and I. G. Czimadia, Kluwer Academic Publishers, Dordrecht, Holland, 1989, pp. 291-346.
29. Kuppermann, A. In *Potential Energy Surfaces and Dynamics Calculations*; Editor D. G. Truhlar; Plenum: New York, NY, 1981, pp. 375-420.
30. Presser, N.; Gordon, R. J. *J. Chem. Phys.* **1985**, *82*, 1291.
31. Zhu, Y.-F.; Arepalli, S.; Gordon, R. J. *J. Chem. Phys.* **1989**, *90*, 183.
32. Robie, D. C.; Arepalli, S.; Presser, N.; Kitsopoulos, T.; Gordon, R. J. *J. Chem. Phys.* **1990**, *92*, 7387.
33. Ehhalt, D. H.; Davidson, J. A.; Cantrell, C. A.; Friedman, I.; Tyler, S. *J. Geophys. Res.* **1989**, *94*, 9831.
34. Ravishankara, A. R.; Nicovich, J. M.; Thompson, R. L.; Tully, F. P. *J. Phys. Chem.* **1981**, *85*, 2498.
35. Bowman, J. M.; Wagner, A. F. *J. Chem. Phys.* **1987**, *86*, 1957.
36. Joseph, T.; Truhlar, D. G.; Garrett, B. C. *J. Chem. Phys.* **1988**, *88*, 6892.
37. Umemoto, H.; Wada, Y.; Tsunushima, S; \, T.; Sato, S. *Chem. Phys.* **1990**, *143*, 333.
38. Wada, Y.; Umemoto, H.; Tsunushima, S.; Sato, S. *J. Chem. Phys.* **1991** *94*, 4896.
39. Gonzalez, A.; Reid, I. D.; Farmer, D. M.; Senba, M.; Fleming, D. G.; Arseneau, D. J.; Kempton, J. R. *J. Chem. Phys.* **1989**, *91*, 6164.
40. Baer, S.; Fleming, D.; Arseneau, D.; Senba, M.; Gonzalez, A., *this volume*.

41. Cantrell, C. A.; Shetter, R. E.; McDaniel, A. H.; Calvert, J. G.; Davidson, J. A.; Lowe, D. C.; Tyler, S. C.; Cicerone, R. J.; Greenberg, J. P. *J. Geophys. Res.* **1990**; *95*, 22455.
42. Tyler, S. C., *this volume*.
43. Lasaga, A. C.; Gibbs, G., *Geophys. Res. Lett.* **1991**, 18, 1217-1220.
44. Gordon, S.; Mulac, W. A. *Int. J. Chem. Kinet.* **1975,** *7 (Symp. 1)*, 289.
45. Smith, I. W. M.; Williams, M. D. *J. Chem. Soc. Faraday Trans. 2* **1985,** *81*, 1849.
46. Bossard, A. R.; Singleton, D. L.; Paraskevopooulos *Int. J. Chem. Kinet.* **1988**, *20*, 609.
47. Dransfeld, P.; Lukacs, J.; Wagner, H. Gg. *Z. Naturforsch A.* **1986,** *41*, 1283.
48. Singleton, D. L.; Paraskevopoulos, G.; Irwin, R. S. *J. Phys. Chem.* **1991**, *95*, 694.
49. Stanton, C. T.; Garland, N. L.; Nelson, H. H. *J. Phys. Chem.* **1991,** *95*, 1277.
50. Wagner, A. F.; Harding, L. B., *this volume*.
51. Zabarnick, S.; Fleming, J. W.; Lin, M. C. *Chem. Phys.* **1991**, *150*, 109.
52. Wine, P.; Nicovich, J. M.; Hynes, A. J., *this volume*.

RECEIVED May 14, 1992

Theoretical Studies of Isotope Effects in Chemical Reactions

Chapter 2

Variational Transition-State Theory with Multidimensional, Semiclassical, Ground-State Transmission Coefficients
Applications to Secondary Deuterium Kinetic Isotope Effects in Reactions Involving Methane and Chloromethane

Donald G. Truhlar[1-3], Da-hong Lu[1,3,4], Susan C. Tucker[1,3,5], Xin Gui Zhao[2,3,6], Angels Gonzalez-Lafont[1,3,7], Thanh N. Truong[1,3,8], David Maurice[1,3,9], Yi-Ping Liu[1,3], and Gillian C. Lynch[1,3]

[1]Department of Chemistry, [2]Chemical Physics Program, and [3]Supercomputer Institute, University of Minnesota, Minneapolis, MN 55455-0431

This article has two parts. The first provides an overview of variational transition state theory with multidimensional semiclassical ground-state transmission coefficients. The second provides an update of recent applications to three secondary deuterium kinetic isotope effects in gas-phase C_1 reactions, in particular:

$$H + CD_3H \rightarrow H_2 + CD_3$$
$$*Cl^- + CD_3Cl \rightarrow CD_3*Cl + Cl^-$$

and

$$*Cl(D_2O)^- + CH_3Cl \rightarrow CH_3*Cl + Cl(D_2O)^-$$

where *Cl denotes a labeled chlorine atom.

Molecular modeling techniques have allowed for significant progress in the quantitative treatment of kinetic isotope effects (KIEs) for atom-diatom reactions. Both accurate quantum mechanical calculations and generalized transition state theory approaches have been used, and the former have been used to test the latter (1-9).

[4]Current address: Department of Chemistry, Fitchburg State College, 160 Pearl Street, Fitchburg, MA 01420
[5]Current address: Department of Chemistry, University of California, Davis, CA 95616
[6]Current address: Fuel Science Program, Department of Materials Science and Engineering, 210 Academic Projects Building, Pennsylvania State University, University Park, PA 16802
[7]Current address: Unidad Quimica Fisica, Departamento Quimica, Universidad Autonoma de Barcelona, Bellaterra 08193 (Barcelona), Spain
[8]Current address: Department of Chemistry, University of Houston, Houston, TX 77204-5641
[9]Current address: Department of Chemistry, University of California, Berkeley, CA 94720

Some critical tests against experiment are also available (for one especially relevant example see *10*). The extension of accurate molecular modeling techniques for the chemical dynamics of gas-phase reactions from atom-diatom collisions to polyatomic collisions is a challenge addressed in the present chapter. Our own work in this area has been focused on the generalized transition state theory approach in which quantized variational transition state theory, in particular in either the canonical variational theory (CVT) or improved canonical variational theory form (*11-16*), is combined with a transmission coefficient based on multidimensional semiclassical tunneling calculations and the ground-state transmission coefficient approximation (*13-20*). The present article reviews some of our work for the following set of C_1 reactions:

Reaction		References
(R1)	$H + CH_4 \rightarrow H_2 + CH_3$	(*21-29*)
(R2)	$^*Cl^- + CH_3Cl \rightarrow CH_3{}^*Cl + Cl^-$	(*30-34*)
(R3)	$^*Cl(H_2O)^- + CH_3Cl \rightarrow CH_3{}^*Cl + Cl(H_2O)^-$	(*31-34*)

Note that $^*Cl$ denotes a labeled chlorine atom. We will discuss secondary deuterium KIEs for CH_3/CD_3 isotopic substitution in reaction R1 (*27*) and R2 (*30,32-34*) and secondary deuterium KIEs for H_2O/D_2O substitution in R3 (*32-34*). Although the original papers consider a wide range of temperatures in all cases, we restrict our discussion here to 300-700 K for R1 and to 300 K for R2 and R3. We have also calculated secondary deuterium KIEs for CH_3/CD_3 substitution in the reverse of R1 (*27*), in $D + CD_3H \rightarrow HD + CD_3$ (*27*), and in R3 (*32-34*) and for both CD_3 and D_2O substitution in the reaction of $^*Cl(H_2O)_2^-$ with CH_3Cl (*31-33*), but that work is not reviewed here. Similar techniques have also been applied to calculate rate constants for three other reactions of methane, $Cl + CH_4 \rightarrow HCl + CH_3$ (*29*), $OH + CH_4 \rightarrow H_2O + CH_3$ (*35*), and $CF_3 + CH_4 \rightarrow CF_3H + CH_3$ (*36*), but secondary KIEs have not been calculated yet for these reactions. We have also calculated primary kinetic isotope effects for reactions R1 (*23*) and $CF_3 + CH_4 \rightarrow CF_3H + CH_3$ (*36*), but these will not be discussed here.

An important lesson of the work reviewed here is that interpretations of kinetic isotope effects based on conventional theory are often not borne out by the more complete treatments, which are the focus of this review. We will start out by reviewing the conventional theory, and then we will contrast it to the improved theory.

Conventional theory

Kinetic isotope effects are usually interpreted in terms of conventional transition state theory (TST, *37-39*), and tunneling is typically invoked only to explain "anomalous" (large) KIEs. The TST rate expression is

$$k = \kappa\sigma \frac{\tilde{k}T}{h} \frac{Q^{\neq}}{\Phi^R} e^{-V^{\neq}/RT} \qquad (1)$$

where κ is the tunneling transmission coefficient (often set equal to unity), σ is a symmetry factor (number of equivalent reaction paths), $\tilde{k}$ is Boltzmann's factor, T is temperature, h is Planck's constant, $Q^{\neq}$ is the transition state partition function with the zero of energy at $V^{\neq}$ and one degree of freedom (the reaction coordinate) fixed, Φ^R is the reactant partition function with zero of energy at classical equilibrium for reactants, $V^{\neq}$ is the saddle point potential energy with the zero of energy at classical equilibrium for reactants, and R is the gas constant. All symmetry factors are omitted from rotational partition functions $Q^{\neq}$ and Φ^R but are included in σ. The per-site KIE is defined as

$$\eta = \frac{k_H/k_D}{\sigma_H/\sigma_D} \qquad (2)$$

where k_H denotes the rate coefficient for the unsubstituted reactants, and k_D denotes the rate coefficient when one or more H is changed to D. It is informative to factor η values calculated from (1) as

$$\eta = \eta^{\neq}_{tun} \eta^{\neq} \qquad (3)$$

where the tunneling contribution is

$$\eta^{\neq}_{tun} = \kappa_H/\kappa_D \qquad (4)$$

and the conventional TST KIE is

$$\eta^{\neq} = \frac{Q^{\neq}_H/\Phi^R_H}{Q^{\neq}_D/\Phi^R_D}. \qquad (5)$$

If tunneling is included in conventional TST treatments it is almost always treated as one-dimensional. Then $\eta^{\neq}_{tun} \cong 1$ for secondary KIEs in most organic reactions since this kind of substitution usually has only a small effect on the reduced mass for motion along the reaction coordinate; however in some cases participation of secondary (i.e., non-transferred) hydrogens in the imaginary-frequency normal mode at the saddle point has been invoked in order to use conventional TST to explain anomalously high secondary KIEs that have been observed experimentally (40-42). (In conventional one-dimensional models, participation of such secondary hydrogens in the reaction coordinate is *necessary* in order to give a significant isotopic dependence to κ.)

In practice one assumes that rotations and vibrations are separable. Then $\eta^{\neq}$ may be further factored into translational, rotational, and vibrational contributions:

$$\eta^{\neq} = \eta_{trans} \eta^{\neq}_{rot} \eta^{\neq}_{vib}. \qquad (6)$$

The traditional way to interpret α-deuterium KIEs in S_N2 reactions focuses on the vibrational contribution $\eta^{\neq}_{vib}$, and in particular on the bending vibrations at the reactive carbon. For a reaction in which the hybridization at the reactive carbon changes from sp^3 to sp^2, the bending force constants are assumed to be reduced, in which case the zero point energy requirement decreases as the system transforms from reactant to transition state. This zero point effect is greater for H than for D, and it is assumed to dominate the KIE; thus k_H is expected to be greater than k_D (38,39,43). Conversely, the extent to which k_H exceeds k_D is often used (see, e.g., 44,45) as a tool for mechanistic analysis; a larger k_H/k_D may be associated with greater sp^2 character at the transition state (an S_N1-like reaction.) The conventional interpretation has been summarized by Saunders (39): "A larger k_H/k_D indicates a transition state in which the out-of-plane bend of the α-hydrogen is less encumbered than in the reactants." Wolfsburg and Stern extended this argument to reactions of the type $CH_3X + X \rightarrow CH_3XX^{\neq}$, again emphasizing the HCX bending force constant (46). We will see that our calculations indicate that many modes contribute to k_H/k_D so the interpretation is not so clear.

Generalized transition state theory

In the version of generalized transition state theory that we have developed and that we use to analyze KIEs, we still retain the concept of a transition state, but we do not necessarily identify it with the saddle point (highest-energy point on the minimum-energy path from reactants to products). The fundamental property upon which the generalization is based results from the equivalence of the transition state rate expression in classical mechanics to the one-way flux of an equilibrium ensemble of reactants through a hypersurface in configuration space (i.e., a system with one degree of freedom fixed). If we retain the assumption of reactant equilibrium, which is apparently often reasonable, even for fast reactions, under typical experimental conditions, as discussed elsewhere (47), then the one-way flux is an upper bound to the net rate and becomes exact when the transition state is a perfect dynamical bottleneck, i.e., when all trajectories cross the transition state at most once, which is called the no-recrossing assumption (for a pedagogical discussion see 16). Although the variational bound and convergence to the exact result as the variational optimization is improved are both lost in a quantum mechanical world (48), our working procedure is to vary the location of the transition state hypersurface to minimize the rate at a given temperature (i.e., for a canonical ensemble), even though we use quantized vibrational partition functions. Since the reaction coordinate is missing (i.e., fixed) in the generalized transition state vibrational partition function, quantum effects on the reaction coordinate are included in a transmission coefficient which is based on a semiclassical multidimensional tunneling (MT) correction to the CVT prediction for the ground state of the transition state and hence is called κ^{MT}. This leads to canonical variational theory with a semiclassical-tunneling ground-state transmission coefficient; the resulting rate constant expression is written

$$k^{CVT/MT} = \kappa^{MT} \sigma \frac{\tilde{k}T}{h} \frac{Q^{CVT}}{\Phi^R} e^{-V^{CVT}/RT} \qquad (7)$$

where Q^{CVT} is the generalized transition state partition function at the optimized location with zero of energy at the classical equilibrium geometry of this generalized transition state, and V^{CVT} is the potential energy at this point. Again all symmetry factors are in σ.

Details of the practical implementation of these equations are given elsewhere (especially 13). A critical aspect is the way in which the variational transition state definition and optimization is made practical. To search for the best variational transition state as an arbitrary (3N-4)-dimensional hypersurface (for an N-atom system) corresponding to fixed center of mass and fixed reaction coordinate would require, first, that one is able to calculate quantized partition functions (Q^{CVT}) for arbitrarily shaped hypersurfaces and, second, that one is able to perform a multidimensional optimization of a large number of variables required to define such a surface. Instead we define a one-parameter sequence of physically motivated generalized transition states orthogonal to a reaction path (the distance along the reaction path is the reaction coordinate s, and the parameter is the value of s at which the generalized transition state, i.e., the hypersurface, intersects s) for which the partition function may be calculated by standard methods, and we optimize the transition state within this sequence. In particular the reaction path is taken to be the minimum-energy path (MEP) in mass-scaled cartesian coordinates (49-53), and the shape of the hypersurface off the MEP is defined simply in terms of internal coordinates or cartesian vibrations. Although it is not clear *a priori* that the best transition state in this set is good enough, we have obtained very accurate results for cases where the theory can be tested against accurate quantum mechanics (*1-8*) and reasonable results in other cases, and so we accept this as good enough. [Further discussion of the shape of the generalized transition state hypersurfaces off the MEP is provided elsewhere (*28*), where in particular we emphasize that this choice is equivalent to choosing a definition of s for points off the MEP.] In the present calculations we also make the harmonic approximation for all vibrational partition functions.

Since, as discussed above, our extension of variational transition state theory to the quantum mechanical world is not rigorous, the usefulness of the approximation scheme must be tested by comparison to accurate quantum dynamics, which is only feasible at present for atom-diatom reactions, or to experiment. Since comparison to experiment is clouded by uncertainties in the potential energy function, we have relied on comparison to accurate quantum dynamics. Extensive comparisons of this type have been carried out (*1-8,12-20*), and they generally confirm the accuracy of our quantized approach. Thus we hope that applications to polyatomics, such as provided here, may be viewed as testing the potential rather than the dynamical theory. One should not forget though that the theory may be less accurate for new reactions than for the reactions for which it has been tested.

The most important question to be addressed in this regard is recrossing. Variational transition state theory does not include recrossing for the best variational transition state. In principle the transmission coefficient should account for such effects, i.e., for classical trajectories that cross the variational transition state in the direction of products but do not proceed to products or do not proceed directly to products without returning to the variational transition state (*12, 13, 54-57*). Recently, Hase and coworkers (*58*) have carried out trajectory calculations for reaction R2 starting from quantized energy distributions in the interaction region [this is called a quantized Keck calculation (*13*) in our earlier work]. They calculated a recrossing transmission coefficient of about 0.13 ± 0.07 at the conventional transition state for $T = 200-2000$ K (no systematic temperature dependence was observed). It is not known whether this effect would be smaller from the present potential function or if the dividing surface had been variationally optimized rather than placed at the conventional transition state. Ryaboy (*59*) has also suggested that recrossing effects may be important in S_N2 reactions. In the present work we account for recrossing of the conventional transition state to the extent that the canonical variational transition state theory rate is lower than the conventional one, but we do not include possible recrossing of the variational transition state. Inclusion of such effects may lower the predicted rate constants, and it may change the KIEs as well, but the estimation is beyond the scope of the present study. We note, however, that although classical recrossing effects are sometimes large, such effects are also sometimes (*12*, but not always, *60*) negligible in the real quantum mechanical world even when they are very significant in a classical mechanical world. The successes of quantized variational transition state theory with multidimensional tunneling corrections in the cases where it can be tested against accurate quantum dynamics (*1-8*) gives us some hope that it provides useful approximations for more complicated systems as well.

The final point to be discussed is quantum mechanical tunneling effects on the transmission coefficient. Our calculations recognize that at low energy, where tunneling is most important, all reactive flux is funneled through the ground state of the transition state. Thus we base our transmission coefficient on that state. Furthermore, in order to make the calculations practical for systems with many degrees of freedom, we perform them semiclassically. To calculate transmission coefficients accurately for the ground state of the transition state requires taking account of the multidimensional nature of the tunneling process. First of all, the optimum tunneling path is not the MEP. The MEP has the lowest barrier, but the optimum tunneling path involves a compromise between a low barrier and a short path. Since MEPs in mass-scaled cartesian coordinate systems are curved, tunneling paths tend to "cut the corner" (*18,19,50,61-69*) to shorten the tunneling distance. The extent to which this occurs is a function of the curvature of the MEP when it is plotted in mass-scaled cartesians. In the small-curvature case *(18,50,61-65,70)*, corner cutting, though quantitatively very important for the transmission coefficient, is relatively mild in terms of coordinate displacements. In this limit corner cutting tunneling has the nature of a negative bobsled effect due to a negative internal centrifugal potential arising from the curvature of the MEP. In this case accurate transmission coefficients can be calculated from a single path with an effective potential obtained by adding the local zero point energy to the potential energy along the MEP. (The sum is called the vibrationally adiabatic ground-state potential curve, where "vibrational adiabaticity" refers to the fact that the vibrational motion is in the local ground state.) Both the path and the effective potential depend on system masses. In the large-curvature case *(19,66-70)* corner cutting is severe, many paths must be considered, and the effective potentials are vibrationally adiabatic over part but not all of the paths. (In the nonadiabatic region the vibrational motion is not in the ground state, either physically or in the representation we use, but the semiclassical results are still referenced to the CVT results for the ground-state reaction.) The distribution of paths and the effective potentials again depend on system mass. Thus in neither limit does the mass enter so simply as in $\eta_{tun}^{\neq}$; consequently tunneling may be more important in the KIE than is conventionally assumed. In general one should consider tunneling into excited states in the exoergic direction in LCG3 calculations, but in the cases considered here, tunneling into excited states contributes at most a few percent.

We have developed a method, called the least-action ground-state (LAG) approximation *(13,20)*, based on optimized tunneling paths, that is valid for small, medium, and large reaction path curvatures. We have also developed a centrifugal-dominant small-curvature semiclassical adiabatic ground-state (CD-SCSAG or, for short, small-curvature tunneling, SCT) approximation *(18,21,70)* and a large-curvature ground-state (called LCG3 or, for short, large-curvature tunneling, LCT) approximation *(13,21,26,70)*, which are simpler and are expected to be accurate in the corresponding limits. Fortunately it is not usually necessary to perform full LAG optimizations; one obtains reasonable accuracy by simply carrying out both CD-SCSAG and LCG3 calculations and accepting whichever transmission coefficient is larger (since usually, though not always, both methods, at least where we have been able to test them against accurate quantum dynamics, underestimate the tunneling, and full LAG calculations have never been found to yield significantly more tunneling than the largest of the SCT and LCT limiting calculations). This is what we do here. In particular, although the systems discussed here were previously treated by the original *(18)* small-curvature approximation or by comparing this to LCG3 calculations considering only tunneling into the product ground state *(26)*, all calculations have been updated for this chapter by performing CD-SCSAG calculations and LCG3 calculations including all energetically accessible final states in the exoergic direction for reaction R1 (and its isotopically substituted and reverse versions) and the ground final state for R2 and R3 (and their isotopically substituted versions), using methods discussed elsewhere *(70)*, and accepting the larger of the two transmission coefficients at each temperature. The larger of the two results is simply labeled MT, which denotes multidimensional tunneling. In most cases the CD-SCSAG method yields the larger result, the exception being the unisotopically substituted version of reaction R2 with surface S. Individual transmission coefficients differ from those calculated by the original SCSAG method by as much as 31% for the cases considered here, but none of the kinetic isotope effects differs by more than 6% from those calculated with that method.

Factorization of the Kinetic Isotope Effect

For interpretative purposes we now require a new factorization of the per-site KIE, e.g., (3) is replaced by

$$\eta = \eta_{tun}\eta_{PF}\eta_{pot} \qquad (8)$$

where

$$\eta_{tun} = \kappa_H^{MT}/\kappa_D^{MT} \qquad (9)$$

$$\eta_{PF} = \frac{Q_H^{CVT}/\Phi_H^R}{Q_D^{CVT}/\Phi_D^R} \qquad (10)$$

and

$$\eta_{pot} = e^{-\left(V_H^{CVT}-V_D^{CVT}\right)/RT} \qquad (11)$$

Thus there is a critical distinction between the conventional TST and variational theory treatments. In conventional TST the transition state is at the same location (same structure) for both isotopes, and so the potential energy contributions cancel. Thus $\eta^{\neq} = \eta_{PF}^{\neq}$ and $\eta_{pot}^{\neq} = 1$. In variational TST this is not true. As a corollary, although the conventional TST rate constant may be computed from reactant and saddle point force fields that are both independent of mass, in variational TST the force fields used for the transition state partition function are different for the H and D versions of the reaction as a consequence of the change in structure.

For interpretive purposes we factor the partition function contribution η_{PF} to the variational TST value of the KIE into three further factors (71,72):

$$\eta_{PF} = \eta_{trans}\eta_{rot}\eta_{vib} \qquad (12)$$

where the factors are due to translational, rotational, and vibrational partition functions. (Electronic partition functions are assumed to cancel, which is an excellent approximation.)

Applications

H + CD$_3$H → H$_2$ + CD$_3$. The formalism reviewed above was applied (27) to the reaction of H with methane using a semiglobal analytic potential energy function, denoted J1, that was calibrated (23) previously. The form is an analytical function of 12 internal coordinates based on a functional form used earlier by Raff (73). The calibration (23) was based on the vibrational frequencies at the equilibrium geometries of CH$_3$ and CH$_4$ and at the saddle point, the heat of reaction at 0 K, the forward and reverse rate constants at 667K, and the equilibrium constant at 1340 K (74-81). The surface was also tested by calculating the primary KIEs which agree very well with experimental (82-84) values. The secondary deuterium KIE for CD$_3$H at 300K and its factorization are given in Table I. (In all cases, when we tabulate KIEs, they are the ratio of the rate constant for the isotopically unsubstituted reaction to that for the reaction of the isotopically substituted reagent under consideration.) In this table, TST/WT and TST/IPT denote conventional TST with one-dimensional transmission coefficients calculated respectively by the Wigner $\hbar^2$ tunneling correction (85) and by fitting the potential energy along the MEP to an infinite parabola (86,87); the latter is sometimes called the Bell tunneling formula, but is here denoted infinite-parabola tunneling.

Table I shows several interesting results for this abstraction reaction. Conventional TST predicts a small KIE in the normal direction ($k_H/k_D > 1$), primarily because of rotational effects which are partly cancelled by an inverse vibrational effect. The conventional tunneling correction increases this KIE by only 3%. Variational transition state theory leads to a much larger KIE for the same potential energy function, and the vibrational contribution is now normal (> 1). Furthermore the tunneling contribution

Table I. KIEs and factors for $H + CD_3H \rightarrow H_2 + CD_3$ at 300 K

	η_{tun}	η_{trans}	η_{rot}	η_{vib}	η_{pot}	η
TST	...	1.01	1.27	0.83	...	1.07
TST/WT	1.01	"	"	"	...	1.08
TST/IPT	1.03	"	"	"	...	1.10
CVT	...	"	1.27	1.03	0.92	1.22
CVT/MT*	1.24	"	"	"	"	1.51

*denotes row used for further comparisons in Table V.

does not cancel out when its multidimensional aspects are included. Next we discuss the reasons why the variational effect on the overbarrier KIE is so large, raising the prediction from 1.07 to 1.22, and why the multidimensional tunneling contribution raises the predicted KIE still further, from 1.22 to 1.51.

The location of the variational transition state is determined mainly by competition between the potential energy $V_{MEP}(s)$ along the MEP and the zero point energy $\varepsilon_{int,2}^G(s)$ of one mode, v_2 of the transition state. The frequency of the mode that correlates with this mode and $V_{MEP}(s)$ are given for four critical points along the MEP for the unsubstituted reaction in Table II. As this mode changes from a reactant stretch to a transition state stretch, it reaches a minimum of 1417 cm^{-1} shortly before the saddle point. By the time the saddle point is reached it is already increasing rapidly toward the product value of 4405 cm^{-1} (the harmonic stretching frequency of H_2 according to the J1 potential energy function). Thus, as the potential energy slowly begins to decrease from its stationary value at the saddle point, v_2 is already raising rapidly. By the time the variational transition state has been reached, V has decreased only 0.1 kcal/mol, but $\varepsilon_{int,2}^G$ has increased 0.6 kcal/mol. At this point the free energy of activation is a maximum; beyond this the potential energy decreases faster than the zero point energy and generalized entropy of activation components increase. This behavior, which is illustrated in Figure 1, actually follows from simple bond-order arguments and was predicted qualitatively correctly in this way for a three-body model of this reaction years earlier (88).

Figure 2 shows the contributions of the various vibrational modes to the predicted KIEs at 300 K as calculated both at the saddle point and at the canonical variational transition state. (For degenerate modes the contribution shown is the total from both components.) Although v_2 is the primary reason why the variational transition state is displaced from the saddle point, its own contribution to the KIE is the same for either location within 1%. The contributions of three other modes, however, differ by more than 7% (computed from the unrounded values), and this is the primary reason for the variational effect shown in Table I.

Figure 2 shows that there are indeed large contributions from the CH_3 bending and deformation modes (the modes at 1131-1379 cm^{-1}), as anticipated in the conventional discussions reviewed above, but the contributions of the other modes are by no means negligible, especially the H-H-C bend at 592 cm^{-1}. Furthermore two of the three CH_3 bending and deformation modes are among the modes whose contributions to the KIE are most sensitive to the variational optimization of the transition state.

The mass dependence of κ^{MT} appears to have two important sources. First, just including the mass-dependent zero point energies in the vibrationally adiabatic ground-state potential curve raises η_{tun} from the value of 1.03 in the Bell formula to 1.04. Second, accounting for corner cutting amplifies the transmission coefficient by a factor of 2.3 for H + H-CH_3 and by a factor of 2.0 for H + H-CD_3. This raises η_{tun} to 1.24.

Note that the imaginary-frequency normal mode at the saddle point has a frequency of 989i cm^{-1} for H-H-CH_3 and 981i cm^{-1} for H-H-CD_3. This small difference indicates that it is not necessary, as assumed in previous work (40-42) to have the C-D motions participate in the imaginary-frequency normal mode at the saddle point in order to have a significant secondary α-D kinetic isotope effect on the tunneling factor.

In order to test our model further, we can compare to experiment for two secondary kinetic isotope effects in this system, in particular, for $k_{CH_3 + H_2}/k_{CD_3 + H_2}$ and $k_{CH_3 + D_2}/k_{CD_3 + D_2}$. In both cases, the experimental value must be obtained indirectly by multiplying several other experimental ratios of rate constants. Based on the paper by Shapiro and Weston (82), we can use the scheme:

$$\frac{k_{CH_3+H_2}}{k_{CD_3+H_2}} = \frac{k_{CH_3+H_2}}{k_{CH_3+D_2}} \frac{k_{CH_3+D_2}}{k_{CH_3+A}} \frac{k_{CH_3+A}}{k_{CD_3+A-d_6}} \frac{k_{CD_3+A-d_6}}{k_{CD_3+H_2}} \quad (13a)$$

$$= R_1 R_2 R_3 R_4 \quad (13b)$$

and

Table II. Reaction-path parameters for H + CH_4 → H_2 + CH_3

location	$V_{MEP}(s)$ (kcal/mol)	v_2 (cm^{-1})	nature of mode v_2
H + CH_4	0.0	3027[a]	H-C stretch
saddle point	13.0	1720	H-H-C stretch
variational transition state[b]	12.9	2195[a]	H-H-C stretch
H_2 + CH_3	2.8	4405[a]	H-H stretch

[a]This is the frequency of the mode that correlates with mode v_2 of the transition state (see Figure I).
[b]300 K

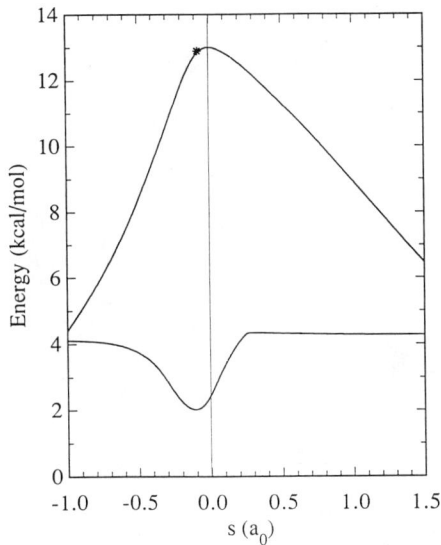

Figure 1. $V_{MEP}(s)$ and $\varepsilon_{int,2}^G(s)$ as functions (upper and lower curves, respectively) of s for reaction R1. The thin vertical line at s = 0 identifies the saddle point along the reaction coordinate, and the * identifies the V_{MEP} at the variational transition state at 300 K.

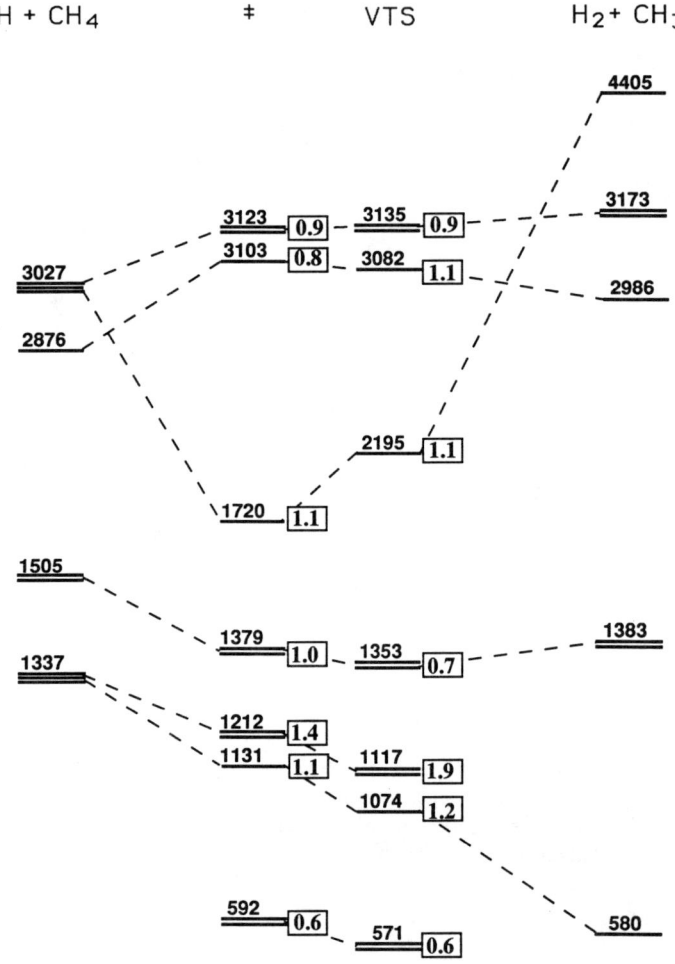

Figure 2. Frequencies (in cm^{-1}) and contributions to secondary KIEs (unitless numbers in boxes) for vibrational modes of (from left to right) H + CH$_4$, CH$_5^{\neq}$, CH$_5^*$, and H$_2$ + CH$_3$, where $\neq$ denotes the saddle point, and * denotes the variational transition state at 300 K.

$$\frac{k_{CH_3+D_2}}{k_{CD_3+D_2}} = \frac{k_{CH_3+D_2}}{k_{CH_3+A}} \frac{k_{CH_3+A}}{k_{CD_3+A-d_6}} \frac{k_{CD_3+A-d_6}}{k_{CD_3+H_2}} \frac{k_{CD_3+H_2}}{k_{CD_3+D_2}} \quad (14a)$$

$$= R_2 R_3 R_4 R_5 \quad (14b)$$

where the ratios are defined R_1, R_2, ... in the order they appear in the line above. Note that A denotes acetone and A-d_6 denotes perdeuterated acetone. Shapiro and Weston measured R_1, R_2, R_4, and R_5, and using their values for these ratios and the value recommended in the review by Kerr and Parsonage (89) for R_3 yields a set of experimental results that we will call SW-KP. For comparison we also computed the experimental results using the values recommended by Kerr and Parsonage for R_1, R_2, and R_3 (89,90) combined with Shapiro and Weston's results for R_4 and R_5; these experimental values are labeled KP-SW. For a third set of values we used the more recent results of Arthur and Newitt (91) for k_{CH_3+A} in R_2 and for k_{CH_3+A} and $k_{CD_3+A-d_6}$ in R_3, again combined with the Kerr-Parsonage recommendations for R_1 and for $k_{CH_3+D_2}$ in R_2 and with Shapiro and Weston's values for R_4 and R_5; these experimental results are labelled AN-KP-SW. A fourth method, suggested by Weston (92), is to combine Shapiro and Weston's values for R_1, R_2, R_4, and R_5 with a value of R_3 calculated from the data analysis of Arthur and Newitt. This has the advantage that R_3 appears in both final ratios in the same way and thus does not affect the *relative* secondary KIEs for the H_2 and D_2 reactions. These results are called SW-AN. Comparison of these sets of experimental results provides an estimate of the error due to the uncertainty in R_1, R_2, and R_3. In addition to the variations in the table, which are due to R_1, R_2, and R_3, all values have an *additional* error contribution of 9-17% due to Shapiro and Weston's error estimates for R_4 and R_5. Table III also gives the theoretical results of Schatz et al. (81) and our calculations for three potential energy surfaces, J1, J2, and J3, all from Joseph et al. (23). For $k_{CH_3+H_2}/k_{CD_3+H_2}$, all calculations agree with experiment within the experimental uncertainty. For the D_2 reaction, the secondary deuterium kinetic isotope effects, $k_{CH_3+D_2}/k_{CD_3+D_2}$, are 19-31% smaller at 400 K, but the theoretical results are 8-10% smaller.

We do not know how reliable the experimental results are; it would be valuable to have experimental results obtained by an independent method. Such measurements would provide a valuable test of the theoretical predictions that the secondary kinetic isotope effect is only about 10% smaller for $CH_3 + D_2$ than for for $CH_3 + H_2$. There is, however, one clue (92) to the source of the disagreement between theory and experiment. As evidenced in Fig. 11 of the paper of Shapiro and Weston (82), the $CD_3 + H_2/D_2$ primary KIE, which is R_5, is low compared to the primary KIEs for $CH_3 + H_2/D_2$ and $CF_3 + H_2/D_2$. The same trend appears in the HD/DH primary KIEs. A low value for R_5 will change the secondary KIE for $CH_3 + D_2/CD_3 + H_2$ but does not affect the one for $k_{CH_3+H_2}/k_{CD_3+H_2}$. Thus it seems likely that there is some systematic error in the CD_3 experiments. As indicated in the derivation (82) of R_5, there are several corrections related to the H isotopic impurity in the acetone-d_6 used. Perhaps this is the source of error (92).

$^*Cl^- + CD_3Cl \rightarrow CD_3{}^*Cl + Cl^-$. For the S_N2 exchange reaction of $^*Cl^-$ with CH_3Cl our results are based on an 18-dimensional semiglobal analytic potential energy function calibrated (30) to fit *ab initio* electronic structure calculations (30,93,94) of properties of the saddle point and ion-dipole complex, of energies, charges, and force fields on the MEP, and of energies at selected points off the MEP, to fit the correct long-range force law, and to fit one critical experimental datum, namely the rate constant for

Table III. KIEs for reverse of R1 and for analogous reaction with D_2

		theory				experiment			
KIE	T(K)	SWD	J1	J2	J3	SW-KP	KP-SW	AN-KP-SW	SW-AN
$k_{CH_3+H_2}/k_{CD_3+H_2}$	400	0.68[a] 0.68[b]	0.67[a] 0.82[c]	0.75 0.89	0.75 0.87	0.85	0.74	0.75	0.88
	500	0.77 0.77	0.77 0.87	0.84 0.94	0.83 0.92	0.86	0.86	0.82	0.85
	600	0.83 0.83	0.84 0.91	0.89 0.96	0.88 0.95	0.87	0.94	0.88	0.83
	700	0.87 0.87	0.87 0.94	0.92 0.99	0.92 0.97	0.88	1.00	0.92	0.81
$k_{CH_3+D_2}/k_{CD_3+D_2}$	400	0.68 0.68	0.67 0.74	0.75 0.81	0.74 0.80	0.59	0.60	0.60	0.61
	500	0.77 0.77	0.76 0.81	0.83 0.88	0.83 0.87	0.72	0.76	0.73	0.70
	600	0.84 0.84	0.82 0.87	0.88 0.93	0.88 0.92	0.82	0.89	0.83	0.78
	700	0.88 0.88	0.86 0.90	0.92 0.95	0.92 0.94	0.90	1.00	0.91	0.83

[a] Upper entry: conventional transition state theory with unit transmission coefficient. [b] Lower entry in SWD column: final result of Ref. 79 as obtained by conventional transition state theory with Wigner transmission coefficient. [c] Lower entry in J1, J2, and J3 columns: final results of this study as obtained by canonical variational theory with the CD-SCSAG transmission coefficient. All vibrational partition functions are calculated in the harmonic approximation for all calculations in the table.

the unsubstituted reaction at 300 K (95), which was used to adjust the *ab initio* barrier height. The final surface is called S because of this semiempirical adjustment.

A second set of calculations (33) was performed using the "direct dynamics" (96) approach. In this approach we do not parameterize an explicit potential energy function; rather we adjust the parameters in a semiempirical molecular orbital approximation to reproduce selected reaction features and the value of the potential energy function is defined implicitly by the energy yielded by the molecular orbital theory at any selected geometry. We selected the neglect-of-diatomic-differential-overlap (NDDO) level (97-99) of semiempirical molecular orbital theory, and we used the Austin model 1 (AM1) general parameter set (100,101) as our starting point. Then we readjusted two parameters (in particular, the one-electron, one-center atomic core matrix elements U_{pp} for Cl and C) specifically to energetic features assumed to be important for this S_N2 reaction. The resulting NDDO calculations with specific reaction parameters are called NDDO-SRP (33).

Table IV compares the KIEs for the CD_3Cl reaction (i.e., the ratio of the rate constant for the reaction in the section heading above to the rate constant with all protiums) and their factorizations for the two potential energy functions (PEFs). The results are remarkably similar, which tends to confirm the general correctness of both potential energy functions. In addition, the variational and conventional transition state theory results agree very well, and the tunneling contribution cancels out within one per cent (although the tunneling contribution is not negligible; κ_H^{MT} = 1.31 for both S and NDDO-SRP). Thus we can use any of the levels of dynamical theory in this case, and the same is true, within about 1%, for the KIEs in the microhydrated versions of this S_N2 reaction that are discussed below. Thus in the rest of the discussion of S_N2 reactions we will explicitly consider only the results at the most reliable (CVT/MT) level.

Table IV. KIEs and factors for $^*Cl^- + CD_3Cl \rightarrow CD_3{}^*Cl + Cl^-$ at 300 K

PEF	dynamics	η_{tun}	η_{trans}	η_{rot}	η_{vib}	η_{pot}	η
S	TST	...	1.04	1.22	0.76	...	0.96
	CVT	...	"	1.22	0.76	1.00	0.96
	CVT/MT*	1.00	"	"	"	"	0.96
NDDO-SRP	TST	...	"	1.23	0.75	...	0.96
	CVT	...	"	1.23	0.75	1.00	0.96
	CVT/MT	1.00	"	"	"	"	0.96

*denotes row used for further comparisons in Table V.

Table V shows a breakdown of the vibrational contribution to the KIE into contributions from three frequency ranges (for comparison, similar breakdowns for the other two KIEs discussed in this paper are also shown). As expected from the conventional interpretation of KIEs in S_N2 reactions (38,39,43-45,102), the contribution of the middle-frequency CH_3 bend and deformation modes is significant and > 1. However the low-frequency modes also contribute significantly, and the largest percentage deviation from unity comes from the high-frequency C-H stretches, which are almost universally ignored in discussions of secondary deuterium KIEs in the literature of the field. Perhaps some of the low KIEs observed experimentally which have been interpreted in terms of leaving group participation in tightening of the CH_3 bends and deformations are really caused to larger extent by stronger C-H bonds at the sp^2 center. Certainly the correlated electronic-structure frequencies on which our saddle point C-H contributions are based are at least semiquantitatively reliable for the present case, and there can be no question that this effect exists and is very important.

At the finer level of detail exhibited in Table V, the NDDO-SRP results for $\eta_{vib,mid}$ and $\eta_{vib,high}$ differ from the presumably more accurate one calculated from surface S by 20-25% (from the NDDO-SRP calculations, we get 0.84 for $\eta_{vib,low}$, 1.01 for $\eta_{vib,mid}$, and 0.88 for $\eta_{vib,high}$), although as seen in Table IV, the differences eventually cancel. Nevertheless we are quite encouraged by the NDDO-SRP results since this method is very much easier to apply. In later work on CF$_3$ + CH$_4$ (Liu et al., 36), we are varying more parameters and paramaterizing to frequencies as well as energies, and we believe we can obtain even better potential energy functions by this method.

Table V. KIEs and factorization of η_{vib} at 300 K[a]

Reactants	$\eta_{vib,low}$	$\eta_{vib,mid}$	$\eta_{vib,high}$	η_{vib}	η
H + CD$_3$H	0.55	1.74	1.08	1.03	1.42
Cl$^-$ + CD$_3$Cl	0.85	1.26	0.71	0.76	0.96
Cl(D$_2$O)$^-$ + CH$_3$Cl	0.40	1.67	0.90	0.60	0.87

[a]In all cases the mode classification is based on the isotopically unsubstituted case. For the abstraction reaction the low/mid and mid/high borders are taken as 600 and 1700 cm^{-1}, respectively, and for the S$_N$2 reactions they are taken as 515 and 1900 cm^{-1}.

*Cl(D$_2$O)$^-$ + CH$_3$Cl → CH$_3$*Cl + Cl(D$_2$O)$^-$. When we add water of hydration, the PEF has three parts: solute, solvent, and solute-solvent interaction. We have used several different choices (31,33,34), and three of the combinations are summarized in Table VI along with the KIEs (i.e., the ratio of the rate constant for the reaction in the section heading above to the rate constant for all protiums) and their factorization. In the first calculations the solute was again treated by our semiempirical potential energy function S, the solute-solvent potentials were extended versions of the Kistenmacher-Popkie-Clementi (103) and Clementi-Cavallone-Scordamaglia (104) potentials, denoted eKPC and eCCS, respectively, and the intramolecular water potential was taken from Coker, Miller, and Watts (CMW, 105). The second calculation was based on the NDDO-SRP method with no further changes in any parameters. The third is like the first except it is based on new chloride-water and intramolecular water potentials discussed in the next paragraph.

In carrying out calculations nos. 1 and 2 of Table VI we noted that the solvent KIEs are very sensitive to the low-frequency vibrations associated with the coupling of the solute to the solvent. Therefore, in work carried out in collaboration with Steckler (106), we calibrated a new potential energy function for Cl(H$_2$O)$^-$. The parameters of the new potential were determined to improve agreement with experiment for the dipole moment of water, with new extended-basis-set correlated electronic structure calculations for D$_e$, for the geometry and frequencies of the complex, for the individual and total vibrational contributions to the equilibrium isotope effect for Cl(H$_2$O)$^-$ + D$_2$O $\rightleftarrows$ Cl(D$_2$O)$^-$ + H$_2$O, and for the energy at a geometry close to the saddle point geometry for reaction R$_3$, which is shown in Figure 3, and with the *ab initio* calculations of Dacre

Table VI. KIEs and factors for $^*Cl(D_2O)^- + CH_3Cl \rightarrow CH_3{}^*Cl + Cl(D_2O)^-$ at 300 K

No.	solute	PEF Cl-H$_2$O	PEF CH$_3$-H$_2$O	PEF H$_2$O	η_{tun}	η_{trans}	η_{rot}	η_{vib}	η_{pot}	η
1.	S	eKPC	eCCS	CMW	1.00	1.03	1.41	0.72	1.00	1.04
2.	NDDO-SRP	NDDO-SRP	NDDO-SRP	NDDO-SRP	1.00	"	1.43	0.63	1.00	0.93
3.*	S	ZGTS	eCCS	mKH	1.00	"	1.41	0.60	1.00	0.87

*denotes row used for further comparisons in Table V.

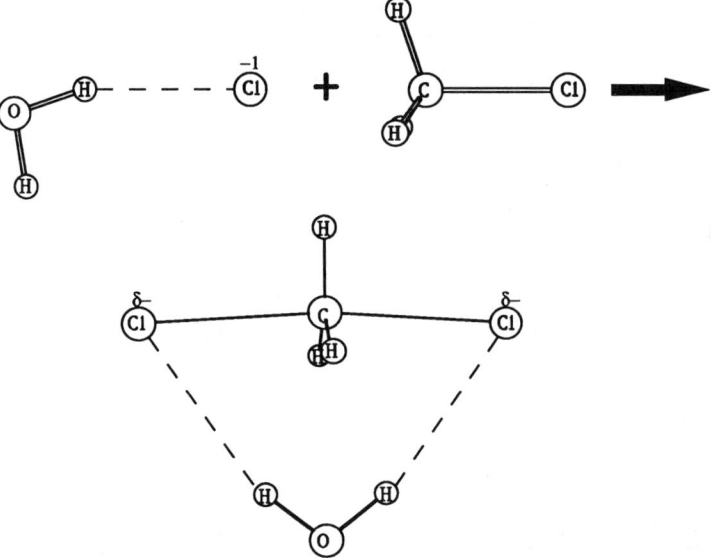

Figure 3. Structures of reactants and transition state for Cl⁻(H$_2$O) + CH$_3$Cl → ClCH$_3$Cl⁻(H$_2$O)$^\ne$, based on the ZGTS chloride–water potential and the mKH intramolecular water potential.

(*107*) for the interaction energy of Cl⁻ with water at 369 widely distributed reaction sites. The chloride-water part of the interaction potential is totally new, and the intramolecular water part is a modified version (denoted mKH) of the accurate force field fit to spectroscopic data by Kauppi and Halonen (*108*), the modification being a change in the cubic O-H stretching force constants to get the red shift correct in the $Cl(H_2O)^-$ complex.

Table VI shows that calculations nos. 2 and 3 are in encouraging agreement with each other, and both predict an inverse KIE. The change from calculation no. 1 to no. 3 can be accomplished in 3 steps, and it is instructive to do so, as follows: First we change the CMW water potential to the original Kauppi-Halonen one, and this changes η from 1.04 to 0.97. Then we change the eKPC chloride-water potential to our new one based on the correlated $Cl(H_2O)^-$ calculations, and this decreases η further to 0.89. Finally, using the mKH water potential instead of the original Kauppi-Halonen one lowers η to 0.87.

Table V shows the more detailed factorization of η_{vib}. The biggest difference from the NDDO-SRP potential is in $\eta_{vib,low}$ (0.40 vs. 0.46), and the biggest difference from calculations no. 1 is in $\eta_{vib,high}$ (0.90 vs. 1.00). As a final comment, we note that the standard theory (*109,110*) of solvent kinetic isotope effects in the bulk is based on "water structure breaking" by ions, by which one means that the water librations are in a looser force field in the first hydration shell than in bulk (*111*). In our reaction, as seen in Figure 3, charge is delocalized at the transition state, so presumably, in bulk water, the water molecules would be in a tighter force field at the transition state than at reactants. Thus the bulk solvent KIE would be inverse. We also find an inverse result for a single water molecule, where water structure breaking is obviously impossible. Thus water structure breaking certainly cannot be the sole reason for the inverse KIEs observed in bulk solvent kinetic isotope effects.

Concluding remarks

Since the time when Shiner (*102*) summarized the interpretation of secondary kinetic isotope effects in an earlier ACS Symposium, there has been considerable progress. Both our force fields and our dynamics techniques have improved, and the reliability of our interpretations should correspondingly be much higher.

The techniques presented here are applicable to systems with many degrees of freedom. For example, the reaction $^*Cl(H_2O)_2^- + CH_3Cl$, which we have treated (*31,33*), has 12 atoms, and these techniques have been applied elsewhere to kinetic isotope effects in an embedded cluster with 29 non-fixed atoms (*112*), an intramolecular process with 7 atoms (*113*), a bimolecular neutral reaction with 9 atoms (*36*), and an intramolecular process with 13 atoms (*114*).

Acknowledgments

The authors are grateful to Bruce Garrett, Al Wagner, and Ralph Weston for helpful comments. This work was supported in part by the U. S. Department of Energy, Office of Basic Energy Sciences.

References

1. Truhlar, D. G.; Hase, W. L.; Hynes, J. T. *J. Phys. Chem.* **1983**, *87*, 2664, 5523E.
2. Garrett, B. C.; Truhlar, D. G. *J. Chem. Phys.* **1984**, *81*, 309.
3. Garrett, B. C.; Truhlar, D. G.; Schatz, G. C. *J. Amer. Chem. Soc.* **1986**, *108*, 2876.

4. Zhang, J. Z. H.; Zhang, Y.; Kouri, D. J.; Garrett, B. C.; Haug, K.; Schwenke, D. W.; Truhlar, D. G. *Faraday Discussions Chem. Soc.* **1987**, *84*, 371.
5. Lynch, G. C.; Truhlar, D. G.; Garrett, B. C. *J. Chem. Phys.* **1989**, *90*, 3110.
6. Lynch, G. C.; Halvick, P.; Truhlar, D. G.; Garrett, B. C.; Schwenke, D. W.; Kouri, D. J. Z. *Naturforsch.*, **1989**, *44a*, 427.
7. Schatz, G. C.; Amaee, B.; Connor, J. N. L. *J. Chem. Phys.*, **1990**, *93*, 5544.
8. Garrett, B. C.; Truhlar, D. G. *J. Phys. Chem.*, in press.
9. Bowman, J. M. *J. Phys. Chem.* **1991**, *95*, 4960.
10. Garrett, B. C.; Truhlar, D. G.; Bowman, J. M.; Wagner, A. F.; Robie, D.; Arepalli, S.; Presser, N.; Gordon, R. J. *J. Amer. Chem. Soc.* **1986**, *108*, 3515.
11. Garrett, B. C.; Truhlar, D. G. *J. Chem. Phys.* **1979**, *70*, 1593.
12. Truhlar, D. G.; Garrett, B. C. *Accounts Chem. Res.* **1980**, *13*, 440.
13. Truhlar, D. G.; Isaacson, A. D.; Garrett, B. C. In *Theory of Chemical Reaction Dynamics*; Baer, M., Ed.; CRC Press: Boca Raton, 1985; Vol. 4, p. 65.
14. Truhlar, D. G.; Garrett, B. C. *Annu. Rev. Phys. Chem.* **1984**, *35*, 159.
15. Truhlar, D. G.; Garrett, B. C. *J. Chim. Phys.* **1987**, *84*, 365.
16. Tucker, S. C.; Truhlar, D. G. In *New Theoretical Concepts for Understanding Organic Reactions*; Bertrán, J., Csizmadia, I. G., Eds.; Kluwer: Dordrecht, 1989, p. 291.
17. Garrett, B. C.; Truhlar, D. G.; Grev, R. S.; Magnuson, A. W. *J. Phys. Chem.* **1980**, *84*, 1730.
18. Skodje, R. T.; Truhlar, D. G.; Garrett, B. C. *J. Phys. Chem.* **1981**, *85*, 3019.
19. Garrett, B. C.; Truhlar, D. G.; Wagner, A. F.; Dunning, T. H. *J. Chem. Phys.* **1983**, *78*, 4400.
20. Garrett, B. C.; Truhlar, D. G. *J. Chem. Phys.* **1983**, *79*, 4931.
21. Truhlar, D. G.; Brown, F. B.; Steckler, R.; Isaacson, A. D. In *The Theory of Chemical Reaction Dynamics*; Clary, D. C., Ed.; Reidel: Dordrecht, 1986, p. 285.
22. Steckler, R.; Dykema, K. J.; Brown, F. B.; Hancock, G. C.; Truhlar, D. G.; Valencich, T. *J. Chem. Phys.* **1987**, *87*, 7024.
23. Joseph, T.; Steckler, R.; Truhlar, D. G. *J. Chem. Phys.* **1987**, *87*, 7036.
24. Garrett, B. C.; Redmon, M. J.; Steckler, R.; Truhlar, D. G.; Baldridge, K. K.; Bartol, D.; Schmidt, M. W.; Gordon, M. S. *J. Phys. Chem.* **1988**, *92*, 1476.
25. Baldridge, K. K.; Gordon, M. S.; Steckler, R.; Truhlar, D. G. *J. Phys. Chem.* **1989**, *93*, 5107.
26. Garrett, B. C.; Joseph, T.; Truong, T. N.; Truhlar, D. G. *Chem. Phys.* **1989**, *136*, 271; **1990**, *140*, 207E.
27. Lu, D.-h.; Maurice, D.; Truhlar, D. G. *J. Amer. Chem. Soc.* **1990**, *112*, 6206.
28. Natanson, G. A.; Garrett, B. C.; Truong, T. N.; Joseph, T.; Truhlar, D. G. *J. Chem. Phys.* **1991**, *94*, 7875.
29. Gonzalez-Lafont, A.; Truong, T. N.; Truhlar, D. G. *J. Chem. Phys.*, in press.
30. Tucker, S. C.; Truhlar, D. G. *J. Amer. Chem. Soc.* **1990**, *112*, 3338.
31. Tucker, S. C.; Truhlar, D. G. *J. Amer. Chem. Soc.* **1990**, *112*, 3347.
32. Zhao, X. G.; Tucker, S. C.; Truhlar, D. G. *J. Amer. Chem. Soc.* **1991**, *113*, 826.
33. Gonzalez-Lafont, A.; Truong, T. N.; Truhlar, D. G. *J. Phys. Chem.* **1991**, *95*, 4618.
34. Zhao, X. G.; Lu, D.-h.; Liu, Y.-P.; Lynch, G. C.; Truhlar, D. G., to be published.
35. Truong, T. N.; Truhlar, D. G. *J. Chem. Phys.* **1990**, *93*, 1761.
36. Liu, Y.-P.; Lu, D.-h.; Gonzalez-Lafont, A.; Truhlar, D. G.; Garrett, B. C., unpublished.
37. Johnston, H. S. *Gas Phase Reaction Rate Theory*, Ronald Press: New York, 1966.

38. Melander, L.; Saunders, W. H. Jr. *Reaction Rates of Isotopic Moleucles* (John Wiley & Sons: New York, 1980).
39. Saunders, W. H. Jr. In *Investigation of Rates and Mechanisms of Reactions (Techniques of Chemistry*, 4th ed., Vol. VI), Bernasconi, C. F., Ed.; John Wiley & Sons: New York, 1986, Part I, p. 565.
40. Ostović, D.; Roberts, R. M. G.; Kreevoy, M. M. *J. Amer. Chem. Soc.* **1983**, *105*, 7629.
41. Huskey, W. P.; Schowen, R. L. *J. Amer. Chem. Soc.* **1983**, *105*, 5704.
42. Saunders, W. H., Jr. *J. Amer. Chem. Soc.* **1985**, *107*, 164.
43. Streitwieser, A.; Jagow, R. H.; Fahey, R. C.; Suzuki, S. *J. Amer. Chem. Soc.* **1958**, *80*, 2326.
44. Scheppele, S. E. *Chem. Rev.* **1972**, *72*, 511.
45. Bentley, T. W., Schleyer, P. v. R. *Adv. Phys. Org. Chem.* **1977**, *14*, 1.
46. Wolfsberg, M.; Stern, M. J. *Pure Appl. Chem.* **1964**, *8*, 325.
47. Lim, C; Truhlar, D. G. *J. Phys. Chem.* **1986**, *90*, 2616.
48. Truhlar, D. G. *J. Phys. Chem.* **1979**, *83*, 199.
49. Shavitt, I. University of Wisconsin Theoretical Chemistry Laboratory Report WIS-AEC-23, Madison, 1959.
50. Marcus, R. A. *J. Chem. Phys.* **1966**, *45*, 4493.
51. Marcus, R. A. *J. Chem. Phys.* **1968**, *49*, 2610.
52. Truhlar, D. G.; Kuppermann, A. *J. Amer. Chem. Soc.* **1971**, *93*, 1840.
53. Fukui, K. In *The World of Quantum Chemistry,* Daude, R., Pullman, B., Eds.; Reidel: Dordrecht, The Netherlands, 1974, p. 113.
54. Keck, J. C. *Adv. Chem. Phys.* **1967**, *13*, 85.
55. Garrett, B. C.; Truhlar, D. G. *J. Phys. Chem.* **1979**, *83*, 1052; **1983**, *87*, 4553E.
56. Garrett, B. C.; Truhlar, D. G. *J. Phys. Chem.* **1980**, *84*, 805.
57. Garrett, B. C.; Truhlar, D. G.; Grev, R. S. *J. Phys. Chem.* **1981**, *85*, 1569.
58. Cho, Y. J.; Vande Linde, S. R.; Hase, W. L., to be published.
59. Ryaboy, V. M. *Chem. Phys. Lett.* **1989**, *159*, 371.
60. Truhlar, D. G.; Garrett, B. C. *Faraday Discuss. Chem. Soc.* 1987, *84*, 464.
61. Marcus, R. A. *J. Chem. Phys.* **1964**, *41*, 610.
62. Coltrin, M. E.; Marcus, R. A. *J. Chem. Phys.* **1977**, *67*, 2609.
63. Garrett, B. C.; Truhlar, D. G. *J. Phys. Chem.* **1979**, *83*, 200.
64. Garrett, B. C.; Truhlar, D. G. *Proc. Natl. Acad. Sci. USA* **1979**, *76*, 4755.
65. Skodje, R. T.; Truhlar, D. G.; Garrett, B. C. *J. Chem. Phys.* **1982**, *77*, 5955.
66. Babamov, V. K.; Marcus, R. A. *J. Chem. Phys.* **1978**, *74*, 1790.
67. Bondi, D. K.; Connor, J. N. L.; Garrett, B. C.; Truhlar, D. G. *J. Chem. Phys.* **1983**, *78*, 5981.
68. Garrett, B. C.; Abusalbi, N.; Kouri, D. J.; Truhlar, D. G. *J. Chem. Phys.* **1985**, *83*, 2252.
69. Kreevoy, M. M.; Ostović, D.; Truhlar, D. G.; Garrett, B. C. *J. Phys. Chem.* **1986**, *90*, 3766.
70. Lu. D.-h.; Truong, T. N.; Melissas, V. S.; Lynch, G. C.; Liu, Y.-P.; Garrett, B. C.; Steckler, R.; Isaacson, A. D.; Rai, S. N.; Hancock, G. C.; Lauderdale, J. G.; Joseph, T.; Truhlar, D. G. *Computer Phys. Commun.* to be published.
71. Garrett, B. C.; Truhlar, D. G.; Magnuson, A. W. *J. Chem. Phys.* **1982**, *76*, 2321.
72. Tucker, S. C.; Truhlar, D. G.; Garrett, B. C.; Isaacson, A. D. *J. Chem. Phys.* **1985**, *82*, 4102.
73. Raff, L. M. *J. Chem. Phys.* **1974**, *60*, 2220.
74. Kurylo, M. J.; Hollinden, G. A.; Timmons, R. B. *J. Chem. Phys.* **1970**, *52*, 1773.

75. *JANAF Thermochemical Tables*, 2nd ed.; Stull, D. R., Prophet, H., Eds.; U.S. Government Printing Office: Washington, 1971.
76. Shaw, R. *J. Phys. Chem. Ref. Data* **1978**, *7*, 1179.
77. Sepehrad, A.; Marshall, R. M.; Purnell, H. *J. Chem. Soc. Faraday Trans. 1* **1979**, *75*, 835.
78. Walch, S. P. *J. Chem. Phys.* **1980**, *72*, 4932.
79. Schatz, G. C.; Walch, S. P.; Wagner, A. F. *J. Chem. Phys.* **1980**, *73*, 4536.
80. Sana, M.; Leroy, G.; Villareves, J. L. *Theoret. Chim. Acta* **1984**, *65*, 109.
81. Schatz, G. C.; Wagner, A. F.; Dunning, T. H. *J. Chem. Phys.* **1984**, *88*, 221.
82. Shapiro, J. S.; Weston, R. E., Jr. *J. Phys. Chem.* **1972**, *76*, 1669.
83. Rodriguez, A. E.; Pacey, P. D. *J. Phys. Chem.* **1986**, *90*, 6298.
84. Kobrinsky, P. C.; Pacey, P. D. *Can. J. Chem.* **1974**, *52*, 3665.
85. Wigner, E. P. *Z. Phys. Chem.* **1932**, *B19*, 203.
86. Bell, R. P. *The Tunnel Effect in Chemistry*, Chapman and Hall: London, 1980, pp. 60-63.
87. Skodje, R. T.; Truhlar, D. G. *J. Phys. Chem.* **1981**, *85*, 624.
88. Garrett, B. C.; Truhlar, D. G. *J. Amer. Chem. Soc.* **1979**, *101*, 4534.
89. Kerr, J. A.; Parsonage, M. J. "Evaluated Kinetic Data on Gas-Phase Hydrogen Transfer Reactions of Methyl Radicals," Butterworths, London, 1976.
90. The value recommended by Kerr and Parsonage for $k_{CH_3 + H_2}$ is also recommended in Tsang, W.; Hampson, R. F. *J. Phys. Chem. Ref. Data* **1986**, *15*, 1087.
91. Arthur, N. L.; Newitt, P. *J. Can J. Chem.* **1985**, *63*, 3486.
92. Weston, R., personal communication.
93. Chandrasekhar, J.; Smith, S. F.; Jorgensen, W. L. *J. Amer. Chem. Soc.* **1985**, *107*, 154.
94. Tucker, S. C.; Truhlar, D. G. *J. Phys. Chem.* **1989**, *93*, 8138.
95. Barlow, S. E.; Van Doren, J. M.; Bierbaum, V. M. *J. Amer. Chem. Soc.* **1988**, *106*, 7240.
96. Truhlar, D. G.; Gordon, M. S. *Science* **1990**, *249*, 491.
97. Pople J. A.; Santry, D.; Segal, G. *J. Chem. Phys.* **1965**, *43*, S129.
98. Pople, J. A.; Beveridge, D. J. *Approximate Molecular Orbital Theory* (McGraw-Hill, New York, 1970).
99. Dewar, M. J. S.; Thiel, W. *J. Amer. Chem. Soc.* **1977**, *99*, 4899, 4107.
100. Dewar, M. J. S.; Zoebisch, E. G.; Healy, E. F.; Stewart, J. J. P. *J. Amer. Chem. Soc.* **1985**, *107*, 3902.
101. Dewar, M. J. S.; Zoebisch, E. G. *J. Mol. Struc. (Theochem.)* **1988**, *180*, 1.
102. Shiner, V. J., Jr. *ACS Symp. Ser.* **1975**, *11*, 163.
103. Kistenmacher, H.; Popkie, H.; Clementi, E. *J. Chem. Phys.* **1973**, *59*, 5842.
104. Clementi, E.; Cavallone, F.; Scordamaglia, R. *J. Amer. Chem. Soc.* **1977**, *99*, 5531.
105. Coker, D. F.; Miller, R. E.; Watts, R. O. *J. Chem. Phys.* **1985**, *82*, 3554.
106. Zhao, X. G.; Gonzalez-Lafont; Truhlar, D. G.; Steckler, R. *J. Chem. Phys.* **1991**, *94*, 5544.
107. Dacre, P. D. *Mol. Phys.* **1984**, *51*, 633.
108. Kauppi, E.; Halonen, L. *J. Phys. Chem.* **1990**, *94*, 5779.
109. Swain, C. G.; Bader, R. F. *Tetrahedron* **1960**, *10*, 182, 200.
110. Schowen, R. L. *Progr. Phys. Org. Chem.* **1972**, *9*, 275.
111. Collins, K. D.; Washabaugh, M. W. *Quart. Rev. Biophys.* **1985**, *18*, 323.
112. Truong, T. N.; Truhlar, D. G. *J. Chem. Phys.* **1988**, *88*, 6611.
113. Truong, T. N.; McCammon, J. A. *J. Amer. Chem. Soc.*, submitted.
114. Lynch, G. C.; Liu, Y.-P.; Lu, D.-h.; Truong, T. N.; Truhlar, D. G.; Garrett, B. C., unpublished.

RECEIVED October 3, 1991

Chapter 3

HN_2 and DN_2 Resonance Spectra
Scattering and Stabilization Calculations

Hiroyasu Koizumi[1], George C. Schatz[1], and Joel M. Bowman[2]

[1]Department of Chemistry, Northwestern University, Evanston, IL 60208–3113
[2]Department of Chemistry, Emory University, Atlanta, GA 30322

In this paper we have calculated the energies and lifetimes of several of the lowest energy vibrational levels (resonances) of HN_2 and DN_2 in their ground rotational states. These results are based on a recent global potential surface that was derived from large scale configuration interaction calculations. This surface indicates that HN_2 and DN_2 are unstable to dissociation, but their lowest few energy levels are relatively long lived. We have calculated the energies of these resonances using stabilization methods and using a coupled channel scattering method. The results from these calculations are in reasonable agreement. The HN_2 and DN_2 lifetimes were also obtained from the scattering calculations. The longest HN_2 lifetime is a few ns, while that for DN_2 is estimated to be close to 10^{-4} s.

HN_2 is an unstable molecule that has been postulated *(1)* to play an important role in the thermal De–NO_x process (*i.e.*, removal of NO from exhaust gases by reaction with NH_3) through the reaction

$$NH_2 + NO \longrightarrow \begin{cases} HN_2 + OH \\ N_2 + H + OH \\ N_2 + H_2O \end{cases}.$$

However, its importance in this and other reactions depends on its unimolecular decay lifetime, and this lifetime is highly uncertain. Kinetic modelling studies *(1)* indicate that a 10^{-4} s lifetime is needed to fit De–NO_x measurements, but HN_2 has not been observed experimentally in kinetics studies that should have been able to detect a species with that lifetime *(2)*. An attempt to observe it by neutralizing HN_2^+ in a beam was also unsuccessful *(3)*.

The lifetime and vibrational states of HN_2 have also been the subject of theoretical studies *(3-13)*. In all recent studies it has been concluded that the HN_2 minimum lies above the H + N_2 asymptote, with a 10-20 kcal/mol barrier to dissociation. For example, Curtiss *et al. (10)* used third-order and fourth order Møller-Plesset perturbation theory to calculate a 10.5 kcal/mol barrier, and they estimated a ground state lifetime (based on tunnelling through a one-dimensional parabolic barrier) of 7×10^{-11} s. Walch *et al. (11)* used CASSCF/CCI (complete active space self consistent field/externally contracted configuration interaction) calculations with two different basis sets to calculate barriers of 12.2 kcal/mol and 11.3 kcal/mol. Fits to these barriers using an Eckart potential yielded ground state lifetimes in the range 9×10^{-11} - 6×10^{-9} s.

Walch *(12)* has recently extended the calculations of Ref. 11 using slightly different basis sets to define the HN_2 surface globally. The points from these calculations have subsequently been fit to an analytical function by Koizumi *et al. (13)* with a root mean-square error of 0.08 kcal/mol, and coupled channel scattering calculations were performed to determine lifetimes of the lowest 11 vibrational states associated with the ground rotational state of HN_2. For the largest basis set that Walch used, the HN_2 minimum is 3.8 kcal/mol above H + N_2 on the fitted surface, with a barrier to dissociation of 11.4 kcal/mol. The lifetime of the ground vibrational state of HN_2 is predicted to be 3×10^{-9} s by the scattering calculations, which is consistent with earlier theoretical predictions *(11)* but not with kinetic modelling.

In this paper we extend the HN_2 studies of Koizumi *et al. (13)* in two ways. First, the spectrum of vibrational state energies and lifetimes has been extended to the DN_2 isotope. This species is expected to have a longer lifetime, and if it is long enough (say $>10^{-6}$ s) it may be possible to detect it using the methods of Ref. 4 or by spectroscopic measurements. One previous study of DN_2 *(3)* based on Møller-Plesset perturbation theory found a lifetime 50 times longer than for HN_2. This ratio is, however, very strongly dependent on potential surface *(3)*. Second, we have characterized both the HN_2 and DN_2 vibrational states by stabilization methods *(14)*. These calculations provide independent estimates of the state energies, and they also enable us to assign quantum numbers to the states. In addition, the stabilization calculations are useful in guiding the scattering calculations, particularly with respect to locating narrow resonances.

Potential Surface

A full description of the HN_2 potential surface is given in Ref. 13. Here we only consider the more accurate of the two analytical fits that were developed in that paper (denoted "surface 2"). Table 1 summarizes the properties of the HN_2 minimum and H···N_2 saddle point on this surface, including vibrational frequencies for both HN_2 and DN_2. A contour plot of the surface is presented in Fig. 1. In this plot we show how the surface varies with the Jacobi coordinates R (H–N_2 separation) and r (N–N separation). For each set of R, r, the energy has been minimized with respect to the Jacobi angle γ (the angle between R and r). The angle θ in Table 1 is related to γ by $\theta = \pi - \gamma$. Its value doesn't change significantly between the minimum and saddle point, but at larger R the energy minimizes at $\theta = 0$, meaning that the reaction path has C_{2v} symmetry at long range.

Figure 1 and Table I also show that the N–N stretch distance decreases significantly when HN_2 dissociates, from 2.250 a_0 at the minimum to 2.139 a_0 at the saddle point and to 2.0897 a_0 at infinite separation. There is also a significant increase in the N–N stretch frequency, 1826 cm^{-1} at the minimum, 1990 cm^{-1} at the saddle point, and 2323 cm^{-1} at infinite separation.

Table I. HN_2 and DN_2 Stationary Point Properties[a]

	H(D)N$_2$ Minimum	H(D)···N$_2$ Saddle Point
r (a$_0$)	2.250	2.139
R (a$_0$)	2.673	3.360
θ (deg)	49.1	45.5
r$_{NH}$ (a$_0$)	1.966	2.703
θ$_{NNH}$ (deg)	117.1	119.4
E (kcal/mol)	3.8	15.2

Frequencies (cm^{-1}):

HN$_2$:		
ω$_1$(H–N str)	2653	1667i
ω$_2$(N–N str)	1826	1990
ω$_3$(bend)	1047	749
DN$_2$:		
ω$_1$	1926	1230i
ω$_2$	1815	1958
ω$_3$	807	595

a. Based on fitted surface labelled "bs2" from Ref. 13.

The effect of deuterium substitution on the N–N stretch frequency ω_2 is small as might be expected while the H–N stretch (ω_1) and bend (ω_3) frequencies are reduced by roughly $2^{-\frac{1}{2}}$. One consequence of this is that ω_1, ω_2, and $2\omega_3$ are close in value for DN_2. This will lead to noticeable triplet structure in the vibrational energies.

Scattering and Stabilization Calculations

The coupled channel scattering calculations were done using a standard Jacobi-coordinate based propagation method (the nonreactive portion of a code that is described in Ref. 15). Total angular momentum J was taken to be zero in all calculations. (We have done a few calculations which indicate that the resonance spectrum for J > 0 is related to that for J = 0 according to the usual asymmetric top expression.) The vibration-rotation basis set considered for both H + N$_2$ and D + N$_2$ consisted of six asymptotic vibrational states (v = 0-5) and even rotational states j between 0 and 38 (i.e., 20 states) for each v (120 states total). Selected calculations using a basis of odd j's (which because of the symmetry of the potential are not coupled to the even j basis) produced resonance energies and widths identical to the even j basis for the lowest few energy resonances, indicating, as expected, that there is no coupling between resonances localized in one or the other of the symmetry equivalent HN_2 and N_2H wells. Selected calculations with a larger basis set (160 states) indicate that the ground and first excited resonance energies are converged (to 3 figures) with respect to further increases in basis size.

Convergence of the scattering results with respect to other parameters in the calculation was studied, including the number of Legendre terms used in expanding the potential, the region of R, r coordinates used to define the wavefunction, and the integration step sizes used for both r and R integrations. Masses used were $N = 14.00$ u, $H = 1.008$ u, and $D = 2.014$ u.

Two different kinds of stabilization calculations were done. The method used for the first (method 1) has been described extensively by Bowman and coworkers *(16)* in extensive applications to HCO. In this method, a direct product basis is defined in Jacobi coordinates that consists of contracted harmonic oscillator functions in the stretch coordinates (R,r) and contracted Legendre polynomials in the angle γ. A nonlinear parameter in the stabilization calculation, α, was used to scale the frequency of the harmonic oscillator basis for the R coordinate in determining stable eigenvalues.

The second stabilization calculation (method 2) also uses a basis defined in the Jacobi coordinates, but this basis consists of a Fourier sin series in R and r, and Legendre polynomials in γ. A sequential diagonalization-truncation calculation *(17,18)* was performed, first with the r variable, then R, and finally γ. For R and r, the secular equation was recast in a discrete variable representation (DVR) using equally spaced grid points according to a recent prescription given by Marston and Balint-Kurti *(19)*. A similar DVR *(20)* was used for γ. 50 grid points were used for each of the three Jacobi variables. The r grid covers the range $(1.5-3.0 \, a_0)$.

Stabilization of vibrational eigenvalues was determined by varying the grid range over which the R coordinate calculation was done. Calculations used the R intervals $(1.8, 3.5 \, a_0)$ and $(1.8, 3.7 \, a_0)$. Eigenvalues stable to three figures or better (i.e., <10 cm^{-1}) were taken as approximations to the resonance energies for the lowest 5-6 states of each isotope. Stability of higher states was poorer, sometimes only two figures.

Results

Scattering Calculations. Figure 2 presents the transition probabilities for $H + N_2$ and $D + N_2$ elastic scattering starting from the N_2 ground state (*i.e.*, $v = 0, j = 0 \rightarrow v' = 0, j' = 0$) as a function of the total energy E (measured relative to separated $H + N_2$ with N_2 at its equilibrium geometry). This elastic probability is only one of a large number of state-to-state transition probabilities that we have calculated, but its energy dependence is typical. This is demonstrated in Figs. 3 and 4, where we plot the $v = v' = 0$ transitions: $(j \rightarrow j') = (0 \rightarrow 10), (10 \rightarrow 10),$ and $(16 \rightarrow 16)$ for $H + N_2$ and $D + N_2$. In all the curves we see a slowly varying background probability upon which are superimposed sharp oscillations that are due to resonances. The same resonances

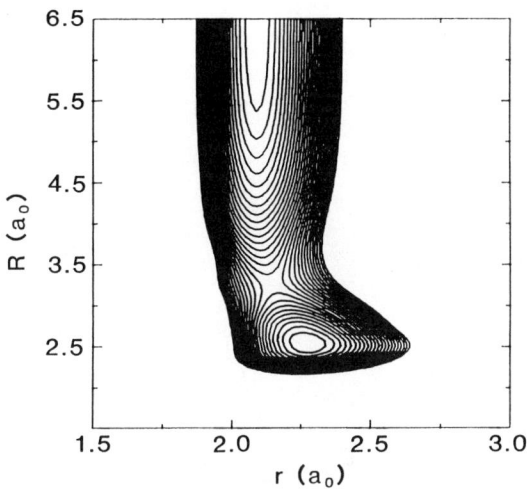

Figure 1. Contours of the HN_2 potential energy surface as a function of the Jacobi coordinates R and r for an angle γ that locally minimizes the energy. Contours are in 1 kcal/mol increments in the range 1-29 kcal/mol.

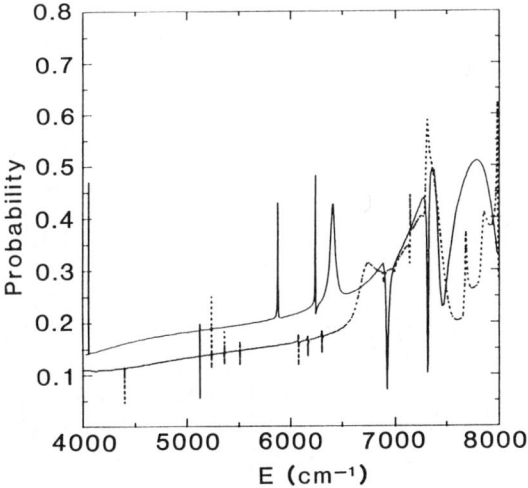

Figure 2. Ground state elastic transition probability ($v = 0$, $j = 0 \to v' = 0$, $j' = 0$) for $H + N_2$ (solid) and $D + N_2$ (dotted) as a function of the total energy E.

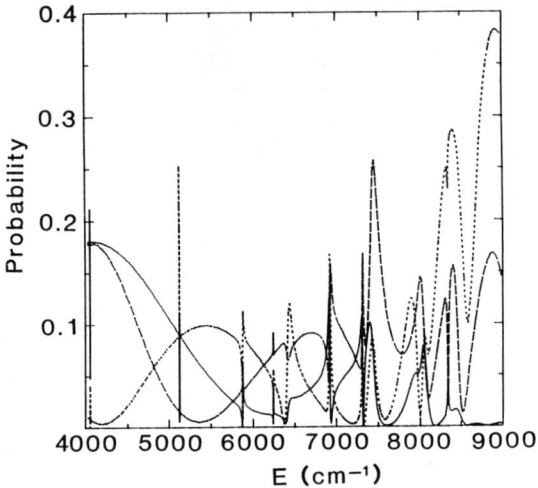

Figure 3. H + N_2 transition probabilities (vj) → (v'j'): (0,0) → (0,10) (solid), (0,10) → (0,10) (dotted), and (0,16 → (0,16) (dashed) versus energy E.

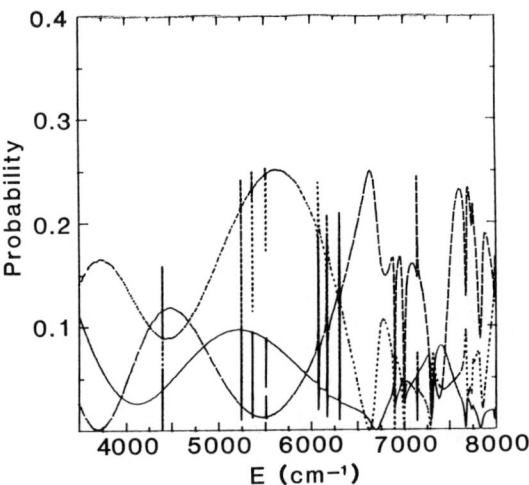

Figure 4. D + N_2 transition probabilities analogous to Fig. 3.

are seen in nearly all the curves for each isotope, implying significant coupling of each resonance to all the asymptotic rotor states. The widths of the resonances increase rapidly with increasing energy, which makes sense given that the barrier energy is 5316 cm^{-1}, and the addition of zero point energy raises this to 6686 cm^{-1} for HN$_2$ (higher than the lowest 5 resonances) and 6593 cm^{-1} for DN$_2$ (higher than the lowest 8 resonances).

The energies (E_{sc}), widths (Γ) and lifetimes ($\tau = \hbar$/width) of the lowest 12 resonances for HN$_2$ are listed in Table II, and the lowest 11 resonances for DN$_2$ in Table III. Both Tables also include harmonic oscillator (HO) estimates of the resonance energies and stabilization results $E_{st}^{(1)}$ and $E_{st}^{(2)}$ that will be discussed below. The harmonic, stabilization, and scattering energies are in reasonable correspondence, but for HN$_2$ we note that the (020) and (101) states are apparently not resolved in the scattering calculations (only one peak is seen in Fig. 2 where two closely spaced energy levels are predicted by the bound state methods).

One important omission from Table III is the scattering result for the DN$_2$ ground state. The lifetime listed there is based on a one-dimensional tunnelling calculation using an Eckart function that was constructed as indicated in Ref. 11 (see also Ref. 20) with the barrier width chosen to fit the lifetime for the (001) state. The lifetime thus obtained, 9×10^{-5} s, corresponds to an energy width of 6×10^{-8} cm^{-1}. Other choices of the Eckart barrier width (fitting to other resonance state results) yielded results within an order of magnitude of the value reported (thereby providing an estimate of the reliability of the Eckart estimate). We have been unable to find this resonance in our scattering calculations but we are confident that it is there. It is not difficult to estimate where it should be located to four significant figures, but our width estimate suggests that we need 11-12 figures to resolve it in the scattering calculations, and this is not a trivial task.

Figure 5 presents a semilog plot of resonance lifetime versus energy for HN$_2$ and DN$_2$. This figure shows a typical energy dependence for unimolecular decay (rapid at low energy, slow at high energy) with HN$_2$ lifetimes that are generally shorter than DN$_2$.

Stabilization Results. The stabilization energies in Tables II and III are in generally good agreement with each other and with the scattering results, but there are also some differences. The best agreement is for DN$_2$, where $E_{st}^{(1)}$ and $E_{st}^{(2)}$ are within 3 cm^{-1} for the first 4 states (and within 10 cm^{-1} for all states except (011) and (004)). The scattering results are also in reasonable correspondence, although it is disappointing that the difference between stabilization and scattering is never smaller than 20 cm^{-1}. For HN$_2$, $E_{st}^{(1)}$ and $E_{st}^{(2)}$ don't agree as well as for DN$_2$ (differences of 10-30 cm^{-1} for states other than (011)) but $E_{st}^{(2)}$ and E_{sc} are relatively close (few cm^{-1}) for the lower energy states.

Stabilization plots (energy eigenvalue versus nonlinear parameter α) are presented for method (1) in Fig. 6 for HN$_2$ and Fig. 7 for DN$_2$. In these plots, resonance states should show up as eigenvalues that are not dependent on α, while nonresonant eigenvalues vary rapidly, showing avoided crossings with the resonant states. Even the resonance energies vary somewhat with α, and as a result, these plots provide a good indication of possible errors in determining the $E_{st}^{(1)}$ values in Tables

Table II. HN$_2$ Resonance Energies, Widths, and Lifetimes

Assignment	E_{HO} (cm^{-1})	$E_{st}^{(1)}$ (cm^{-1})	$E_{st}^{(2)}$ (cm^{-1})	E_{sc} (cm^{-1})	Γ (cm^{-1})	τ (s)
(000)	4092	4045	4055	4053	2×10^{-3}	3×10^{-9}
(001)	5139	5116	5129	5124	3×10^{-2}	2×10^{-10}
(010)	5918	5769	5791	5872	6	9×10^{-13}
(002)	6186	6223	6243	6243	4	1×10^{-12}
(100)	6745	6384	6493	6412	65	8×10^{-14}
(011)	6965	6793	6823	6928	16	3×10^{-13}
(003)	7233	7294	7315	7315	16	3×10^{-13}
(020)	7744		7501		80	
(101)	7792		7565	7452		6×10^{-14}
(012)	8012			8001	40	1×10^{-13}
(004)	8280			8340	24	2×10^{-13}
(110)	8571			8469	80	6×10^{-14}
(102)	8839			8888	160	3×10^{-14}

Table III. DN$_2$ Resonance Energies, Widths, and Lifetimes

Assignment	E_{HO} (cm^{-1})	$E_{st}^{(1)}$ (cm^{-1})	$E_{st}^{(2)}$ (cm^{-1})	E_{sc} (cm^{-1})	Γ (cm^{-1})	τ (s)
(000)	3603	3562	3565		6×10^{-8} [a]	9×10^{-5} [a]
(001)	4410	4375	4376	4396	4×10^{-5}	1×10^{-7}
(002)	5217	5205	5206	5237	5×10^{-2}	1×10^{-10}
(010)	5418	5313	5314	5360	0.3	2×10^{-11}
(100)	5529	5428	5436	5507	0.2	3×10^{-11}
(003)	6024	6025	6017	6073	0.8	7×10^{-12}
(011)	6225	6159	6114	6170	2	2×10^{-12}
(101)	6336	6209	6219	6299	0.3	2×10^{-11}
(004)	6831	6857	6775	6888	3	2×10^{-13}
(012)	7032	6969	6969	6985	16	3×10^{-12}
(102)	7143		7122	7138	3	2×10^{-12}

a. Based on Eckart barrier calculation with parameters (notation of Ref. 11) $V_1 = 11.30$ kcal/mol, $V_2 = 15.56$ kcal/mol, $\omega^* = 1570i$ cm^{-1}.

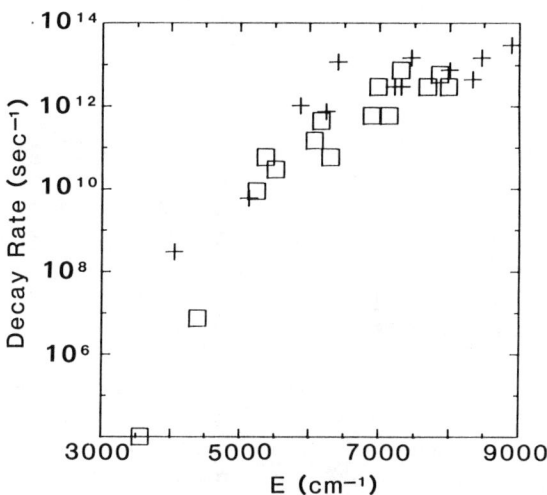

Figure 5. Semilog plot of lifetime versus energy E, showing results for HN_2 (pluses) and DN_2 (squares).

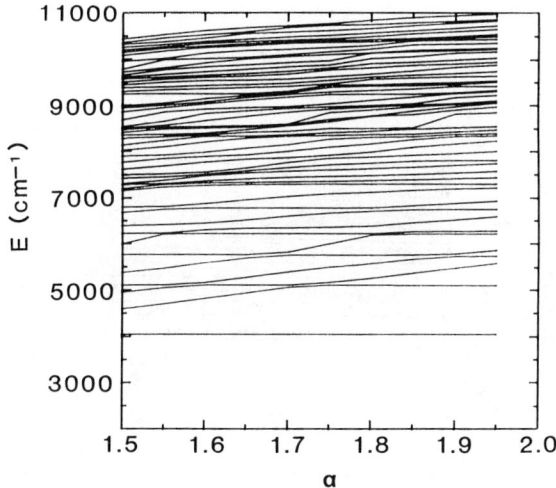

Figure 6. Stabilization plot of energy versus nonlinear parameter α for HN_2.

II and III. For DN_2, Fig. 7 indicates that the lowest 5 energy levels are very stable, so these energy levels should be accurately determined (perhaps suggesting that the scattering results are not fully converged). For HN_2 there is a good deal of uncertainty in determining energy levels, easily tens of cm^{-1} for states other than the ground state. This is consistent with the level of agreement between the stabilization and scattering results.

Conclusion

The most important results in this paper refer to the vibrational energy levels and lifetimes of HN_2 and DN_2. Our stabilization and scattering calculations agree reasonably well with respect to resonance energies, thus providing confidence in the correctness of each calculation, and enabling us to assign quantum numbers to the states. The energies are for the most part similar to the predictions of the harmonic approximation, but we do find perturbations due to anharmonicity.

The DN_2 ground state lifetime is estimated to be 9×10^{-5} s. This is, as expected, substantially longer than the 3×10^{-9} s HN_2 ground state lifetime, and it is long enough to make DN_2 observable in a number of experiments. For HN_2 we estimated earlier *(13)* that the likely uncertainty in our ground state lifetime estimate was about an order of magnitude due to errors in the potential surface. For DN_2, this uncertainty is increased by roughly an order of magnitude due to the approximate evaluation of the ground state lifetime. Experimental tests of our DN_2 lifetime estimate would be extremely important at this point as they would provide a sensitive test of both the HN_2 and DN_2 estimates.

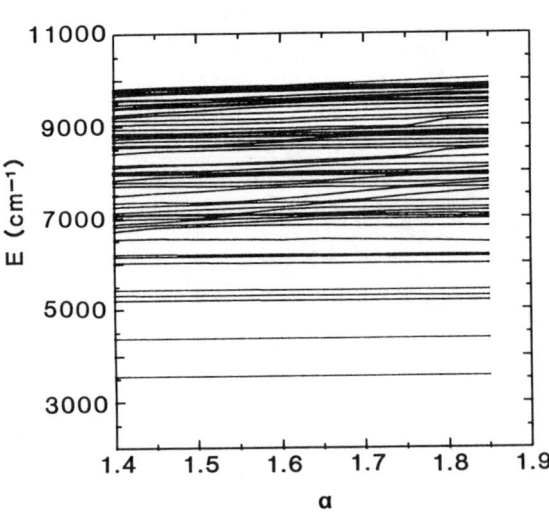

Figure 7. Stabilization plot similar to Fig. 5 but for DN_2.

Acknowledgments

HK and GCS acknowledge partial support of NASA grant NCC2-682 and NSF grant CHE-9016490. JMB acknowledges partial support of NSF grant CHE-8723042. Computations were done at the National Center for Supercomputing Applications (NCSA). HK and GCS thank Professor Z. Bačić for providing unpublished results used in testing the method 2 stabilization calculation.

Literature Cited

1. Miller, J. A.; Kee, R. J.; Westbrook, C. K. *Annu. Rev. Phys. Chem.* **1990**, *41*, 345.
2. Miller, J. A.; Bowman, C. T. *Prog. Energy Combust. Sci.* **1989**, *15*, 287.
3. Unfried, K. G.; Glass, G. P.; Curl, R. F. *Chem. Phys. Lett.* **1990**, *173*, 337.
4. Selgren, S. F.; McLaughlin, P. W.; Gellene, G. I. *J. Chem. Phys.* **1989**, *90*, 1624.
5. Lathan, W. A.; Curtiss, L. A.; Hehre, W. J.; Lisle, J. B.; Pople, J. A. *Progr. Phys. Org. Chem.* **1974**, *11*, 175.
6. Vasudevan, K.; Peyerimhoff, S. D.; Buenker, R. J. *J. Mol. Struct.* **1975**, *29*, 285.
7. Baird, N. C. *J. Chem. Phys.* **1975**, *62*, 300.
8. Baird, N. C.; Kathpal, H. B. *Can. J. Chem.* **1977**, *55*, 863.
9. Casewit, C. J.; Goddard III, W. A. *J. Am. Chem. Soc.* **1980**, *102*, 4057.
10. Curtiss, L. A.; Drapcho, D. L.; Pople, J. A. *Chem. Phys. Lett.* **1984**, *103*, 437.
11. Walch, S. P.; Duchovic, R. J.; Rohlfing, C. M. *J. Chem. Phys.* **1989**, *90*, 3230.
12. Walch, S. P. *J. Chem. Phys.* **1990**, *93*, 2384.
13. Koizumi, H.; Schatz, G. C.; Walch, S. P. *J. Chem. Phys.*, in press.
14. Taylor, H. S., *Adv. Chem. Phys.* **1970**, *18*, 91.
15. Schatz, G. C.; Hubbard, L. M.; Dardi, P. S.; Miller, W. H. *J. Chem. Phys.* **1984**, *81*, 231.
16. Gazdy, B.; Bowman, J. M.; Cho, S.-W.; Wagner, A. F. *J. Chem. Phys.* **1991**, *94*, 4192; Gazdy, B.; Bowman, J. M., in **Advances in Molecular Vibrations and Dynamics**, (JAI Press, Greenwich, 1991) in press.
17. Bačić, Z.; Light, J. C., *Annu. Rev. Phys. Chem.* **1989**, *40*, 469.
18. Bowman, J. M.; Gazdy, B., *J. Chem. Phys.* **1991**, *94*, 454.
19. Marston, C. C.; Balint-Kurti, G. G., *J. Chem. Phys.* **1989**, *91*, 3571.
20. Bačić, Z.; Light, J. C., *J. Chem. Phys.* **1986**, *85*, 4594.
21. Miller, W. H., *J. Am. Chem. Soc.* **1979**, *101*, 6810.

RECEIVED September 4, 1991

Chapter 4

Isotope Effects in Addition Reactions of Importance in Combustion
Theoretical Studies of the Reactions $CH + H_2 \rightleftarrows CH_3^* \rightleftarrows CH_2 + H$

Albert F. Wagner and Lawrence B. Harding

Chemistry Division, Argonne National Laboratory, Argonne, IL 60439

Ab initio electronic structure characterizations of the addition reaction path for the title reaction are described. Variational RRKM calculations employing the reaction path properties are then used to compute thermal rate constants for comparison to kinetics measurements on the title reactions and its isotopic variation.

In combustion chemistry, both addition and abstraction reactions have important roles to play. However, addition reactions, and their reverse, dissociation reactions, often provide a more complex chemistry because of the presence of long-lived metastable adducts that can be stabilized in the presence of third-body collisions or eliminate to produce a variety of products. Addition reactions frequently involve barrier-less reaction paths, often involve close competition in adduct decay between atomic or molecular elimination, and often can involve electronic excitation in the reactive chemistry. The particular reaction

$$CH(^2\Pi) + H_2 \rightleftarrows CH_3^* \rightleftarrows CH_2(^3B_1) + H \quad (1a)$$
$$\downarrow [M]$$
$$CH_3 \quad (1b)$$

where M is a third body and the * indicates metastability, demonstrates all of these features. Consequently this reaction has been the subject of several different types of experimental studies and a variety of theoretical studies as well.

Experimental kinetics studies have measured the thermal rate constants for addition in both directions (1-14). Isotope effects on three variants of reaction (1) have been performed (6 -10):

$$CD + D_2 \rightleftarrows CD_3^* \rightarrow CD_2 + D \quad (2a)$$
$$\rightarrow CD_3 \quad (2b)$$

$$CH + D_2 \rightleftarrows CHD_2^* \rightarrow CD_2 + H \quad (3a)$$
$$\rightarrow CHD_2 \quad (3b)$$

$$CH + D_2 \rightleftharpoons CHD_2^* \rightarrow CD + HD \quad (3c)$$
$$\rightarrow CHD + D \quad (3d)$$

$$CD + H_2 \rightleftharpoons CHD_2^* \rightarrow CDH + H \quad (4a)$$
$$\rightarrow CDH_2 \quad (4b)$$
$$\rightarrow CH + HD \quad (4c)$$
$$\rightarrow CH_2 + D \quad (4d)$$

The thermal dissociation rate constant (15) for reaction (1) has also been measured. Non-kinetics measurements using molecular beams (16-18) have characterized the state-resolved reactive and inelastic cross sections for reactions (1) and (3). Spectroscopy of CH_3 and all its isotopic variants have been measured with special attention (19,20) to the out-of-plane umbrella motion which has quartic characteristics.

Reaction (1) has also been the subject of several theoretical studies and is quite accessible to reasonably rigorous theory due to the few number of electrons involved. Several of the earliest electronic structure studies pointed out that reaction (1) is the simplest reaction with a non-least-motion pathway (21), i.e., CH does not insert into H_2 along a C_{2v} path. Several *ab initio* electronic structure studies have mapped out the general nature of either the $CH + H_2$ (21-22) or the CH_2+H (23) reaction path. However, these pioneering studies employed a relatively small basis set and modestly correlated wave function, resulting in limited accuracy in the reaction path characterization. All previous dynamics studies have focused on the calculation of rate constants. The earliest study (6) did not have the benefit of electronic structure characterizations of the reaction path. The other (23) did employ computed reaction path characteristics but examined only the high pressure limit of the reverse reaction (-1a).

Recently, a new series of theoretical studies (24, 25), of which this is the third, have begun to examine the reaction path characteristics (via electronic structure calculations) and the kinetics [via Rice-Ramsperger-Kassel-Marcus (RRKM) calculations] of reactions (1) - (4). The first study in this series, hereafter called Paper I, gave a description of the Multi-Reference Singles and Doubles Configuration Interaction (MRSDCI) electronic structure method used and the resulting characterization of the fragments CH, CH_2, CH_3, H_2, and the planar CH_2+H addition reaction path at the harmonic level. The high-pressure limiting rate constant of reaction (-1a) was also provided. The second study in this series, hereafter called Paper II, provided a preliminary harmonic description of the $CH+H_2$ reaction path and of the anharmonic out-of-plane motion along both reaction paths.

In this paper, a brief review of the electronic structure theory method used in the work will be provided. A fuller description of the out-of-plane anharmonic motion will be discussed. Variationally RRKM theory will then be applied to produce rate constants for comparison to the experimental results on reactions (1) - (4).

Details of Electronic Structure Calculations

As fully described in Papers I and II, all MRSDCI electronic structure calculations in this work were performed with the COLUMBUS program system (26). The multi-reference wavefunction, the configuration interaction, and the basis set will now be briefly reviewed.

The standard multi-reference wavefunction used in these studies is a FORS/CASSCF type (27) with 1 inactive orbital (the C(1s) orbital) and 7 active molecular orbitals correlating the remaining 7 electrons. This wave function, written in abbreviated form as (7mo/7e), results in 784 configuration state functions (CSFs), an expansion which is well within the capabilities of modern MCSCF methodology. At planar geometries, the above reference space reduces to the direct-product wave function (6mo/6e)x(1mo/1e), because the radical orbital must remain singly occupied in order for the total wave function to possess the required A" symmetry. This roughly halves the reference expansion length to 364 CSFs.

The singles and doubles configuration interaction calculations are based on either the (7mo/7e) or (6mo/6e)x(1mo/1e) reference wavefunction. Configuration interaction calculations performed with both reference wavefunctions at selected geometries of $CH+H_2$ showed differences that never exceeded 0.1 kcal/mole. All energy changes due to out-of-plane motion from planar geometries consistently use the (7mo/7e) reference in Papers I and II and also here.

In papers I and II, a correlation-consistent polarized valence triple zeta (cc-pVTZ) orbital basis set (28) was used to characterize the planar reaction path and equilibrium reactant and adduct properties at largely a harmonic level. In this paper, extensive non-planar, non-harmonic calculations along both reaction paths will be presented. Because of the expense of such calculations, a smaller basis set, cc-pVDZ, was used. This basis set has dissociation energies for $CH_3 \rightarrow CH+H_2$ and the $CH_3 \rightarrow CH_2+H$ about 3 and 5 kcal/mole smaller, respectively, than the energies for the cc-pVTZ basis set which in turn are very close to experiment (as will be discussed later). However, the computed reactant frequencies differ by no more than 50 cm^{-1}, the reactant geometries are essentially identical, and the variation from the asymptotic value of the potential energy along the reaction path differs by no more than a few tenths of a kcal/mole between the two basis sets. Consequently, previous cc-pVTZ and current cc-pVDZ MRSDCI calculations will be carefully mixed together in the reaction path characterizations used in the rate constant calculations. The absolute energetics separating reactants, products, and adducts will be taken from the cc-pVTZ calculations.

In addition to the MRSDCI energies, the use of various multireference Davidson corrections (30) has also been examined. Such empirical corrections are used to estimate the contributions of higher-order excitations. However, since the reference space itself is quite large in the cc-pVTZ calculations, these corrections appear to overestimate the importance of higher-order corrections and were not used. In the cc-pVDZ calculations, these corrections were used and generally gave comparable energies to cc-pVTZ where comparisons between the two calculations were made.

Reaction Path Results

In order to fully describe the kinetics, the reactants, products, adduct, and all reaction paths between them must be described. All reactant, product, and adduct theoretical descriptions are taken from Paper I and summarized in Table I. The frequencies listed in the table are harmonic frequencies. The harmonic representation will be used for all vibrational motion except out-of-plane motion along the reaction path. As indicated in Table I, the agreement between theory and experiment is quite satisfactory.

The energetics separating adduct, reactants, and products are all taken from the cc-pVTZ calculations of Paper I and are summarized in Table II. The energies listed in the table all include zero-point corrections, i.e., they are enthalpies at 0 K. In computing the zero-point corrections for the theoretical entries to the table the *harmonic* frequencies in Table I are used. Two theoretical entries are listed: one labelled "ab initio" and the other "adjusted". The ab initio column is directly calculated by the electronic structure calculations described in Paper I. By comparison with the experimental JANAF entries (31) in the table, the directly computed values are one or two kcal/mole below the nominal experimental values and only outside the error bars for experiment in the case of the dissociation energy for $CH_3 \rightarrow CH_2+H$. The adjusted column represents allowing slight changes in the computed energetics in optimizing the agreement in the computed and measured rate constants (to be described later). These show only one kcal/mole or less adjustments. The small variations between ab initio values and adjusted or experimental values testifies to the intrinsic accuracy of the electronic structure calculations and implies no more than a 1% underestimation of the dissociation energies of CH_3.

Table I. Calculated and observed properties of H_2, CH, CH_2, and CH_3

	r_e(Å)	θ_e(deg.)	harmonic frequency (cm^{-1})			
H_2: theory	0.745		4185			
exp't[a]	0.7412		4401			
CH: theory	1.125		2719			
exp't[b]	1.120		2859			
CH_2: theory	1.082	133.38	3153	3361	1143	
exp't[c]	1.0766	134.037	2985	3205	963	
CH_3: theory	1.0821	120.000	3126	3231,3231	1446,1446	520.0
exp't[d]	1.0790	120.000	3270	3285-3297	1436-1447	495-545

[a] B. Rosen, Spectroscopic Data Relative to Diatomic Molecules (Pergamon Press, New York, 1970).
[b] G. Herzberg and J. W. C. Jones, *Astrophys. J.* **158**, 399 (1969).
[c] Bunker, P. R.; Jensen,Per; Kraemer, W. P.; Beardsworth, *J. Chem. Phys.* **1986**, 85, 3724.
[d] Schatz, G. C.; Wagner, A. F.; Dunning, Jr., T. H.; *J. Physical Chem.* **1984**, 88, 221.

Table II. Calculated and observed reaction enthalpies at 0° K

process	theory[a] (ab initio)	theory[a] (adjusted)	experiment (Ref. 31)
$CH_3 \rightarrow CH + H_2$	102.7	104.0	105.6 ± 4.4 kcal/mole
$CH_3 \rightarrow CH_2 + H$	106.7	107.75	108.3 ± 1.2
$CH_2 + H \rightarrow CH + H_2$	4.0	3.75	2.7 ± 5.2

[a] Zero-point corrections made using harmonic frequencies in Table I.

The harmonic characterization of the planar reaction paths for $CH+H_2$ or CH_2+H addition have been described in Papers I and II and in paper II a cursory characterization of the out-of-plane motion in both paths was presented. Extensive characterizations of the out-of-plane motion have resulted in a somewhat more precise description for $CH+H_2$ but a substantially revised description of the entire reaction path in the case of CH_2+H. This channel is found to have a *non*-planar reaction path, even though CH_3, both in experiment and in the calculations, has a planar equilibrium structure. The results of these new calculations will be presented below. A more detailed account of these results is in preparation.

CH + H₂ Reaction Path

As discussed in Paper II, the CH+H₂ reaction path is planar in the kinetically important region but does not follow the least motion C_{2v} pathway where the CH bond inserts along the perpendicular bisector of the HH bond. Rather, at large distance of separation along the reaction path, the CH bond is aligned nearly parallel with the HH bond but displaced in the parallel direction with the C atom approximately located on the bisector of the HH bond, as in $|\cdots|$ where the line on the left is CH while the line on the right is HH. Unlike this schematic representation, the CH and HH bonds are not perfectly parallel to each other. The CH bond has a slightly acute and the HH bond has a slightly obtuse angle with respect to the vector R between C and the center-of-mass of H₂ and the dotted line in the schematic. Only well down into the potential well does the reaction path incorporate the angular motion that leads to the symmetric configuration of CH_3.

The out-of-plane motion along the reaction path can be thought of as the dihedral motion of the H-C-(center-of-mass of H₂) plane relative to the C-H₂ plane, i.e., the twirling of the CH bond about the vector R. Schematically, a 180° variation in this dihedral angle would correspond to $|\cdots| \rightarrow |\cdots|$. If all the other degrees of freedom perpendicular to the reaction path are allowed to relax during the much slower dihedral motion, then the barrier to this internal rotation would occur at the symmetric geometry where the dihedral angle is exactly at 90° and the HH bond is perpendicular to the R vector. Calculations of this relaxed barrier to hindered rotation have been carried out as a function of R at the cc-pVDZ level. Test calculations at one value of R suggest a simple constant plus cosine fit to potential change as a function of dihedral angle is reliable. The resulting sinusoidal fits in the vicinity of the reaction bottleneck (as discussed later) are displayed as a function of the dihedral angle in Figure 1 for various values of R-R_e where R_e is the value R assumed by CH_3 at equilibrium, i.e., 1.02 a_o. Generally, the barrier to hindered rotations is small, rising above a kcal/mole only for fairly deep penetration along the reaction path. The increase in the potential with deviations for planar configurations is always much smaller than the decrease in the potential with motion along the reaction path. Consequently, even at a dihedral angle of 90°, the potential at every value of R is still attractive relative to the CH+H₂ asymptote. The results in Figure 1 are similar to the limited results presented in Paper II that are based on cc-pVTZ calculations.

Given the force constant for harmonic displacements from planarity provided by the sinusoidal fits in Figure 1 and the reduced moments of inertia for H₂ and CH spinning about the R vector, the harmonic frequency for the torsional approximation to out-of-plane motion can be calculated. This frequency, along with the two other harmonic frequencies that arise out of free rotation of the reactants, i.e., in-phase and out-of-phase bending motion in the plane (presented in Paper II), are displayed in Figure 2 as a function of R-R_e. The figure has a logarithmic scale to illustrate the approximate exponential growth of all the frequencies that correlates to the free rotations of the reactants. An exponential growth in these types of frequencies with progress along the addition reaction path is often assumed (32) with an exponential factor of approximately 1.0 Å^{-1}. In this case, the factor is between 1.1 to 1.3 Å^{-1}. The small size of the out-of-plane frequency relative to the other two frequencies is another measure of how small the barrier to hindered out-of-plane rotation is. This suggests that a free-rotor, rather than torsional, approximation to this motion may be closer to the hindered rotor model.

CH₂ + H Reaction Path

As presented in Papers I and II, the reaction path for H+CH₂ has C_{2v} symmetry leading to the equilibrium planar structure of CH_3. The deviation from planarity was

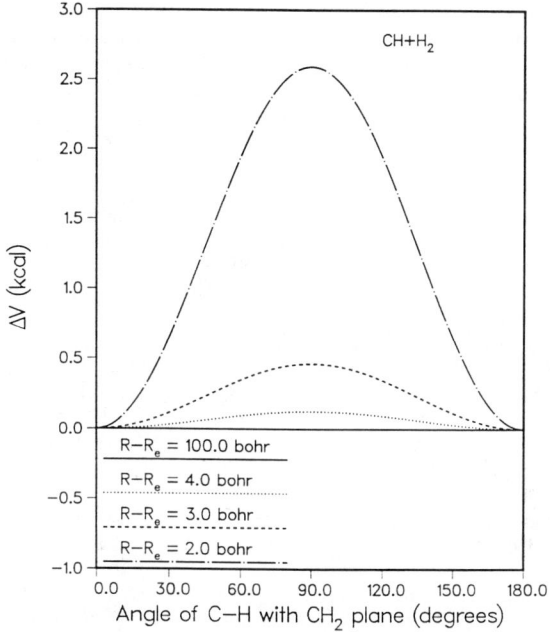

Figure 1. Plot of the variation of the potential energy surface with out-of-plane rotation angle at fixed distances of $R-R_e$ for $CH+H_2$ (see text).

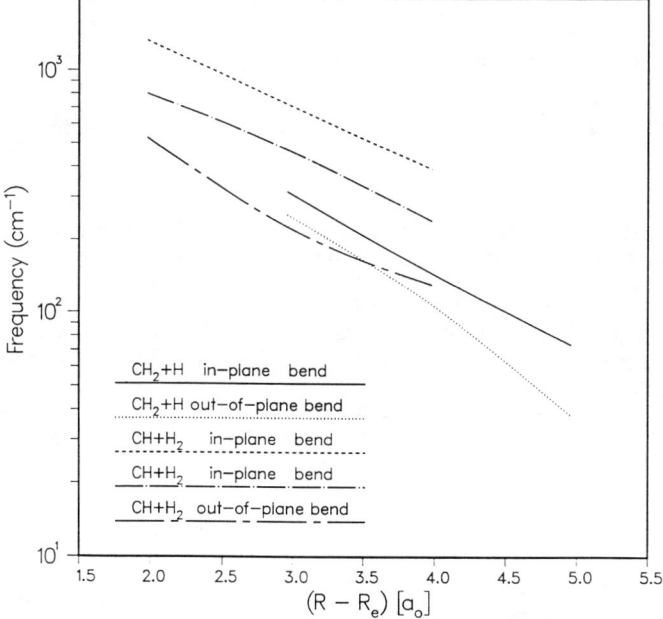

Figure 2. Plot of the in-plane and out-of-plane frequencies for motion perpendicular to the minimum energy path along the $CH+H_2$ or $H+CH_2$ minimum energy path as a function of $R-R_e$ (defined differently for each path [see text]).

briefly examined in Paper II. The out of plane motion corresponds to changes in the angle θ formed by the attacking H, the C, and midpoint between the two H's in CH_2. With this angle at 180°, the system is planar with the CH_2 splayed away from the attacking H, placing H+CH_2 in a •···⟨ or "Y" spatial configuration. If the angle is set to 0°, the system is still planer but an energetically less favorable ⟨···• or "arrow" configuration is formed. Calculations discussed in Paper II sampled a sparse grid of intermediate values of this angle and determined a monotonic rise in the potential as the angle decreased from 180° to 0°. This grid was too sparse to detect a *decrease* in the potential with decreases in the angle in the vicinity of 180°, i.e., a non-planar reaction path. In this regard, Merkle et al. (23) with a lower quality calculations did detect a non-planar reaction path.

More complete explorations of the out-of-plane geometries with the pVDZ calculations indicate that the reaction path is decidedly nonplanar except in the immediate vicinity of the equilibrium CH_3. In Figure 3, the change in the potential energy ΔV with variation in θ is displayed as a function of the progress along the reaction path. That progress is measured by R-R_e where R here is the distance of attacking H from C and R_e is the value of R at the CH_3 equilibrium (i.e., 2.039 a_0). The values of R-R_e are selected to be in the vicinity of the reaction path bottleneck (see below). (Note that the symbol R-R_e, used for *both* the CH+H_2 and H+CH_2 reactions paths as a measure of progress along the path, has a different definition for each path and assumes different values in the kinetically relevant region.) In Figure 3, the most favorable angle of H atom attack varies from about 100° at the larger distances to about 130° at closer distances along the reaction path, where 180° is the planar configuration. The barrier separating the above the plane (i.e., <180°) and below the plane (i.e., >180°) attack becomes more pronounced at closer distances but is always less than a kcal/mole and, at still closer distances, will decrease to zero as the planar CH_3 geometry asserts itself.

While the non-planarity in H+CH_2 is surprising, the more kinetically relevant feature of Figure 3 is the high barrier to complete rotation posed by the energetically unfavorable ⟨···• configuration. This barrier is on the order of seven times larger than the barrier between above and below the plane attack. Figure 3 and Figure 1 are on the same scale and show that the barrier hindering out-of-plane internal rotation in H+CH_2 (Figure 3) is about ten times higher than that hindering out-of-plane internal rotation in CH+H_2 (Figure 1) *at the common value of R* (i.e., due to the difference in R_e values between the two channels, a value of R-R_e of 4.0 bohr in Figure 1 corresponds to that R value where R-R_e has a value of 3.0 bohr in Figure 3). However, different values of R are in the kinetically important region for the different paths. The rate constants calculations described later indicate the kinetically relevant range of values of R are those given in Figures 1 and 3 for the same temperature range. The shorter values of R correspond to the high end of the common temperature range and the larger values of R to the low end of the temperature range. In that sense the barriers to out-of-plane motion show approximately the same variation over the common temperature range. Like the CH+H_2 path, even at the barrier at 0°, the potential at every value of R along the H+CH_2 pathway is still attractive relative to the asymptote.

For the convenience of the kinetics calculations discussed below, the potential energy variations shown in Figure 3 were approximated by a single minimum potential of the form:

$$\Delta V(\theta) = .5(V_0 - V_{180})(1 + \cos(\theta)) + V_{180} \tag{5}$$

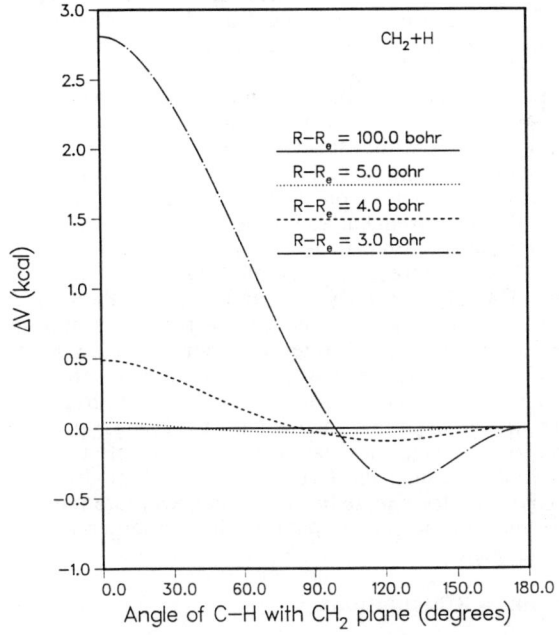

Figure 3. Plot of the variation of the potential energy surface with out-of-plane rotation angle at fixed distances of R-R$_e$ for H+CH$_2$ (see text).

Here the height of the barrier at 0° and at 180° is, respectively, V_0 and V_{180}. In the fitting, the higher and more kinetically limiting barrier at 0° was fixed at its true value while the value at 180° was selected to minimize the least squares error between the fitted potential and the actual potential in Figure 3. The resulting value of V_{180} was always negative, indicating the effect of the out-of-plane reaction path. Treatment of the out-of-plane rotation as single sinusoidal variation about a planar reaction path whose energy is corrected by the negative value of V_{180} allows the use of standard hindered rotor representations (33) in the kinetics calculations with negligible error in the rate constant (less than 10% by auxiliary calculations).

The reduced moment for the out-of-plane rotation is constructed from the lowest moment of inertia of the CH_2 fragment and the moment of the attacking H about the same CH_2 axis used for the lowest CH_2 moment. With the reduced moment and the fitted ΔV in equation (5), a harmonic frequency can be determined and compared to the other harmonic frequencies for motion perpendicular to the reaction path. In computing these other frequencies, the true out-of-plane reaction path is used, producing frequency values slightly different from those displayed in Papers I and II. The resulting comparison of frequencies is displayed in Figure 2 along with the analogous frequencies for $CH+H_2$ previously discussed. As the figure shows, the out-of-plane frequency is smaller, but not much smaller, than the other frequency evolving out of free rotations of the fragments. Both frequencies are growing approximately exponentially with respect to progress along the reaction path with an exponential factor similar to that in $CH+H_2$. Comparison of the frequencies for the two paths show the out-of-plane frequencies have a nearly common R dependence. On the other hand, the other frequencies for the two paths show larger values for the $CH+H_2$ path than for the $H+CH_2$ path at a given value of $R-R_e$.

Because $CH+H_2$ has three frequencies arising out of free rotation of the fragments while $H+CH_2$ has only two and because there is not a complete predominance in size of frequencies of one path over the other, it is not clear from the frequencies alone which channel is tighter, i.e., more sterically hindered. The most relevant feature of steric hindrance shown in Figure 2 is the absolute value of $R-R_e$ for the two channels. For the $H+CH_2$ path the values are larger, indicating that, in the kinetically important regions of the reaction path, the $H+CH_2$ path is more extended, with consequently lower frequencies and higher moments of inertia, and in that sense is less sterically hindered than the $CH+H_2$ path. Although both paths show no barriers, the $H+CH_2$ path for simple bond fission goes more rapidly down hill than the $CH+H_2$ molecular rearrangement path, leading to larger values of $R-R_e$ in the kinetically relevant regions.

Details of Dynamics Calculations

Given the calculated addition reaction path energetics, structure, and harmonic frequencies or hindered rotational barriers reported above or in Papers I and II, the thermal addition rate constants for reactions (1) - (4) can be calculated with variational RRKM theory (34,35). RRKM theory uses a steady state expression for the temperature and pressure dependent rate constant for reactant loss. This expression involves rates for complex formation, e.g., $CH+H_2 \rightarrow CH_3^*$, complex decay back to reactants, e.g., $CH_3^* \rightarrow CH+H_2$, complex decay on to products, e.g., $CH_3^* \rightarrow CH_2+H$, and stabilization of the complex, e.g., $CH_3^*+M \rightarrow CH_3+M$ (from which the pressure dependence arises). While empirical or adjustable stabilization rates must be provided, RRKM theory provides a statistical, transition-state-theory-like expression for all the other rates in terms of a ratio of partition functions at the transition state and at the reactants times a rate of crossing through the transition state. A variational RRKM approach is required for there are no potential energy barriers to addition along either pathway to locate the transition state and therefore its location must be varied along the reaction path until the most constraining, and therefore optimum, location (i.e., the reaction bottleneck) is found. There are five features to the RRKM calculations reported here that are described below.

First, a canonical, rather than micro-canonical, variational RRKM theory was carried out. This means that the reaction bottleneck was located for each reaction path (including all isotopic variations) as a function of temperature. A more rigorous location of the bottleneck as a function of total energy E and total angular momentum J has not yet been done.

Second, explicit summation over the total angular momentum was included in the calculations in order to be sensitive to the changes in extension of the geometries of the reaction bottlenecks for the different reaction paths. The quantum number K for the projection of J on the principal axis of CH_3^* was treated as active.

Third, the influence of the buffer gas M was treated in an analytic way (36) that approximates the rigorous Master Equation description of the effect of the buffer gas. In this approximate approach, an effective rate constant for buffer-gas collision-induced stabilization of metastable CH_3^* to thermalized CH_3 is derived from a gas kinetic rate constant (using approximate Lennard-Jones parameters (37)) modified by a scaling constant γ that is a function of ΔE_{tot}, the average energy lost per collision of CH_3^* with M. The scaling constant is derived from analytic solutions to a simplified Master Equation model in which only one unimolecular decay route (not two as in CH_3) is available. There is no direct measurements of ΔE_{tot} and in the calculations it is used as a temperature and isotopically independent adjustable constant. The significant approximations used in the treatment of the buffer gas are probably partially remedied by the adjustable value of ΔE_{tot}.

Fourth, the treatment of the out-of-plane motion was determined by preliminary calculations of the high pressure limit to CH_3 addition along both pathways. Since the high pressure limiting rate constant is reached only at pressures where every adduct formed is stabilized, it is sensitive only to the addition reaction path characteristics, not to any competing pathways for adduct unimolecular decay or to the properties of the adduct itself. For $CH+H_2$, negligible variation in the rate constant (<5%) between an free or hindered rotor treatment of the out-of-plane motion was found while both treatments differed substantially from a harmonic torsional treatment. This is consistent with the very low barriers to hindered rotation seen in Figure 1. In the RRKM calculations, a free rotor treatment was used throughout. For $H+CH_2$, a similar agreement between the three models was seen at lower temperatures but, between about 1000 K to 3000 K, the hindered rotor treatment becomes more similar to the torsional treatment. This is also consistent with the high barriers to hindered rotation seen in Figure 3. In the RRKM calculations, a hindered rotor treatment was used throughout.

Fifth and last, while the variation of energy, structure, and frequencies were not changed from the calculated values, as was described above, the asymptotic, zero-point corrected dissociation energies were varied to obtain best agreement with experiment. For the addition rate constants discussed below, the only asymptotic energetics important is the *difference* between the energies for CH_3 dissociation to $CH+H_2$ and $H+CH_2$. As listed in Table II, the adjustment decreased the directly computed value by 0.25 kcal/mole. The exact value of the dissociation energy for any one channel is only sensitive to the thermal dissociation rate constant. Although that rate constant will not be discussed here, an increase of 1.0 kcal/mole brings the computed dissociation rate constant into good agreement with the measured rate (15). This is the value used in the calculations discussed below and the adjusted value listed in Table II. The resulting adjusted values lie within the JANAF error bars (31). The very minor changes between directly computed and adjusted values testifies to the accuracy of the electronic structure calculations.

Kinetics Results and the Comparison to Experiment

The results presented here will concentrate on the isotope effects in thermal rate constants, in particular the addition rate constants of $CH+H_2$ and its isotopic variants as written in reactions (1) - (4). Not only is the experimental record available for comparison for these particular reactions but they show all the most important isotope effects. In principle, both temperature and pressure variations of the addition rate constant can be examined. However, the isotope effects are most clearly brought out with the temperature variation of the rate constant at a fixed pressure of buffer gas. The experimental record is the most complete in this regard also and in all cases argon was the buffer gas used. Therefore the theoretical results will focus on the temperature variation of the rate constant at a fixed argon pressure. The results obtained below are similar to those obtained by Berman and Lin (6) who did not have access to reaction path characterizations from electronic structure calculations.

Discussion of: (1) the pressure dependence of the addition rate constant, (2) the addition rate constant from the $H+CH_2$ direction, and (3) the thermal dissociation rate constants will be presented elsewhere in a fuller discussion of all the kinetics. In the case of $H+CH_2$ addition or CH_3 dissociation rate constants, no isotope effects have been measured.

$CH+H_2$ and Isotopic Variants

The $CH+H_2$ addition rate constants have received the most experimentally complete study of any of the thermal rate constants of the CH_3 system and its isotopic analogs. Three different rate constant measurements (6, 7, and 9) as a function of inverse temperature at a fixed pressure of argon of either 4 or 100 torr are represented in Figure 4. The calculated rate constants for two different values of ΔE_{tot} are also displayed in the figure. The comparison between the theory and experiment is excellent. Both display a decline of the rate constant with increasing temperature for the lower temperatures. Theory confirms that this decline is due to the dominance of buffer-gas stabilized CH_3 as the reaction product. As the temperature increases, the excess energy content of the initially formed metastable CH_3^* increases and the buffer gas can not effectively drain away the excess energy in competition with unimolecular decay back to the reactants. Consequently, the rate constant declines with temperature. However, at the higher temperatures, the dominant reaction product becomes $H+CH_2$, which is 3.75 kcal/mole endoergic (see Table II) and thus is activated by higher temperatures. A decrease in the fixed pressure dramatically lowers the rate constant for themalized CH_3 formation but does not effect substantially the rate constant for $H+CH_2$ formation, a process that can proceed, at high enough temperature, at zero pressure. The low pressure measurements of Becker et al. (9) in particular most clearly expose the atomic bond fission product channel over the largest range of inverse temperature space.

The value of ΔE_{tot} most optimal for agreement to experiment for $CH+H_2$ and also for the other isotopic variations lies between about -50 to -75 cm^{-1}. This is somewhat lower than measurements of this quantity for other, generally much larger, metastable adducts being stabilized by argon (38). However, CH_3 is a small adduct with only one low frequency vibrational mode and is therefore not too similar to systems for which measurements have been performed. At least one other small adduct with no low frequency vibrational modes, HCO, has required similar low values for ΔE_{tot} in order to obtain optimal agreement between theory and experiment for $H+CO \rightleftharpoons HCO$ thermal rate constants (39).

The $CD+D_2$ addition rate constants have been studied by two separate experiments (6, 10) and the resulting addition rate constant as a function of inverse temperature is displayed in Figure 5 for a fixed pressure of argon at 100 torr. The

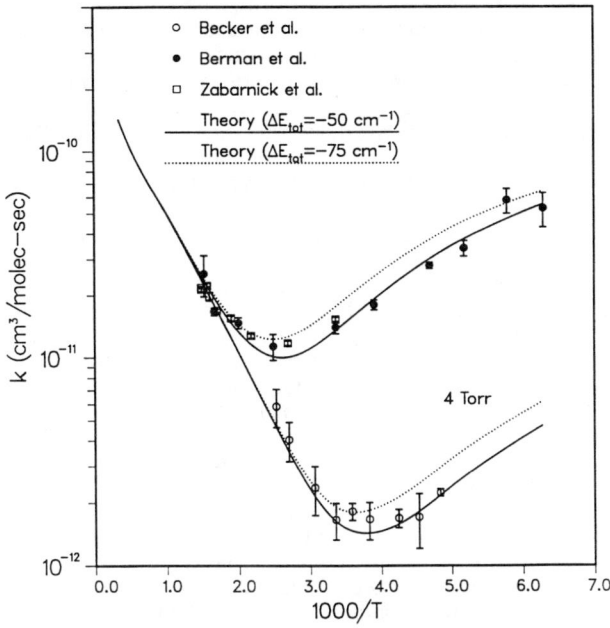

Figure 4. The thermal addition rate constant for CH+H$_2$ as a function of temperature at a fixed 100 torr pressure of Ar buffer gas.

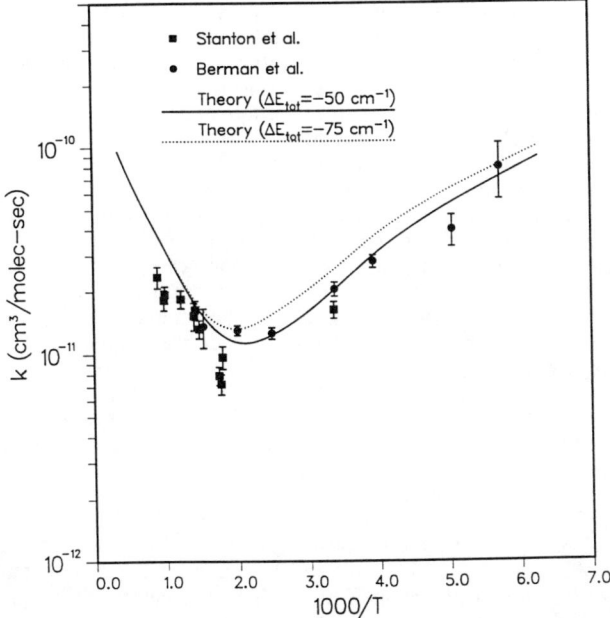

Figure 5. The thermal addition rate constant for CD+D$_2$ as a function of temperature at a fixed 100 torr pressure of Ar buffer gas.

theoretical results, using the same values of ΔE_{tot} used in $CH+H_2$, are also displayed on the figure. The theory and the experiment are in good agreement and both indicate that the rate constant has a temperature dependence very similar to that of $CH+H_2$.

At a fine level of comparison between theory and experiment, it would appear that the high temperature increase in the rate constant due to the production of $D+CD_2$ is occurring at a slightly lower temperature than that indicated by the experiments of Stanton et al. (10). The variation in the calculations for different values of ΔE_{tot} indicates that only massive and unrealistic changes in this value from that used for the $CH+H_2$ system could effect an improvement between theory and experiment. The agreement with experiment could be improved by a slight (~0.25 kcal/mole) increase in the endothermicity for $CD+D_2 \rightarrow CD_2+D$. However such an increase would also push to higher temperatures the rate constant increase in $CH+H_2$ to such an extent that the calculated rate constant would fall below the error bars at high temperature in the Becker et al. experiment (9). If one accepts that the theory is correctly determining small variations between isotopes, then the results in Figures 5 and 6 could be interpreted as indicating a small inconsistency between the experiments of Becker et al. and Stanton et al. A further interpretation would be that the experiments of Stanton et al. may slightly underestimate the rate constant at higher temperatures. This interpretation is relevant to the following isotopic variation.

For $CH+D_2$, the only rate constant measurements over a variety of temperatures is by Stanton et al. at a fixed pressure of 20 torr of argon. The results, along with the theoretical calculations, are displayed in Figure 6. As can be seen from the figure, the temperature dependence of the rate constant changes dramatically from that of the fully protonated or deuterated system. This is due to the fact that this isotopic variant has two additional product channels, as detailed in reaction (3). One of those products, $CD+HD$, is an exoergic isotope exchange that requires no pressure or temperature activation. The theoretical calculations show that at the lower temperatures in the figure, this process completely dominates. There is almost no influence of pressure on the results, and, consequently, almost no influence of ΔE_{tot}, because this isotope exchange can proceed at zero pressure. The rate constant at low temperatures does not fall with temperature, as in $CH+H_2$ or $CD+D_2$, because adduct stabilization is not an important channel. In this lower temperature region, the theory and experiment agree. At higher temperatures, the theory predicts that $H+CD_2$ becomes a product channel competitive with isotope exchange. (The other triatomic product channel, $D+CHD$, is 1.5 kcal/mole more endoergic and over the temperature range displayed is always a trace product.) This endoergic channel is activated by temperature giving rise to the increase of the rate at higher temperatures for the same reasons found in the other isotopic variants. The experimental results show no clear evidence of this increase in the rate constant and tends to grow somewhat noisier at the higher temperatures. As in the case of $CD+D_2$, the atomic bond fission channel could be made more endoergic, delaying its appearance in the temperature dependence of the addition rate constant to much higher temperatures. However, even massive changes in the endoergicity would lower but not eliminate the disagreement between theory and experiment at high temperatures. Since the atomic bond fission channels and the isotope exchange channel compete with each other, the calculations indicate that elimination of the atomic bond fission channel by artificially increasing its endoergicity will also have the effect of increasing the rate constant for isotope exchange. While the overall rate will decline, the calculated rate constant will still lie above the measured rate constant at the higher temperatures. An interpretation of this discrepancy is that the Stanton et al. measurements tend to somewhat underestimate the rate at higher temperatures in both $CH+D_2$ and in $CD+D_2$.

For $CD+H_2$, the measured rate constant consists of one value at room temperature in 100 torr of argon (6). This result and the calculated rate constant over the full temperature range are displayed in Figure 7. The calculations fall below the measured

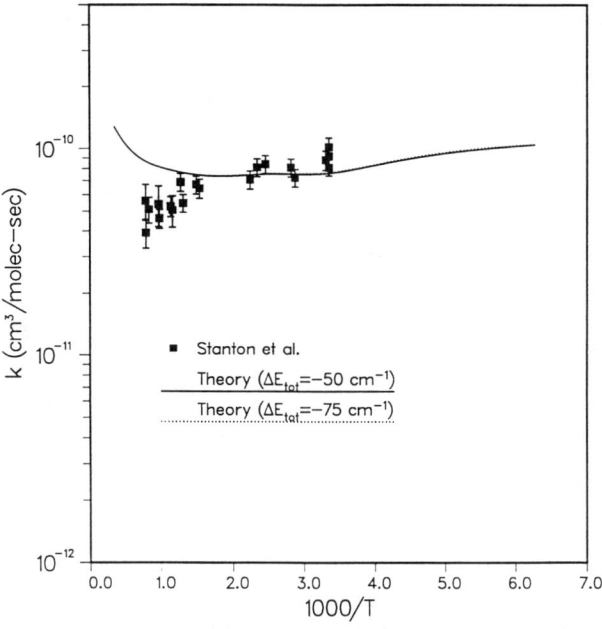

Figure 6. The thermal addition rate constant for CH+D$_2$ as a function of temperature at a fixed 20 torr pressure of Ar buffer gas.

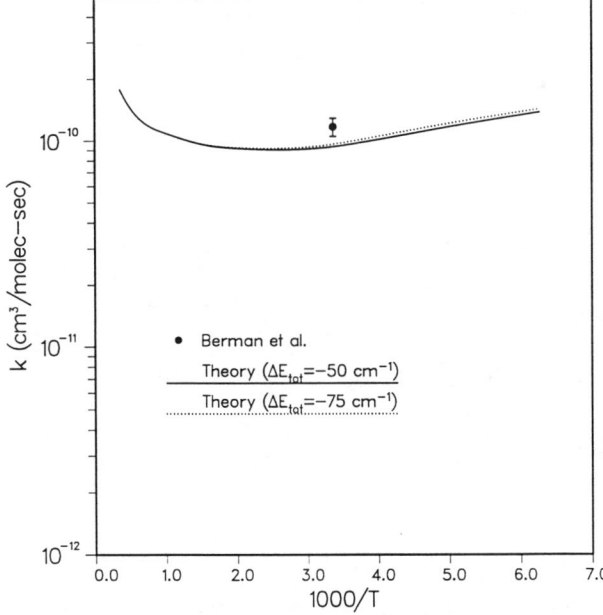

Figure 7. The thermal addition rate constant for CD+H$_2$ as a function of temperature at a fixed 100 torr pressure of Ar buffer gas.

value by about 10% to 20%. Consideration of the scatter found in the measured data as a function of temperature in the other isotopic variants (Figures 4 - 6) suggests this small difference is probably not significant. The overall theoretical temperature dependence is very similar to that of $CH+D_2$. Both $CD+H_2$ and $CH+D_2$ have four product channels, i.e., reactions (4) and (3), respectively, and have similar temperature dependences for the same reason. In $CD+H_2$, the simple isotope exchange to $CH+HD$ is slightly endoergic. At quite low temperatures then, the product is dominated by CDH_2 product formation. In $CH+D_2$, at such low temperatures, the isotope exchange process dominates.

The remaining two other isotopic variants of $CH+H_2$, namely $CH+HD$ and $CD+HD$, have never been examined experimentally. These two sets of reactants sample reactions (3) and (4) from a direction different from that written. No theoretical results will be presented, but the temperature dependence at fixed pressure is very similar to that of $CH+D_2$ and $CD+H_2$ for related reasons.

Conclusion

Isotopic variations in the addition thermal rate constants for $CH+H_2$ have been calculated and compared to experiment using variational RRKM theory and an *ab initio* electronic structure characterization of the reaction paths $CH+H_2$ and $H+CH_2$ and their isotopic variants. In general the agreement is good. Both theory and experiment clearly distinguish the change in character between the mixed isotope additions, i.e., $CH+D_2$, and the fully protonated or deuterated additions. The consistency of agreement over all the isotopic variations suggests several of the experimental measurements may have somewhat underestimated the rate constant.

Acknowledgments. This work was performed under the auspices of the Office of Basic Energy Sciences, Division of Chemical Sciences, U.S. Department of Energy, under Contract W-31-109-Eng-38. The large scale computing resources required for this project were provided in part through a Grand Challenge grant under the auspices of the Office of Basic Energy Sciences, Office of Scientific Computing, U.S. Department of Energy.

Literature Cited

1. Bohland, T.; Temps, F. Ber. Bunsenges. *Phys. Chem.* **1984**, 88, 459.
2. Grebe, J.; Homann, K. H. Ber. Bunsenges. *Phys. Chem.* **1982**, 86, 581.
3. Frank, P.; Bhaskaran, K. A.; Just, Th. *J. Phys. Chem.* **1986**, 90, 2226.
4. Lohr, R.; Roth, P. Ber. Bunsenges. *Phys. Chem.* **1981**, 85, 153.
5. Peeters, J.; Vinckier, C *Comb. (International) Symp.* **1974**, *15*, 969.
6. Berman, M. R.; Lin, M. C. *J. Chem. Phys.* **1984**, 81, 5743.
7 Zabarnick, S.; Fleming, J. W.; Lin, M. C. *J. Chem. Phys.* **1986**, 85, 4373.
8. Becker, K.H.; Engelhardt, B.; Wiesen, P.; Bayes, K. D. *Chem. Phys. Lett.* **1989**, *154*, 342.
9. Becker, K. H.; Kurtenbach, R.; Wiesen, P. *J. Phys. Chem.* **1991**, 95, 2390.
10. Stanton, C.T.;Garland, N.L.; Nelson, H.H. *J. Phys. Chem.* **1991**, 95, 1277.
11. Braun, W.; McNesby, J. R.; Bass, A. M. *J. Chem. Phys.* **1967**, 46, 2071.
12. Butler, J. E.; Goss, L.P.; Lin, M. C.; Hudgens, J.W. *Chem. Phys. Lett.* **1979**, 63, 104.
13. Anderson, S. M.; Freedman, A.; Kolb, C. E. *J. Phys. Chem.* **1987**, 91, 6272.
14. Bosnali, M. W.; Perner, D. Z. *Naturforsch.* **1971**, 26a, 1768.
15. Roth, P.; Barner, U.; Lohr, R. Ber. Bunsenges. *Phys. Chem.* **1979**, 83, 929.
16. Liu, K.; Macdonal, G. *J. Chem. Phys.* **1988**, 89, 4443.

17. Liu, K.; Macdonal, G. *J. Chem. Phys.* **1990**, *93*, 2431.
18. Liu, K.; Macdonal, G. *J. Chem. Phys.* **1990**, *93*, 2443.
19. Jacox, M. E. *J. Mol. Spectrosc.* **1977**, 66, 272.
20. Holt, P. L.; McCurdy, K. E.; Weisman, R. B.; Adams, J. S.; Engel, P.S. Ibid. **1984**, 81, 3349.
21. Brooks, B. R.; Schaefer, III, H.F. *J. Chem. Phys.* **1977**, 67, 5146.
22. Dunning, Jr., T. H.; Harding, L. B.; Bair, R. A.; Eades, R. A.; Shepard, R. L. *J. Phys. Chem.* **1986**, 90, 344.
23. Merkel, A.; Zulicke, L. *Molec. Phys.***1987**, 60, 1379.
24. Aoyagi, M.; Shepard, R.; Wagner, A. F.; Dunning, Jr., T. H.; Brown, F. B. *J. Phys. Chem.* **1990**, *94*, 3236.
25. Aoyagi, M.; Shepard, R.; Wagner, A. F. *Intl. J. Suprecomp. Appl.* **1991**, *5*, 72.
26. Shepard, R.; Shavitt, I.; Pitzer, D. C.; Pepper, M.; Lischka, H.; Szalay, P. G.; Ahlrichs, R.; Brown, F. B.; Zhao, J.-G. *Int. J. Quantum Chem.* **1988**, S22, 149.
27. Shepard, R. in *Ab Initio Methods in Quantum Chemistry II, Advances in Chemical Physics*, K. P. Lawley, Ed. (Wiley, New York, 1987), Vol. 69, pp. 63-200.
28. Dunning, Jr., T. H. *J. Chem. Phys.* **1989**, 90, 1007.
29. Aoyagi, M.; Dunning, Jr., T. H.(to be published)
30. Shavitt, I.; Brown, F. B.; Burton, P. G. *Int J. Quantum Chem.* **1987**, 31, 507.
31. Chase, Jr., M.W.; Davies, C. A.; Downey, Jr., J. R.; Frurip, D. J.; McDonald, R. A.; Syverus, A. N. *J. Phys. Chem.* Ref. Data **1985**, 14, 1211.
32. Cobos, C.J.; Troe, J. *J. Chem. Phys.* **1985**, 83, 1010.
33. Pitzer, K. S.; Gwinn, W. D. *J. Chem. Phys.* **1942**, *10*, 428.
34. Robinson, P. J.; Holbrook, K. A. Unimolecular Reactions : Wiley-Interscience, New York, 1972.
35. Truhlar, D. G.; Hase, W. L.; Hynes, J. T. *J. Phys. Chem.* **1983**, 87, 2264.
36. Troe, J. *J. Chem. Phys.* **1977**, *66*, 4745.
37. Hippler, H.; Troe, J.;Wendelken, H. J. *J. Chem. Phys.* **1983**, *78*, 6709.
38. Dove, J. E.; Hippler, H.; Troe, J. *J. Chem. Phys.* **1985**, *82*, 1907.
39. Timonen, R. S.; Ratajczak, E.; Gutman, D; Wagner, A. F. *J. Phys. Chem.* **1987**, *91*, 5325.

RECEIVED September 4, 1991

Experimental Studies of Isotope Effects in Reactions of Neutral Systems

Chapter 5

Hydrogen Isotope Effects in Chemical Reactions and Photodissociations

B. Katz[1] and R. Bersohn[2]

[1]Department of Chemistry, Ben Gurion University of the Negev, Beersheba, Israel
[2]Department of Chemistry, Columbia University, New York, NY 10027

A review is presented of recent results on the cross section and energy release in the exchange collisions of fast hydrogen atoms with deuterium containing molecules and reactions in which neutral atoms attack HD. The cross sections for the $H + D_2$ reaction at different energies are in excellent agreement with theoretical values based on classical trajectories on an ab initio surface. With the use of polarized light to generate H atoms the reactions $H + MD_4 \longrightarrow D + MD_3H$ (M=C,Si) are proven to proceed via an inversion mechanism. A review is also presented of recent results on the atomic H/D ratios and energy release observed in the photodissociation of partially deuterated molecules such as HDO, HCCD and CH_nD_{4-n} (n=1,2,3). There is always a preference for H atom release either because its wave function is more diffuse or it tunnels more easily or it forms slightly weaker bonds than does D.

I. Hydrogen Isotope Effects in Exchange Reactions and Attack of Neutral Atoms on HD

Experimental Methods. The experiment is essentially a pump and probe experiment(1). The pump is an excimer laser which produces H atoms by photodissociation or a dye laser which produces excited atoms and the probe is a laser which excites hydrogen atoms to the 2p state at a wavelength of 121.6 nm. The latter light is generated by a nonlinear four wave mixing process. In earlier work (1) third harmonic generation was used in which 364.8 nm light generated by an excimer pumped dye laser was focused in krypton gas(2). More recently the development of barium metaborate (BBO) crystals has permitted the use of a more efficient four wave mixing process. Light at 212.6 nm is focused into a Kr cell virtually exciting a two photon transition in the Kr atom. Simultaneously 845 nm light is focused at the same point generating by a 2 $\omega_1 - \omega_2$ process the desired 121.6 nm light. The second moment of the fluorescence excitation curve yields $<v_z^2>$ where z is the axis of the probing light. By adjusting the polarization of the pumping light, one can obtain all three averages, $<v_x^2>$, $<v_y^2>$ and $<v_z^2>$. In this way one determines the anisotropy (if any) of the velocity distribution and the total kinetic energy.

Exchange Reaction: H + D$_2$. Bimolecular exchange reactions in the gas phase are of fundamental importance in molecular dynamics. This is even more true when hydrogen atoms appear in the reaction as both reactants and products. The simpler the reaction the less difficult it is to construct a potential surface for the reaction and to use it for calculating observables such as cross sections and translational energy release. Recently (3) we have measured the absolute cross sections of some of the elementary reactions involving the exchange of H for D. The H + D$_2$ reaction and its isotopic variants, the simplest of all A + BC reactions, have been extensively investigated recently. The groups of Valentini(4) and Zare(5) have recently shown that the HD product has little rotational excitation at low collision energies. As the collision energy is increased higher J states start to be populated. The reason is that at higher collision energies the attacking H atoms can react at larger angles away from the linear configuration which is the lowest energy path of this reaction. For the same reason the cross section becomes larger as the collision energy is increased above the threshold value. There is fine agreement between the cross sections measured at four different energies(6) and theoretical values based on classical trajectories run on an <u>ab initio</u> potential surface(7).(Figure 1) We have another (unpublished) indication that at higher collision energies the reaction path is no longer colinear. At collision energies of about 1.2 eV we found some correlation between the velocities of reactant H and product D atoms. However at collision energies of about 2.2 eV we could not observe any such correlation.

Exchange Reaction: H + DCCD and C$_2$D$_4$

The exchange reactions of fast(1 eV) H atoms with the unsaturated molecules DCCD and C$_2$D$_4$ might be expected to proceed by an addition followed by internal vibrational redistribution and eventual unimolecular ejection of a D atom. Our main finding is that no long lived complex is involved.(8) The evidence is that the product D atoms from collisions of H with DCCD or CH$_2$CCD have almost the same energy, which is about 40% of the initial H atom kinetic energy. Also, for DCCD, the velocities of the leaving D atoms were correlated with the velocities of the incident H atoms. At low energies the initial expectations of complex formation must still be true.

Exchange Reaction: H + MD$_4$, H + MHD$_3$, M = C,Si

Almost 30 years ago several groups (9,10) studied the reaction of "hot" (fast) tritium atoms with CH$_4$ and CD$_4$ and concluded that abstraction to form HT had a lower threshold than exchange to

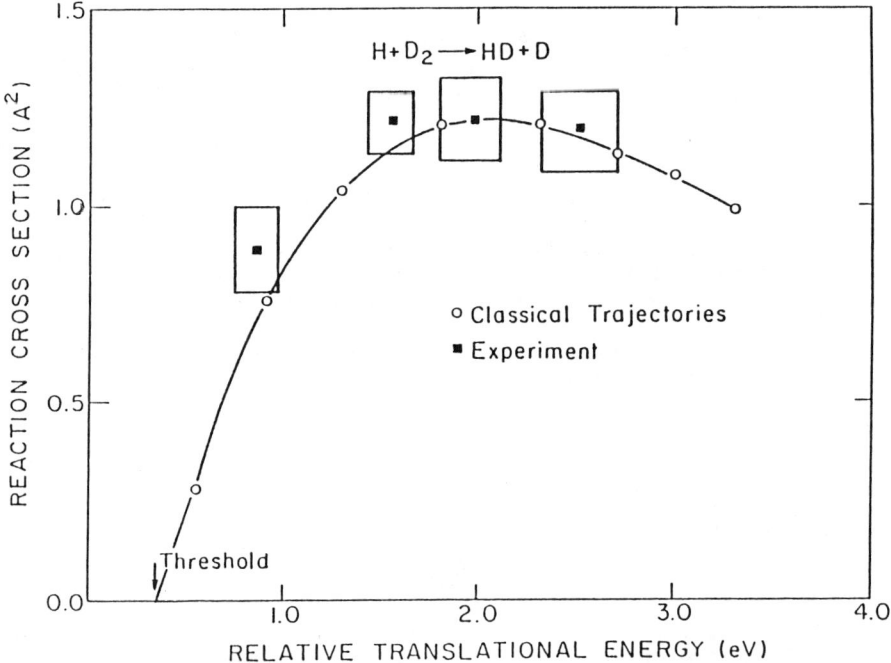

Figure 1. Cross section (Å^2) as a function of relative kinetic energy (eV) for the H + D$_2$ reaction.

form CH_3T or CD_3T. The barrier to exchange was about 1.5 eV. The mechanism of the exchange, that is, whether it was a direct three center process or an inversion involving all atoms of the molecule could not be determined. Today using lasers it is possible to reexamine this reaction and to prove definitively that at least at 2.1 eV the exchange involves an inversion.

All exchange reactions with optically active molecules of the form CHXYZ where X,Y,Z are heavier groups result in retention of configuration which means that inversion does not take place. Nevertheless when all four ligand atoms are hydrogen atoms the barrier to inversion should be much lower. This is particularly true for SiD_4 which is a floppier molecule than CD_4 with weaker bonds (92 kcal/mol for Si-D versus 98 kcal/mol for C-D. For both CD_4 and SiD_4 three observables of the reaction were measured at an average 2.1 eV collision energy, the rate coefficient, the average kinetic energy of the product D atoms and the alignment of the velocity of the product D atoms when the reactant H atom velocity was aligned.(11,12)

The H atoms were generated by photodissociating H_2S at 193.3 nm. The H atoms so produced have a spread of energies (13) but the average energy is 2.1 eV. While 63% of the H atoms have the maximum speed of 2.09×10^6 cm/s, the average speed is 1.94×10^6. For the H + SiD_4 reaction the rate coefficient found was $(0.70 \pm 0.07) \times 10^{-10}$ molec^{-1} s^{-1}. Dividing by the average relative speed of 1.94×10^6 cm/s, one obtains a cross section of $(0.36 \pm 0.03) Å^2$. Table I lists this and other cross sections measured by the same method.

Table I. Cross Sections for Hydrogen Isotope Exchange Reactions

Molecule	Relative Energy (eV)	Cross Section ($Å^2$)
HD	0.82	0.48±0.05
HD	1.47	0.60±0.04
HD	1.86	0.68±0.09
D_2	0.87	0.88±0.12
D_2	1.57	1.28±0.09
D_2	1.97	1.27±0.14
D_2	2.70	1.25±0.12
DCCD	1.0	1.69±0.22
CH_3CCD	1.0	0.50±0.15
C_2D_4	1.0	1.85±0.20
C_3D_6	1.0	1.10±0.14
CD_4	2.1	0.084±0.015
CH_3D	2.1	0.040±0.015
SiD_4	2.1	0.36±0.03
SiH_3D	2.1	0.17±0.03

Two important points emerge from the cross section data. The cross sections of SiD_4 and $SiDH_3$ are about a factor of five larger than those of the corresponding methanes. This suggests, as one would assume from its smaller force constants, that the barrier for exchange in silane is much lower than in methane. Also the cross sections for MDH_3 were about one half of those for MD_4. This is an argument for the inversion mechanism. If the reaction had been a front side attack without inversion, the cross section for exchange would have been expected to be proportional to the number of deuterons.

The strongest argument for an inversion mechanism is derived from observation of a velocity alignment of the D atoms when the H atom velocities are aligned. When a hydride molecule is dissociated by polarized light, the angular distribution of the velocities is given by:

$$f(\theta_{v,E}) = (1/4\pi)\{1 + \beta_H P_2(\cos\theta_{v,E})\} \quad (1)$$

where $\theta_{v,E}$ is the angle between the velocity of the H atom fragment v and the electric vector of the dissociating light. β_H is the anisotropy parameter of the H atom velocity distribution. In order to obtain the angular distribution of the velocities of the product D atoms, we need to integrate over all initial directions of the H atom velocities. Using the Legendre polynomial addition theorem, one finds

$$\langle P_2(\cos\theta_{v,E})\rangle = \langle P_2(\cos\theta_{v,v'})\rangle P_2(\cos\theta_{v',E}) \quad (2)$$

v' is the velocity of the D atom, $\theta_{v,v'}$ is the angle between the H and D atom velocities and $\theta_{v',E}$ is the angle between the D atom velocity and the electric vector. Thus the angular distribution of the velocity of the D atom is

$$f(\theta_{v',E}) = (1/4\pi)\{1 + \beta_D P_2(\cos\theta_{v',E})\} \quad (3)$$

where $\beta_D = \beta_H \langle P_2(\cos\theta_{v,v'})\rangle \quad (4)$

Therefore the product angular distribution will be anisotropic only if both the reactant angular distribution and the differential cross section are anisotropic i.e. if both factors in Eq.(4) are different from zero. Assuming that the anisotropy parameter is independent of the speed, the velocity distribution function can be factored into a product of a speed distribution function and an angular distribution function. By averaging over the distribution function given above, one can show that the anisotropy parameters β_H and β_D can be measured from the second moments of the fluorescence excitation curves as shown in Figure 2. The average squares of the velocity component perpendicular and parallel to the E vector are:

$$\langle v_\parallel^2\rangle = \langle v^2\rangle\{1/3 + 2\beta/15\} \quad (5a)$$
$$\langle v_\perp^2\rangle = \langle v^2\rangle\{1/3 - \beta/15\} \quad (5b)$$

Solving these equations one finds that the average kinetic energy is

$$m\langle v^2\rangle/2 = m\{\langle v_\parallel^2\rangle + 2\langle v_\perp^2\rangle\}/2 \quad (6)$$

and the anisotropy parameter is

$$\beta = 5\{\langle v_\parallel^2\rangle - \langle v_\perp^2\rangle\}/\langle v^2\rangle. \quad (7)$$

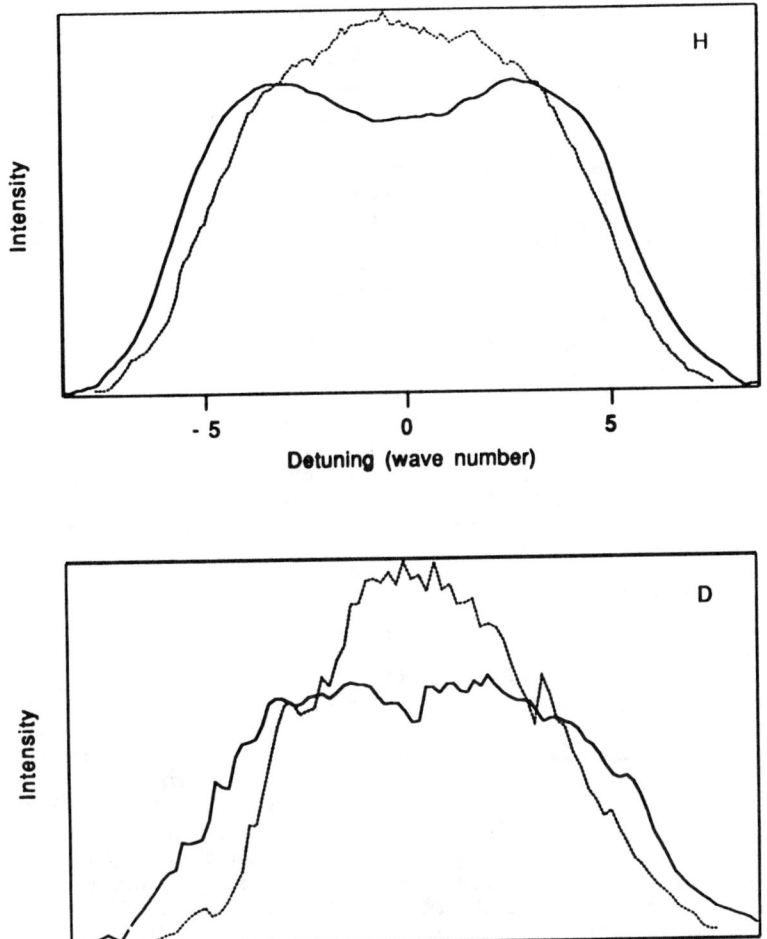

Figure 2. H and D atom fluorescence excitation spectra taken 100 ns after irradiation of a mixture of SiD_4 and H_2S with polarized 193 nm light. The solid and dotted lines are the spectra obtained when the probing laser is perpendicular and parallel, respectively, to the E vector of the dissociating light. All four curves have equal areas. (Reproduced with permission from ref. 12. Copyright 1991 American Institute of Physics.)

As an example of the above equations, for H_2S dissociated by polarized 193 nm light $\beta_H = -0.82\pm0.08$. For the D product atoms from the reaction of H atoms with SiD_4 we obtained $\beta_D = -0.72\pm0.35$. The reaction anisotropy may be defined as $<P_2(\cos\Theta_{v,v'})> = \beta_D/\beta_H$. For H + SiD_4 the anisotropy is 0.88 which leads to a typical scattering angle of $16°$. For H + CD_4, β_D was -0.53 ± 0.28 implying a typical angle of $25°$. The finding that the velocities of the H and D atoms are in the same direction is consistent with the inversion mechanism. The alignment experiment rules out the possibility that the mechanism is a direct frontal attack by the H atom on the Si-D or C-D bond. Such a mechanism would have resulted in larger scattering angles making the reaction anisotropy small or even negative. In summary, the findings that a) the cross section of H + MDH_3 is about twice per channel that of H + MD_4, b) the D product is ejected in the same direction as the reactant H atoms and c) most of the available energy remains as product D translation all support the inversion mechanism.

Isotopic Competition: F+HD. The $F + H_2$ reaction has been the subject of numerous studies because it is an important reaction in a chemical laser. Potential surfaces have been calculated for it.(14,15) The reaction is exothermic by about 32 Kcal/mol and has a barrier for reaction of about 1 Kcal/mol.(14) Therefore many sorts of trajectories will be reactive which makes it hard to describe with a simple picture. Because we have a reliable method of measuring the H and D fluorescence excitation spectra, we studied the branching ratio of the reaction of F atoms with HD:

F + HD --> FD + H   (8a)
F + HD --> FH + D   (8b)

The ratio of the yields of reaction (8a) to reaction (8b) was measured at three different collision energies. Reaction (8b) has a larger rate constant than (8a). This effect is an aspect of dynamical stereochemistry(26). The three ratios obtained were 0.75 ± 0.05, 0.94 ± 0.03 and 0.94 ± 0.03 at relative energies of 4.5 ± 2.2, 6.05 ± 2.50 and 8.3 ± 3.3 Kcal/mol respectively. We found that the dominant effect giving rise to the unequal branching ratio is the shift of the center of mass of HD from its center of charge. While the H/D ratio observed for reaction with room temperature HD is less than one, for the J=0 state of HD the ratio is predicted to be greater than one(15).

For a rotationless HD molecule there are three reasons for a preferential attack of the fluorine atom on the D end of the molecule. First, the cone of approach (see Figure 3) of the F atom along the coordinate to the center of mass(R) will be wider near the D end than near the H end of the HD molecule. Secondly, calculations show(15) that within the cone near the D end the reactivity is greater than near the H end because of reorientation of the HD by the F atom towards the colinear configuration. Figure 3 shows that the force **F** is the same at either end of HD; the force is repulsive because the system is approaching the barrier. For the D end this force acts to reorient the HD towards the F atom while for the H end the force acts to rotate the HD away from the F atom. Thirdly, recrossing favors the exit channel DF + H over HF + D. Trajectory

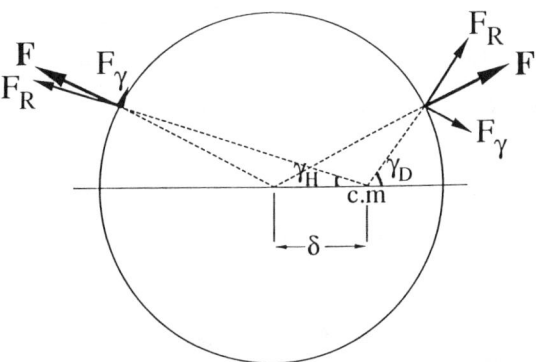

Figure 3. The reorientation effect due to mass asymmetry of HD is the distance between the center of mass and the center of charge of the HD molecule. The F atoms shown as points on the circumference of the circle are approaching either end of the barrier at the same height (i.e., at the same angle with respect to the center of charge) and will have different angles γ_H and γ_D with respect to the center of mass. For a given approach angle to the center of charge, the forces at the two ends are qualitatively different. The force F is the same at either end and is repulsive because the F atoms are enroute to the barrier. The force F is resolved into a component F_R along the vector R to the center of mass and F_γ perpendicular to R. For the heavier D end, the force acts so as to favorably reorient HD towards the D atom. For the lighter H end, the force acts in the opposite direction. (Reproduced with permission from ref. 15. Copyright 1991 American Institute of Physics.)

calculations showed that because of the mass combinations, recrossing of F atoms attacking the H end is more frequent than when the D atom is attacked(15).

The preceding discussion has applied to a rotationless HD reactant. However, when the HD molecule is at room temperature as it was in our experiments, higher J values are populated. Muckerman has pointed out that since the HD rotates about its center of mass, the H atom will sweep out a larger circle than the D atom while rotating(16). This will cause a preferential attack on the H by the F atom. This effect definitely changes the branching ratio. An additional explanation is that anisotropies in most potential surfaces for the reaction cause the D end to be recessed more than the H end of the rotating HD and consequently more HF is produced.(see Fig. 6 of Ref.15) A measurement of this isotope effect on rotationally cold HD is needed to test the above ideas.

II. Hydrogen Isotope Effects in Photodissociations

In a molecule containing one or more hydrogen atoms there can be an isotope effect in that the quantum yield for the process

$$MH + h\nu \longrightarrow M + H \qquad (9a)$$

is different from that of the process

$$MD + h\nu \longrightarrow M + D \qquad (9b)$$

A well known example is that of OH and OD in their A^2 excited state. At certain energies the OH spectral lines are broadened by predissociation whereas those of OD remain relatively sharp. The H atom can more rapidly tunnel from the bound A state to a repulsive curve. Thus at about the same energy of excitation, OH has a lower fluorescence yield and a higher photodissociation yield.

We have studied molecules containing two or more chemically equivalent hydrogen atoms in which the quantum yields of the processes

$$MHD + h\nu \longrightarrow MD + H \text{ and} \qquad (10a)$$
$$MHD + h\nu \longrightarrow MH + D \qquad (10b)$$

were compared as well as the quantum yields of

$$MH_2 + h\nu \longrightarrow MH + H \text{ and} \qquad (11a)$$
$$MD_2 + h\nu \longrightarrow MD + D. \qquad (11b)$$

Table II lists all the data known to us on the photodissociation of partially deuterated molecules. What stands out is the fact that the H/D ratio is always significantly greater than one. However, when one examines the properties of the excited states of individual molecules, it turns out that the specific explanation of the H/D ratio is different for almost every case. The important factors are whether the state excited initially is repulsive or bound, the difference of zero-point energies between the two isotopes and the relative rates of internal conversion, intersystem crossing, and radiative decay.

Table II. Normalized Hydrogen to Deuterium Atom Ratio and Kinetic Energy Release in the Photodissociation of Partially Deuterated Molecules

Molecule	Wavelength (nm)	H/D*	$\langle E_T \rangle$ (kcal/mol)	$\langle f_T \rangle$	Ref.
HDO	157	4±1	56.1	0.88	3,17
CHDO		2 to 4			18
HCCD	193.3	2.1±0.2	8.0±1.4	0.47	19
HCCD	193.3	2.85±0.3			20
CH_3D	121.6	1.96±0.19	50.0±4.7	0.38	21
CH_2D_2	121.6	2.0±0.2	51.4±3.0	0.39	21
CHD_3	121.6	1.89±0.18	48.7±8.4	0.37	21
t-CHDCHD	193.3	1.2±0.1	7.3±1.3	0.18	22
CH_2CD_2	193.3	2.2±0.2	7.3±1.3	0.18	22
PhenylCH_2D	193.3	1.8±0.1	7.8±0.3	0.13	23
C_4H_4CHD	193.3	3.9±0.5	12.2±1.6	0.17	24

* H/D is the ratio of H atoms to D atoms divided by the atomic ratio in the parent molecule, e.g. 1/3 for CHD_3. $\langle E_T \rangle$ is the average kinetic energy of the H atom. For the ethylenes it is the average kinetic energy of H from C_2H_4. $\langle f_T \rangle$ is the average fraction of the available energy released as kinetic energy of the H atom.

HDO The first absorption band of water centered at 60,000 cm^{-1} is due to a transition between two singlet states, the upper one being repulsive(25). Both states correlate at infinite separation to the fragments $H(^2S_{1/2})$ and $OH(^2\Pi_i)$. This pair of fragments has eight electronic states. Besides the two singlets there are six states which are the components of two triplet states. When H_2O is dissociated by 63,570 cm^{-1} (157.3 nm) photons(17), the OH radicals are found in the v=0,1,2 states in the ratios 1:1:0.58 and the rotational temperature is comparatively low, 475 K. This low rotational energy implies only weak torques act on the upper surface; in other words the bond angle does not change much on excitation. The bond angle can therefore be neglected in a first order calculation and the potential surface is a function, approximately, of only two coordinates. The average fraction, $\langle f_T \rangle$ of available energy released as translation is 0.88. The large release of kinetic energy is a characteristic of a descent on a steep repulsive potential.

The vibrational excitation occurs because, while the photon does not supply enough energy to release two hydrogen atoms, initially the symmetric vibration is excited as well as the antisymmetric motion. The potential surface has a barrier in the symmetric coordinate but the potential energy decreases when just one of the OH bonds is elongated. A strong isotope effect in the photodissociation of HDO was predicted by theoretical calculations(27,28). The physical explanation is that electronic excitation is accompanied by vibrational excitation of both the OH and OD bonds but the more diffuse wave function of the former extends further into the region of no return. An H/D ratio of 4±1 was found(18) for HDO dissociated by 157 nm light in agreement with a theoretical prediction (28). However, an even larger ratio was observed when HDO was excited with four quanta in the O-H stretch(29). There is even a strong effect with only one quantum of the O-H stretch excited (30).

HDCO Formaldehyde, H_2CO can dissociate into the radical pair $H(^2S_{1/2})$ and $HCO(\tilde{X}^2A_1')$ which have four electronic states. One is a singlet which correlates to the ground state and the other three are components of a repulsive triplet. The singlet state excited in the transition is a n, singlet which has relatively large matrix elements connecting it with the triplet state. Dissociation was shown(18) to be preceded partly by internal conversion to the ground state and partly by intersystem crossing to the triplet state. The H/D ratio from HDCO measured by Chuang et al.(18) as a function of energy was useful in proving these points.

HCCD When HCCD is photodissociated with 51,733 cm^{-1} (193.3 nm) light an H/D ratio was reported as 2.85±0.3 by Cool et al.(20) and 2.1±0.2 by Satyapal et al.(19). The dynamics of the photodissociation are not well understood but recent data suggest that dissociation is both from an excited singlet state and an excited triplet state. The fragment C_2H has been detected both in its ground $\tilde{X}^2$ state and in an excited $\tilde{A}^2$ state indicating dissociation on two excited state surfaces (31). The evidence against dissociation from a hot equilibrated ground state are the facts that a) the translational energy distribution does not peak at zero energy(34), b) the translational energy release is high ($<f_T>=0.5$), c) CH_3CCD yields only D atoms although the C-H bond is considerably weaker and d) DCCD and CH_3CCD yield D atoms of nearly the same kinetic energy(4).

Baldwin et al.(33) showed that the H atoms dissociated in the energy region 46,300 to 49,750 cm^{-1} (216 to 201 nm) had isotropic velocity distributions. This is consistent with a lifetime longer than a rotation period but could have occurred because the H atoms may be released at angles not too far from 54.7° so that the anisotropy might almost vanish. They also showed however that in the $\tilde{A} \leftarrow \tilde{X}$ $2_0^1 v_0^0$ band at 48,636 cm^{-1} rotational lines could be resolved but had widths corresponding to lifetimes of the order of 10-20 ps. This is clear evidence for a predissociation. It is a long enough time for intersystem crossing to take place so that dissociation may well occur on a triplet surface.

CH_nD_{4-n} Recently the dynamics of the photodissociation of methane have been studied using 121.6 nm light(21). More precisely the channel leading to hydrogen atoms was studied but not the hydrogen molecule + methylene channel. However CD_4 and CH_4 yielded the same concentrations of D and H atoms respectively showing that the quantum yield for this channel was independent of isotope. The data in Table II on the H/D ratios measured in the dissociation of the three partially deuterated methanes are quite consistent. A hydrogen atom has twice the probability of escaping from an excited methane as a deuterium atom.

To understand this preference for H atom elimination we must consider the mechanism of dissociation. A ground state CH_4

molecule can dissociate to an H atom and an $\tilde{X}^2A_2$" CH_3 radical. This pair has four electronic states, one correlating to the ground state and the other three to a repulsive triplet state. Thus the first absorption band of methane being a strongly allowed transition must bring the molecule to an excited singlet state which dissociates either to an $\tilde{A}^2A_1$ 'CH_3 + H(1s) or to a ground state CH_3 and H(2s or 2p). The thresholds for these processes are 82,400 and 118,300 cm^{-1} respectively. The first absorption band of methane, a continuum begins around 70,000 cm^{-1} and extends at least as far as 90,000 cm^{-1}.

The average kinetic energy of the H atom released was 50 kcal/mol (18,700 cm^{-1}). This fact shows that dissociation does not take place from the initially excited electronic state because there is not enough energy available to break the C-H bond, excite electronically a methyl radical and release so much kinetic energy. There must be a crossover to the triplet state or the ground singlet state. It is not clear which is the final state. A strong argument against the triplet state is that the spin-orbit coupling is very weak and the dissociation time is very short. The photon energy is 235 kcal/mol as compared to a C-H bond energy of 103 kcal/mol. On the other hand, the translational energy release, 50 kcal/mol is 38% of the available energy which is very large for a dissociation taking place from a hot ground state. A possible explanation is that the surface crossing occurs at an inner crossing where the C-H bond is highly compressed, and subsequently there is incomplete vibrational energy exchange. The H atom, having a more diffuse wave function than a D atom will have a faster rate of crossing to the repulsive region of the final surface.

CH_2CD_2 and trans-CHDCHD The dynamics of the hydrogen atom channel in the photodissociation of ethylene has recently been reported for 193 and 157 nm light.(7) The fraction of available energy released as translational energy was only 18% suggesting that internal conversion to the ground state precedes dissociation. A Rice-Rampsberger-Kassel-Marcus (RRKM) calculation to be described below predicts that the average lifetime of ethylene excited by a 193 nm photon is about 30 ps.

The isotope effects observed at 193 nm are revealing. CD_2CH_2 exhibits an H/D ratio of 2.2±0.2 whereas that from trans-CHDCHD is 1.2±0.1. The difference is nicely explained by the RRKM theory. The RRKM expression for the rate constant for decomposition is

$$k(E^*) = Q^+W(E^+)/[hQ^*N(E^*)] \tag{12}$$

where E^+ and E^* are the internal energies of the activated complex and the excited molecule respectively. Q^+ and Q^* are the respective partition functions. $W(E^+)$ is the density of states of the activated complex and $N(E^*)$ is the number of states of the excited molecule.

The assumptions of the theory are that the wag, rock and twist frequencies involving the leaving atom become much softer. This effect is relatively more important for the higher frequency H atom vibrations. Thus the CH_2 group of CH_2CD_2 breaks up faster than the CD_2 group. On the other hand in trans-CHDCHD the transition states are nearly the same. The moment of inertia for the transition state in which the D leaves is larger than the moment of inertia for the transition state in which the H leaves. This effect tends to cancel the effect of the larger zeropoint energy in the C-H bond which weakens it relative to the C-D bond. Thus the H/D ratio is close to one.

$C_6H_5CH_2D$ When toluene is excited at 193 nm, a hydrogen atom is released in an average time of about 300 ns-ten thousand times longer than for the smaller molecule, ethylene at the same energy.(8) This is clearly a dissociation from a vibrationally hot ground state. When $C_6H_5CH_2D$ is dissociated at 193 nm, the experimental H/D ratio is 3.63±0.10 which implies that each C-H bond dissociates faster than an equivalent C-D bond by a factor of 1.82. The major reason for this effect, which is again interpretable by an RRKM model, is that the zeropoint energy of the C-H bond is larger than that of the C-D bond.

Isotope Effects and Potential Surfaces The preference for H atom dissociation over D atom dissociation is linked to the surface on which the process takes place. When excited in the lowest energy absorption band, water dissociates on a repulsive singlet surface, formaldehyde and acetylene on excited singlet and triplet surfaces, and more complex molecules such as ethylene and toluene on the ground state surface. Theoretical work has been carried out for the extreme cases. For the simplest case of HOD a full quantum treatment has been carried out for the first excited state.(25,26) A wave packet is observed to have a larger amplitude in the H exiting channel than in the D exiting channel. For complex molecules such as ethylene and toluene internal conversion to the ground state is much faster than dissociation. The latter process is appropriately treated by a statistical model such as RRKM. The photodissociation of molecules such as acetylene and methane has not been treated with a rigorous theory. Methane, having only ten electrons and high symmetry may be easier to study than acetylene with fourteen electrons and a rich variety of structures in its upper states.

Acknowledgments The experiments reviewed on the inversion mechanism for silane and methane isotope exchange were supported by the US National Science Foundation. All other work reviewed was supported by the US Department of Energy.

Literature Cited

1. Johnston,G.W.;Park,J;Satyapal,S.;Shafer,N.;Katz,B.; Tsukiyama,K.;Bersohn,R.Acc.Chem.Res.**1990**,23,232
2. Wallenstein,R.;Opt.Commun.**1980**,33,119
3. Shafer,N;Satyapal,S.;Bersohn,R.J.Chem.Phys.**1989**,99,6807
4. Gerrity,D.F.;Valentini,J.J.J.Chem.Phys.**1983**,79,5202
5. Blake,R.S.;Rinnen,K.;Kliner,D.A.;Zare,R.N. Chem.Phys.Lett.**1988**,153,365,371
6. Johnston,G.W.;Katz,B.;Tsukiyama,K.;Bersohn,R. J.Phys.Chem.**1987**,91,5445
7. Schechter,I.;Levine,R.D.Inter.J.Chem.Kinet.**1986**,18,1026

8. Johnston,G.W.;Satyapal,S.;Bersohn,R.;Katz,B.
   J.Chem.Phys.**1990**,_92_,308
9. El-Sayed,M.A.;Estrup,P.J.;Wolfgang,R.;
   J.Phys.Chem.**1958**,_62_,1356
10. Chou,C.C.;Rowland,F.S.J.Chem.Phys.**1969**,_50_,2763,5133
11. Katz,B.;Park,J.;Satyapal.S.;Tasaki,S.;Bersohn,R.
    Chattopadhyay,A.;Yi,W.;Disc.Farad.Soc. 91 in press
12. Chattopadhyay,A.;Tasaki,S;Bersohn,R.;Kawasaki,M.
    J.Chem.Phys.**1991**,_95_,1033
13. Xie,X.;Schneider,L.;Wallmeier,H.;Boettner,R,;
    Welge,K.H.;Ashfold,M.W.R.J.Chem.Phys.**1990**,_92_,1608
14. Schaefer,H.F.;J.Phys.Chem.**1985**,_89_,5336
15. Johnston,G.W.;Korenweitz,H.;Schechter,I.;Persky,A.;Katz,B.;
    Bersohn,R.;Levine,R.D.J.Chem.Phys.**1991**,_94_,2749
16. Muckerman,J.T.J.Chem.Phys.**1971**,_54_,1155
17. Andresen,P.;Schinke,R. in Molecular Photodissociation
    Dynamics, Ashfold,M.N.R.;Baggott,J.E.eds.,Royal
    Society of Chemistry,London,1987 p.107
18. Chuang,M.C.;Foltz,M.;Moore,C.B.J.Chem.Phys.**1987**,_87_,3895
19. Satyapal,S.; Bersohn,R. J.Phys.Chem.,in press
20. Cool,T.A.;Goodwin,A.;Otis,C.J.Chem.Phys.**1990**,_93_,3714
21. Yi,W.;Chattopadhyay,A.;Tasaki,S.;Bersohn,R.
    J.Chem.Phys.,submitted
22. Satyapal,S.;Johnston,G.W.;Bersohn,R.;Oref,I.
    J.Chem.Phys.**1990**,_93_,6398
23. Park,J.;Bersohn,R.;Oref,I;J.Chem.Phys.**1990**,_93_,5700
24. Yi,W.;Chattopadhyay,A.;Bersohn,R.J.Chem.Phys.**1991**,94,5994
25. Staemmler,V.;Palma,A.Chem.Phys. **1985**,93,63
26. Levine,R.D.J.Phys.Chem.**1990**,_94_,8872
27. Engel,V.;Schinke,R.J.Chem.Phys.**1988**,_88_,6831
28. Zhang,J.;Imre,D.;Frederick,J.J.Phys.Chem.**1989**,_83_,1840
29. Vander Wal,R.L.;Scott,J.L.;Crim,F.F.J.Chem.Phys.**1990**,_92_,803
30. Bar,I.;Cohen,Y.;David,D.;Rosenwaks,S.;Valentini,J.J.
    J.Chem.Phys.**1990**,_93_,2146
31. Fletcher,T.R.;Leone,S.R.J.Chem.Phys.**1989**,_90_,871
32. Wodtke,A.M.;Lee,Y.T.J.Phys.Chem.**1985**,_89_,4744
33. Baldwin,D.P.;Buntine,M.A.;Chandler,D.W.J.Chem.Phys.**1990**,_93_,6578

RECEIVED September 4, 1991

Chapter 6

Isotope Effects at High Temperatures Studied by the Flash or Laser Photolysis–Shock Tube Technique

J. V. Michael

Chemistry Division, Argonne National Laboratory, Argonne, IL 60439

During the past five years, the flash or laser photolysis-shock tube (FP or LP-ST) technique has been used to measure absolute thermal bimolecular rate constants in a previously difficult temperature range, ~700-2500 K. The technique is described. Protonated and deuterated versions of six reactions have been studied to date. The reactions are $C_2H(C_2D) + C_2H_2(C_2D_2)$, $O + C_2H_2 (C_2D_2)$, $H(D) + O_2$, $H(D) + H_2O(D_2O)$, $O + H_2(D_2)$, and $D(H) + H_2(D_2)$. These results are reviewed. In many cases the high temperature results can be combined with lower temperature results, and the experimental isotope effects can then be determined over a very large range of temperature. For one of the cases to be discussed, namely the isotope effect between $D + H_2$ and $H + D_2$, the range of temperature is from ~200-2000 K. This large range then gives an unprecedented opportunity for experimental comparison to theoretical predictions of isotope effects since data now exist (a) at low temperatures where quantum mechanical tunneling predominates and (b) at high temperatures where tunneling is unimportant.

The flash or laser photolysis-shock tube (FP or LP-ST) technique for studying thermal bimolecular reaction rates was originally envisioned by Burns and Hornig (1). Following this pioneering work, Zellner and coworkers (2,3) studied three OH-radical with molecule reactions. The use of atomic resonance absorption spectroscopy (ARAS) for atomic detection in such experiments is relatively recent and started about five years ago (4,5). Subsequently, the technique has been used on about twenty reactions (6) many of which are isotopic variations of the same reaction. These cases will be reviewed in this article.

Since the method is useful at high temperatures, it can and has been used, along with lower temperature data sets, to extend the temperature range of a specific reaction thereby giving an accurate understanding of the rate behavior over a very large temperature range. For reactions in which H-atoms are abstracted and which have relatively high activation energies, the technique can be used in a temperature range where tunneling is relatively unimportant. Hence, the measured activation energy relates directly to the barrier height on the potential energy surface for the given reaction. This feature of the results has recently been discussed in detail (7).

When absolute determinations are used for individual isotopic modifications of the same reaction, the derived isotope effect is calculated from the ratio of the absolute rate constant values. With any gas phase chemical kinetics method, this procedure is never as accurate as classical relative methods that are based on product analysis in systems where both isotopic reactions are simultaneously occuring. With the FP or LP-ST technique, the absolute accuracy of the results is typically between ±15 to 25%. Taking the square root of the sum of variances, the ratio value will be accurate to ±21 to 35%. When isotope effects approach unity at high temperature, it will be difficult with this technique alone to assess whether an isotope effect significantly different from unity actually exists. This is the reason that combinations of data sets with lower temperature results are desirable because the continuous changes in the kinetic isotope effect can be documented from low to high temperature.

Experimental

The FP or LP-ST technique has been described previously (*1-7*), and therefore, only a brief description of the method will be given here. Figure 1 shows a schematic diagram of the apparatus. The shock tube is of general design (*8*) and consists of a driven section that is separated from a driver section by a thin Al diaphragm. He is used as the driver gas, and the driven or test gas is predominantly Ar with small quantities of added source molecule and reactant molecule. The source molecule is chosen so that on photolysis it will photodissociate to give the transient species that will subsequently be spectroscopically measured as it reacts with the reactant molecule. In some cases the source molecule and the reactant molecule are the same; eg., H + NH_3 (*9*) or H + H_2O (*10*). However, in most cases, two different molecules are used, and accurate determinations of their compositions in premixtures in Ar are necessary. This is accomplished with capacitance manometric measurement.

Experiments are performed behind reflected shock waves where the hot gas is effectively stagnant and not flowing. Flash or laser photolysis occurs after the reflected shock wave has gone past the spectroscopic observation station, the ARAS photometer system. Transient species are observed radially across the shock tube. Reflected shock pressure and temperature are kept sufficiently low so that concurrent thermal decomposition is minimized. Therefore, the initial transient species concentration will be totally controlled by photolysis, and its subsequent decay will be totally controlled by bimolecular reaction. Diffusion out of the viewing zone is negligibly slow on the time scale of the experiment. This experiment is then an adaptation of the well known static kinetic spectroscopy experiment with the reflected shock serving as a source of high temperature and density; i. e., shock heating is equivalent to a pulsed furnace.

Pressure transducers, mounted at equal intervals along the shock tube, are used to accurately measure the incident shock wave velocity. Temperature and density in the reflected shock wave regime are calculated from incident shock velocities through well known relations and correction procedures (*5, 8, 11*) that take boundary layer formation into account. Since the initial mole fractions of the source and reactant components are known, the absolute concentrations of both species can then be determined in the reflected shock wave regime.

In the ARAS adaptation of the method, atomic species are spectroscopically monitored as a function of time. H- (*6,7,9,10*), D- (*12,13*), O- (*14,15*), and N-atom (*16,17*) reactions have been studied by the technique. Beer's law holds if absorbance, (ABS), is kept low. Then, (ABS) $\equiv -\ln(I/I_o)$ (where I and I_o are transmitted and incident intensities of the resonance light, respectively) is proportional to the atomic concentration; i. e., $(ABS)_t = \sigma[A]_t l$. σ is the effective cross section for

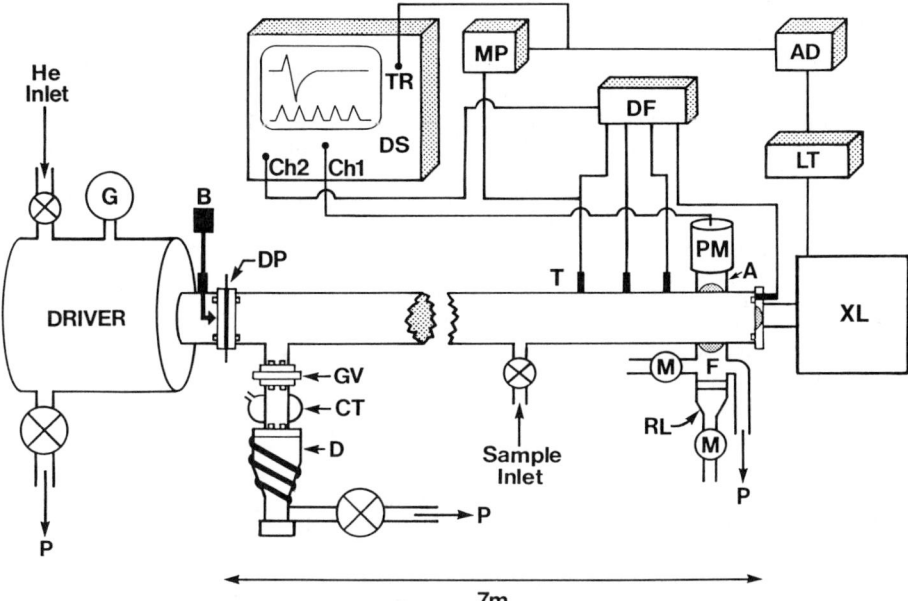

Figure 1. Schematic diagram of the apparatus. P - rotary pump. D - oil diffusion pump. CT - liquid nitrogen baffle. GV - gate valve. G - bourdon gauge. B - breaker. DP - diaphragm. T - pressure transducers. M - microwave power supply. F - atomic filter. RL - resonance lamp. A - gas and crystal window filter. PM - photomultiplier. DS - digital oscilloscope. MP - master pulse generator. TR - trigger pulse. DF - differentiator. AD - delayed pulse generator. LT - laser trigger. XL - excimer laser.

resonance absorption by atom, A, and l is the path length. If the temporal behavior of species A is controlled by a bimolecular reaction, A + R, where R is the stable reactant molecule, then the rate of depletion of [A] will be given by the product of the bimolecular rate constant (k_{bim}), [R], and [A]. If [R]>>[A] then the decay of A-atoms will follow pseudo-first-order kinetics with the decay constant being given as $k_{1st} = k_{bim}$[R]. Because (ABS)$_t$ is proportional to [A]$_t$, observation of the temporal dependence of (ABS) is sufficient to determine k_{1st}. Since [R] is known from the mole fraction and the final thermodynamic conditions as determined from the initial pressure and temperature and the shock strength, a value for k_{bim} can be deduced from each experiment. Figure 2 shows a typical example of raw data and the derived first-order plot. The negative slope of the first-order plot (k_{1st}) is obtained by linear least squares analysis, and the value of k_{bim} is determined by dividing by [R]. The results from many experiments are then usually displayed as Arrhenius plots, and, if curvature is not apparent in the results, a simple linear least squares line is derived from the composite set in order to describe the rate behavior over the experimental temperature range. It is also possible to carry out experiments with varying total density thereby measuring the pressure dependence of k_{bim} if such pressure dependence exists. This allows termolecular reactions to be studied with the method.

Results and Discussion

The FP or LP-ST results for six reactions are presented in Table I as Arrhenius expressions. The one standard deviation accuracy of the results and the temperature range of applicability is also given. The ratio of the results on isotopic modifications of the same reaction gives the high temperature kinetic isotope effect. To date, data have been obtained on three addition-elimination reactions (18-20) and on three H-atom abstraction reactions (10,12,13,15,21,22).

$C_2H + C_2H_2 \rightarrow C_4H_2 + H$ **and** $C_2D + C_2D_2 \rightarrow C_4D_2 + D$. Rate constants for these reactions have been measured with the LP-ST technique over the temperature range, ~1230 to 1500 K (20). The results are shown in Table I, and the kinetic isotope effect, KIE, is given by the ratio of k_H to k_D. The temperature independent result is 1.39 ± 0.40 indicating that an isotope effect different from unity is indeterminate. The absolute rate constants in both cases are fast, being about one half of the collision rate. The products of the reaction would strongly indicate that the reaction is a simple addition-elimination reaction, and therefore, the isotope effect would be secondary. Undoubtedly the initially formed adduct is vibrationally excited well above the dissociation energy for the forward process to diacetylene and H-atoms, and therefore a large isotope effect would not be expected. This conclusion is corroborated by the experimental result.

$O + C_2H_2 \rightarrow$ **Products and** $O + C_2D_2 \rightarrow$ **Products.** Absolute rate constants have been measured for these reactions between ~850 and 1950 K (18). Even though there are a significant number of lower temperature results for the protonated case, thereby allowing for an evaluation over the extended temperature range, 200 to 2500 K (18,23), comparable data do not exist for the deuterated case. Therefore, the kinetic isotope effect can be evaluated from only the FP-ST data at high temperature. The Arrhenius expressions that describe the results are presented in Table I, and the KIE is,

$$\text{KIE} = 1.03 \exp(26 \text{ K/T}). \quad (1)$$

Equation (1) gives 1.06 for 850≤T≤1950 K with an error of ~±30%, and this indicates that the isotope effect is unity within experimental error.

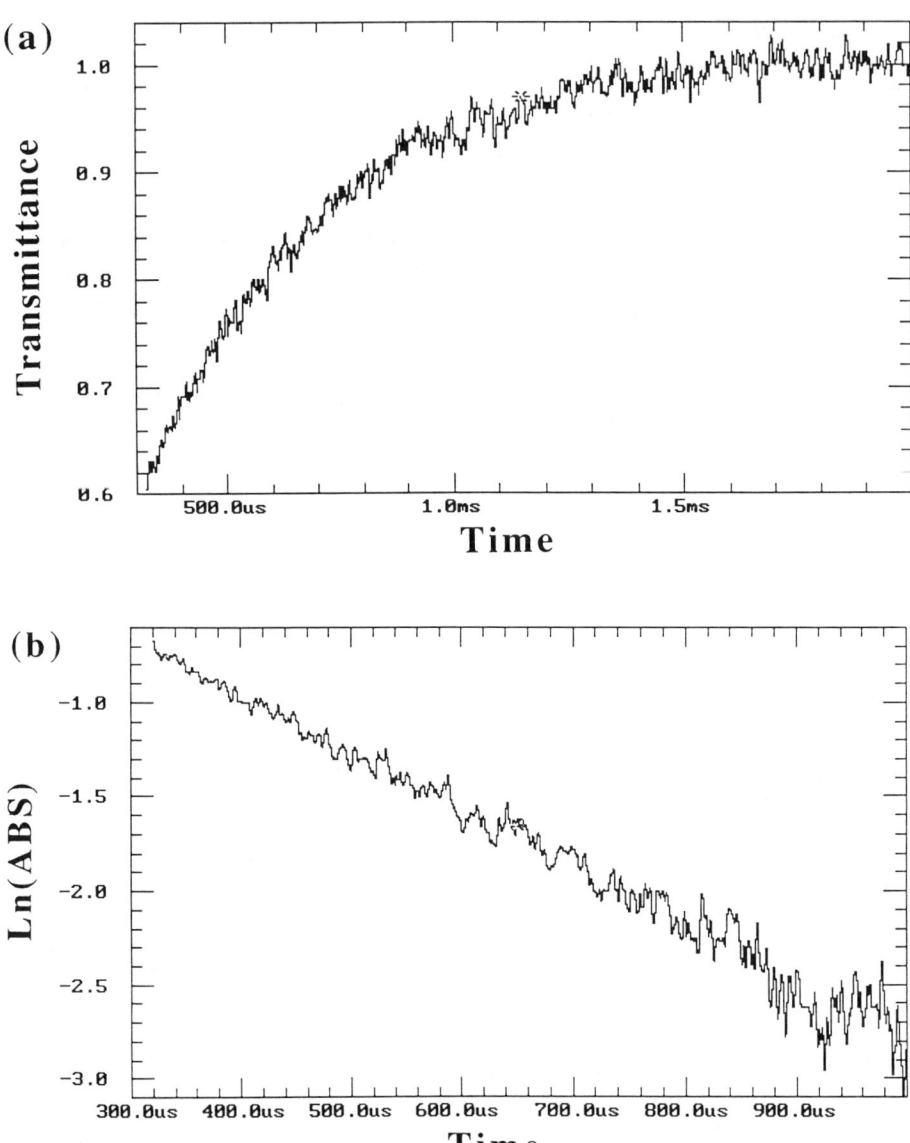

Figure 2. (a) H-atom transmittance as a function of time after laser photolysis in the reflected shock wave region. (b) First-order plot of $\ln(ABS)_t$ against time that is obtained from the record in panel (a); $k_{1st} = 3012$ s^{-1} in an H + O$_2$ experiment where [O$_2$] = 1.58 x 10^{15} cm^{-3} at 1697 K. Division gives the value for k_{H+O_2} at 1697 K. (Reproduced with permission from ref. 19. Copyright 1991 American Institute of Physics.)

Table I: Rate Constant Expressions of the Form, $k = AT^n \exp(B/T)$, from FP- or LP-ST Protonated and Deuterated Studies

Reaction	A cm^3 molecule^{-1} s^{-1}	n	B/K	Comments and References
Addition-elimination reactions:				
$C_2H + C_2H_2$	2.50(-10)[a]	0	0	1230 - 1475K, ±28%, (20)
$C_2D + C_2D_2$	1.80(-10)	0	0	1230 - 1700K, ±28%, (20)
$O + C_2H_2$	1.78(-10)	0	-2714	820 - 1921K, ±20%, (18)
$O + C_2D_2$	1.73(-10)	0	-2740	857 - 1980K, ±23%, (18)
$H + O_2$	1.15(-10)	0	-6917	1100 - 2050K, ±27%, (19)
$D + O_2$	1.09(-10)	0	-6937	1050 - 2300K, ±27%, (19)
H-atom abstraction reactions:				
$H + H_2O$	4.58(-10)	0	-11558	1246 - 2297K, ~±25%, (10)
[b]$H + H_2O$	1.56(-15)	1.52	-9249	250 - 2297K, ~±25%, (12)
$D + D_2O$	2.90(-10)	0	-10815	1285 - 2261K, ±27%, (12)
$O + H_2$	3.10(-10)	0	-6854	880 - 2495K, ±16%, (21)
[b]$O + H_2$	8.44(-20)	2.67	-3167	297 - 2495K, ±20%, (21)
$O + D_2$	3.22(-10)	0	-7293	825 - 2487K, ±17%, (15)
[b]$O + D_2$	2.43(-16)	1.70	-4911	343 - 2487K, ±16%, (15)
$D + H_2$	3.76(-10)	0	-4985	655 - 1979K, ±28%, (13)
[b]$D + H_2$	4.00(-18)	2.29	-2627	250 - 1979K, ±30%, (13)
$H + D_2$	3.95(-10)	0	-5919	724 - 2061K, ±25%, (22)
[b]$H + D_2$	1.69(-17)	2.10	-3527	256 - 2160K, ±23%, (22)

[a]parentheses denotes the power of ten; i. e., 2.50 x 10^{-10}. [b]evaluated with data reviewed in the indicated references.

The reaction has been discussed in terms of RRKM theory (*18,24*) in which the O-atom adds to the double bond in C_2H_2 initially forming a vibrationally hot species. Two major processes for the reaction have been documented (*see 18*),

$$O + C_2H_2 \rightarrow HCCO + H, \qquad (2)$$

$$O + C_2H_2 \rightarrow CH_2 + CO, \qquad (3)$$

both of which have been included in the RRKM description. Reaction (2) is a simple addition-elimination process whereas reaction (3) involves addition with subsequent intramolecular rearrangement followed by decomposition. Since the vibrational energy of the adduct is high and the isotope effects are secondary, a significant isotope effect should not be expected for this reaction from either channel. This conclusion is corroborated by the experimental results.

$H + O_2 \rightarrow OH + O$ and $D + O_2 \rightarrow OD + O$. Rate constants for these reaction have recently been measured with the same LP-ST apparatus (*19*), and the kinetic isotope effect between ~1100 and 2100 K is the ratio of Arrhenius expressions given in Table I.

$$KIE = 1.06 \exp(20 \text{ K/T}) \qquad (4)$$

Equation (4) gives a nearly constant value of 1.1 over the temperature range. The accuracy of this value is ~±35%, and therefore an isotope effect significantly different from unity is not indicated.

There are earlier less direct determinations of the isotope effect giving values of 0.84-0.99 (800-1000 K) (*25*), 4.0-5.3 (1000-2200 K) (*26*), and 1.2-1.7 (1000-2500 K) (*27*). The last study (*27*) supersedes the earlier study (*26*) from the same laboratory, and therefore, there is experimental agreement that the kinetic isotope effect in this case is not large.

Theoretical calculations on these two reactions have been carried out in order to assess the magnitude of the KIE. The simplest calculation is a conventional transition state theory calculation (CTST) with the bending frequency taken as a parameter (*19*). The double many body expansion potential energy surface (DMBE IV) on which this calculation is based is from Varandas and coworkers (*28*), and the resulting estimate is 0.69-0.89 (1100-2200 K). A more sophisticated quasi-classical trajectory (QCT) calculation has been given by Miller (*29,30*). The potential energy surface (*31*) on which these calculations are based is clearly not as accurate as that of Varandas and coworkers (*28*). This theory gives a KIE of 1.07-1.35 (1100-2200 K). Lastly, with the new more accurate potential energy surface (*28*), Varandas and coworkers (*32*) have carried out QCT calculations that indicate values of 0.75-1.06 for the temperature range, 1000-2500 K. Regardless of sophistication, all methods agree that a significant isotope effect does not exist in the higher temperature range, and this theoretical conclusion agrees with experiment within experimental error.

$H + H_2O \rightarrow OH + H_2$ and $D + D_2O \rightarrow OD + D_2$. Absolute rate constants for these reactions have been measured with the FP-ST technique (*10,12*) between ~1250 and 2300 K. The results are presented in Table I. Even though an evaluation for the protonated reaction has been made from 250 to 2300 K the data base for the deuterated reaction is much less extensive, and therefore, the kinetic isotope effect is only derivable from the higher temperature FP-ST results. The values from Table I give,

$$KIE = 1.58 \exp(-743 \text{ K/T}), \quad (5)$$

for the experimental temperature range. Between 1250 and 2300 K, the experimental isotope effect would vary from 0.87 to 1.14 with an estimated error of ~±38%. Because of the error, an isotope effect different from unity cannot be determined.

There have been several theoretical attempts to estimate thermal rate constants for the protonated reaction. The most significant is that of Issacson and Truhlar (33) who used a fit that was suspect (34) to an *ab initio* potential energy calculation (35). Subsequently, E. Kraka and T. H. Dunning, Jr. (Dunning, T. H., Jr., Pacific Northwest Laboratory, personal communication, 1990) have re-determined the potential surface, and the saddle point properties for both the protonated and deuterated cases have been used to estimate the absolute rate constants and the kinetic isotope effect with conventional transition state theory including Wigner tunneling (CTST/W) (12). In these calculations the *ab initio* potential energy surface was slightly scaled in order to reproduce the known exoergicity of the reaction and also to give agreement with the reverse protonated reaction, $OH + H_2$. This scaling procedure is well within the accuracy of the *ab initio* calculation (±2 kcal mole^{-1}) and is therefore not in contradiction to it. The database for the reverse reaction is large, and the data for both the forward and reverse rate constants can be combined through equilibrium constants to give consistency, thereby showing that the system is microscopically reversible. Therefore, the rate constant for the protonated case can be evaluated over a very large temperature range, and this evaluation is given in Table I. The CTST/W calculation agrees quite well with the evaluation being high by only ~15-25% over the entire temperature range, 250-2297 K. A calculation for the deuterated case with this successful model predicts a substantial isotope effect over the entire temperature range. Since the database for the deuterated case is not extensive, only the theoretical prediction in the higher temperature range is given here. The CTST/W predictions for KIE at 1250 and 2300 K are 1.89 and 1.49, respectively. This compares to the respective experimental values of 0.87 and 1.14. Even taking the uncertainty into account, the experimental values for the KIE are lower than the theoretical estimates probably indicating that CTST/W is too simple a theory. Additional theoretical calculations perhaps with variational transition state theory (VTST) might resolve this experimental to theoretical discrepancy.

$O + H_2 \rightarrow OH + H$ and $O + D_2 \rightarrow OD + D$. Data for these reactions have been extensively reviewed (36-39). However, conclusions about the rate constants have been mostly based on model fits to complex reaction mechanisms. Recently, both the protonated and deuterated reactions have been studied by the direct FP-ST technique (15,21), and the results are given in Table I. Inspection of these results alone shows that the difference in apparent activation energies is ~0.9 kcal mole^{-1} indicating that a primary isotope exists for this reaction. However, in these cases, there are additional lower temperature studies (21,40-44) with which the FP-ST data can be compared. The studies by Pirraglia et al. (21) and Gordon and coworkers (42,44) are the most notable, and, in both cases, the FP-ST data sets are in good agreement over the common range of temperature overlap. The lower and higher temperature results can then be combined and evaluations can be made over a very large temperature range. These evaluations are also shown in Table I. The KIE can then be evaluated as the ratio,

$$KIE = 3.47 \times 10^{-4} \, T^{0.97} \exp(1744 \text{ K/T}), \quad (6)$$

for the temperature range, 350 to 2500 K.

The KIE has also been measured between 390 and 1420 K by Marshall and Fontijn (*43*) with the HTP (high temperature photochemistry) technique. These results, at intermediate temperatures, overlap both the FP-ST results (*15,21*) and those of Gordon and coworkers (*42,44*). The combined results are plotted in Figure 3 along with estimates of the experimental accuracies in both low and high temperature ranges. The one standard deviation error is based on the results of Gordon and coworkers whose results are generally accurate to between ±5-10% giving an error of ~±15% in the low temperature KIE. Similarly the derived error at high temperatures is dominated by the FP-ST results both of which are accurate to ~±15-20%. This gives an error of ~±24% in the ratio.

Theoretical estimates of the absolute rate constants for both reactions are extensive. This case has served as a test case for modern theories of chemical kinetics. There are three notable calculations by Bowman et al. (*45*), Garrett and Truhlar (*46*), and Joseph et al. (*47*), all of which have used the same *ab initio* potential surface or an analytic fit to it (*48-50*). The first calculation uses the CEQB (co-linear exact quantum with adiabatic treatment of the bend) method, and the latter two calculations are variational transition state theoretical (VTST) estimates. The overall endoergicity has been slightly adjusted from the *ab initio* result so as to agree with the known value, and, in the latest VTST calculation (*47*), the saddle point energy has been slightly scaled upward by 0.45 kcal mole^{-1} in order to better agree with the results for the $O + H_2$ reaction. It should be noted that the *ab initio* energy calculation at the saddle point is only accurate to a few kcal mole^{-1}, and energy scaling within this range is acceptable and does not contradict the *ab initio* calculation. Both the CEQB and VTST methods have included modern models for quantum mechanical tunneling. The actual comparisons of the calculations with the evaluated experimental absolute rate constant results are excellent, being different from experiment by no more than one standard deviation over the entire temperature range. However, the theories do slightly overestimate tunneling at low temperatures. Evenso, these calculations represent an important confirmation of modern theories and indicate the importance of quantum mechanical tunneling at low temperatures. This conclusion is corroborated by experimental and theoretical branching ratio results (*51-54*) for the O + HD reaction.

It is well known that the magnitude of the KIE for reactions involving H- or D-atom abstraction is another sensitive measure of the phenomenon of quantum mechanical tunneling (*53,54*). Since the theoretical values for the rate constants for each reaction are well represented by the abovementioned theories, the theoretical estimates of the KIE should also be in substantial agreement with the experimental result shown in Figure 3. The comparisons of the predictions of Bowman et al. (CEQB) (*45*) and Joseph et al. (VTST) (*47*) are shown in Figure 3 along with experiment. Both calculations are lower than experiment, but the disagreement is not serious, particularly if the comparison is made at the experimental two standard deviation level (95% confidence).

In the present work, a CTST/W calculation that is based on the same *ab initio* potential energy surface has been carried out, and the predicted KIE is shown in Figure 3. The saddle point energy was scaled upward by an additional 0.5 kcal mole^{-1} from the value adopted by Joseph et al. (*47*). Therefore, the total increase in the *ab initio* saddle point barrier height is 0.95 kcal mole^{-1}, a value still well within the accuracy of the original calculation (*48-50*). Such a simple model gives remarkably good agreement with the $O + D_2$ thermal rate constant in Table I; however, the rate constant prediction slightly diverges above experiment in the low temperature range. The prediction for $O + H_2$ is poorer in the low temperature range, giving lower values relative to experiment than the Table I evaluation. The ratio,

KIE, then becomes substantially more flat than experiment. The reason for this behavior is simply that the ultra-simple Wigner formula is not adequate and underestimates the extent of quantum mechanical tunneling. This has been noted before for these cases (53,54). Lastly, it should be pointed out that even though this CTST/W KIE estimate is worse than the CEQB or VTST estimates, it is still only slightly outside the two standard deviation error of the experimental evaluation.

$D + H_2 \rightarrow HD + H$ and $H + D_2 \rightarrow HD + D$. FP-ST experiments have been carried out on these reactions over the temperature range, ~700 to 2000 K (13,22), and the results are summarized in Table I. By inspection, the FP-ST data alone show a difference in apparent activation energies of 1.86 kcal mole^{-1} indicating the existence of a primary isotope effect. In these cases, a large number of precedent studies exist because of the historical importance of these reactions in gas phase chemical kinetics. The absolute rate constant experimental work of Le Roy and coworkers (55-58) and Westenberg and de Haas (59) in the 1960's and 70's is most notable. There is a recent study by Jayaweera and Pacey (60) that has extended the temperature range for $H + D_2$ down to 256 K. All of these data can be combined to give evaluations for both reactions over the large temperature range, ~250 to 2000 K, and these evaluations are given in Table I. The kinetic isotope effect over this temperature range is then calculated as the ratio,

$$KIE = 0.237 \, T^{0.19} \exp(900 \, K/T). \tag{7}$$

Equation (7) is plotted in Figure 4. The errors indicated in this figure were roughly estimated from the accuracy of the low temperature results (55-60), ±16%, and from that of the high temperature FP-ST results (13,22), ±38%.

When the absolute rate constant evaluations for both reactions are compared to theoretical calculations, the agreement is generally good. In this case the theoretical potential energy surface is known with such high accuracy that adjustments of vibration frequencies or total electronic binding energy are not possible; i. e., the calculation does not allow for parameterization. The accurate *ab initio* potential energy for this case comes from the work of Liu (61) and Siegbahn (62) as fitted by Truhlar and Horowitz (63). It is commonly called the LSTH potential energy surface. The LSTH surface has then been used in VTST calculations to estimate the thermal rate behavior for $D + H_2$ and $H + D_2$, and the results are in fairly good agreement with the experimental evaluations over the entire temperature range (64,65). Similarly, Sun and Bowman (66) have carried out CEQB calculations with the same surface, and the results are also in good agreement with the evaluations. Following this work, Varandas et al. (67) have calculated additional *ab initio* points and have used the DMBE method to obtain an even more accurate representation of the potential energy surface. Subsequently, Garrett et al. (68) have calculated rate constants with the new DMBE potential energy surface for the $D + H_2$ reaction by the VTST method. These results show improvement over the original LSTH calculation particularly in the low temperature region where tunneling is dominating. Lastly, Michael, Fisher, Bowman, and Sun (69) have presented CEQB results that are based on the DMBE potential surface, and these give agreement with both experimental evaluations that are excellent over the temperature range, ~200 to 2000 K. For completeness, a simple CTST/W calculation has also been presented that is based on the DMBE potential surface (13,22), and these "simplest" calculations are in very good agreement with the evaluations over the temperature range, 300 to 2000 K. Calculations on both reactions in the 200 to 300 K range diverge from the data giving estimates that are too low. This is not a suprising result since the Wigner tunneling

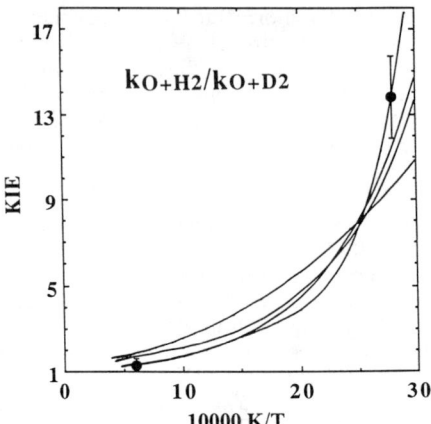

Figure 3. Kinetic isotope effect results for the $O + H_2/O + D_2$ system. The line with symbols is the experimental result as calculated from refs. 15, 21, 42, 43, and 44, as described in the text. The indicated error at low temperatures, $1\sigma \cong \pm 15\%$, is derived from the ratio of experimental results given in refs. 42 and 44 with attendant errors. The error at high temperatures, $1\sigma \cong \pm 24\%$, is derived from the two FP-ST studies, refs. 15 and 21. The three other lines are theoretical calculations. Starting on the right hand ordinate and reading down from top to bottom, these are from refs. 45, 47, and the CTST/W calculation (described in the text), respectively.

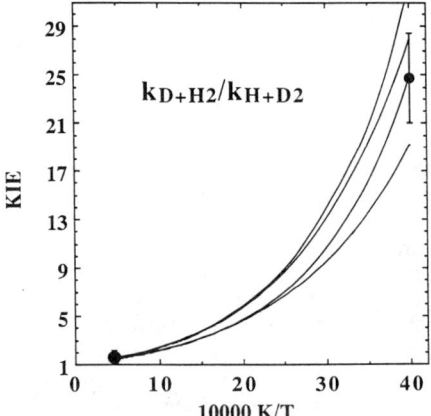

Figure 4. Kinetic isotope effect results for the $D + H_2/H + D_2$ system. The line with symbols is the experimental result as calculated from the evaluated expressions in Table I. The indicated error at low temperatures, $1\sigma \cong \pm 16\%$, is derived from the ratio of experimental results given in refs. 56 and 60 with attendant errors. The error at high temperatures, $1\sigma \cong \pm 38\%$, is derived from the two FP-ST studies, refs. 13 and 22. The three other lines are theoretical calculations. Starting on the right hand ordinate and reading down from top to bottom, these are from CEQB (ref. 69), CTST/W (refs. 13 and 22), and VTST (refs. 64 and 65), respectively.

model is known to be inadequate; however, these tunneling factors above 300 K are approximately correct.

Since the abovementioned theoretical calculations are in relatively good agreement with the absolute data for both reactions, the ratios should give a good representation of the KIE. Figure 4 shows the predictions of the kinetic isotope effect from the VTST-LSTH calculations of Garrett and Truhlar (*64,65*), the CEQB-DMBE calculations of Michael et al. (*69*), and the CTST/W-DMBE calculations of Michael and Fisher (*13*) and Michael (*22*). When these predictions are compared to the experimental result, the agreement is quite good, particularly at the two standard deviation level. This is true in both calculations where tunneling was adequately explained (*64,65,69*); however it is also true in the CTST/W calculations where the Wigner method failed to give adequate tunneling corrections. As can be seen in Figure 4, the extent of negative deviation was fortuitously the same for both reactions thereby giving an accurate value for the ratio.

Conclusions

Data for six thermal bimolecular reactions have been discussed. Kinetic isotope effects for the three addition-elimination cases are not signficantly different from unity. These results are in agreement with theoretical ideas and calculations since in all cases the isotope effect should be secondary. By constrast, the kinetic isotope effects for the three abstraction reactions are primary. Theoretical calculations have therefore predicted significant primary isotope effects, and these have been documented in two out of the three cases. The failure is the $H + H_2O/D + D_2O$ case and indicates that further theoretical work may be needed. However, the two other cases are model cases in theoretical chemical kinetics. The predicted isotope effects agree with experiment to within two standard deviations over very large temperature ranges. There is no doubt but that these results indicate the importance of quantum mechanical tunneling, and such corrections should routinely be applied in all H-atom abstraction calculations where the electronic energy barrier is large. Lastly, the relatively good results with even the simplest theory, CTST/W, suggests that the simplest theory should always be used first. This realization amounts to an approximate corroboration of the procedures of Bigeleisen (*70*). It should further be noted that in cases where the potential energy is not known with high accuracy, the energy scaling methods need not be the same for the different dynamical methods of calculation (eg. CEQB vs. VTST vs. CTST). If agreement with experiment can be obtained by an energy scaling procedure that is not outside the uncertainty in an *ab initio* potential surface then the calculation will not be in contradiction to the surface. As increasingly accurate *ab initio* results become available, this situation will no doubt change, and a preferred method for dynamical calculation will emerge.

Acknowledgements

This work was supported by the U. S. Department of Energy, Office of Basic Energy Sciences, Division of Chemical Sciences, under Contract No. W-31-109-ENG-38.

Literature Cited

1. Burns, G.; Hornig, D. F. *Can. J. Chem.* **1960**, *38*, 1702.
2. Ernst, J.; Wagner, H. Gg.; Zellner, R. *Ber. Bunsen-Ges. Phys. Chem.* **1978**, *82*, 409.
3. Niemitz, K. J.; Wagner, H. Gg.; Zellner, R. *Z. Phys. Chem. (Frankfurt am Main)* **1981**, *124*, 155.

4. Michael, J. V.; Sutherland, J. W.; Klemm, R. B. *Int. J. Chem. Kinet.* **1985**, *17*, 315.
5. Michael, J. V.; Sutherland, J. W. *Int. J. Chem. Kinet.* **1986**, *18*, 409.
6. For a review, see Michael, J. V. *Prog. Energy Combust. Sci.*, in press.
7. Michael, J. V. *Adv. Chem. Kin. and Dynamics*, in press.
8. Bradley, J. N. *Shock Waves in Chemistry and Physics*, Wiley, New York, NY, 1962.
9. Michael, J. V.; Sutherland, J. W.; Klemm, R. B. *J. Phys. Chem.* **1986**, *90*, 497.
10. Michael, J. V.; Sutherland, J. W. *J. Phys. Chem.* **1988**, *92*, 3853.
11. Michael, J. V.; Fisher, J. R. In *Current Topics in Shock Waves, Seventeenth International Symposium on Shock Waves and Shock Tubes;* Kim, Y. W., Ed.; American Institute of Physics, New York, NY, 1989; pp. 210-215.
12. Fisher, J. R.; Michael, J. V. *J. Phys. Chem.* **1990**, *94*, 2465.
13. Michael, J. V.; Fisher, J. R. *J. Phys. Chem.* **1990**, *94*, 3318.
14. Sutherland, J. W.; Michael, J. V.; Klemm, R. B. *J. Phys. Chem.* **1986**, *90*, 5941.
15. Michael, J. V. *J. Chem. Phys.* **1989**, *90*, 189.
16. Davidson, D. F.; Hanson, R. K. *Int. J. Chem. Kinet.* **1990**, *22*, 843.
17. Koshi, M.; Yoshimura, M.; Fukuda, K.; Matsui, H.; Saito, K; Watanabe, M.; Imamura, A.; Chen, C. *J. Chem. Phys.* **1990**, *93*, 8703.
18. Michael, J. V.; Wagner, A. F. *J. Phys. Chem.* **1990**, *94*, 2453.
19. Shin, K. S.; Michael, J. V. *J. Chem. Phys.* **1991**, *95*, 262.
20. Shin, K. S.; Michael, J. V. *J. Phys. Chem.* **1991**, *95*, 5864.
21. Sutherland, J. W.; Michael, J. V.; Pirraglia, A. N.; Nesbitt, F. L.; Klemm, R. B. In *Twenty-first Symposium (International) on Combustion*, The Combustion Institute, Pittsburgh, PA, 1986, pp. 929-939.
22. Michael, J. V. *J. Chem. Phys.* **1990**, *92*, 3394.
23. Bohn, B; Stuhl, F. *J. Phys. Chem.* **1990**, *94*, 8010.
24. Harding, L. B.; Wagner, A. F. *J. Phys. Chem.* **1986**, *90*, 2974.
25. Kurzius, S. C.; Boudart, M. *Combust. Flame* **1969**, *21*, 477.
26. Chiang, C.; Skinner, G. B. In *Twelfth International Symposium on Shock Tubes and Shock Waves;* Lifshitz, A.; Rom, J., Eds.; Magnes, Jerusalem, Israel, 1980; pp. 629-639.
27. Pamidimukkala, K. M.; Skinner, G. B. In *Thirteenth International Symposium on Shock Waves and Shock Tubes*, SUNY Press, Albany, NY, 1981; pp. 585-592.
28. Pastrana, M. R.; Quintales, L. A. M.; Brandão, J.; Varandas, A. J. C. *J. Phys. Chem.* **1990**, *94*, 8073, and references cited therein.
29. Miller, J. A. *J. Chem. Phys.* **1981**, *74*, 5120; ibid. **1986**, *84*, 6170.
30. Miller, J. A. *J. Chem. Phys.* **1981**, *75*, 5349.
31. Melius, C. F.; Blint, R. *Chem. Phys. Lett.* **1979**, *64*, 183.
32. Varandas, A. J. C.; Brandão, J.; Pastrana, M. R. *J. Chem. Phys.*, submitted.
33. Issacson, A. D.; Truhlar, D. G. *J. Chem. Phys.* **1982**, *76*, 1380.
34. Schatz, G. C.; Elgersma, H. *Chem. Phys. Lett.* **1980**, *73*, 21.
35. Walch, S. P.; Dunning, T. H., Jr. *J. Chem. Phys.* **1980**, *72*, 1303.
36. Baulch, D. L.; Drysdale, D. D.; Horne, D. G.; Lloyd, A. C. *Evaluated Kinetic Data for High Temperature Reactions, Vol. 1: Homogeneous Gas Phase Reactions of the H_2-O_2 System*, Butterworths, London, 1973, p. 49.
37. Cohen, N.; Westberg, K. *J. Phys. Chem. Ref. Data* **1983**, *12*, 531.
38. Warnatz, J. In *Combustion Chemistry*; Gardiner, W. C., Jr., Ed.; Springer-Verlag, New York, NY, 1984, Chap. 5.
39. Tsang, W.; Hampson, R. F. *J. Phys. Chem. Ref. Data* **1986**, *15*, 1078.
40. Clyne, M. A. A.; Thrush, B. A. *Proc. R. Soc. London* **1963**, *A275*, 544.

41. Westenberg, A. A.; de Haas, N. *J. Chem. Phys.* **1967**, *47*, 4241.
42. Presser, N.; Gordon, R. J. *J. Chem. Phys.* **1985**, *82*, 1291.
43. Marshall, P.; Fontijn, A. *J. Chem. Phys.* **1987**, *87*, 6988.
44. Zhu, Y-F.; Arepalli, S.; Gordon, R. J. *J. Chem. Phys.* **1989**, *90*, 183.
45. Bowman, J. M.; Wagner, A. F.; Walch, S. P.; Dunning, T. H., Jr. *J. Chem. Phys.* **1984**, *81*, 1739.
46. Garrett, B. C.; Truhlar, D. G. *J. Chem. Phys.* **1984**, *81*, 309.
47. Joseph, T.; Truhlar, D. G.; Garrett, B. G. *J. Chem. Phys.* **1988**, *88*, 6982.
48. Walch, S, P.; Dunning, T. H., Jr.; Raffenetti, R. C.; Bobrowicz, F. W. *J. Chem. Phys.* **1980**, *72*, 406.
49. Walch, S. P.; Wagner, A. F.; Dunning, T. H., Jr.; Schatz, G. C. *J. Chem. Phys.* **1980**, *72*, 2894.
50. Lee, K. T.; Bowman, J. M.; Wagner, A. F.; Schatz, G. C. *J. Chem. Phys.* **1982**, *76*, 3563, 3583.
51. Robie, D. C.; Arepalli, S.; Presser, N.; Kitsopoulos, T.; Gordon, R. J. *Chem. Phys. Lett.* **1987**, *134*, 579.
52. Garrett, B. C.; Truhlar, D. G.; Bowman, J. M.; Wagner, A. F.; Robie, D.; Arepalli, S.; Presser, N.; Gordon, R. J. *J. Am. Chem. Soc.* **1986**, *108*, 3515.
53. Bowman, J. M.; Wagner, A. F. *J. Chem. Phys.* **1987**, *86*, 1967.
54. Wagner, A. F.; Bowman, J. M. *J. Chem. Phys.* **1987**, *86*, 1976.
55. Ridley, B. A.; Schulz, W. R.; Le Roy, D. J. *J. Chem. Phys.* **1966**, *44*, 3344.
56. Mitchell, D. N.; Le Roy, D. J. *J. Chem. Phys.* **1973**, *58*, 3449.
57. Schulz, W. R.; Le Roy, D. J. *Can. J. Chem.* **1964**, *42*, 2480.
58. Schulz, W. R.; Le Roy, D. J. *J. Chem. Phys.* **1965**, *42*, 3869.
59. Westenberg, A. A.; de Haas, N. *J. Chem. Phys.* **1967**, *47*, 1393.
60. Jayaweera, I. S.; Pacey, P. D. *J. Phys. Chem.* **1990**, *94*, 3614.
61. Liu, B. *J. Chem. Phys.* **1973**, *58*, 1925.
62. Siegbahn, P.; Liu, B. *J. Chem. Phys.* **1978**, *68*, 2457.
63. Truhlar, D. G.; Horowitz, C. J. *J. Chem. Phys.* **1978**, *68*, 2466.
64. Garrett, B. C.; Truhlar, D. G. *Proc. Natl. Acad. Sci.* **1979**, *76*, 4755.
65. Garrett, B. C.; Truhlar, D. G. *J. Chem. Phys.* **1980,** *72*, 3460.
66. Sun, Q.; Bowman, J. M. *J. Phys. Chem.* **1990**, *94*, 718.
67. Varandas, A. J. C.; Brown, F. B.; Mead, C. A.; Truhlar, D. G.; Blais, N. C. *J. Chem. Phys.* **1987**, *86*, 6258.
68. Garrett, B. C.; Truhlar, D. G.; Varandas, A. J. C.; Blais, N. C. *Int. J. Chem. Kinet.* **1986**, *18*, 1065.
69. Michael, J. V.; Fisher, J. R.; Bowman, J. M.; Sun, Q. *Science* **1990**, *249*, 269.
70. Bigeleisen, J. *J. Chem. Phys.* **1949**, *17*, 675.

RECEIVED September 4, 1991

Chapter 7

Deuterium Substitution Used as a Tool for Investigating Mechanisms of Gas-Phase Free-Radical Reactions

P. H. Wine, A. J. Hynes, and J. M. Nicovich

Physical Sciences Laboratory, Georgia Tech Research Institute, Georgia Institute of Technology, Atlanta, GA 30332

> Results are presented and discussed for a number of gas phase free radical reactions where H/D isotope effects provide valuable mechanistic insights. The cases considered are (1) the reactions of OH, NO_3, and Cl with atmospheric reduced sulfur compounds, (2) the reactions of OH and OD with CH_3CN and CD_3CN, and (3) the reactions of alkyl radicals with HBr and DBr.

A major focus of modern chemical kinetics research is on understanding complex chemical systems of practical importance such as the atmosphere and fossil fuel combustion. In these applications, accurate information on reaction mechanisms (i.e., product identities and yields) as well as reaction rate coefficients is often critically important. Since detailed experimental kinetic and mechanistic information for every reaction of importance in a complex chemical system is often an unrealizable goal, it is highly desirable to develop a firm theoretical understanding of well studied reactions which can be extrapolated to prediction of unknown rate coefficients and product yields.

In recent years it has become apparent that many reactions of importance in atmospheric and combustion chemistry occur via complex mechanisms involving potential energy minima (i.e., weakly bound intermediates) along the reaction coordinate. The OH + CO reaction is one of the best characterized examples (*1*). While theoretical descriptions can sometimes be employed to rationalize experimental observations (*1-3*), a theoretical framework does not yet exist for predicting complex behavior. In this paper we discuss some experimental studies carried out in our laboratory over the last several years which were aimed at characterizing the kinetics and mechanisms of a number of complex chemical reactions of practical interest. Mechanistic details were deduced in part from studies of the effects of temperature, pressure, and [O_2] on reaction kinetics and from direct observation of reaction products. However, studies of H/D isotope effects were also employed as a tool for deducing reaction mechanisms; information obtained from the isotope effect studies is highlighted in the discussion.

The chemical processes we have chosen for discussion are (1) the reactions of OH, NO_3, and Cl with atmospheric reduced sulfur compounds (2) the reactions of OH and OD with CH_3CN and CD_3CN, and (3) the reactions of alkyl radicals with HBr and DBr. The experimental methodology employed to investigate the above reactions involved coupling generation of reactant radicals by laser flash photolysis with time resolved detection of reactants and products by pulsed laser induced fluorescence (OH and OD), atomic resonance fluorescence (Cl and Br), and long path tunable dye laser absorption (NO_3).

The Reactions of OH, NO_3, and Cl with Atmospheric Reduced Sulfur Compounds

Dimethylsulfide (CH_3SCH_3, DMS) emissions into the atmosphere from the oceans are thought to account for a significant fraction of the global sulfur budget (*4*). It has been suggested that DMS oxidation in the marine atmosphere is an important pathway for production of cloud condensation nuclei and, therefore, that atmospheric DMS can play a major role in controlling the earth's radiation balance and climate (*5*). Hence, there currently exists a great deal of interest in understanding the detailed mechanism for oxidation of atmospheric DMS.

It is generally accepted that the OH radical is an important initiator of DMS oxidation in the marine atmosphere (*4*). Several years ago, we carried out a detailed study of the kinetics and mechanism of the OH + DMS reaction (*6*). We found that OH reacts with DMS via two distinct pathways, one of which is only operative in the presence of O_2, and one of which is operative in the absence or presence of O_2 (see Figure 1). The rate of the O_2-dependent pathway increases with increasing [O_2], increases dramatically with decreasing temperature, and shows no kinetic isotope effect, i.e., CH_3SCH_3 and CD_3SCD_3 react at the same rate. These observations indicate that the O_2-dependent pathway involves formation of a weakly bound adduct which reacts with O_2 in competition with decomposition back to reactants.

$$OH + CH_3SCH_3 + M \rightleftarrows (CH_3)_2SOH + M \qquad (1,-1)$$

$$(CH_3)_2SOH + O_2 \rightarrow products \qquad (2)$$

The absence of a kinetic isotope effect strongly suggests that none of the three elementary steps in the above mechanism involve breaking a C-H bond.

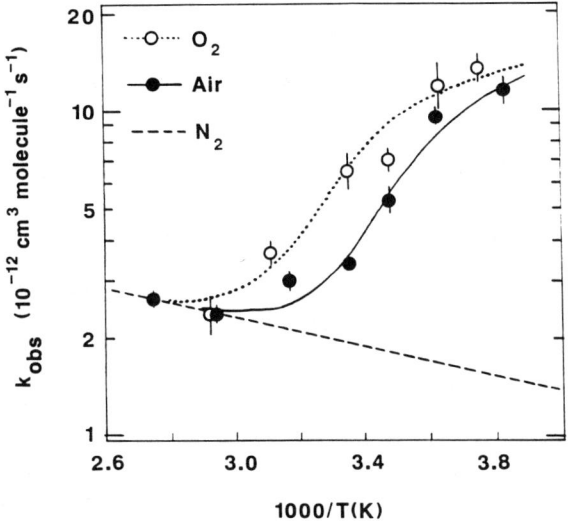

Figure 1. Arrhenius plots for the OH + CD_3SCD_3 reaction in 700 Torr N_2, air, and O_2. k_{obs} ≡ the slope of a plot of the pseudo-first order OH decay rate versus the CD_3SCD_3 concentration under conditions where the adduct $(CD_3)_2SOH$ is removed much more rapidly than it is formed. (Reproduced from reference 62. Copyright 1987 American Chemical Society.)

The O_2-independent channel for the OH + DMS reaction proceeds with a 298K rate coefficient of 4.4×10^{-12} cm^{3}molecule^{-1}s^{-1}; in one atmosphere of air, the O_2-independent channel is dominant at T > 285K while the O_2-dependent channel dominates at lower temperatures (6). We find that the rate of the O_2-independent channel is pressure independent but increases slightly with increasing temperature (small positive activation energy). Furthermore, the O_2-independent channel displays a significant kinetic isotope effect, $k_H/k_D \sim 2.3$ at 298K. Based upon the observed positive activation energy and significant isotope effect, we have postulated (6) that the O_2-independent pathway is a direct hydrogen abstraction reaction, i.e., there is no potential energy minimum (corresponding to an OH-DMS adduct) on the potential energy surface connecting reactants with products.

$$OH + CH_3SCH_3 \rightarrow CH_3SCH_2 + H_2O. \tag{3}$$

Interestingly, Domine et al. (7) have recently observed production of C_2H_5 + CH_3SOH from the reaction of OH with $C_2H_5SCH_3$ at low pressure and in the absence of O_2, although the branching ratio for production of C_2H_5 + CH_3SOH remains rather uncertain. By analogy, Domine et al.'s results suggest that the O_2-independent pathway in OH + DMS may involve cleavage of the relatively weak C-S bond rather than the C-H bond.

$$OH + CH_3SCH_3 \rightarrow [(CH_3)_2SOH^{\ddagger}] \rightarrow CH_3 + CH_3SOH \tag{4}$$

If the O_2-independent pathway for OH + DMS is reaction 4 rather than reaction 3, then the H/D isotope effect we have observed (6) would, to our knowledge, be the largest secondary isotope effect known for a gas phase reaction. Clearly, direct determination of the product yields from the O_2-independent channel of the OH + DMS reaction could have a major impact not only on our understanding of atmospheric sulfur oxidation, but also on our understanding of chemical reactivity in general and kinetic isotope effects in particular.

In coastal marine environments where NO_x levels are relatively high, it is generally believed that NO_3 can compete with OH as an initiator of DMS oxidation (4). The 298K rate coefficient for the NO_3 + DMS reaction is known to be about 1×10^{-12} cm^{3}molecule^{-1}s^{-1} (8-13) and a significant negative activation energy has been reported (12). The reaction of NO_3 with DMS could proceed via direct H or O atom transfer or via formation of long-lived adduct.

$$NO_3 + CH_3SCH_3 \rightarrow CH_3SCH_2 + HNO_3 \tag{5}$$

$$NO_3 + CH_3SCH_3 \rightarrow (CH_3)_2SO + NO_2 \tag{6}$$

$$NO_3 + CH_3SCH_3 + M \rightleftarrows (CH_3)_2SONO_2 + M \tag{7,-7}$$

Attempts to detect NO_2 as a reaction product have been unsuccessful (9,12) suggesting that O atom transfer via either a direct mechanism or via adduct decomposition is unimportant. As pointed out by Atkinson et al. (8), the NO_3 + DMS reaction is several orders of magnitude faster than the known rates of H-abstraction of, for example, relatively weakly bound aldehydic hydrogens by NO_3; this fact, coupled with the observed negative activation energy (12), strongly suggests that the NO_3 + DMS reaction does not proceed via a direct H-abstraction pathway. By the process of elimination, it is generally accepted that the initial step in the NO_3 + DMS reaction is adduct formation, i.e., reaction (7).

In a recent study of the kinetics of NO_3 reactions with organic sulfides (13), we observed a large kinetic isotope effect for the NO_3 + DMS reaction; at 298K NO_3 reacts with CH_3SCH_3 a factor of 3.8 more rapidly than with CD_3SCD_3. The observed isotope effect, coupled with the observation that at 298K $C_2H_5SC_2H_5$ reacts with NO_3 a factor of 3.7 more rapidly than does CH_3SCH_3, clearly demonstrates that the adduct decomposes via a process which involves C-H bond cleavage. A very recent chamber study by Jensen et al. (14) confirms the magnitude of our reported isotope effect and reports quantitative observation of HNO_3 as a reaction product.

$$(CH_3)_2SONO_2 + M \rightarrow CH_3SCH_2 + HNO_3 + M \qquad (8)$$

As we discuss elsewhere (13), the postulate that the NO_3 + DMS reaction proceeds via reactions 7, -7, and 8 appears to be consistent with all available product data.

It is interesting to compare and contrast kinetic and mechanistic findings for the NO_3 + DMS reaction, with those for the reaction of OH with CH_3SH. Like NO_3 + DMS, the OH + CH_3SH reaction becomes faster with decreasing temperature (15-18), suggesting that the initial step in the mechanism is adduct formation.

$$OH + CH_3SH + M \rightleftarrows CH_3S(OH)H + M \qquad (9)$$

Also, as appears to be the case for NO_3 + DMS, the OH + CH_3SH reaction is known to give H-abstraction products with unit yield (19).

$$CH_3S(OH)H + M \rightarrow CH_3S + H_2O \qquad (10)$$

Hence, there are important similarities between the NO_3 + DMS and OH + CH_3SH reactions. However, there are also important differences. First, at 298K the OH + CH_3SH reaction is about 30 times faster than the NO_3 + DMS reaction. Secondly, while NO_3 + DMS displays a large H/D kinetic isotope effect (see above), isotope effects in OH reactions with CH_3SH, CD_3SH, and CH_3SD are minimal (17,18). These reactivity differences can be rationalized by postulating that decomposition of $(CH_3)_2SONO_2$ to products competes relatively unfavorably with decomposition back to reactants (i.e. $k_{-7} \gg k_8$), whereas decomposition of $CH_3S(OH)H$ to products is much faster than decomposition back to reactants (i.e. $k_{-9} \ll k_{10}$). Hence, the rate of the adduct → product step, which should be sensitive to isotopic substitution, strongly influences the overall rate of the NO_3 + DMS reaction but does not influence the overall rate of the OH + CH_3SH reaction.

Recently in our laboratory we have investigated the kinetics of chlorine atom reactions with CH_3SH, CD_3SD, H_2S, and D_2S (20) as a function of temperature. There have been no previous reports of the temperature dependence of the Cl + CH_3SH rate coefficient and no previous kinetics studies of Cl reactions with CD_3SD or D_2S. Nesbitt and Leone (21,22) have shown that, at 298K, the Cl + CH_3SH reaction occurs at a gas kinetic rate (k ~ 1.84 x 10^{-10} $cm^3 molecule^{-1} s^{-1}$) and that k_{11}/k_{12} ~ 45.

$$Cl + CH_3SH \rightarrow CH_3S + HCl \quad (11)$$

$$Cl + CH_3SH \rightarrow CH_2SH + HCl \quad (12)$$

Several kinetics studies of the Cl + H_2S reaction have been reported (21,23-27) with published 298K rate coefficients spanning the range (4.0 - 10.5) x 10^{-11} $cm^3 molecule^{-1} s^{-1}$. Two temperature dependence studies (26,27) both conclude that the Cl + H_2S rate coefficient is temperature independent. Internal state distributions in the HCl product of Cl + H_2S and Cl + CH_3SH (28,29) and the SH product of Cl + H_2S (29) have also been reported.

Arrhenius plots for reactions of Cl with H_2S, D_2S, CH_3SH, and CD_3SD are shown in Figure 2. Arrhenius expressions derived from our data are as follows (units are 10^{-11} $cm^3 molecule^{-1} s^{-1}$; errors are 2σ, precision only):

Cl + H_2S: $\quad$ k = (3.60 ± 0.26) exp [(210 ± 20)/T], 202-430K

Cl + D_2S: $\quad$ k = (1.65 ± 0.27) exp [(225 ± 45)/T], 204-431K

Cl + CH_3SH, CD_3SD: k = (11.9 ± 1.7) exp [(151 ± 38)/T], 193-430K

Kinetic data for CH_3SH and CD_3SD were indistinguishable so one Arrhenius expression incorporating all data is presented. One important aspect of our results is that all reactions are characterized by small but well-defined negative activation energies, suggesting that long range attractive forces between S and Cl are important in defining the overall rate coefficient. Our interpretation of observed kinetic isotope effects follows the same arguments as employed above in the comparison of NO_3 + DMS with OH + CH_3SH. In the case of the Cl + CH_3SH reaction, adduct decomposition to products is apparently fast compared to adduct decomposition back to reactants whereas in the case of the Cl + H_2S reaction the two adduct decomposition pathways occur at competitive rates. This argument seems

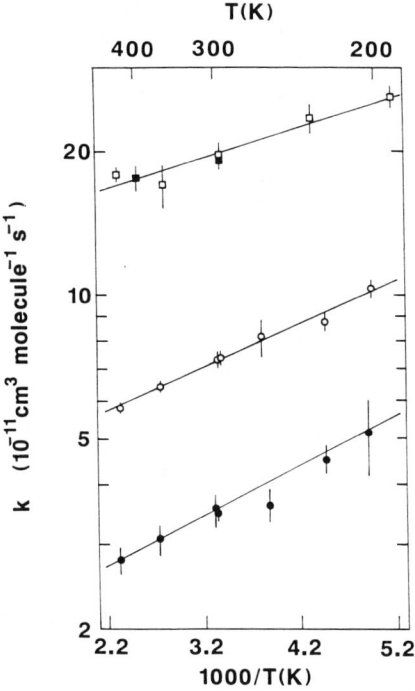

Figure 2. Arrhenius plots for the reactions of chlorine atoms with H_2S (○), D_2S (●), CH_3SH (□), and CD_3SD (■). Error bars are 2σ and represent precision only. Solid lines are obtained from linear least squares analyses which yield the Arrhenius parameters given in the text.

reasonable since we expect H_2SCl to be a less strongly bound species than $CH_3S(Cl)H$, thus making adduct decomposition back to reactants considerably more rapid for $Cl + H_2S$ than for $Cl + CH_3SH$. The relative stabilities of the adducts can be predicted based on the facts that a methyl group releases electron density to the sulfur atom more efficiently than does a hydrogen atom (30), and that the ionization potential of CH_3SH is about 1 ev lower than the ionization potential of H_2S (31), thus facilitating the formation of a more stable charge transfer complex in the $Cl + CH_3SH$ case.

The Reactions of OH and OD with CH_3CN and CD_3CN

Acetonitrile (CH_3CN) is present at significant levels in both the troposphere and the stratosphere, and has been implicated in stratospheric ion chemistry (32-35). Reaction with OH is generally thought to be a major atmospheric removal mechanism for acetonitrile (35). Early studies of the kinetics of the $OH + CH_3CN$ reaction demonstrated that $k(298K) \sim 2 \times 10^{-14}$ $cm^3 molecule^{-1}s^{-1}$ and that $E_{act} \sim 2$ kcal $mole^{-1}$ (36-41); it has generally been thought that reaction proceeds via a direct H-abstraction mechanism (40-42).

We recently carried out a detailed study of the hydroxyl reaction with acetonitrile which demonstrates that the reaction mechanism is considerably more complex than previously thought (43). The kinetics of the following four isotopic variants were investigated:

$$OH + CH_3CN \rightarrow products \tag{13}$$

$$OH + CD_3CN \rightarrow products \tag{14}$$

$$OD + CH_3CN \rightarrow products \tag{15}$$

$$OD + CD_3CN \rightarrow products \tag{16}$$

All four reactions were studied at 298K as a function of pressure and O_2 concentration, while reactions 13 and 14 were also studied as a function of temperature.

Experiments which employed N_2 buffer gas gave some results which appear inconsistent with the idea that reactions 13 - 16 occur via direct H (or D) abstraction pathways. First, rate coefficients for reactions 13 and 14 (but not reactions 15 and 16) increase with increasing pressure over the range 50 - 700 Torr; the largest increase, nearly a factor of two, is observed for reaction 14. Second, observed isotope effects on the (high pressure limit) 298K rate coefficients are not as would be expected for an H (or D)-abstraction mechanism. Measured 298K rate coefficients in units of 10^{-14} $cm^3 molecule^{-1}s^{-1}$ are $k_{13} = 2.48 \pm 0.38$, $k_{14} = 2.16 \pm 0.11$, $k_{15} = 3.18 \pm 0.40$, and $k_{16} = 2.25 \pm 0.28$ (errors are 2σ). If the dominant reaction pathway is H (or D) abstraction we would expect reactions 13 and 15, which break C-H bonds, to be faster by a factor of two or more than reactions 14 and 16, which break C-D bonds. Observed differences in reactivity are quite small, although reaction 15 does appear to be somewhat faster than the other reactions.

The observed kinetics in the absence of O_2 can best be reconciled with a complex mechanism which proceeds via formation of an energized intermediate, i.e.

$$OH + CH_3CN \underset{k_{-a}}{\overset{k_a}{\rightleftarrows}} \text{energized complex} \overset{k_b}{\rightarrow} \text{products}$$

$$\overset{M, k_c}{\longrightarrow} \text{adduct}$$

Such an energized intermediate could decompose to produce $CH_2CN + H_2O$ or other products, decompose back to reactants, or be collisionally stabilized at sufficiently high pressures. Hence, the reaction proceeds at a finite rate at low pressure but shows an enhancement in the rate as the pressure is increased. Such a mechanism is well documented for the important atmospheric reactions of OH with CO and HNO_3 (*44*) and has recently been observed in our laboratory for the Cl + DMS reaction (*45*). The pressure, temperature, and isotopic substitution dependences of the elementary rate coefficients k_a, k_{-a}, k_b, and k_c interact to produce the observed complex behavior.

Perhaps the most conclusive evidence that the OH + CH_3CN reaction proceeds, at least in part, via formation of an intermediate complex comes from experiments carried out in reaction mixtures containing O_2. Observed OH temporal profiles in the presence of CH_3CN and O_2 are non-exponential and suggest that OH is regenerated via a reaction of O_2 with a product of reaction 13. Two possibilities are as follows:

$$OH + CH_3CN \rightarrow CH_2CN + H_2O \tag{13a}$$

$$CH_2CN + O_2 \rightarrow \rightarrow OH + \text{other products} \tag{17}$$

$$OH + CH_3CN + M \rightarrow \text{adduct} + M \tag{13b}$$

$$\text{adduct} + O_2 \rightarrow \rightarrow OH + \text{other products} \tag{18}$$

In the mixed-isotope experiments, we observe that OD is regenerated from OD + CH_3CN + O_2 and that OH is regenerated from OH + CD_3CN + O_2; these findings conclusively demonstrate that an important channel for the hydroxyl + acetonitrile reaction involves formation of an adduct which lives long enough to react with O_2 under atmospheric conditions, and also places considerable constraints on possible adduct + O_2 reaction pathways. A plausible set of elementary steps via which OH can be regenerated in the OH + CD_3CN + O_2 reaction is shown in Figure 3. The mechanism involves OH addition to the nitrogen atom, followed by O_2 addition to the cyano carbon atom, isomerization, and decomposition to D_2CO + DOCN + OH. Further studies are needed to establish whether or not OD as well as OH is generated from OH + CD_3CN + O_2 and whether or not OH as well as OD is generated from OD + CH_3CN + O_2. Further studies are also needed to directly detect end products of the adduct + O_2 reactions(s).

The Reactions of Alkyl Radicals with HBr and DBr

The thermochemistry and kinetics of alkyl radicals are subjects of considerable importance in many fields of chemistry. Accurate evaluation of alkyl radical heats of formation are required for determination of primary, secondary, and tertiary bond dissociation energies in hydrocarbons, for establishing rates of heat release in combustion, and for relating unknown "reverse" rate coefficients to known "forward" values. Kinetic data for numerous alkyl radical reactions are needed for modeling hydrocarbon combustion.

Recent direct kinetic studies (46-51), primarily by Gutman and coworkers (46-49), strongly suggest that alkyl + HX reactions have negative activation energies, a finding which seems counter-intuitive for apparently simple hydrogen abstraction reactions. It should be noted, however, that one recent direct study (52) reports much slower rate coefficients compared to other direct studies (46,48,50,51) and positive activation energies for the reactions of t-C_4H_9 with DBr and DI.

Motivated initially by the desire to obtain improved thermochemical data for sulfur-containing radicals of atmospheric interest, we developed a method for studying radical + HBr(DBr) reactions by observing the appearance kinetics of product bromine atoms (53). We have recently applied the same experimental approach to investigate the kinetics of the following reactions (54):

$$CH_3 + HBr \rightarrow Br + CH_4 \tag{19}$$

$$CD_3 + HBr \rightarrow Br + CD_3H \tag{20}$$

$$CH_3 + DBr \rightarrow Br + CH_3D \tag{21}$$

$$C_2H_5 + HBr \rightarrow Br + C_2H_6 \tag{22}$$

$$C_2H_5 + DBr \rightarrow Br + C_2H_5D \tag{23}$$

$$t\text{-}C_4H_9 + HBr \rightarrow Br + (CH_3)_3CH \tag{24}$$

$$t\text{-}C_4H_9 + DBr \rightarrow Br + (CH_3)_3CD \tag{25}$$

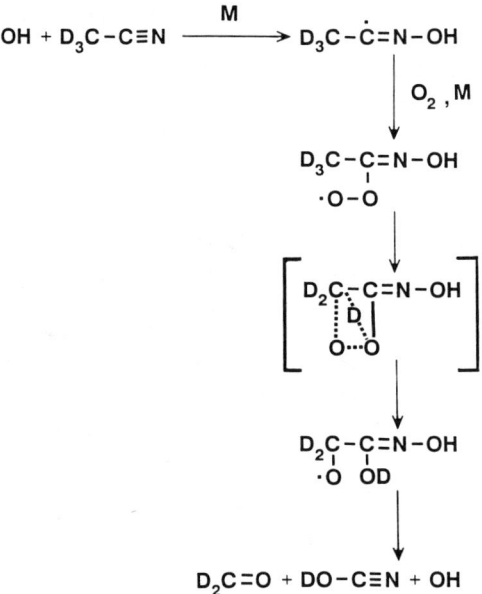

Figure 3. Plausible set of elementary steps for the reaction $OH + CD_3CN + O_2 \rightarrow D_2CO + DOCN + OH$. Adduct decomposition to products is shown as a single step; in reality, it probably occurs via two sequential steps with either D_2CO or OH coming off before the other. (Reproduced from reference 43. Copyright 1991 American Chemical Society.)

The isotope effect studies were motivated by a recent theoretical investigation of the t-C_4H_9 + HI, DI reactions (55) which suggests that negative activation energies for alkyl + HX reactions should be accompanied by inverse kinetic isotope effects, i.e., k_H/k_D < 1.

In Table I our results (54) are compared with other available direct kinetic data for reactions 19 - 25. The negative activation energies and fast rate coefficients for alkyl + HBr reactions reported by Gutman and coworkers (46,47,49) are confirmed in our study. In fact, the activation energies derived from our data are consistently a little lower, i.e., more negative, than those reported by Gutman and coworkers and the 298K rate coefficients obtained in our study are consistently more than a factor of two faster than those reported by Gutman and coworkers. Our 298K rate coefficient for the t-C_4H_9 + HBr reaction exceeds the values reported by Russell et al. (46) and Richards et al. (50) by a factor of 2.7, but is in excellent agreement with the value reported by Seakins and Pilling (51); interestingly, the experimental technique employed by Seakins and Pilling was very similar to the technique employed in our study. Our 298K rate coefficient for the t-C_4H_9 + DBr reaction exceeds the value reported by Richards et al. (50) by a factor of 2.7 and exceeds the value reported by Müller-Markgraf et al. (52) by more than two orders of magnitude. As discussed in some detail by Gutman (56), the probable source of error in the Müller-Markgraf et al. study (52) is neglect of heterogeneous loss of t-C_4H_9 in their data analysis.

Traditionally, hydrogen transfer reactions such as R + HX → RH + X have been thought of as "direct" metathesis reactions with a barrier along the reaction coordinate and a single transition state located at the potential energy maximum. Rationalization of observed negative activation energies for R + HX reactions has centered around the postulate that product formation proceeds via formation of weakly bound R···XH complexes (45-48,55). As shown by Mozurkewich and Benson (57), if the transition state leading from reactants to complex (TS1) is loose and the transition state leading from complex to products is both tighter and lower in energy compared to TS1, then a negative activation energy for the overall reaction should be observed. McEwen and Golden (55) have carried out a two channel RRKM calculation that models the t-C_4H_9 + HI(DI) reactions as proceeding through a weakly bound complex; they were able to reproduce the kinetics results of Seetula et al. (48) for t-C_4H_9 + HI assuming complex binding energies as low as 3 kcal mole^{-1}. Probably the most interesting aspect of McEwen and Golden's study is the fact that models which were capable of reproducing experimentally observed (48) k(T) values for t-C_4H_9 + HI also predicted an inverse kinetic isotope effect (KIE), i.e., t-C_4H_9 + DI was predicted to be faster than t-C_4H_9 + HI. The predicted inverse KIE results from the fact that the transition state leading from complex to products becomes looser with lower vibrational frequencies associated with deuterium substitution. Contrary to McEwen and Golden's prediction for t-C_4H_9 + HI, we observe normal KIE's for CH_3, C_2H_5, and t-C_4H_9 reactions with HBr. Richards et al. (50) also observe a normal KIE for the t-C_4H_9 +

Table I. Comparison of our results (reference 54) with other direct determinations of alkyl + HBr(DBr) rate coefficients.[a]

Reaction	Exptl Method[b]	Range of T	A[c]	-E/R[c]	k_i(298K)[d]	Reference
CH_3 + HBr	LFP - PIMS	296 - 532	0.87	160 ± 110	1.49	47
	LFP - RF	257 - 422	1.36	233 ± 23	2.97	54
CD_3 + HBr	LFP - IRE	298			4.7	60
	LFP - RF	297			3.35	54
CH_3 + DBr	VLPP	608 - 1000	0.32	0 ± 500	0.32	61
	LFP - RF	267 - 429	1.07	130 ± 55	1.66	54
C_2H_5 + HBr	LFP - PIMS	295 - 532	1.0	410 ± 110	3.96	47
	LFP - RF	259 - 427	1.33	539 ± 78	8.12	54
C_2H_5 + DBr	LFP - RF	298 - 415	(0.92)	(580)	6.44	54
$t-C_4H_9$ + HBr	LFP - PIMS	296 - 532	0.99	700 ± 110	10.4	46
	LFP - DLA	297			10	50
	LFP - RF	298			32	51
	LFP - RF	297 - 429	1.07	963 ± 152	27.1	54
$t-C_4H_9$ + DBr	VLPP	295 - 384	(8.3)	(-1180)	0.16	52
	LFP - DLA	297			8	50
	LFP - RF	298 - 415	(1.03)	(919)	22.5	54

a. Units are T, E/R: degrees K; A, k_i(298K): 10^{-12} cm^{3}molecule^{-1}s^{-1}.
b. LFP: laser flash photolysis; PIMS: photoionization mass spectrometry; RF: resonance fluorescence; IRE: infrared emission; VLPP: very low pressure pyrolysis; DLA: diode laser absorption; VLPΦ: very low pressure photolysis.
c. Parentheses indicate Arrhenius parameters which are based on experiments at only two temperatures.
d. Calculated from Arrhenius parameters when temperature dependent data were obtained. Error limits not quoted due to inconsistencies in methods used by different groups to arrive at uncertainties; most values of k_i(298K) have absolute accuracies in the 15-30% range.

HBr reaction. It does appear, however, that the magnitude of the KIE is reduced as the activation energy becomes more negative, i.e., the observed KIE is largest for R = CH_3 and smallest for R = t-C_4H_9. Chen et al. have recently calculated a potential energy surface for the CH_3 + HBr reaction at the G1 level of theory and deduced the existence of a <u>hydrogen bridged</u> complex which is bound by 0.28 kcal mole^{-1} and is formed without activation energy (58). They have also calculated rate coefficients for CH_3 + HBr, CH_3 + DBr, and CD_3 + HBr from RRKM theory with corrections for tunneling evaluated using the Wigner method (59). Their calculated isotope effects agree quantitatively with our measured isotope effects, a result which lends strong support to the idea that the methyl-HBr complex is hydrogen-bridged rather than bromine-bridged.

Summary

Experimental kinetic data have been presented and discussed for a number of reactions where H/D isotope effects provide valuable mechanistic insights. For the reactions of atmospheric free radicals with reduced sulfur compounds, isotope effect studies provide information not only about C-H or S-H bond cleavage versus other reactive pathways but also on the relative rates of adduct decompositions back to reactants versus on to products. For the reaction of hydroxyl radicals with acetonitrile, isotope effect studies conclusively demonstrate the intermediacy of a long-lived adduct and also provide site-specific information which places important constraints on the detailed mechanism for hydroxyl generation from the adduct + O_2 reaction. For the CH_3 + HBr reaction, comparison of observed and theoretical isotope effects supports the view that reaction proceeds via formation of a very weakly bound, hydrogen-bridged addition complex. In one case considered, namely the O_2-independent channel for the OH + CH_3SCH_3 reaction, there exist potential problems in relating experimental observations (6,7) to existing prejudices concerning the nature of kinetic isotope effects.

Acknowledgments

Support for the work described in this paper has been provided by grants ATM-8217232, ATM-8600892, ATM-8802386, and ATM-9104807 from the National Science Foundation and grant NAGW-1001 from the National Aeronautics and Space Administration.

Literature Cited

1. Brunning, J.; Derbyshire, D. W.; Smith, I. W. M.; Williams, M. D., *JCS Farad. Trans. 2* **1988**, *84*, 105, and references therein.
2. Mozurkewich, M.; Lamb, J. J.; Benson, S. W. *J. Phys. Chem.* **1984**, *88*, 6435.

3. Lamb, J. J.; Mozurkewich, M.; Benson, S. W. *J. Phys. Chem.* **1984**, *88*, 6441.
4. see for example, Toon, O. B.; Kasting, J. F.; Turco, R. P.; Liu, M. S. *J. Geophys. Res.* **1987**, *92*, 943.
5. Charlson, R. J.; Lovelock, J. E.; Andreae, M. O.; Warren, S. G. *Nature* **1987**, *326*, 655.
6. Hynes, A. J.; Wine, P. H.; Semmes, D. H. *J. Phys. Chem.* **1986**, *90*, 4148.
7. Domine, F.; Ravishankara, A. R.; Howard, C. J. *J. Phys. Chem.*, in press
8. Atkinson, R.; Pitts, J. N., Jr.; Aschmann, S. M. *J. Phys. Chem.* **1984**, *88*, 1584.
9. Tyndall, G. S.; Burrows, J. P.; Schneider, W.; Moortgat, G. K. *Chem. Phys. Lett.* **1986**, *130*, 463.
10. Wallington, T. J.; Atkinson, R.; Winer, A. M.; Pitts, J. N., Jr. *J. Phys. Chem.* **1986**, *90*, 4640.
11. Wallington, T. J.; Atkinson, R.; Winer, A. M.; Pitts, J. N., Jr. *J. Phys. Chem.* **1986**, *90*, 5393.
12. Dlugokencky, E. J.; Howard, C. J. *J. Phys. Chem.* **1988**, *92*, 1188.
13. Daykin, E. P.; Wine, P. H. *Int. J. Chem. Kinet.* **1990**, *22*, 1083.
14. Jensen, N. R.; Hjorth, J.; Lohse, C.; Skov, H.; Restelli, G. *J. Atmos. Chem.*, in press.
15. Atkinson, R.; Perry, R. A.; Pitts, J. N., Jr. *J. Chem. Phys.* **1977**, *66*, 1578.
16. Wine, P. H.; Kreutter, K. D.; Gump, C. A.; Ravishankara, A. R. *J. Phys. Chem.* **1981**, *85*, 2660.
17. Wine, P. H.; Thompson, R. J.; Semmes, D. H. *Int. J. Chem. Kinet.* **1984**, *16*, 1623.
18. Hynes, A. J.; Wine, P. H. *J. Phys. Chem.* **1987**, *91*, 3672.
19. Tyndall, G. S.; Ravishankara, A. R. *J. Phys. Chem.* **1989**, *93*, 4707.
20. Nicovich, J. M.; van Dijk, C. A.; Kreutter, K. D.; Wine, P. H., to be published.
21. Nesbitt, D. J.; Leone, S. R. *J. Chem. Phys.* **1980**, *72*, 1722.
22. Nesbitt, D. J.; Leone, S. R. *J. Chem. Phys.* **1981**, *75*, 4949.
23. Braithewaite, M.; Leone, S. R. *J. Chem. Phys.* **1978**, *69*, 840.
24. Clyne, M. A. A.; Ono, Y. *Chem. Phys. Lett.* **1983**, *94*, 597.
25. Clyne, M. A. A.; MacRobert, A. J.; Murrels, T. P.; Stief, L. J. *JCS Farad. Trans. 2* **1984**, *80*, 877.
26. Nava, D. F.; Brobst, W. D.; Stief, L. J. *J. Phys. Chem.* **1985**, *89*, 4703.
27. Lu, E. C. C.; Iyer, R. S.; Rowland, F. S. *J. Phys. Chem.* **1986**, *90*, 1988.
28. Dill, B.; Heydtmann, H. *Chem. Phys.* **1978**, *35*, 161.
29. Agrawalla, B. S.; Setser, D. W. *J. Phys. Chem.* **1986**, *90*, 2450.
30. see for example, Morrison, R. T.; Boyd, R. N. *Organic Chemistry*, 2nd edition. Allyn and Bacon, Inc., Boston, MA, **1966**.

31. Lias, S. G.; Bartmess, J. E.; Liebman, J. F.; Holmes, J. L.; Levin, R. D.; Mallard, W. G. *J. Phys. Chem. Ref. Data* **1988**, *17*, Supplement I.
32. Arnold, F.; Böhringer, H.; Henschen, G. *Geophys. Res. Lett.* **1978**, *5*, 653.
33. Arijs, E.; Nevejans, D.; Ingels, J. *Nature* **1983**, *303*, 314.
34. Schlager, H.; Arnold, F. *Planet. Space Sci.* **1986**, *34*, 245.
35. Arijs, E.; Nevejans, D.; Ingels, J. *Int. J. Mass Spectrom. Ion Processes* **1987**, *81*, 15.
36. Harris, G. W.; Kleindienst, T. E.; Pitts, J. N., Jr. *Chem. Phys. Lett.* **1981**, *80*, 479.
37. Fritz, B.; Lorenz, K.; Steinert, W.; Zellner, R. in *Proceedings of the 2nd European Symposium on the Physico-Chemical Behavior of Atmospheric Pollutants.* Varsino, B., Angeletti, G., Eds. D. Reidel: Boston, MA, **1982**.
38. Zetszch, C.; Bunsekelloquium; Battelle Institut: Frankfurt, **1983**.
39. Kurylo, M. J.; Knable, G. L. *J. Phys. Chem.* **1984**, *88*, 3305.
40. Rhasa, D.; Diplomarbeit, Gottingen, FRG, **1983**.
41. Poulet, D.; Laverdet, G.; Jourdain, J. L.; LeBras, G. *J. Phys. Chem.* **1984**, *88*, 6259.
42. Atkinson, R. *J. Phys. Chem. Ref. Data* Monograph 1, **1989**.
43. Hynes, A. J.; Wine, P. H. *J. Phys. Chem.* **1991**, *95*, 1232.
44. DeMore, W. B.; Sander, S. P.; Golden, D. M.; Molina, M. J.; Hampson, R. F.; Kurylo, M. J.; Howard, C. J.; Ravishankara, A. R. *Chemical Kinetics and Photochemical Data for Use in Stratospheric Modeling*, Evaluation No. 9, JPL publication 90-1, **1990**, and references therein.
45. Nicovich, J. M.; Wang, S.; Stickel R. E.; Wine, P. H., to be published.
46. Russell, J. J.; Seetula, J. A.; Timonen, R. S.; Gutman, D.; Nava, D. F. *J. Am. Chem. Soc.* **1988**, *110*, 3084.
47. Russell, J. J.; Seetula, J. A.; Gutman, D. *J. Am. Chem. Soc.* **1988**, *110*, 3092.
48. Seetula, J. A.; Russell, J. J.; Gutman, D. *J. Am. Chem. Soc.* **1990**, *112*, 1347.
49. Seetula, J. A.; Gutman, D. *J. Phys. Chem.* **1990**, *94*, 7529.
50. Richards, P. D.; Ryther, R. R.; Weitz, E. *J. Phys. Chem.* **1990**, *94*, 3663.
51. Seakins, P. W.; Pilling, M. J. *J. Phys. Chem.*, **1991**, *95*, 9874.
52. Müller-Markgraf, W.; Rossi, M. J.; Golden, D. M. *J. Am. Chem. Soc.* **1989**, *111*, 956.
53. Nicovich, J. M.; Kreutter, K. D.; van Dijk, C. A.; Wine, P. H. *J. Phys. Chem.*, in press.
54. Nicovich, J. M.; van Dijk, C. A.; Kreutter, K. D.; Wine, P. H. *J. Phys. Chem.*, **1991**, *95*, 9890.
55. McEwen, A. B.; Golden, D. M. *J. Mol. Struct.* **1990**, *224*, 357.
56. Gutman, D. *Acc. Chem. Res.* **1990**, *23*, 375.
57. Mozurkewich, M.; Benson, S. W. *J. Phys. Chem.* **1984**, *88*, 6429.

58. Chen, Y.; Tschuikow-Roux, E.; Rauk, A. *J. Phys. Chem.*, **1991**, *95*, 9832.
59. Chen, Y.; Rauk, A.; Tschuikow-Roux, E. *J. Phys. Chem.*, **1991**, *95*, 9900.
60. Donaldson, D. J.; Leone, S. R. *J. Phys. Chem.* **1986**, *90*, 936.
61. Gac, N. A.; Golden, D. M.; Benson, S. W. *J. Am. Chem. Soc.* **1969**, *91*, 309.
62. Hynes, A. J.; Wine, P. H. in *The Chemistry of Acid Rain. Sources and Atmospheric Processes*; Johnson, R. W.; Gordon, G. E., Eds.; American Chemical Society Symposium Series 349, Washington, DC, **1987**, pp. 133-141.

RECEIVED December 17, **1991**

Chapter 8

Kinetic Isotope Effects in Gas-Phase Muonium Reactions

Susan Baer, Donald Fleming, Donald Arseneau, Masayoshi Senba, and Alicia Gonzalez[1]

Tri University Meson Facility (TRIUMF) and Department of Chemistry, University of British Columbia, Vancouver, British Columbia V6T 2A3, Canada

> The study of the reaction dynamics of muonium (Mu), an ultralight isotope of hydrogen (^{m}Mu/^{m}H ≈ 1/9), provides a sensitive measure of mass effects in chemical reactions. The remarkable mass difference between Mu and the other hydrogen isotopes produces large kinetic isotope effects, providing a rigorous test of calculated potential energy surfaces (PES) and reaction rate theories. The low Mu mass also necessitates careful consideration of quantum effects, i.e. tunneling in the reaction coordinate. A review of recent results in gas phase Mu chemistry is presented, including comparison with relevant H chemistry and calculated PESs, where available. The magnitude and direction of the kinetic isotope effect is shown to be a sensitive function of the PES, particularly the height and position of the saddle point.

Kinetic isotope effects have been extensively studied in previous years to provide information about proposed potential energy surfaces and theories of reaction rates on those surfaces.[1] The study of hydrogen isotope effects has proved particularly interesting due to the large mass differences between the different isotopes, leading to correspondingly large differences in rate constants; e.g., substitution of a deuterium atom for a protium atom in a molecule will have a much larger effect on the relevant vibrational frequency than will substitution of a ^{13}C atom for a ^{12}C. In addition, the probability of quantum tunneling is greatly increased by the low mass of H compared to heavier atoms and therefore the study of hydrogen isotope effects can provide insight into this fundamental process.

[1]Current address: Loker Hydrocarbon Research Institute, University of Southern California, Los Angeles, CA 90089

The three common isotopes of hydrogen are protium (H), deuterium (D), and tritium (T), each possessing 1 proton, 1 electron, and 0, 1, or 2 neutrons respectively. One novel isotopic form of hydrogen is the radioactive species muonium (Mu), comprized of one muon (μ^+) and one electron. The muon, which is an elementary particle of the lepton family, possesses roughly one-ninth the mass of a proton, yielding Mu/H mass ratio of 0.113. Although not an isotope by the textbook definition (variation in neutron number), muonium behaves like a hydrogen isotope in that it shares essentially identical electronic characteristics and differs only in mass. This can be seen by the comparison between the properties of the hydrogen isotopes with Mu, given in Table I. Despite large mass variations, the electronic properties of these species, e.g. their ionization potentials, are virtually identical, confirming that muonium can be treated as an ultralight hydrogen isotope.

The light mass of the muon (ca. 200 times the mass of an electron), however, raises the question of the validity of the Born-Oppenheimer approximation in the atomic and molecular interactions of muonated species. It is well known that this approximation begins to break down as the mass difference between the nucleus and the electron becomes small, and the comparisons in Table I suggest that this could be of concern in the case of Mu. Theoretical calculations of one-electron problems,[2,3] however, (comparing for example, HD$^+$ and HMu^{+3}), indicate that the Born-Oppenheimer approximation remains valid for muon interactions.

The main advantage in the study of Mu reaction kinetics over that of traditional hydrogen isotopes lies in the remarkable range and magnitude of possible isotope effects it affords. Within the Born-Oppenheimer approximation, isotopic species share a common potential energy surface; any differences that arise in their respective reaction rates depend on mass effects only. The true surface must be able to account for the behavior of all isotopes, no matter how light or how heavy. With the inclusion of Mu, the available mass ratio range of hydrogen isotopes is increased from 3 to 27. This unprecedented mass range therefore provides a uniquely sensitive probe of reaction dynamics and of the underlying potential energy suface, particularly near threshold.

Perhaps the most propitious result of the low Mu mass is the greatly enhanced predilection for quantum tunneling of Mu relative to H, enabling observation of tunneling effects at easily accessible tempertures (e.g. 100-200 K). This has facilitated experimental observation of tunneling regimes in some exothermic reactions where the experimental activation energy approaches zero,[4] indicative of threshold tunneling.[5] Another important advantage inherent in the study of Mu lies in the experimental technique. Muonium atoms are easy to form (provided a pion source is available) and Mu events are individually monitored, thereby eliminating the radical-radical interactions that often plague H experiments.[6,7] The experimentally obtained Mu rate constants may therefore be more accurate than those of their heavier atom counterparts and can, in principle, be used to predict H atom reaction rates, providing an accurate potential energy surface is available. Although this application clearly lies in the future for most reactions, as calculation methods become

Table I. Comparison of the properties of Mu with H, D, and T.

	mass/m_H	reduced mass/m_e	IP(eV)[a]	Bohr radius/a_o(H) [a]
Mu	0.1131	0.9952	13.541	1.0043
H	1	0.9995	13.598	1
D	1.998	0.9997	13.602	0.9998
T	2.993	0.9998	13.603	0.9996

a) Calculated from difference in reduced mass, based on vapor values for the H atom.

faster and more accurate, the study of Mu reactivity may be used more in this predictive capacity.

The experimental technique also imposes a significant constraint, however, on the study of muonium isotope effects. The muon is a radioactive particle with a half-life of 2.2 μs, limiting the long term observation of muonium. Hence, very small rate constants are difficult to measure. On the other hand, measurement of very fast reactions can be limited by the formation time of thermal Mu (up to 100 ns at 1 atm in some gases),[8] and the time resolution of the detection technique (≈1 ns). Therefore any chemistry we wish to observe typically occurs within a 0.1-10 μs time window. These time constraints limit the observation of product molecules and secondary reactions of Mu. (Such observations can be facilitated, however, using resonant techniques.) Most reaction studies to date have been of primary reactions of free Mu atoms, the subject of this review paper.

Experimental Technique

The reactivity of Mu atoms can be monitored by the technique of muon spin rotation (μSR), which relies on the fact that the Mu is formed 100% spin polarized. Polarization can be lost in a number of ways including reaction to form some diamagnetic product. In a magnetic field, this loss of polarization and therefore the disappearance of free Mu atoms can be monitored. The μSR technique has been well described elsewhere[9-11] and will not be addressed in great detail here. The following discussion is intended to elucidate the basic features necessary to understand and evaluate the experimental results presented below.

Formation and Decay of Muonium Positive muons with high kinetic energy (ca. 4 MeV) are produced 100% longitudinally spin polarized from the parity violating decay of positive pions. Muon production therefore requires a source of pions. All the experiments described in this paper were performed at the TRIUMF accelerator in Vancouver, Canada.[12]

When high energy muons enter a reaction chamber filled with some inert bath gas, they are thermalized by collisions. During the thermalization process they undergo a series of cyclic charge exchange reactions with the bath gas, M, as indicated in R1.[8,10,11,13]

$$\mu^+ + M \rightleftharpoons Mu + M^+ \qquad \text{R1}$$

Below some threshold energy, which depends on the ionization potential of M, R1 can no longer occur and the relative amounts of μ^+ and Mu are fixed. These final μ^+ and Mu yields depend reproducibly on the bath gas.[8,13] Therefore, when studies of Mu chemistry are desired, it is important to choose a bath gas, such as Ar or N_2, which produces a large amount of Mu. Since the μ^+ is spin 100% polarized, muonium can be thought of as being produced equally in two forms: "singlet" muonium, ^SMu, where the spin of the muon and the electron are paired; and "triplet" muonium, ^tMu, where the two spins are unpaired.[9,11,13] The ^SMu state is, in fact, a superposition of |1,0⟩ and |0,0⟩ hyperfine states and thus is quickly depolarized due to the strong hyperfine interaction (ν_o = 4463 Hz) between the μ^+ and the e^- spins. Classically, it can be considered as total spin S = 0 in a weak magnetic field and thus as diamagnetic. If thermalization takes place in a short time period, the ^{t}Mu atoms will retain their spin polarization. Loss of polarization will occur if the time spent as ^{S}Mu during the charge exchange process is long relative to the characteristic period of the hyperfine interaction between the μ^+ and the electron in ^{S}Mu ($1/\nu_o$ = 0.22 ns). Since the time between ^{S}Mu collisions is directly related to the collision frequency, loss of spin polarization can be avoided through use of sufficiently high pressures (~1 atm) of the bath gas.

Muonium Spin Rotation (μSR) When placed in a transverse magnetic field, the muon spin precesses with a characteristic frequency that depends on both the strength of the field and the environment of the muon. Diamagnetic species such as μ^+ and MuH precess with a frequency of 13.6 kHz/G, while paramagnetic ^{t}Mu precesses much faster--1.39 MHz/G in low fields. Regardless of its form, the muon decays with a mean life of 2.2 μs, emitting a positron preferentially along its spin axis. Thus, a positron counter placed in the plane of spin rotation displays a wave-like signal that varies in amplitude with the characteristic frequency corresponding to the precession of the muon spin. An example is shown in Figure 1a, for μ^+ stopping in N_2 at ≈1 atm pressure. This figure is a histogram of millions of collected positron events (timed relative to when the corresponding muon entered the reaction chamber). The characteristic decay that arises from the 2.2 μs lifetime of the muon has been substracted out for clarity. The solid line is a computer fit to the data. The frequency of the precession signal indicates the environment of the muon (μ^+, Mu, etc.), while the amplitude, often called asymmetry, indicates the number of muons that are in that environment. In an inert bath gas, like N_2, the slight damping in the amplitude of the precession signal over time is simply due to some spin dephasing arising from small inhomogeneities in the magnetic field. The rate at which this occurs is called the background relaxation rate, λ_o.

In the presence of a dilute reactant, chemical reaction of the Mu atom can occur. When a diamagnetic product molecule is formed, the muon spin begins to precess at its characteristic diamagnetic frequency. (Chemical shifts are typically too small to be resolved by the μSR technique.) Since this frequency is ca. 100 times slower than that of Mu, and, moreover since these diamagentic products are formed at random times and thereby possess no coherent phase, they are not experimentally observeable and the overall Mu ensemble is seen to relax exponentially.[10] This process is illustrated in Figure 1b where the initial amplitude of the precession signal is quickly damped as the reaction occurs.[14] The overall relaxation rate, λ, is related to the bimolecular rate constant, k, as shown in Equation (1), where [X] is the concentration of the reactant.

$$\lambda = \lambda_o + k[X] \qquad (1)$$

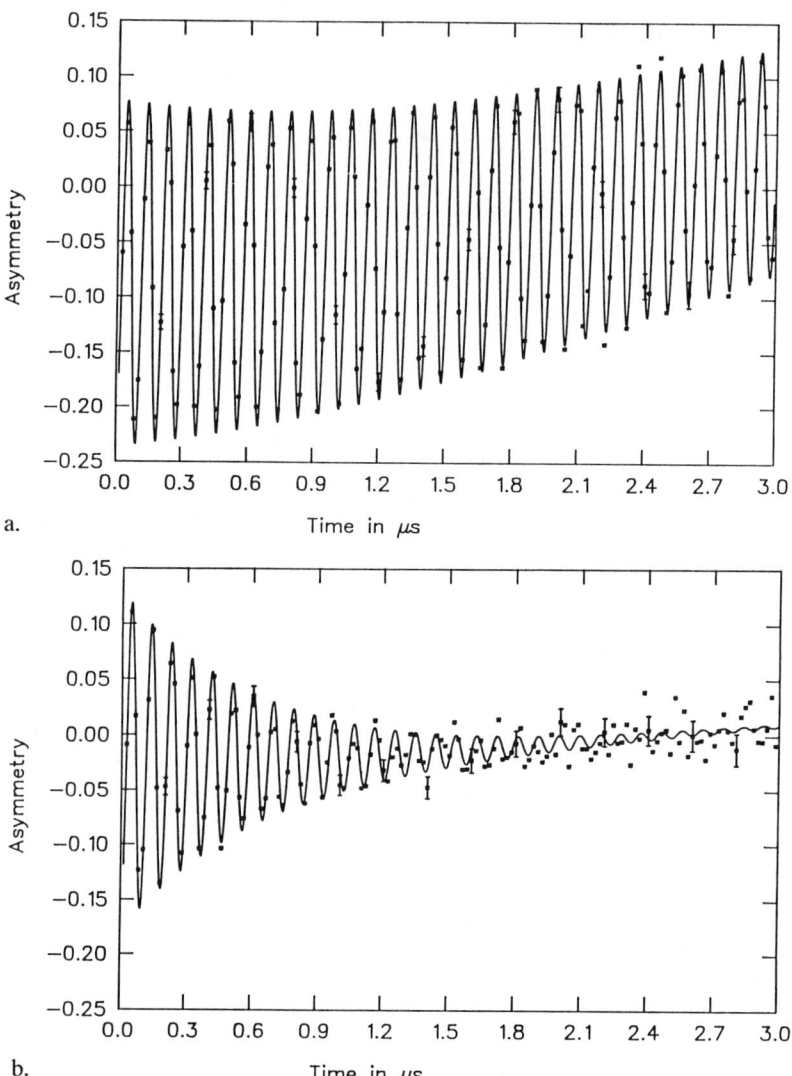

Figure 1. The μSR signal at 200 K for Mu precession in a transverse magnetic field of 7.7G for a) 1000 torr pure N_2 and b) 1000 torr N_2 in the presence of 1% additional C_2D_4. The solid line is a chi-squared fit to the data, yielding the indicated relaxation rates. (Reproduced with permission from Ref. 15. © 1990 American Institute of Physics).

By measurement of λ at a series of concentrations of X, the kinetic quantity of interest, the reaction rate constant, can be extracted.

Some addition reactions of Mu result in the formation of a paramagnetic, product (e.g., MuC_2H_4 in the presence of a bath gas[15]). This type of reaction also results in a decay of the Mu signal that can be described by (1).[16] The environment of the muon is still paramagnetic, but the coupling between the μ^+ and the electron is greatly reduced relative to Mu due to the greater distance bewteen the μ^+ and the e^- spin density in the free radical. Although such muonium radicals are the subject of intense investigation in their own right,[9,11,17] only the kinetics of the addition step will be considered in this paper.

As noted above, the μSR technique monitors the reaction of each Mu atom individually. Each time a muon enters the reaction chamber a clock is started, which is only stopped after a positron has been detected or a fixed amount of time greater than several muon lifetimes has elapsed. If another muon enters the chamber during that interval, both muon events are rejected and do not contribute to the final histogram. In this way, the possibility of distorting the experimental results by competing Mu interactions is eliminated. As mentioned above, this has frequently been a problem in kinetic studies of the corresponding H reactions.[6,7]

Muonium Kinetic Isotope Effects

Despite the growing sophistication of molecular beam experiments and measurements of state-to-state reaction cross sections,[18,27] the study of bulk chemical kinetics, as manifest by thermal reaction rate constants, remains important. First, rate constants provide some measure, albeit indirect, of <u>absolute</u> cross sections, rarely reported in beam experiments. Second, only thermal rate constants test the nature of the potential energy surface near the reaction threshold, thus usually providing the best determination of both the position and height of the potential barrier. The measurement of kinetic isotope effects is of vital importance to this point, particularly in the case of the ultralight hydrogen isotope, Mu, with its pronounced sensitivity to dynamical mass effects.

This section addresses the specific effects of the low muonium mass on reaction rates; i.e. what kind of kinetic isotope effects (k_{Mu}/k_H) do we expect for Mu atom reactions? Consider first the relative energies for an atom-molecule reaction of either Mu or H. Since isotopic reactions share a common reaction potential surface, energetic differences must arise from differences in translations or internal mode vibrations and rotations of the respective transition states. The lower mass of the Mu atom causes an increase in the vibrational frequencies and therefore the zero point energy of the transition state. The zero point energies of the reactants are, of course, independent of isotopic substitution. The resulting rise in the reaction activation energy causes a decrease in the muonium reaction rate; i.e. $k_{Mu}/k_H < 1$. This effect can be quite large, particularly for an endothermic reaction, such as Mu + H_2,[19,20] possessing a "late" potential barrier. For such a reaction, the reduced mass of the transition state resembles the reduced mass of the products more closely than that of the reactants, resulting in a maximum difference in zero point energies and thereby enhancing the rate of H abstraction relative to Mu abstraction.

A second muonium isotope effect, mentioned previously, is the enhanced possibility of tunneling. Since tunneling probability depends exponentially on the square root of the mass, Mu, with one-ninth the mass of H, has a much greater chance of tunneling at a given temperature,[21,22] leading to a corresponding increase in the muonium reaction rate constant; i.e. $k_{Mu}/k_H > 1$. Quantum tunneling is most important for exothermic reactions, such as $Mu + F_2$,[4] which tend to exhibit "early" barriers.

A third mass effect, arising from translational motion in simple collision theory, is that the average velocity of the reactive particle at a given temperature is inversely proportional to the square root of its mass. This implies that Mu has three times larger velocity than H, causing a three-fold increase in the Mu encounter frequency. For certain reactions this increased encounter frequency translates directly into an enhanced reaction rate, causing an isotope ratio of up to $k_{Mu}/k_H = 2.9$. A further effect of this enhanced velocity is that the duration of the Mu collision is shorter than its H counterpart. This effect can be important in reactions where steric effects and/or molecular orientation play a role since the Mu atom may not have time to "find" the proper orientation. This effect therefore tends to decrease the relative muonium reaction rate: $k_{Mu}/k_H < 1$.

The magnitude of the final kinetic isotope effect, k_{Mu}/k_H, is an interplay of all the above, and a sensitive function of the potential energy surface; in particular, the position of the saddle point on that surface. A wide range of kinetic isotope effects has been observed: from $k_{Mu}/k_H \approx 0.08$ for the reaction of Mu(H) with H_2,[19] to $k_{Mu}/k_H \approx 100$ at low temperatures for the reaction of Mu(H) with F_2.[4] These reactions are discussed below.

Transition State Theory The effect of muonium substitution on reaction rates can be expressed in terms of transition state theory.[4,15,20,23,24] The expression for the thermally averaged rate constant, k(T), for the reaction A + B → products, is given in Equation (2), where k_B is Boltzmann's constant; h is Planck's constant; Γ_t is the transmission or tunneling coefficient; $Q^\ddagger$, Q_A, and Q_B are the partition functions for the transition state, A, and B, respectively; and E_{va} is the vibrationally adiabatic barrier to reaction. (The vibrationally adiabatic barrier equals the difference in energy between the ground vibrational state of the transition state and that of the reactants.)

$$k(T) = \Gamma_t \frac{k_B T}{h} \frac{Q^\ddagger}{Q_A Q_B} e^{-E_{va}/k_B T} \qquad (2)$$

The kinetic isotope effect, k_{Mu}/k_H, can then be expressed as Equation (3) where ΔE_{va} is the difference in the vibrational adiabatic barriers, and the prime denotes the muonated species.

$$\frac{k_{Mu}}{k_H} = \frac{\Gamma_t'}{\Gamma_t} \frac{Q_H}{Q_{Mu}} \frac{Q^{\ddagger'}}{Q^\ddagger} e^{-\Delta E_{va}/k_B T} \qquad (3)$$

By expanding the relevant partition functions and expressing $\Delta E_{va}^{\ddagger}$ in terms of the vibrational frequencies of the transition state, $\nu_i^{\ddagger}$'s (excluding the reaction coordindate), the kinetic isotope effect can be written as (4), where m equals mass, µ equals reduced mass, and I_A, I_B, and I_C refer to the moments of inertia of the transition state.

$$\frac{k_{Mu}}{k_H} = \frac{\Gamma_t'}{\Gamma_t} \left(\frac{m_H}{m_{Mu}}\right)^{3/2} \left(\frac{\mu^{\ddagger'}}{\mu^{\ddagger}}\right)^{3/2} \left(\frac{I_A'I_B'I_C'}{I_A I_B I_C}\right)^{\ddagger 1/2}$$

$$\prod_i^{3n-7} \left(\frac{1-\exp(-h\nu_i^{\ddagger}/k_B T)}{1-\exp(-h\nu_i^{\ddagger'}/k_B T)}\right) \exp\left[-h/2k_B T \sum_i (\nu_i^{\ddagger'} - \nu_i^{\ddagger})\right] \quad (4)$$

This expression can be simplified by use of the Redlich-Teller product theorem, Equation (5), where m_i are the masses of the atoms comprising the molecule with mass M.

$$\left(\frac{M'}{M}\right)^{3/2} \left(\frac{I_A'I_B'I_C'}{I_A I_B I_C}\right)^{1/2} = \prod_i^n \left(\frac{m_i'}{m_i}\right)^{3/2} \prod_i^{3n-6} \frac{\nu_i'}{\nu_i} \quad (5)$$

After factoring out the imaginary frequencies $\nu^{\ddagger}$, corresponding to the reaction coordinate and substituting $u_i = h\nu_i/k_B T$, k_{Mu}/k_H can be writen as Equation (6).

$$\frac{k_{Mu}}{k_H} = \frac{\Gamma_t'}{\Gamma_t} \frac{\nu^{\ddagger'}}{\nu^{\ddagger}} \prod_i^{3n-7} \frac{u_i' \sinh(u_i/2)}{u_i \sinh(u_i'/2)} \quad (6)$$

Using the harmonic oscillator approximation, the isotopic frequency ratio, $\nu^{\ddagger'}/\nu^{\ddagger}$, can be written as $(u^{\ddagger}/u^{\ddagger'})^{1/2}$ where $\mu^{\ddagger}$ refers to the reduced mass of the reaction complex at the geometry corresponding to the transition state. The rate constant ratio can therefore be written as Equation (7).

$$\frac{k_{Mu}}{k_H} = \frac{\Gamma_t'}{\Gamma_t} \left(\frac{\mu^{\ddagger}}{\mu^{\ddagger'}}\right)^{1/2} \prod_i^{3n-7} \frac{u_i' \sinh(u_i/2)}{u_i \sinh(u_i'/2)} \quad (7)$$

This equation is composed of three separate ratios, which can be considered independently. The first is the strongly temperature dependent ratio of the transmission coefficients, Γ_t'/Γ_t, which includes contributions from both barrier recrossing and quantum tunneling. Although rigorous quantum formulations exist,[25] transition state theory is essentially a classical theory and does not a-priori include quantum tunneling. The transmission coefficient is therefore introduced as an ad-hoc factor to incorporate quantum tunneling into the formalism. There has been significant debate on the methods of calculating these transmission coefficients; and measurements of Mu isotope effects have been important in evaluating the accuracy of these methods.[22,26,27] The second ratio, $(\mu^{\ddagger}/\mu^{\ddagger'})^{1/2}$, is the square root of the reduced mass of the two transition states. It is important to emphasize that this term depends on the geometry of the transition state and therefore on the position of the reaction barrier. This term, therefore, contains information about the reaction dynamics and is often called the primary isotope effect.[28] For an endothermic reaction with a correspondingly late barrier, the reduced mass of the transition state resembles more closely that of the products and the ratio $(\mu^{\ddagger}/\mu^{\ddagger\prime})^{1/2}$ approaches 1. For an exothermic reaction, on the other hand, with an early barrier to reaction, the reduced mass of the transition state resembles that of the reactants. In this limit, the ratio $(\mu^{\ddagger}/\mu^{\ddagger\prime})^{1/2}$ approaches 2.9, the velocity difference mentioned above.

The third ratio, $\Pi(u_i^{\prime}\sinh(u_i/2))/(u_i \sinh(u_i^{\prime}/2))$, appears complicated at first, but can be simply considered as containing all the information about the internal modes of the transition state other than the reaction coordinate (e.g., for a bimolecular exchange reaction, the symmetric stretch and bend modes). For this reason, it is often called a secondary isotope effect.[28] For an early barrier, where these internal modes have very low frequencies, the value of this ratio approaches 1. For a late barrier, on the other hand, vibrational modes have much higher frequencies and are strongly affected by isotopic substitution. The secondary isotope effect approaches zero and thus can have a dramatic effect on the kinetics.

In conclusion, for an endothermic reaction with a late barrier: the ratio of the transmission coefficients should approach 1 since tunneling is not expected to be important (barrier recrossing could reduce this ratio); the primary kinetic isotope effect approaches 1; and the secondary isotope effect approaches 0. The net effect can therefore be a pronounced negative isotope effect: $k_{Mu}/k_H \ll 1$, as observed for Mu + H_2[19] and Mu + HCl.[47] On the other hand, for an exothermic reaction with a correspondingly early barrier: the secondary isotope effect approaches 1; the primary isotope effect approaches 2.9 (simply the ratio of the mean velocities for Mu and H); and the ratio of the transmission coefficients, $\Gamma_t^{\prime}/\Gamma_t$, containing all the interesting information on the reaction dynamics will be greater than or equal to 1. The net effect, which will likely be temperature dependent due to the ratio of the transmission coefficients, is therefore a positive isotope effect: $k_{Mu}/k_H \geq 2.9$.

Mu + H_2 (D_2)

The bimolecular exchange reaction of H + H_2 and its deuterated variations have been extensively studied over the last 60 years both from an experimental[29-31] and theoretical perspective.[29,32-34] The persistent interest in this reaction stems both from its fundamental interest as the simplest of all bimolecular exchange reactions and from the fact that it is the only reaction for which the potential energy surface has been essentially exactly calculated.[34] Because an accurate surface is available, this reaction provides an excellent test of different theories of reaction dynamics. Since the errors in these theories are most accentuated under conditions where quantum interactions become important (e.g. very low temperature, low mass), the study of the isotopic reaction, Mu + H_2 ⟶ MuH + H, where quantum tunneling may play an enhanced role, is of considerable interest. This has led to several theoretical studies of the Mu + H_2 system in recent years,[29,32,35] as well as experimental measurements of the reaction rate constants for Mu + H_2 and Mu + D_2 in this laboratory.[19,20]

The Mu + H_2 reaction is endothermic by the difference in the zero point energy between MuH and H_2: 31.8 kJ/mol (for Mu + D_2, 38.5 kJ/mol).[20] Therefore, both high temperatures and high pressures of $H_2(D_2)$ were necessary in order to observe the reaction within the muon lifetime.

A comparison between the measured reaction rates and Arrhenius parameters for Mu + H_2 and H + H_2 is given in Table II. The most striking result is the large negative isotope effect: $k_{Mu}/k_H \approx 0.08$. This value is temperature independent.[19,20,29] As mentioned above, this can be qualitatively understood in terms of a late potential barrier for the endothermic Mu + H_2 reaction, which results in an increase in the zero point energy of the transition state. Since the reactant energy is unchanged, this leads to a considerably larger activation energy (E_a) for the Mu + H_2 reaction, and correspondingly much slower reaction rates. The roughly three-times slower reaction rate of Mu + D_2 relative to Mu + H_2 occurs for the same reason: an increase in the zero point energy of the transition state relative to the reactants.[20] This same effect is observed for H + D_2 versus H + H_2 (Table II).[31]

A comparison of the Mu + H_2 and Mu + D_2 experimental results with theory is shown in the Arrhenius plot from ca. 500-850 K, given in Figure 2. The dot-dashed line refers to the variational transition state theory calculations of Garrett and Truhlar using the least action ground state approximation to compute the tunneling correction.[26] The barrier height of the potential energy surface has been adjusted slightly in the calculation. As can be seen from the figure, this method works very well at high temperatures, but starts to break down slightly at lower temperatures for both Mu + H_2 and Mu + D_2. Tunneling is predicted to become important for these reactions around 500 K,[32,35] so the fitting problems in the transition state theory calculations at lower temperatures are likely due to inadequate treatment of tunneling effects. (As noted above, Mu reactivity provides an extremely sensitive measure of the correct tunneling path to be used.) This disagreement between the experimental data and the transition state theory calculations would be expected to become increasingly pronounced at lower temperatures where tunneling begins to dominate. The slowness of the reaction relative to the μ^+ lifetime, however, prohibited its measurement at these temperatures. The dashed line in Figure 2 refers to the "exact" coupled states 3-dimensional quantum calculation by Schatz,[32] which shows excellent agreement with the data over the entire temperature range. (There are no adjustable parameters in this calculation.) This suggests that under conditions where tunneling can be important, such as in Mu reactions and/or at low temperature, the quantum calculations are significantly superior to even sophisticated transition state theory calculations. It would be interesting to see the comparison between Mu and H reaction rates extended to more recent quantum calculations of the H + H_2 system.[33]

Table II. Comparison between Mu + H_2 and H + H_2 at 745K.

Reaction	ΔH(kJ/mol)	E_a(kJ/mol)	k(cm^{3}molecules^{-1}s^{-1})
1)[a] H + H_2	0	35.6±2.1	40 x 10^{-14}
2)[a] H + D_2	4.2	39.3±1.3	14.4 x 10^{-14}
3)[b] Mu + H_2	31.8	55.6±0.8	3.3 x 10^{-14}
4)[b] Mu + D_2	38.5	61.5±1.7	1.0 x 10^{-14}

$k_3/k_1 = 0.08$ $k_4/k_2 = 0.07$

a) Data from Ref. 31
b) Data from Ref. 19

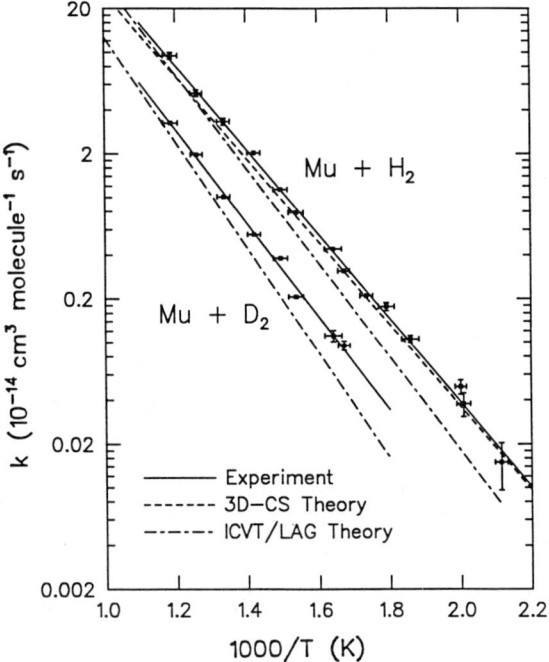

Figure 2. Arrhenius plots for Mu+H_2 and Mu+D_2. The solid lines are fits to the experimental data from Ref. 19. The dashed line gives the result of 3-dim. quantum mechanical calculations.[32] The dot-dashed lines refer to variational transition state theory calculations on the same surface.[26] (Reproduced with permission from Ref. 19. © 1987 American Institute of Physics).

Mu + X_2

The reactions of Mu with the halogens (F_2, Cl_2, and Br_2)[4] are also simple abstraction reactions like Mu + H_2 but differ markedly in their reaction energetics. As shown in R2-R4, all three reactions are very exothermic and therefore very different kinetic isotope effects from those in the Mu + H_2 reaction are expected.

Mu + F_2 ——→ MuF + F	ΔH = -369.0 kJ/mol	R2
Mu + Cl_2 ——→ MuCl + Cl	ΔH = -155.2 kJ/mol	R3
Mu + Br_2 ——→ MuBr + Br	ΔH = -142.3 kJ/mol	R4

These experiments were performed in a similar manner to those for Mu + H_2 described above, except that, due to the exothermicities involved, high pressures and temperatures were not required. In each case, the reactions Mu + X_2 were measured from ca. 150-500 K, depending on the vapor pressure of X_2, and at ca. 1 atm N_2 pressure.

The measured rate constants and activation energies for R2-R4 are given in Table III, along with the kinetic isotope effects at 298 K. The kinetic isotope effects for these reactions are all ≥ 2.9, as predicted for an exothermic reaction with an early barrier. The early position of the barrier along the reaction coordinate has been confirmed by calculations of HX_2 potential energy surfaces.[21,36] As discussed above, the kinetic isotope effect for a highly exothermic reaction consists of a simple mass contribution of 2.9 and a tunneling contribution given by the ratio of the transmission coefficients (barrier recrossing effects are not significant for exothermic reactions). The extent of the tunneling contribution can therefore be gauged by dividing the experimentally determined kinetic isotope effect by 2.9, yielding the ratio of the transmission coefficients in Table III. These values are all greater than one, demonstrating that quantum tunneling is indeed enhanced for Mu relative to H, even at room temperature. At 250 K, this ratio is further enhanced, equaling 8.0 for Mu + F_2 and 2.1 for Mu + Cl_2.

Further evidence of quantum tunneling can be found in the comparison of the experimentally obtained Arrhenius activation energies E_a(Mu) and E_a(H). The Arrhenius activation energy can be defined by Equation (8),[37]

$$E_a = \langle E^* \rangle - \langle E \rangle \quad (8)$$

where $\langle E^* \rangle$ is the average energy of those collisions leading to reaction and $\langle E \rangle$ is the average energy of all molecules. Classically, for an <u>exothermic</u> reaction, E_a is expected to depend only slightly on temperature and/or on isotopic substitution,[4,38] as observed for H + X_2.[39-41] The Arrhenius plots for these reactions yield essentially straight lines over the measured temperature range. This classical behavior is not observed, however, in the reactions of Mu with the halogens, as shown dramatically by the

Table III. Comparison between the reactions of Mu and H with the halogens.

	F_2[a]	Cl_2[b]	Br_2[c]
$k_{298}(Mu)$[d] (10^{-11} cm^{3}molecules^{-1}s^{-1})	2.62±0.06	8.50±0.14	56.0±0.9
$k_{298}(H)$ (10^{-11} cm^{3}molecules^{-1}s^{-1})	0.16±0.01	2.10±0.10	6.5±0.5
k_{Mu}/k_H (298K)	16.4	4.0	8.6
Γ_{Mu}/Γ_H (298K)	5.7	1.4	3.0
Γ_{Mu}/Γ_H (250K)	8.0	2.1	4.2
$E_a(Mu)$[e] (kJ/mol)	3.1±0.3	2.7±0.2	−0.40±0.08
$E_a(H)$[e] (kJ/mol)	9.2±0.3	5.0±0.4	5.6±0.5

a) H data from Ref. 39.
b) H data from Ref. 40.
c) H data from Ref. 45.
d) Mu data from Ref. 4.
e) From Arrhenius fits over 250-500 K.

Arrhenius plots for reactions R2-R4, given in Figure 3. Both the reactions with F_2 and Cl_2 show a pronounced curvature at low temperatures. (Mu + Br_2 displays quite different behavior, as discussed below.) In fact, the rate constant of Mu + F_2 appears to approach a temperature independent regime below ca. 150 K. This dramatic decrease in the apparent activation energy of the reaction is indicative of Wigner threshold tunneling, which has been hypothesized to occur when the ratio of the de Broglie wavelength of the tunneling particle to the thickness of the barrier is much greater than 1.[5] This can be expressed in terms of temperature, T, as given by Equation (9),

$$T \ll h^2/2\mu k_B t^2 \tag{9}$$

where t is the thickness of the barrier and μ is the reduced mass of the particle. Below this threshold, everything tunnels through the barrier and therefore further decreases in temperature do not enhance the tunneling rate. From the potential energy surface for Mu+ F_2 given in Ref. 21, T need only be below ~500 K for this threshold to be attained, a condition easily met by the experiment. (For H + F_2, the same condition requires temperatures below 50 K.[36]) This is believed to be the first experimental indication of Wigner threshold tunneling in the gas phase.

Variational transition state theory calculations of the Mu + F_2 and Mu + Cl_2 reaction rates[21] do not account well for the observed Arrhenius behavior. For both reactions, the calculations overestimate the rate constants at high temperature and underestimate them at low temperature, as shown in Figure 4 for the reaction of Mu(H) + F_2. The corresponding calculations for H + F_2 and H + Cl_2[21] give much better agreement with the experimental data,[39,40] suggesting that errors in the calculated potential surface cannot entirely account for the poor agreement of the theory and experiment in the muonium reactions. In particular, the underestimation of the calculated rate constants for Mu at low temperatures suggests that quantum tunneling is not properly accounted for in the calculation. This is consistent with the Mu + H_2 calculations, described above; although the "exact" surface for Mu + H_2 is known, the transition state theory calculations predicated on a given tunneling path[36] did not give as good agreement with the experimental data as the 3-D quantum calculations[32] at low temperatures. The overestimation of the calculated rate constants for Mu + F_2 and Mu + Cl_2 at high temperatures, however, likely arises at least in part from an error in the potential energy surface. In both reactions (H_2 and the halogens) the study of muonium isotope effects has provided a stringent test of reaction rate theories.

The Arrhenius behavior of Mu + Br_2 is quite different than that of the other halogens, as seen in Figure 3; Mu + Br_2 displays a negative temperature dependence over the measured temperature range. This type of behavior, which is commonly observed in ion-molecule and radical-radical reactions, suggests that the energetic barrier to reaction is either non-existent or lies below the reactants in energy.[42] An Arrhenius fit over the temperature range 200-400 K yields a negative activation energy of -0.040 ± 0.08 kJ/mol, consis-

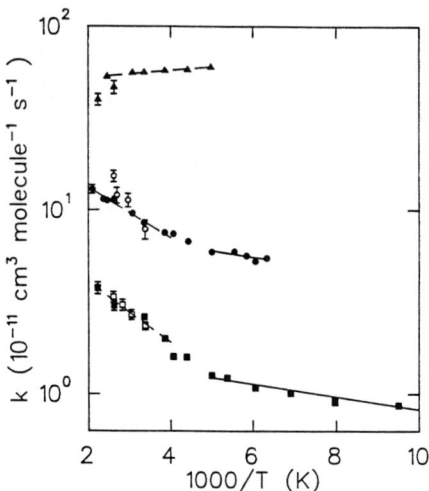

Figure 3. Arrhenius plots for reactions R2 (squares), R3 (circles), and R4 (triangles). Solid points are from Ref. 4; open points are from Ref. 24. Dashed lines are experimental Arrhenius fits over 250-500 K for R2 and R3 and 200-400 K for R4, while solid lines are fits over 100-200 K for R2 and 160-200 K for R3. (Reproduced with permission from Ref. 4. © 1989 American Institute of Physics).

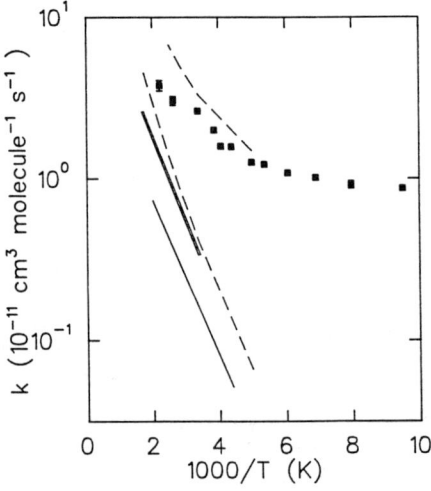

Figure 4. Comparison of theory and experiment for Arrhenius plots of the Mu + F_2 reaction. The Mu data are given by the squares.[4] The H data are given by the solid lines (thin solid line from Homann et al.; thick solid line from Albright et al.).[39] The dashed lines are the transition state theory calculations for Mu + F_2 (upper) and H + F_2 (lower).[21] (Reproduced with permission from Ref. 4. © 1989 American Institute of Physics).

tent with atomic beam studies of H + Br_2, which also exhibit a negative energy dependence.[32] Theoretical work on both H + Br_2 and Mu + Br_2 has been unable to account satisfactorily for the observed behavior.[21,44]

The temperature dependence of the H(D) + Br_2 reaction rate has been recently directly measured by a Lyman-α resonance technique.[45] The activation energy for abstraction by H is found to be *positive* (5.6 ± 0.5 kJ), in contrast to the Mu results (-0.40 ± .08 kJ) and also in conflict with the energy dependence of the H + Br_2 molecular beam data.[43] A transition state theory calculation in Ref. 45 on a modified LEPS surface is able to account for the H(D) + Br_2 kinetic data over the small experimental temperature range, 214-295 K. Transmission coefficients do not differ appreciably from unity and hence tunneling is relatively unimportant. Interestingly, these authors have also calculated the Mu + Br_2 rate on the same surface. They are not able to reproduce the observed negative temperature dependence, but do find transmission coefficients $\Gamma_t(Mu)/\Gamma_t(H)$ greater than 1, even at room temperature, qualitatively consistent with the experimental values in Table III. This is a puzzling aspect of the Mu + Br_2 data. The temperature dependence, as mentioned, suggests *no* barrier, in which case quantum tunneling should have little or no influence on the reaction rate.

The enhancement in the ratio of the transmission coefficients is not understood. It is unlikely to be due to a steric effect in the bimolecular collision since any steric constraints would be expected to slow the Mu reaction relative to H due to the shorter duration of the Mu collision. One possible explanation of the enhancement is the presence of a bound reaction complex, $MuBr_2$, on the potential energy surface. This Van der Wals type complex could be thought of as a weakly bound analogue of ion-molecule collision complexes, which are estimated to be bound by ca. 10-15 kcal/mol.[46] The presence of a bound intermediate species on the reaction surface necessarily implies the existence of an energetic barrier lying *below* the reactants in energy. Such a barrier would slow the reaction rate relative to the collision rate for entropic rather than energetic reasons and could account for the negative temperature dependence observed. Further theoretical studies of both H + Br_2 and Mu + Br_2 are required, however, to establish the existence of such a bound $MuBr_2$ species on the reaction surface.

Mu + HX

The reactivity of Mu with the hydrogen halides, HCl, HBr, and HI has also been investigated.[47] Both the exchange reaction, resulting in formation of MuX, and the abstraction reaction, resulting in formation of MuH, are possible. Due to the zero point energy difference between HX and MuX, however, exchange is endothermic (≈29 kJ/mol) and hence abstraction is the only pathway observed experimentally. The abstraction reactions vary from endothermic for Mu + HCl to highly exothermic for Mu + HI, as shown in R5-R7.

$$Mu + HCl \longrightarrow MuH + Cl \quad \Delta H = 21.8 \text{ kJ/mol} \quad R5$$
$$Mu + HBr \longrightarrow MuH + Br \quad \Delta H = -41.0 \text{ kJ/mol} \quad R6$$

$$\text{Mu} + \text{HI} \longrightarrow \text{MuH} + \text{I} \quad \Delta H = -110.9 \text{ kJ/mol} \quad \text{R7}$$

The corresponding H reactions are all exothermic due to more favorable zero point energy effects, but show the same trend of increasing reaction exothermicity with increasing halogen size.

Because it encompasses both endo- and exothermic reactions, the series given in R5-R7 is particularly interesting to the study of muonium kinetic isotope effects. Based on the discussion above, the magnitude and direction of the k_{Mu}/k_H ratio is expected to change dramatically from the reaction with HCl to the reaction with HI. In addition, since the exothermicities of R6 and R7 are still relatively modest compared to those of $\text{Mu} + X_2$, the magnitude of the kinetic isotope effects are expected to lie between those observed for $\text{Mu} + H_2$ and $\text{Mu} + X_2$. This intermediate regime should be very sensitive to small differences in the isotope effect contributions discussed above.

The reactions of $\text{Mu} + \text{HBr}$ and $\text{Mu} + \text{HI}$ were studied from 150-500 K in an analogous manner to the reactions of Mu with X_2.[4,47] Due to the endothermicity of the reaction of Mu with HCl, only an upper limit for the reaction rate constant was reported at room temperature.

A comparison between the kinetic data for $\text{Mu} + \text{HX}$[47] and $\text{H} + \text{HX}$[48,49] is given in Table IV. As expected, the kinetic isotope effects, k_{Mu}/k_H, vary dramatically, from much less than one for the endothermic $\text{Mu} + \text{HCl}$ reaction to greater than one for $\text{Mu} + \text{HI}$. Although the effect reported for $\text{Mu} + \text{HCl}$ is only an upper limit, its magnitude appears reasonable based on comparison with the reaction of $\text{Mu} + H_2$.[19] The endothermic HCl reaction is not expected to display any contributions from tunneling, as observed for $\text{Mu} + H_2$.

The kinetic isotope effects observed for the reactions of $\text{Mu} + \text{HBr}$ and $\text{Mu} + \text{HI}$ are intriguing. The potential energy surfaces for the corresponding H reactions indicate the presence of an early barrier,[50] and therefore similar behavior to that of the $\text{Mu} + X_2$ reactions is expected.[4] In other words, the Mu reaction rate should be greatly enhanced compared to that of H due to the primary isotope effect of ca. 2.9 and due to tunneling effects in the lower temperature regions.

The observed isotope effects differ considerably from these expectations. Although k_{Mu}/k_H is greater than one, as expected, its magnitude does not differ significantly from the simple velocity factor of 2.9 at any temperature; i.e. $\Gamma_t(\text{Mu})/\Gamma_t(\text{H}) \approx 1$ for both Mu+ HBr and $\text{Mu} + \text{HI}$, suggesting that tunneling does not contribute appreciably to these reactions. Consistent with these observations, Arrhenius plots of the data show no curvature, even at low temperatures, as shown in Figure 5 for the reaction of $\text{Mu} + \text{HBr}$. However, the experimental E_as for the Mu reactions are <u>smaller</u> than for the corresponding H reactions, as shown in Table IV. (This effect is quite dramatic for $\text{Mu} + \text{HI}$ where the E_a is over four times smaller than $\text{H} + \text{HI}$.[52]) These decreases in E_a for the muonated reactions are puzzling. They cannot be explained in terms of zero point energy effects, which would tend to <u>increase</u> E_a in the absence of any tunneling. As noted above, for the reactions of Mu with F_2 and Cl_2, this dramatic decrease in E_a is strongly suggestive of quantum tunneling.

Table IV. Comparison between the reactions of Mu[a] and H with HCl, HBr, and HI.

	HCl[b]	HBr[c]	HI[d]
k_{298}(Mu) (10^{-11} cm^3 molecules^{-1} s^{-1})	<0.0003	1.9±0.06	8.0±0.3
k_{298}(H) (10^{-11} cm^3 molecules^{-1} s^{-1})	0.005	0.60±0.01	1.8±0.1
k_{Mu}/k_H (298K)	<0.06	3.2	4.4
Γ_{Mu}/Γ_H (298K)	---	1.1	1.5
E_a(Mu) (kJ/mol)	---	2.3±0.08	0.7±0.2
E_a(H) (kJ/mol)	14.6±0.2	3.3±0.6	3.0±1.1

a) Mu data from Ref. 47. E_a from Arrhenius fit over 150-500 K.
b) H data from Ref. 48.
c) H data constant from Ref. 49; E_a from Ref. 51.
d) H data from Ref. 52.

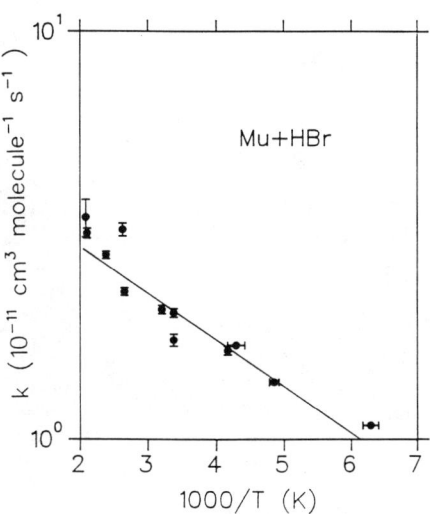

Figure 5. Arrhenius plot for Mu + HBr over 170-480 K.[47]

A possible resolution of this conflict lies in the secondary effect of the enhanced velocity of Mu relative to H: the shorter collision duration mentioned briefly above. This type of dynamic, steric effect has been suggested to be important at epithermal energies where reaction of D with HBr is observed to be enhanced relative to that of H.[53] Since Mu is much lighter and therefore faster-moving, this effect could become important at thermal energies and lead to a corresponding decrease in reaction rate, masking possible rate enhancements due to tunneling. In this case, the final kinetic isotope effect would result from a delicate balance of these contributing effects: the simple velocity factor of 2.9 that enhances the Mu rate, the steric effect that decreases the Mu rate, and quantum tunneling that enhances the Mu rate. One difficulty with this argument lies in the temperature independence of the observed isotope effect. Since tunneling is expected to become more important at low temperatures while steric factors become less important, a gradual enhancement of the isotope effect at low temperatures would be expected. Calculations of the Mu + HX reaction dynamics will be instrumental in clarifying this complex reaction system.

Mu + Ethylene

The addition reaction of Mu with ethylene[15] forming MuC_2H_4 typifies a host of reaction studies of Mu addition to unsaturated bond systems underway at TRIUMF.[17,54] Addition reactions differ from the abstraction reactions discussed above in that collisions with a third body are required to stabilize the reaction product. The well established mechanism is shown in R8, where k_1 is the rate of formation of the complex, k_{-1} is the rate of dissociation, and k_2 is the rate of stabilization.

$$Mu + C_2H_4 \underset{k_{-1}}{\overset{k_1}{\rightleftarrows}} MuC_2H_4^* \xrightarrow{M,\ k_2} MuC_2H_4 \qquad \text{R8}$$

If the excited addition complex does not collide with the bath gas, M, in a time scale faster than its lifetime, $\tau_c = 1/k_{-1}$, it will decompose back to reactants. The reaction rate will therefore depend on the density of M, as shown by the usual definition of the observed rate constant, given in Equation (10).

$$k_{obs} = k_1 k_2 [M] / (k_{-1} + k_2 [M]) \qquad (10)$$

In the limit of high pressure $k_2[M]$ is much greater than k_{-1} and the observed reaction rate, k_{obs}, simply equals the rate of association, k_1. Under conventional experimental conditions, this limit is determined by varying the pressure of the bath gas and observing the onset of the pressure independent regime. In reality, this limit can be difficult to determine unambiguously and often occurs at quite high pressures (>50 atm).[55,56] In contrast, the μSR technique may allow <u>direct</u> determination of the high pressure limit, k_1, at all pressures of M, depending on the lifetime, τ_c, of the activated complex formed.[16] This apparent paradox arises from the rapid rate of spin dephasing of the excited addition product, $MuC_2H_4^*$, due both to intramolecular spin relaxation and mixing between the large number of muon-electron-proton couplings in the radical. If the rate of spin dephasing is much faster than the rate of decay of $MuC_2H_4^*$, which includes both the dissociation rate k_{-1}, and the stabilization rate k_2, as soon as the addition product is formed it will lose polarization and become invisible to µSR detection. Under these circumstances, the observed reaction rate is simply k_1, the rate of association.

This situation is predicted for the reaction of Mu with ethylene at modest N_2 pressures of ca. 1 atm: the rate of spin dephasing is estimated to be $1 \times 10^{10} s^{-1}$, while the rate of dissociation of $MuC_2H_4^*(k_{-1})$ is estimated to be $1 \times 10^9 s^{-1}$ from unimolecular dissociation theory, comparable to the rate of stabilization (k_2). This hypothesis is confirmed by the pressure independence of the reaction rate constant in the moderate pressure range 200-1500 torr.[15]

The addition reaction, R8, is exothermic and therefore is expected to display behavior consistent with an early reaction barrier, analogous to the halogen abstraction reactions discussed above. This is suggested as well by the C_2H_5 potential energy surface.[57] The validity of this expectation is shown in Figure 6, Arrhenius plots of $Mu + C_2H_4$, $H + C_2H_4$, and $D + C_2H_4$ from ca. 150-500 K. While the slope of the H and D reactions is constant in the temperature range examined,[7,58] the Mu line shows a pronounced curvature. At high temperatures ($\approx$500 K) the kinetic isotope effect, $k_{Mu}:k_H:k_D$, approximately equals 3.9:1.4:1, roughly the classically expected value based on the relative velocities of Mu, H and D. The ratio of the transmission coefficients, $\Gamma_t(Mu):\Gamma_t(H):\Gamma_t(D)$, therefore equals 1; there is no tunneling enhancement of the lighter species at high temperature. At low temperature, however, the situation is dramatically different. Whereas H and D continue to behave classically, the Mu rate is greatly enhanced. At 165 K the kinetic isotope effect $k_{Mu}:k_H:k_D$ equals roughly 42:1.4:1, clearly indicating the importance of tunneling for the Mu reaction. Similar behavior for the reaction of Mu with C_2D_4 is observed. There are, as yet, no theoretical calculations for Mu + C_2H_4. It can be noted, however, that theory can not account for even the observed H,D isotope effect in C_2H_4 (see Ref. 15).

Other addition reactions currently being studied include the reaction of Mu with NO in the presence of a bath gas, which has been observed to remain in the low pressure regime at up to 60 atm of N_2.[59] In the low pressure limit, the reaction rate depends termolecularly on the concentrations of Mu, NO, and the bath gas, M. The observed rate constant is given by k_1k_2/k_{-1}, using the same notation as in R8 above. Due to the higher vibrational frequencies of $MuNO^*$ relative to HNO^*, the lifetime τ_c is likely much shorter than the spin relaxation time, resulting in an inverse isotope effect for the termolecular rate constant, $k_{Mu}/k_H = 0.28$.[59-60] Interestingly, in the high pressure limit, where the observed rate constant is given by k_1, the rate of the exothermic addition step, the kinetic isotope effect is predicted to become greater than 1 due to the higher velocity of Mu relative to H. (This is analogous to the Mu + ethylene addition reaction, discussed above.) Thus, the kinetic isotope effect in a termolecular association reaction can display a pressure as well as a temperature dependence.

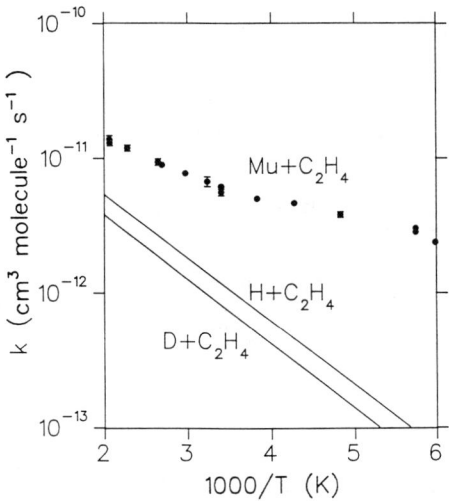

Figure 6. Arrhenius plots for the Mu + C_2H_4 addition reaction (circles) and experimental results for H + C_2H_4 and D + C_2H_4.[7,58] (Reproduced with permission from Ref. 15. © 1990 American Institute of Physics).

Conclusions

The reactions discussed above -- Mu + H_2, Mu + X_2, Mu + HX, Mu + C_2H_4, and Mu + NO -- give an overview of the study of gas phase muonium isotope effects in the last 10 years at the TRIUMF cyclotron. The dramatic mass difference between Mu and the other hydrogen isotopes produces strikingly large kinetic isotope effects that are very sensitive to the nature of the potential energy surface. Small discrepancies between theory and experiment, barely apparent in a deuterium reaction, for example, become much more pronounced in a muonium reaction. This sensitivity makes the study of Mu isotope effects a rigorous test for the accuracy of both calculated potential surfaces and reaction rate theories. For example, the poor agreement of the variational transition state theory calculations[21] with the Mu + F_2 experimental results[4] suggested problems both with the classical formulation of transition state theory and with the semi-empirical potential energy surface for the HX_2 system. Comparisons such as these play an important role in creating and refining molecular models for motion along the potential surface.

The remarkably low mass of the muonium atom also makes it a good candidate for the study of quantum effects, particularly tunneling in the reaction coordinate. Tunneling regimes such as Wigner threshold tunneling, which can only be seen at very low temperatures in the case of H atoms,[5] can, in principle, be probed easily by Mu. The reactions of Mu with F_2, Cl_2, and C_2H_4 all give strong evidence of how dominant tunneling can be, with E_a(Mu) ≈ 0 at temperatures around 150 K. At the present time, good theoretical accounts of the temperature dependencies of these particular reactions have not been found. Development of reaction rate theories which can successfully deal with the tunneling behavior of Mu is not only of importance to our understanding of fundamental muonium chemistry. Such development could also aid in the extrapolation of existing H data down to low temperatures where experimental measurements are difficult. This has potential interest for the chemistry of interstellar space, as well as other low temperature applications.

Finally, due to the μSR technique, the Mu reaction rate measurements are inherently more accurate than H rate measurements. As our knowledge of potential energy surfaces and reaction rate theories advances, Mu studies may be used in a predictive capacity to yield information about their H and D atom counterparts.

Further directions in the study of Mu isotope effects currently in progress in our research group include a study of the excited vibrational distribution of MuF, formed by reaction of Mu with F_2. The corresponding H reaction is known to produce highly vibrationally excited HF (up to υ = 9).[61] Isotopic substitution of Mu for H is expected to result in much lower vibrational excitation of the product,[62] as originally postulated by Polanyi, Schreiber, and Sloan[61] who referred to this as the "light atom anomaly". A comparison of the MuF* distribution at high and low temperatures should yield information on how tunneling (dominant at low temperatures) affects the energy partitioning of the reaction.

Literature Cited

1. Melander, L.; Saunders, W.H. *Reaction Rates of Isotopic Molecules*; Wiley-Interscience: New York, 1980.
2. Garrett, B.C.; Truhlar, D.G.; Melius, C.F. *Phys. Rev. A* **1981**, *24*, 2853.
3. McKenna, D.; Webster, B. *J. Chem. Soc. Faraday Trans. II* **1985**, *81*, 225.
4. Gonzalez, A.C.; Reid, I.D.; Garner, D.M.; Senba, M.; Fleming, D.G.; Arseneau, D.J.; Kempton, J.R. *J. Chem. Phys.* **1989**, *91*, 6164.
5. Takayanagi, T.; Masaki, N; Nakamura, K.; Okamato, M.; Sato, S.; Schatz, G.C. *J. Chem. Phys.* **1987**, *86*, 6133. Takayanagi, T.; Nakamura, K.; Sato, S. *J. Chem. Phys.* **1989**, *90*, 1641.
6. Cowfer J.A.; Michael, J.V. *J. Chem. Phys.* **1975**, *62*, 3504. Michael J.V.; Suess, G.N. *J. Chem. Phys.* **1973**, *58*, 2807. Malins R.J.; Setser, D.W. *J. Chem. Phys.* **1980**, *73*, 5666. Ambidge, P.F.; Bradley J.N.; Whytock, D.A. *J. Chem. Soc. Faraday Trans. I* **1976**, *72*, 1157.
7. Lightfoot P.D.; Pilling, M.J. *J. Phys. Chem.* **1987**, *91*, 3373.
8. Senba, M. *J. Phys. B: At. Mol. Opt. Phys.* **1990**, *23*, 1545; **1989**, *22*, 2027; **1988**, *21*, 3093.
9. Walker, D.C. *Muon and Muonium Chemistry*; Cambridge University Press: Cambridge, 1983, and references included therein.
10. Fleming, D.G.; Garner, D.M.; Vaz, L.C.; Walker, D.C.; Brewer, J.H.; Crowe, K.M. *ACS Adv. Chem. Ser.* **1979**, *175*, 279.
11. Fleming, D.G.; Senba, M. in *Advances in Meson Science*; Nagamine, K.; Nakai, K.; Yamazaki, T. Eds., in press.
12. For a description of the TRIUMF facility, see: Beveridge, J.L.; Doornbos, J.; Garner, D.M. *Hyperfine Interactions* **1986**, *32*, 907.
13. Fleming, D.G.; Senba, M. in *Atomic Physics with Positrons*; Humberston, J.W., Armour, E.A.G. Eds.; Plenum Press: London, 1987, p. 343. Fleming, D.G.; Mikula, R.J.; Garner, D.M. *Phys. Rev. A* **1982**, *26*, 2527.
14. This figure actually refers to formation of a paramagnetic product, MuC_2D_4. As noted in the text, the relaxation process in a weak tranverse magnetic field appears identical whether the product is diamagnetic or paramagnetic. (See also Ref. 16)
15. Garner, D.M.; Fleming, D.G.; Arseneau, D.J.; Senba, M.; Reid, I.D.; Mikula, R.J. *J. Chem. Phys.* **1990**, *93*, 1732.
16. Duchovic, R.J.; Wagner, A.F.; Turner, R.E.; Garner, D.M.; Fleming, D.G. *J. Chem. Phys.* **1991**, *94*, 2794.
17. For example, see: Fleming, D.G.; Kiefl, R.F.; Garner, D.M.; Senba, M.; Gonzalez, A.C.; Kempton, J.R.; Arseneau, D.J.; Venkateswaren, K.; Percival, P.W.; Brodovitch, J.C.; Leung, S.K.; Cox, S.F.J. *Hyperfine Interactions* **1990**, *65*, 767. Cox, S.F.J.; Eaton, G.H.; Magraw, J.E. *Hyperfine Inter-*

actions **1990**, *65*, 773. Roduner, E. *Prog. Rxn. Kinetics* **1981**, *14*, 261.
18. Levine, R.D.; Bernstein, R.B. *Molecular Reaction Dynamics and Chemical Reactivity*; Oxford University: New York, 1987. Lee, Y.T. *Chem. Scr.* **1987**, *27*, 215.
19. Reid, I.D.; Garner, D.M.; Lee, L.Y.; Senba, M.; Arseneau, D.J.; Fleming, D.G. *J. Chem. Phys.* **1987**, *86*, 5578.
20. Garner, D.M.; Fleming, D.G.; Mikula, R.J. *Chem. Phys. Lett.* **1985**, *121*, 80.
21. Steckler, R.; Truhlar, D.G.; Garrett, B.C. *Int. J. Quantum Chem. Symp.* **1986**, *20*, 495. Garrett, B.C.; Steckler, R.; Truhlar, D.G. *Hyperfine Interactions* **1986**, *32*, 779.
22. Lynch, G.G.; Truhlar, D.G.; Garrett, B.C. *J. Chem. Phys.* **1989**, *90*, 3102.
23. Johnston, H.S. *Gas Phase Reaction Rate Theory*; Ronald: New York, 1966; Chap. 13.
24. Garner, D.M. Ph.D. Thesis, Univ. of British Columbia, 1979.
25. Miller, W.H. *J. Chem. Phys.* **1974**, *61*, 1823.
26. Garrett, B.C.; Truhlar, D.G. *J. Chem. Phys.* **1984**, *81*, 309.
27. Truhlar, D.G.; Garrett, B.C.; Hipes,P.G.; Kuppermann, A. *J. Chem. Phys.* **1984**, *81*, 3542. Kuppermann, A. *J. Phys. Chem.* **1979**, *83*, 171.
28. Nikitin, E.E., *Theory of Elementary Atomic and Molecular Processes in Gases*; Clarendon: Oxford, 1974.
29. Valentini, J.J.; Phillips, D.L. in *Bimolecular Collisions* Ashfold, M.N.R., Baggot, J.E., Ed.s; Royal Society of Chemistry: London, 1989; and references included therein.
30. For example see: Kliner, D.A.V.; Zare, R.N. *J. Chem. Phys.* **1990**, *92*, 2107. Rinnen, K.D.; Kliner, D.A.V.; Zare, R.N. *J. Chem. Phys.* **1989**, *91*, 7514. Gerrity, D.P.; Valentini, J.J. *J. Chem. Phys.* **1984**, *81*, 1298. Pratt, G.; Rogers, D. *J. Chem. Soc. Faraday Trans. I* **1976**, *72*, 1589. Götting, R.; Mayne, H.R.; Toennies, J.P. *J. Chem. Phys.* **1984**, *80*, 2230.
31. Michael, J.V., these proceedings. See also: Westenberg, A.A.; de Haas, N. *J. Chem. Phys.* **1967**, *47*, 1393. Quickert, K.A.; LeRoy, D.J. *J. Chem. Phys.* **1970**, *53*, 1325; **1971**, *54*, 5444. Mitchell D.N.; LeRoy, D.J. *J. Chem. Phys.* **1973**, *58*, 3449.
32. Schatz, G.C. in *Theory of Chemical Reaction Dynamics*; Clary, D.C. Ed; Reidel: Dordrecht, 1986. Schatz, G.C. *J. Chem. Phys.* **1982**, *83* 3441.
33. Park T.J.; Light, J.C. *J. Chem. Phys.* **1989**, *91*, 974. Webster, F.; Light, J.C. *J. Chem. Phys.* **1989**, *90*, 300.
34. Siegbahn, P.; Liu, B. *J. Chem. Phys.* **1978**, *68*, 2457. Truhlar, D.G.; Horowitz, C.J. *J. Chem. Phys.* **1978**, *68*, 2466. Blomberg, M.R.A.; Liu, B. *Chem. Phys.*

1985, *82*, 1050. Liu, B. *J. Chem. Phys.* **1984**, *80*, 581.
35. Bondi, D.K.; Clary, D.C.; Connor, J.N.L.; Garrett, B.C.; Truhlar, D.G. *J. Chem. Phys.* **1982**, *76*, 4986. Garrett, B.C.; Truhlar, D.G. *J. Chem. Phys.* **1984**, *81*, 309.
36. Connor, J.N.L.; Jakubetz, W.; Manz, J.; Whitehead, J.C. *J. Chem. Phys.* **1980**, *72*, 6209. Jonathan, N.; Okuda, S.; Timlin, D. *Mol. Phys.* 1972, *24*, 1143. Ding, A.M.G.; Kirsch, L.J.; Perry, D.S.; Polanyi, J.C.; Schreiber, J.L. *J.Chem. Soc. Faraday Discuss.* **1973**, *55*, 252. Blais, N.C.; Truhlar, D.G. *J. Chem. Phys.* **1985**, *83*, 5546. Last, I.; Baer, M. *J. Quantum Chem.* **1986**, *29*, 1067.
37. Furue, H.; Pacey, P.D. *J. Phys. Chem.* **1986**, *90*, 397.
38. Connor, J.N.L.; Lagana, A.; Turfa, A.F.; Whitehead, J.C. *J. Chem. Phys.* **1981**, *75*, 3301.
39. Homann, R.H.; Schweinfurth, H.; Warnatz, J. *Ber. Bunsenges Phys. Chem.* **1977**, *81*, 724. Albright, R.G.; Dodonov, A.F.; Lavrovskaya, G.K.; Morosov, I.I.; Talroze, V.L. *J. Chem. Phys.* **1969**, *50*, 3632.
40. Wagner, H.Gg.; Welzbocher, U.; Zellner, R. *Ber. Bunsenges Phys. Chem.* **1976**, *80*, 902. Bemand, P.P.; Clyne, M.A.A. *J. Chem. Soc. Faraday Trans. II* **1977**, *73*, 394.
41. Jaffe, S.; Clyne, M.A.A. *J. Chem. Soc. Faraday Trans. II* **1981**, *77*, 531.
42. Su, T.; Bowers, M.T. in *Gas Phase Ion Chemistry*; M.T. Bowers, Ed.; Academic Press: New York, 1979, vol. 1.
43. Hepburn, J.W.; Klimek, D.; Liu, K.; Polanyi, J.C.; Wallace, S.C. *J. Chem. Phys.* **1978**, *69*, 4311.
44. Blais, N.C.; Truhlar, D.G. *J. Chem. Phys.* **1985**, *83*, 5546.
45. Wada, Y.; Takayanagi, T.; Umemoto, H.; Tsunashima, S.; Sato, S. *J. Chem. Phys.* **1991**, *94*, 4896.
46. Kebarle, P.; Chowdhury, S. *Chem. Rev.* **1987**, *87*, 513.
47. Arseneau, D.J.; Gonzalez, A.C.; Tempelmann, A.; Kempton, J.R.; Senba, M.; Pan, J.J.; Fleming, D.G. *J. Chem. Phys.*, in press. Tempelmann, A. MSc. Thesis, Univ. of British Columbia, 1990.
48. Miller, J.C.; Gordon, R.J. *J. Chem. Phys.* **1983**, *78*, 3713; **1981**, *75*, 5305.
49. Husain, D.; Slatert, N.K.H. *J. Chem. Soc. Faraday Trans. II* **1980**, *76*, 276.
50. Lynch, G.G.; Truhlar, D.G.; Garrett, B.C. *J. Chem. Phys.* **1989**, *90*, 3102. Tucker, S.C.; Truhlar, D.G.; Garrett, B.C.; Isaacson, A.D. *J. Chem. Phys.* **1985**, *82*, 4102. Garrett, B.C.; Truhlar, D.G.; Magnuson, A.W. *J. Chem. Phys.* **1982**, *76*, 2321. Sudhakaran, M.P.; Raff, L.M. *Chem. Phys.* **1985**, *95*, 165. Schwenke, D.W.; Tucker, S.C.; Steckler, R.; Brown, F.B.; Lynch, G.G.; Truhlar, D.G.; Garrett, B.C. *J. Chem. Phys.* **1989**, *90*, 3110. Umemoto, H.; Nakagawa, S.; Tsunashima, S.; Sato, S. *Chem. Phys.*

1988, *124*, 259. Zhang, Y.C.; Zhang, J.Z.H; Kouri, D.J.; Haug, K; Schwenke, D.W.; Truhlar, D.G. *Phys. Rev. Lett.* 1988, *60*, 2367. Clary, D.C. *J. Chem. Phys.* 1985, *83*, 1685.
51. Umemoto, H.; Wada, Y.; Tsunashima, S.; Sato, S. *J. Chem. Phys.* 1990, *143*, 333.
52. Umemoto, H.; Nakagawa, S.; Tsunashima, S.; Sato, S. *Chem. Phys.* 1988, *124*, 259.
53. Sudhakaran, M.P.; Raff, L.M. *Chem. Phys.* 1985, *95*, 165. Hepburn, J.W.; Klimek, D.; Liu, K.; Polanyi, J.C.; Wallace, S.C. *J. Chem. Phys.* 1978, *69*, 4311.
54. Senba, M.; Gonzalez, A.C.; Kempton, J.R.; Arseneau, D.J.; Fleming, D.G. TRIUMF Report No. TRI-89-2. Fleming, D.G. et al., work in progress.
55. Wagner, A.F.; Bowman, J.M. *J. Phys. Chem.* 1987, *91*, 5314. Lee, K.T.; Bowman, J.M. *J. Chem. Phys.* 1987, *86*, 215.
56. Croce de Cobos, A.E.; Hippler, H.; Troe, J. *J. Phys. Chem.* 1984, *88*, 5083. Hippler, H.; Rahn, R.; Troe, J. *J. Chem. Phys.* 1990, *93*, 6560. Baer, S.; Hippler, H.; Rahn, R.; Siefke, N.; Troe, J. *J. Phys. Chem.*, submitted 1991.
57. Nagase, S.; Fueno, T.; Morokuma, K. *J. Am. Chem. Soc.* 1979, *101*, 5849. Nagase, S.; Kern, C.W. *J. Am. Chem. Soc.* 1980, *102*, 4513. Schlegel, H.B. *J. Phys. Chem.* 1982, *86*, 4878. Schlegel, H.B.; Bhalla, K.C.; Hase, W.L. *J. Phys. Chem.* 1982, *86*, 4883. Michael, J.V.; Suess, G.N. *J. Chem. Phys.* 1973, *58*, 2807.
58. Sugawara, K.; Okazaki, K.; Sato, S. *Chem. Phys. Lett.* 1981, *78*, 259; *Bull. Chem. Soc. Jpn.* 1981, *54*, 2872.
59. Senba, M.; Gonzalez, A.C.; Kempton, J.R.; Arseneau, D.J.; Pan, J.J.; Tempelmann, A.; Fleming, D.G. *Hyperfine Interactions* 1990, *65*, 979.
60. H-atom data from: Campbell, I.M.; Handy, B.J. *J. Chem. Soc. Faraday Trans. I* 1975, *71*, 2097.
61. Polanyi, J.C.; Schreiber, J.L.; Sloan, J.J. *Chem. Phys.* 1975, *9*, 403.
62. Conner, J.N.L.; Jakubetz, W.; Manz, J. *Chem. Phys.* 1978, *28*, 219.

RECEIVED November 13, 1991

Chapter 9

Mass-Independent Isotopic Fractionations and Their Applications

M. H. Thiemens[1]

Institut für Physikalische Chemie der Universität Gottingen, Tammannstrasse 6, D–3400 Gottingen, Germany

There are a wide variety of conventional isotope effects. However, they all have a common feature; they are ultimately dependent upon mass. A new type of isotope effect which is mass independent has been discovered and is detailed in this chapter. There are important aspects of this new effect. First, the fractionation mechanism derives from molecular symmetry and this unique property provides unique insight into reaction features. Second, the specific isotopic fractionation pattern produced in laboratory experiments is observed in meteoritic components. The meteoritic observations had been previously interpreted as deriving from supernova debris, thus, there are cosmochemical applications. Finally, mass independent fractionations have been observed in different stratospheric molecules, thus, there are applications in resolving stratospheric chemical reactions and transformation mechanisms.

Stable isotope ratio measurements have been successfully employed as a diagnostic probe of a wide range of processes. The extent of such studies encompasses, e.g., paleoclimatology, oceanic circulation, atmospheric chemical transformation mechanisms, igneous rock geothermometry and the evolutionary history of the solar system. Such studies are contingent upon fundamental, quantitative knowledge of the physical chemical principles which govern the relevant processes which produce the characteristic isotope variations. Generally, these isotopic variations are attributed to well known, quantitatively detailed processes, e.g. the position of equilibrium in isotope exchange reactions, with its concomitant temperature dependency, diffusion and chemical kinetic effects. The basic tenents of, for example, isotope exchange, are rather well known, with the classic works of (1-2), establishing the basic physical chemical formalisms.

Although the governing principles of conventional isotope effects differ, they all have one common feature: they are ultimately dependent upon mass. For example, isotopic exchange between two isotopically substituted species results

[1]Current address: Department of Chemistry, University of California—San Diego, La Jolla, CA 92093–0317

primarily from the difference in zero point vibrational energy for the isotopomers. The vibrational frequencies for two isotopically substituted species, e.g. υ_1, υ_2 differ by a factor, equation 1:

$$\upsilon_1/\upsilon_2 = (\mu_2/\mu_1)^{1/2} \qquad (1)$$

where μ_1 and μ_2 are the reduced masses. Therefore, the frequency ratio is dependent upon the ratio of the reduced masses. Similarly, an isotopic fractionation deriving from, e.g. a property due to translational energy (Σ_i) of a particle is simply: $\Sigma_i = \frac{1}{2} m V^2$, with the species possessing mass (m) and velocity (V). The ratio of the velocities of two isotopically substituted molecules, i and j, is then $V_i/V_j = (m_j/m_i)^{1/2}$, again, mass dependent.

Under normal conditions (standard temperature and pressure) isotope effects are small since the free energy difference associated with, e.g. isotopic exchange between isotopically substituted species is likewise, small. Partly for this reason, and for sake of comparison of isotopic measurements of natural samples, an isotopic convention was established (3). The delta notation is defined, for example in the case of oxygen isotopes, as, equation (2):

$$\delta^{18}O\ (o/oo) = (R^{18}/R^{18}_{STD} -1)\ 1000 \qquad (2)$$

with $R^{18} = {}^{18}O/{}^{16}O$. The subscript "STD" refers to a conventional standard, which for oxygen, is standard mean ocean water (SMOW). The δ thus expresses the $^{18}O/^{16}O$ ratio (R^{18}) of a sample, in parts per thousand, or per mil, (o/oo), variation with respect to a standard. The choice of the standard in the experiments reported in this chapter is the make up oxygen gas. The $\delta^{17}O$ similarly refers to R^{17} which is the $^{17}O/^{16}O$ ratio.

In the delta notation then, a mass dependent isotopic composition has $\delta^{17}O \approx 0.5\ \delta^{18}O$, since any process producing a change in the $^{18}O/^{16}O$ ratio concomitantly produces half that change in the $^{17}O/^{16}O$ ratio.

Measurements of many natural samples, e.g. terrestrial and lunar rocks and minerals, water (oceanic and meteoric) and air O_2 all possess a $\delta^{17}O/\delta^{18}O \sim 0.5$. Figure 1 displays this relation. A profound deviation from this observation was made (4), who observed that the so called high temperature Ca, Al inclusions in the Allende meteorite had an isotope composition $\delta^{17}O/\delta^{18}O = 1$, rather than 1/2. They suggested, that since chemical isotope effects only produce isotopic fractionations with $\delta^{17}O/\delta^{18}O \approx 0.5$, the observations must reflect intervention of a nuclear process. Specifically, it was suggested that explosive carbon or helium burning, which produces essentially monoisotopic ^{16}O, may be the progenitor of the observed isotopic anomaly. The Allende inclusions thus are alien grains, admixed in varying amounts, to the "normal" ($\delta^{17}O = 0.5\ \delta^{18}O$) solar nebular reservoir, of pure ^{16}O. Given that 1) oxygen is the major element in stony planetesinals and 2) the magnitude of the anomaly is large, the process responsible for generation of the observed isotopic composition must represent a major process in the early history of the solar system. A recent review of mass independent isotopic compositions and the implications for cosmochemistry is given (5).

It was later observed (6) that in the process of ozone formation a mass independent isotopic fractionation arose. In fact, the product ozone was enriched

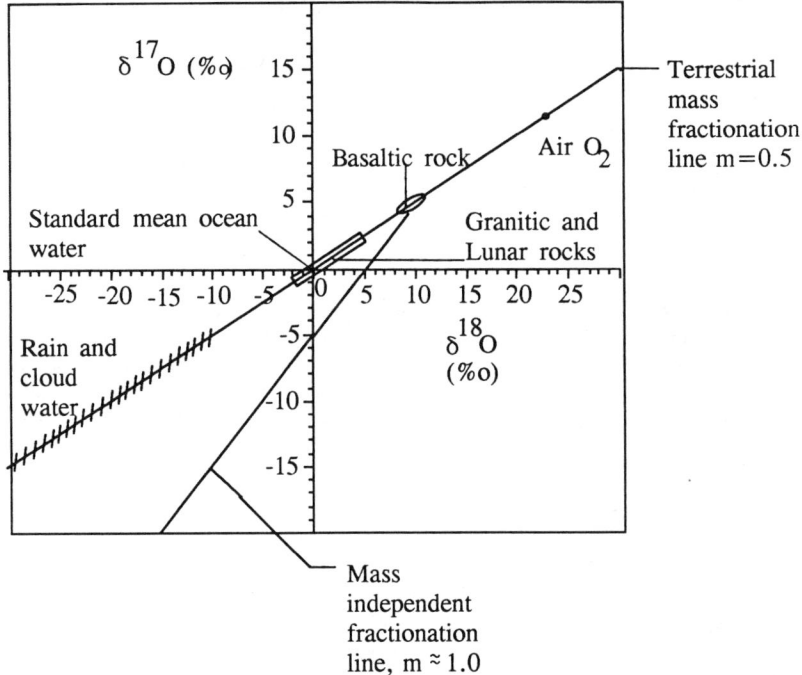

Figure 1. The oxygen isotopic composition of terrestrial, lunar and meteoritic materials.

in the heavy isotopes (^{17}O, ^{18}O) with $\delta^{17}O = \delta^{18}O$, precisely the same fractionation observed in the Allende inclusions, and which was presumed to be supernova ejecta. Thus, there are two significant questions immediately posed: (1) what is the physical-chemical explanation for this anomalous isotope effect and (2) is the process operative in the early solar system, thus accounting for the observed meteoritic observations.

It is now known that mass independent isotopic fractionations manifest themselves in other environments, notably the Earth's atmosphere. A remarkable <u>in situ</u> balloon borne mass spectrometric measurement demonstrated that stratospheric ozone possessed an enormous ^{18}O enrichment, as high as 40% (7). It was later demonstrated that there is an equal ^{17}O, ^{18}O enhancement, ($\delta^{17}O/\delta^{18}O \cong 1$) but of variable magnitude (8). Return sample isotopic measurements have also confirmed the original measurements (9). The details of the stratospheric isotopic measurements are presented in a chapter of this text (10), and the reader is referred to that section. The important aspect as pertains to this chapter is that the isotopic composition of stratospheric ozone is mass independently fractionated with $\delta^{17}O/\delta^{18}O \sim 1$, essentially the same ratio as observed in the lab and meteoritic material, though the magnitude of the $\delta^{18}O$, $\delta^{17}O$ enrichment differs. Even more recently, the oxygen isotopic composition of stratospheric carbon dioxide has been measured and shown that it too possesses a mass independent oxygen isotopic composition, possibly related to the known ozone anomaly (11).

It is now clear that there are many applications of mass independent fractionation processes. As in the case of conventional mass dependent isotope effects, successful application requires a quantitative description of the relevant physical chemical process. At present, the mechanism responsible for the observed mass independent isotopic fractionation process is unclear. This chapter presents the present state of knowledge and details the aspects of the fractionation process which are experimentally known.

The Mass Independent Isotopic Fractionation In Ozone Formation. As discussed, the first demonstration of a chemically produced mass independent isotopic fractionation was in ozone formation, produced by O_2 dissociation, which was originally done by electron impact (6). It was observed that ozone was produced with equal ^{17}O, ^{18}O enrichment (with respect to precursor molecular oxygen). Later experiments (12) led to the suggestion that the observed isotopic fractionation derives from the symmetry of the different isotopomeric species $^{16}O^{16}O^{16}O$, $^{16}O^{16}O^{17}O$ and $^{16}O^{16}O^{18}O$. It was suggested that, owing to the appearance of alternate rotational states for the asymmetric (C_s) isomers, a longer lifetime is expected for this metastable state compared to the C_{2v} ($^{16}O^{16}O^{16}O$) due to its slightly longer lifetime and consequently increased probability of stabilization to a stable, ground state entity. This enhanced lifetime may develop as a result of the higher state density, which is known , in part, to determine the lifetime (13). Thus, the isotopic fractionation would not derive from a property of mass, but rather symmetry. Isotopic fractionations arising from O_2 properties during dissociation may be ruled out both theoretically (14-16) and experimentally (17). It has also been experimentally demonstrated that O_2 dissociation via electron impact in the RF region (6), microwave region (18), and by UV light (19-20) produces ozone in a mass independent manner, with $\delta^{17}O/\delta^{18}O = \sim 1$. In addition, in a recent series of experiments it was demonstrated that O_3 dissociation, and subsequent recombination in the presence of O_2, results in a similar ozone enrichment of the heavy isotopes (21). One of the important aspects of this work is that it clearly relates the source of the anomalous isotopic fractionation process to the O (^{3}P) + O_2 ($^3\Sigma_g$) recombination reaction. Isotopic effects arising from secondary ozone dissociation have also been experimentally eliminated as a source of the anomalous isotopic fractionation. (22) demonstrated that ozone photolysis produces products isotopically depleted in the heavy isotopes with $\delta^{17}O \approx .6\ \delta^{18}O$, essentially as expected on the basis of zero point energy differences.

As discussed in (21), theories for the observed mass independent fractionation process which derive from differential lifetimes for the isotopomers metastable state, O_3^* (12), (22-23) should obey known kinetic unimolecular theory. It is well known that reaction order is pressure dependent, and in particular, a well developed formalism exists which relates pressure dependencies of recombination reactions to unimolecular inter and intra molecular energy rearrangements (24-26). In the particular case of the $O + O_2 \rightarrow O_3$ recombination reaction, experimentally determined falloff curves have been determined (27-28). At third body [M] pressures to $\sim 1.2 \times 10^{20}$ molecules cm^{-3}, the reaction rate obeys third order kinetic relations (28), however, at higher pressures, kinetic fall off towards second order is observed. It has been suggested (27-28) that at pressures in excess of $\sim 10^{20}$ molecules cm^{-3} contributions from a radical complex mechanism during ozone recombination become important. The kinetic fall off behavior may be interpreted by the following equations 3-7 (28):

$$O + M \rightarrow OM \tag{3}$$

$$O_2 + M \rightarrow O_2M \tag{4}$$

$$O_2 + OM \rightarrow O_3 + M \tag{5}$$

$$O + O_2 M \rightarrow O_3 M \tag{6}$$

$$O_2 M + OM \rightarrow O_3 + 2M \tag{7}$$

Kinetic evaluation of the proposed reaction sequence establishes that the fall off to second order is consistent with experimental observations. As discussed by (21), if the observed ^{18}O, ^{17}O enrichment derives from a property of the O_3 metastable state, the enrichment should obey a similar pressure dependency as the observed reaction order fall off since the O_3 excited state becomes decreasingly important with increasing pressure. It was shown that the ^{18}O enrichment, following UV photolysis of O_2, began to decrease at pressures above ~ 450 Torr (20), which was also later confirmed in the O_3 photolysis experiments (21). Comparison of the results from the two laboratories, although experimental procedures differ, give essentially the same result. More recent studies of the pressure dependency of the ^{18}O enrichment to higher pressures demonstrates that the enrichment decreases in the 8 to 45 atmosphere region by 84 o/oo and is absent at ~ 56 atm (29). The behavior of the ^{18}O enrichment in O_3 formation thus does not coherently track the observed kinetic order fall off curves (21), (28). Therefore, it is unlikely that theories involving differential lifetimes of the metastable state are the source of the observed mass independent fractionations.

Although the lifetime hypothesis appears incorrect, the contention that isotopomeric symmetry is a factor mediating the fractionation process, as first suggested by (12), does appear correct. If symmetric distributions influence the fractionation process, then the effect should be dependent upon isotopic abundance. At natural abundances, species such as $^{18}O^{16}O^{18}O$, $^{17}O^{16}O^{17}O$, $^{17}O^{17}O^{16}O$, $^{18}O^{18}O^{16}O$, etc., are too low to contribute to the overall observed fractionation. However, at increased ^{18}O abundances, species such as $^{18}O^{16}O^{18}O$ become significant, and the relative proportion of C_s/C_{2v} species is altered. Thus, if symmetry is at least in part responsible for the mass independent fractionation, the process should be abundance dependent. As demonstrated by (22-23) this is clearly the case; the $\delta^{17}O/\delta^{18}O$ ratio depends upon isotopic abundance, as expected. The absolute dependence upon isotopic abundance has now been determined (30). When the ozone isotopomers of masses 48 to 54 are normalized to mass 48 ozone (pure ^{16}O) it is observed that the symmetric ^{18}O isotopomer (mass 54) possesses no enrichment, and the largest enrichment (20.3%) is at mass 51, which experimentally totally consists of the asymmetric species $^{16}O^{17}O^{18}O$. At the other masses, which possess mixtures of both asymmetric and symmetric species, the enrichments are observed to be significantly smaller.

There now is direct evidence that symmetry influences the fractionation process. Infrared, tunable laser absorption measurements of the abundances of the $^{16}O^{18}O^{16}O$ and $^{16}O^{16}O^{18}O$ species have demonstrated that the asymmetric isotopomeric species possesses 80% of the observed ^{18}O enrichment (31), which is significantly greater than that expected from a purely statistical isotopomeric distribution. A second important observation of these experiments is that the ^{18}O enrichment does not exclusively reside in the asymmetric species. Any theory which quantitatively details the anomalous mass independent fractionation process must also account for this partitioning between symmetric and asymmetric isotopomers. This will be discussed in more detail in a following section. Additionally, experiments which qualitatively support the role of symmetry, specifically those involving both other elements and molecules will be discussed later.

Regarding influencing parameters for the isotopic ozone recombination process, it is known that the fractionation is temperature dependent (21). The ^{18}O enrichment, after the effect of $O + O_2$ exchange is subtracted, linearly varies from ~165 o/oo at 360° K to ~88 o/oo at 140° K. Present theories for the fractionation based upon lifetime differences (12), (22-23) are qualitative and thus possess no temperature dependency. A recent theory (32-33) has, at least partially, addressed the effect of temperature and will be discussed in greater detail.

Present Theoretical State Of Art For Mass Independent Isotopic Fractionation Processes. It is observed by CARS spectroscopic analysis that following ozone photolysis in the Hartley band (230-311 nm), the rotational and vibrational state distributions of the product O_2 ($^1\Delta_g$), that there exists anomalous quantum state distributions. There is a significant enhancement of the even-J states observed for $^{16}O^{16}O$ and not $^{16}O^{18}O$ (34). It was suggested that the even-J state propensity derives from selection rules for the non adiabatic curve crossing for O_2 ($^1\Delta_g$) → O_2 ($^3\Sigma_g$). In an expanded treatment, (35) suggested that this anomalous J-state distribution may relate to the mass independent fractionations observed (6). The symmetry and parity restrictions specifically associated with the O_2 ($^1\Delta_g$) → O_2 ($^3\Sigma_g$) transition and the nuclear spin selections could result in fractionations which are independent of mass. However, the UV-O_3 photolysis of experiments (30), were specifically done by photolysis in the ozone Chappuis bands, where the products are O (^{3}P) and O_2 ($^3\Sigma_g$), with no O_2 ($^1\Delta_g$). Given that the same isotopic results are obtained as are observed in the discharge experiments, the role of O_2 ($^1\Delta_g$) is eliminated. Although the specific mechanism involved (*35*) is ruled out for the observed mass independent fractionation, the suggestion that isotope effects may derive from selection rules arising from electronic state transitions is important, and still potentially relevant, albeit not in the original connotation.

At present, the only theory remaining is that recently developed (*32*) (*36*). It is, therefore, informative to consider it in some detail, particularly as it relates to experimental observations. Based upon kinetic arguments (*33*) it was shown that symmetry alone, during recombination, does not sufficiently account for the observed isotopic enhancements in ozone formation. It was suggested (*36*) that the role of symmetry in the O_3^* energized complex is linked with the process of energy randomization and dissociation. The energy randomization frequency is given by υ (R) and mean dissociation frequency, υ (D), for the energized ozone. A kinetic argument (*32*) suggested that the enrichment of the, e.g. asymmetric ^{18}O, is given by: r = 1 +δ, where δ is, equation 8:

$$\delta = \{\upsilon (D)/\upsilon (R)\} \{1- \exp[-\upsilon (R) /\upsilon (D)]\} \quad (8)$$

However, this relation would lead to essentially a complete enrichment in $^{16}O^{16}O^{18}O$, and none in $^{16}O^{18}O^{16}O$, which is inconsistent with experimental observation (*31*). To accommodate this deficiency, it was suggested that a "flip" occurs, with isotopic, isomeric rearrangement of the metastable O_3 state (*32*). The rearrangement occurs during the stretch associated with the excess energy accompanying the O ~ O_2 association, with the terminal atom sufficiently stretched to initiate bond formation with the other ozone terminal atom. Thus $^{16}O^{16}O^{18}O$ may be converted to $^{16}O^{18}O^{16}O$ in the process, plus all other isotopomeric variations. Inclusion of stabilization rates (*32*) with further kinetic analysis leads to an ^{18}O enhancement equation, equation 9:

$$r (^{50}O_3) = 1 + \Delta (^{50}O_3) = 1 + \frac{5a}{6} \quad (9)$$

with a suggested as being 0.52. Δ is the measured or observed enrichment factor. The value r, however, is based upon <u>in situ</u> stratospheric $^{18}O_3$ measurements. The ratio of the dissociation to randomization frequencies ($\upsilon(D)/\upsilon(R)$) is calculated to be 1.6. A problem with this outcome is that the enrichment observed in laboratory experiments is, $\Delta \approx 0.085$ (*20*), thus, υ (R) would have to be 3.3 x 10^{13} s^{-1} (*32*). Physically, this is nearly an impossible requirement since it is of the order of the greatest vibrational frequency. A more realistic value for υ (R) is obtained for the value $\Delta \approx 0.43$, as observed in stratospheric O_3 via balloon borne measurements (*7-8*). To alleviate this difficulty, it was suggested that a secondary fractionation has occurred, both in the La Jolla and Minneapolis experiments (*6*) (*19-21*) and, in the stratospheric ozone isotopic measurements (*9*). It is specifically suggested that exchange between O_3, which initially possesses an ^{18}O enrichment of ~ 430 o/oo, and molecular oxygen occurs, subsequently reducing the magnitude of the ^{18}O enrichment. The isotopic exchange, equation 10:

$$^{16}O^{16}O^{18}O + {}^{16}O^{16}O \rightarrow {}^{16}O^{16}O^{16}O + {}^{18}O^{16}O \qquad (10)$$

is suggested as occurring in either the gas phase or on the chilled walls of the collection surface. In considering this suggestion, the gas phase exchange may be ruled out simply because it is kinetically too slow. The experimentally determined rate for this reaction, as an upper limit, is 2×10^{-25} cm^3 s^{-1} (37), thus the rate of O_3-O_2 isotopic exchange is too slow to be significant. It is, however, known that the same exchange may occur under conditions where the ozone is condensed in the presence of molecular oxygen (37), thus the possibility that this catalyzed exchange may reduce the ^{18}O enhancement associated with the O_3 recombination step must be considered.

A simple, yet relatively unambiguous argument may be made against participation of catalytic O_3-O_2 isotopic exchange. Figure 2 shows the original results first reported (6). Note, that the best fit line (r = 0.99) passes nearly exactly through the origin, that is, the starting molecular oxygen isotopic composition. The spread in the data derives simply from the extent of reaction in the small (76 cm^3) volume. A major advantage of such a plot is that one may immediately determine if more than one fractionation process has occurred. It should be emphasized that this does not require that they occur either simultaneously or sequentially. In fact, the only restriction is that the processes have different $\delta^{17}O/\delta^{18}O$ fractionation factors. The contention in (32) is that the ozone recombination and energy randomization processes are mass independent (probably $\delta^{17}O \approx \delta^{18}O$) and of a magnitude of ~ 430 o/oo. Catalytical exchange between O_3 and O_2 is suggested as reducing this enrichment to approximately e.g. $\delta^{18}O = 10\%$, which is about the average ^{18}O enrichment experimentally observed by the two groups. Isotope exchange between molecules, gas, liquid or solids is well known theoretically and experimentally to be strictly mass dependent processes, or $\delta^{17}O = 0.5 \, \delta^{18}O$. If these two processes influence the final O_3 isotopic composition, as suggested by (32), then the best fit of the data, by simple vector argument, may not pass through the starting isotopic composition. Figure 2 demonstrates that the data clearly pass through the origin. In fact, one may, by vector addition, determine what the final O_3 isotopic composition should be assuming that there is a fractionation factor $\delta^{17}O = \delta^{18}O = 430$ o/oo associated with O_3 recombination, and a O_3-O_2 isotopic fractionation process, with $\delta^{17}O = 0.5 \, \delta^{18}O$, which produces O_3 with $\delta^{18}O \approx 100$ o/oo. The combination of those two processes would produce O_3 with isotopic composition $\delta^{18}O = 100$ o/oo and $\delta^{17}O = 265$ o/oo! This is clearly not observed. In fact none of the experiments to date in any laboratory produce isotopic fractionations of this sort. Therefore, based upon both kinetic and isotopic arguments, it must be concluded that both the existing laboratory data and the returned stratospheric O_3 measurements reflect the true isotopic composition of gas phase ozone (see, however, theoretical arguments (33a)). As discussed later, experiments involving the O + CO reaction also support this conclusion. It must, however be emphasized that the largest experimentally observed O_3 isotopic composition from both laboratories is ~ 14 % at room temperature, yet the stratospheric O_3 enrichment exceeds 40%. The lower temperatures of the stratosphere, as experimentally shown (21), only lowers the enrichment. The presence of nitrogen as a third body in air proportions does

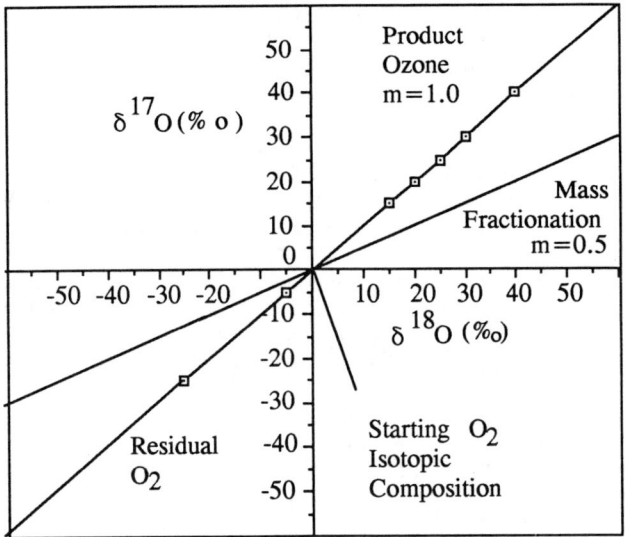

Figure 2. The oxygen isotopic composition of ozone, produced by O_2 dissociation.

not affect the fractionation process (*20*), nor does, apparently, the wavelength of O_2 or O_3 photolysis (*19-21*). Thus, at present, a paradox exists concerning this difference between the magnitude of the stratospheric ^{18}O-O_3 enrichment and laboratory observations. It is clear that future experimental and theoretical work is needed to resolve the present dilemma.

Mass Independent Isotopic Fractionations Observed In Non-Ozone Systems.
It has been discussed (*5*), (*32*), (*36*) that a more extended experimental verification of the role of symmetry could be obtained if other reactions, besides $O + O_2$, were investigated. Such studies thus might provide additional insight into the relative roles of properties such as symmetry and state density. In addition, regarding future cosmochemical applications, it is desirable to ascertain whether the mass independent isotopic fractionation process is restricted to ozone formation and the isotopes of oxygen. With this as motivation, two other reactions have been investigated, $O + CO$ and $SF_5 + SF_5$. Both reactions should in principle, be subject to symmetry contraints and concomitantly produce mass independent isotopic fractionations. In addition, the density of states differ considerably and it may be useful in determining the relevant isotopic fractionations.

The Reaction of $O + CO$. As discussed (*32*), the anomalous isotopic fractionation which occurs in the $O + O_2$ reaction may, in part, derive from symmetry number considerations in an association reaction. The relative rate dependence of the energy randomization process and the activated complex lifetime were suggested as the factors which mediate the mass independent isotopic fractionation process. If this assertion is correct, then the $O + CO$ reaction should also produce a mass independent isotopic fractionation since the relevant isotopic symmetry is the same as for $O + O_2$. In addition, the vibrational energy density of states for $O + CO \leftrightarrow CO_2$ is greater than 200 times that for $O + O_2 \Rightarrow O_3$ (*25*). The state density is well known to affect the lifetime of metastable intermediate species, thus there should be significant differences in the isotopic systematics.

The oxygen isotopic fractionation observed in the reaction of $O + CO$ has now been measured and reported (*38-39*). These studies were carried out in a ~5 liter glass photolysis system. Molecular oxygen was photolyzed at two wavelength regions, 180-260 and 120-160 nm in a stoichiometric excess (~180-400) of CO, thus insuring that the predominant reaction is termolecular $O + CO$ association. A kinetic evaluation (*38-39*) demonstrated that ozone chemistry is insignificant under the experimental conditions.

The experiments demonstrate that a large, mass independent isotopic composition is observed in the product CO_2. Kinetic and isotopic evaluation demonstrates that there is an associated, equilibrium isotope exchange between O atoms and O_2 and CO (*38-39*). The isotopic exchange values observed are consistent with the reduced partition functions calculated by Urey (*1*). Unlike the $O + O_2$ reaction, a pure $^{17}O = ^{18}O$ enrichment in the product CO_2 is not observed, even after substraction of the effects of O-atom exchange with both O_2 and CO. The best fit of the data gives a $\delta^{17}O = 0.8 \; \delta^{18}O$, suggesting simultaneous participation of mass dependent and independent processes. Thus the suggestion that the $O + CO$ reaction should produce a mass independent fractionation is confirmed. However, there are clearly fractionations associated with the reaction which are not accounted for. It was suggested (*36*) that the ratio of the lifetime of the intermediate complex $\tau(D)$ to the energy randomization time $\tau(R)$ should influence the magnitude of the isotopic enrichment. The value $\tau(R)$ for O_3 and CO_2 is essentially the same, and $\tau(D)$ for CO_2 is ~ 200 times that of O_3. Since the enrichment factor (δ) immediately following ozone formation is given by $\delta = \{\upsilon(D)/\upsilon(R)\} \{1- \exp[-\upsilon(R)/\upsilon(D)]\}$ the magnitude of enrichment of ^{18}O, ^{17}O in CO_2 should be significantly different than for O_3. The experimental observation, however, is that it is less. As discussed in (36), this particular reaction has complications not associated with the $O + O_2$ recombination reaction. The CO_2 activated complex must undergo a spin change upon curve crossing, a forbidden process (36), and it is not clear what effect this will have on the isotopic fractionation processes. At present, no theoretical framework exists to address such phenomena.

The more recent, and expanded treatment of the isotopic systematics of the $O + CO$ reaction (32) deals with the complexities in more detail. From the experimental data, it is estimated that $\upsilon(R)$ is about 1.2×10^{12} s^{-1} and $\upsilon(D)$ for carbon dioxide is $\sim 8 \times 10^{10}$ s^{-1}. Thus, the value $\upsilon(D)/\upsilon(R)$ is $\sim 6.7 \times 10^{-2}$ for CO_2, compared to 1.6 for O_3 (for the value derived from the in situ balloon borne mass spectrometric measurements). If the formalism given (32) is correct, the magnitude of fractionation between O_3 and CO_2 should differ considerably since it is determined by the ratio $\tau(D)/\tau(R)$. The magnitudes, however do not differ by more than ~ 2-3 % in the value of $\delta^{18}O$, thus further development of the theory is required. It was suggested (32) that the equations 11, 12:

$$O + O_2 \leftrightarrow O_3^* \quad (11)$$

$$O_3^* + CO \rightarrow CO_2 + O_2 \quad (12)$$

could participate in the observed fractionation, thus transferring the known ozone ^{18}O enrichment to CO_2. Also, the reaction sequence, equations 13, 14:

$$O + CO \leftrightarrow CO_2^* \quad (13)$$

$$CO_2^* + O_2 \leftrightarrow CO_2 + O_2 \quad (14)$$

is suggested as possibly contributing to the observed isotopic fractionations. The participation of reactions (3)-(4) is based upon a rate coefficient for equation 15:

$$O + O_2 + CO \rightarrow CO_2 + O_2 \quad (15)$$

reported (40) as being 1.7×10^{-33} cm^6 s^{-1}. The 20 year old value of Kondratiev, however is likely to be seriously overestimated and needs to be checked.

It was suggested (32) that the observed experimental data may have been complicated by the involvement of the isotopic exchange, equation 16:

$$^{18}O\ C^{16}O + C^{16}O \rightarrow {}^{16}O\ C^{16}O + C^{18}O \quad (16)$$

This exchange was suggested as occurring between the carbon dioxide trapped at liquid nitrogen temperatures on the walls of sprial traps while the carbon monoxide is pumped away. There are three problems with this. First, as explained in (*38*) the carbon dioxide is not trapped in the spiral traps. They serve merely as a secondary collection surface. The actual cryogenic removal is done in the 5.2 liter photolysis reservoir, thus carbon monoxide is never directly pumped over condensed carbon dioxide. More compellingly, however, are control experiments, where carbon dioxide of known isotopic composition, is admixed with carbon monoxide and recollected and its isotopic composition determined. There is no alteration of isotopic composition observed, thus, it may be safely concluded that secondary isotopic exchange between CO and condensed CO_2 has not altered the original isotopic composition of the carbon dioxide.

A final observation may be made regarding possible isotopic exchange between frozen carbon dioxide and a flowing gas. The collection procedures of CO_2 from CO employed (*38-39*) are essentially the same as those typically employed for the separation of atmospheric CO_2 from air for isotopic analysis. In this case, there is an even larger excess of exchangeable gas (O_2) which should significantly alter the CO_2 isotopic composition. However, it has been known for more than 20 years that the $\delta^{18}O$ value of air CO_2 is ~ + 41 o/oo, which is the value for equilibrium between CO_2 and ocean water at 25°C, and as expected. Thus, it is concluded at the present time that there is absolutely no supporting evidence for the secondary exchange of CO_2 with CO in the experiments (*38-39*) and other explanations must be sought. It was suggested (*38-39*) that secondary CO_2 photolysis may account for a portion of the observed linear range of $\delta^{18}O$ for the CO_2. It is observed that the greatest extent of fractionation is for the experiments done by photolysis at the higher energy (~ 130 nm) where the CO_2 absorption cross section is 1.6×10^3 times greater. One may also not rule out the possibility that energy dependent processes may influence the isotopic fractionations. It is clear, however, that in the recombination reaction of $O + CO \rightarrow CO_2$ a significant, mass independent isotopic fractionation is observed. This is as expected on the basis of molecular symmetry. There are, however, contributions from at least two other processes, one of which is an equilibrium isotope exchange process between oxygen atoms and O_2 and CO. There are two important aspects derived from these experiments. First, it further supports the role of symmetry. Secondly, it demonstrates that the mass independent fractionation process is not confined to ozone. For meteorites, this is particularly interesting since ozone is not a likely pre solar nebula molecule. That the mass independent fractionation chemistry may be a more general phenomena is further evidence of its possible participation in early solar system chemistry and the generation of observed meteoritic oxygen isotopic anomalies.

Mass Independent Isotopic Fractionations In Sulfur Isotopes. As a further test of symmetry involvements, another element which may be tested is sulfur. It possesses 4 stable isotopes (^{32}S = 95.02, ^{33}S = 0.75, ^{34}S = 4.21, ^{36}S = 0.017 for percent abundances) with abundances of the rare isotopes sufficiently low to provide a similar test as oxygen. Unfortunately, given that the reaction must be (presumably) gas phase, with the product molecules possessing sulfur at each end, the restrictions are severe. In addition, the mass spectrometric measurements must be performed on SF_6 to observe the ^{33}S, ^{34}S variations. Measurements of SO_2 would not permit this because of isobaric interferences from ^{17}O and ^{18}O. Thus, any product molecule must quantitatively and cleanly be converted to SF_6 for isotopic measurement. The chemistry and mass spectrometry were developed for this and are reported in (43). For sulfur, the $\delta^{34}S$ refers to the $^{34}S/^{32}S$ ratio and $\delta^{33}S$, $^{33}S/^{32}S$. The mass spectrometric precision for both $\delta^{34}S$, $\delta^{33}S$ is $\pm$ 0.03 o/oo, which is required for the experiments. The molecule chosen for the sulfur studies, given the restrictions cited above, was S_2F_{10}. The difficulty with sulfur chemistry is that it is significantly more complex than either O_3 or CO_2. The full details of the experiments and an extended discussion are given in (44); a brief summary of the pertinent points is presented here.

The S_2F_{10} molecule is created by reaction of two SF_5 radicals. These radicals were created by three independent means using photolysis and electron discharge. Molecular fluorine gas was photolyzed to F atoms in the presence of SF_4 to create SF_5. Electrical discharge in CF_4 produced F atoms to react with SF_4, yielding SF_5 radicals, and pure SF_5Cl was dissociated by electron impact, yielding SF_5. In all cases, S_2F_{10} was created and isotopically analyzed for its sulfur isotopic composition. The same results are obtained in all cases and, as discussed in more detail in (44), effectively rule out any significant production role of SF_5 for the mechanism associated with the observed S_2F_{10} isotopic fractionations.

A mass independent isotopic composition (with respect to the precursor molecules) is observed in all experiments, regardless of how SF_5 is created. The isotopic systematics have some similarity to the O + CO experiments. There is strong evidence that there is simultaneous participation of a mass dependent, near equilibrium isotope exchange between SF_5 and SF_4, and a mass independent fractionation associated with the SF_5 + SF_5 recombination reaction, as expected on the basis of symmetry. Fractionations associated with S_2F_{10} dissociation are ruled out on the basis of control experiments which clearly show that it is a mass dependent process, of small ($\delta^{34}S \sim 2$ o/oo) magnitude (44). In the case of S_2F_{10} production, the magnitude of the observed mass independent isotopic fractionation is significantly smaller than for CO_2 or O_3; at most, 3.5 o/oo for $\delta^{34}S$ and $\delta^{33}S$.

The experiments involving sulfur isotopes and S_2F_{10} production provide the following information. First, mass independent isotopic fractionations are not restricted to oxygen. Again, this is important for cosmochemical applications since it is further evidence of the general nature of the process. Second, it is further evidence of the mediation of symmetry factors upon isotopic fractionation processes. Third, the source of mass independent effect is in recombination reactions, e.g. $O + O_2$, $O + CO$ and SF_5 + SF_5. Finally, the magnitude of the mass independent fractionation process is variable, which may ultimately be of importance in the development of a successful theory for the chemical production of symmetry dependent isotopic fractionations.

Present State of Knowledge For Chemical Mass Independent Fractionations. Since the first observation of a chemically produced mass independent isotopic fractionation (6) a great deal has been established. From consideration of all the existing experimental data, we may conclude the following. (1) The fractionation process is related to symmetry. The dependence upon isotopic abundance (23) (30), direct observation of preferential ^{18}O enrichment in the asymmetric $^{16}O^{16}O^{18}O$ species (31), and observation of mass independent fractionations in reactions subject to symmetry constraints; $O + O_2$, (6) $O + CO$, (38) SF_5 + SF_5 (44) all support symmetry involvement. (2) Both the magnitude of the ^{18}O enrichment (21), (29) and the $\delta^{17}O/\delta^{18}O$ ratio are pressure dependent. The magnitude of the enrichment decreases at pressures significantly below 1 atm total pressure and is absent at 56 atm. At pressures to 86 atm, a conventional kinetic isotopic fractionation is observed. (3) The effect is temperature dependent, at least for ozone formation (21). The ^{18}O enrichment in O_3 varies from ~ 16.5% at 360°K to ~ 8.8% at 140°K (21). These three points are relatively secure and any successful theory must account for them.

There are other experimentally observed features which should be considered. First, there is the magnitude of fractionation. In all cases, intervention of secondary O_3 photolysis superimposes a fractionation. It is known that O_3 photolysis preferentially dissociates isotopically light species, thus enriching the residual O_3 in ^{18}O (22). Thus it is nontrivial to determine the precise magnitude of ^{18}O enrichment. It is at most, 16% (20- 21) and at least, 8.5% (20). A typical enrichment is ~ 10%. For O + CO the largest extent of fractionation (after correction for O-CO isotopic exchange) is ~ 8.5% and the lowest, 3% (38-39). This is significantly lower than O_3. In the case of S_2F_{10}, the magnitude is only 3.5% (44). As previously suggested (36), (44) this may possibly indicate a dependence, at least for the magnitude of fractionation, upon the density of states. (36) suggested that the state density is an important factor since it regulates the energy distribution lifetimes. The observations are suggestions of this. The order of the isotopic enrichment is $O_3 > CO_2 > S_2F_{10}$. The density of states follows the same order 20.95, 4529, ~ 10^9 (25) ,(44). This is of course, strictly qualitative in nature, but worth pursuing in the future, both theoretically and experimentally.

General Remarks And Future. It is fairly clear now that chemically produced mass independent isotopic fractionations have a number of unique applications. For chemical physics, multi-isotope studies have been important in establishing and delineating the role of symmetry in recombination reactions. The reactions studied to date include $O + O_2$, $O + CO$ and $SF_5 + SF_5$. The isotopic studies have permitted insights into the chemical dynamics that otherwise would have been unobtainable, or at the least, not looked for by other spectroscopic techniques. A great deal remains to be understood, but the applicability of theory thus obtained renders such development worthwhile. Even more recently, thermal decomposition studies (45-46) have demonstrated what is apparently yet another new isotope effect. Through a series of controlled experiments it is demonstrated that the product O_2 from O_3 decomposition is equally enriched in ^{17}O, ^{18}O. Kinetic isotopic fractionation based upon unimolecular theory (e.g. (25)) predicts that the product molecular oxygen should be depleted in the heavy isotopes and in a mass dependent manner ($\delta^{17}O \cong 0.5\ \delta^{18}O$). The theoretical treatment of the data demonstrates that the effect is apparently derived from the complex, collisional energy transfer process associated with the thermal dissociation. As such, this is not strictly the reverse of the recombination reaction, which entails a higher energy complex and, stabilization. The actual process associated with thermal dissociation is extraordinarily complex, including collisional energy transfer in both an upward (towards dissociation) and downward (stabilization) direction. It is well beyond the scope of this chapter to review the pertinent unimolecular theories, however, it is appropriate to refer to the studies since it is another example of how mass independent isotopic fractionations may be used as a probe of physical chemical phenomena.

From precise isotopic measurements of returned stratospheric samples it is unambiguously clear now that mass independent isotopic compositions are present in at least two species. Ozone possesses a mass independent composition, similar to that observed in laboratory experiments (9). However, the magnitude is apparently extraordinarily variable and not understood. It is of importance to resolve both by laboratory experiments and by further precise return sample measurements the source and mechanism of the enrichment and variability. Until this is achieved, a significant feature of global stratospheric ozone is not understood. It is also now known that stratospheric carbon dioxide possesses a mass independent composition thought to arise from exchange with the O(^{1}D) atom derived from O_3 photolysis (11), (47). This unambiguous link between two of the most important species of the earth's atmosphere could not have been resolved by any other technique other than multi isotope ratio measurements. There are presently only a few relevant laboratory measurements for the multi-isotopic exchange. There is essentially no theory, such as that which exists for ground state species (1-2) for the calculation of isotopic reduced partition functions for the exchange between an electronically excited and a ground state species. This represents a new horizon which has immediate applications in stratospheric chemistry.

Regarding meteoritics, an original criticism directed towards involvement of chemically produced oxygen isotopic fractionations in the early solar system was that the effect was rather specific and was not observed in nature. Both statements are now known to be false. As discussed in detail, it is known that the effect is in fact somewhat general, and a feature of gas phase reactions. It has been demonstrated to occur in several reactions and elements. It is also observed in nature now as well and will be of importance in future stratospheric studies. The occurrence of the chemical fractionation process may also relieve a problem associated with meteoritic observations. While pure ^{16}O admixture could, ostensibly, account for the observations of (4), it is known that many meteorites (e.g. H, L, LL chondrites) possess ^{16}O depletions (e.g. (5)). These of course may not be explicable in terms of a relict grain, ^{16}O rich admixtures. In fact, the ^{16}O depletions exist at the bulk level, not individual inclusions. Thus, a significantly large reservoir is required, in an apparently turbulent pre solar nebula, and one which does not mix with other isotopically distinct reservoirs. A number of such distinct reservoirs (at least 4) must be both created and segregated. Chemical production of the anomalies of course requires one reservoir since, by material balance, both positive and negative ^{16}O reservoirs may be simultaneously created, essentially analogous to that observed in Figure 2.

In the first observations of the meteoritic oxygen isotopic anomalies (4) it was stated that if the ^{16}O anomaly is created in an astrophysical event, such as a supernova, it should correlate with another isotopic anomaly, particularly ^{24}Mg or ^{28}Si excesses. Since it is required that the anomaly be sequestered in grains, one would particularly expect this. As reviewed recently (5) it is striking that, even though the magnitude of the isotopic anomaly is large and in the major element, there is no correlation with any other element across the meteoritic classes. In fact, silicon e.g. appears to be essentially homogenized; the total (mass dependent) isotopic range is a few per mil. As regards to chemical production, it is interesting to note that from what is known at present regarding the fractionation mechanism, oxygen is probably the only element in the periodic chart where these isotopic anomalies should be observable. Reactions such as O + SiO, O + MgO, O + FeO could plausibly be envisioned as reactions which could occur in a presolar nebula and which could produce a mass independent isotopic fractionation (5). No isotopic effect would occur in e.g. Si, Mg or Fe since these atoms are typically co-ordinated by oxygen, and to generate the mass independent fractionation they must be external. Clearly, to forge a quantitative model for the early history of the solar system there are two crucial needs. First, the mechanisms associated with the mass independent fractionation processes must be resolved. As stated in the opening paragraph of the chapter, it is the quantitative formalism of conventional isotope effects which led to the wealth of detailed applications. Secondly, the relevant experiments must be performed. The studies of $O + O_2$, $O + CO$ and $SF_5 + SF_5$ demonstrate that there are significant variations between each reaction, possibly related to the density of states, though not certainly. The difficult task of studying the relevant reactions, such as $O + SiO$ must be done and these studies are presently underway.

As a final comment, as is the case for conventional mass dependent isotope effects, mass independent fractionations may be used as a probe of a wide range of physical phenomena. The time scales of the processes range from occurrences of events on the 10^{-12} sec scale to $>10^9$ years before present. The range in size extends from a.u. (atomic units) to A.U. (astronomical units).

Acknowledgments. The author thanks Jack Kaye for many useful conversations and an anonymous reviewer for comments. NSF (grant ATM-9020462) and the Alexander Von Humboldt Foundation are gratefully acknowledged. Lynn Krebs and Teresa Jackson are both thanked for their help in the preparation of the manuscript.

Literature Cited

1. Urey, H. C.; *J. Chem. Soc.* (London) **1947**, 562-581.
2. Bigeleisen, J.; Mayer, M. G. *Chem. Phys.* **1947**, *15*, 261-267.
3. Craig, H.; *Geochim. Cosmochim. Acta* **1957**, *12*, 133-149.
4. Clayton, R. N.; Grossman, L., Mayeda, T. K. *Science* **1973**, *182*, 485-488.
5. Thiemens, M. H. In *Meteorites and the Early Solar System;* Kerridge, J. F., Matthews, M. S. Eds.; University of Arizona Press, Tucson, AZ 1988, 899-923.
6. Thiemens, M. H.; Heidenreich III, J. *Science* **1983**, *219*, 1073-1075.
7. Mauersberger, K.; *Geophys. Res. Lett.* **1981**, *8*, 935-937.
8. Mauersberger, K.; *Geophys. Res. Lett.* **1987**, *14*, 80-83.
9. Schueler, B.; Morton, J., Mauersberger, K. *Geophys. Res. Let.* **1990**, *17*, 1295-1298.
10. (Mauersberger, K. Chapter in the text (1991)).
11. Thiemens, M. H.; Jackson, T., Mauersberger, K., Schueler, B., Morton, J. *Geophys. Res. Lett.* **1991**, *18*, 669-672.
12. Heidenreich III, J. E.; Thiemens, M. H. *J.Chem. Phys.* **1986**, *84*, 2129-2136.
13. Herzberg, G.; *Electronic Spectra of Polyatomic Molecules;* Molecular Spectra and Molecular Structure III; Van Nostrand Reinhold Company: New York, NY, 1954; Vol. III, 1-745.
14. Kaye, J. A.; Strobel, D. F. *J. Geophys. Res.* **1983**, *88*, 8447-8452.
15. Navon, O.; Wasserburg, G. J. *Earth Planet. Sci. Lett.* **1985**, *73*, 1-16.
16. Kaye, J. A.; *J. Geophys. Res.* **1986**, *91*, 7865-7874.
17. Heidenreich III, J. E.; Thiemens, M. H. *Geochim. Cosmochim. Acta* **1985**, *49*, 1303-1306.

18. Bains-Sahota, S. K.; Thiemens, M. H. *J. Phys. Chem.* **1987**, *91*, 4370-4374.
19. Thiemens, M. H.; Jackson, T. *Geophys. Res. Lett.* **1987**, *6*, 624-627.
20. Thiemens, M. H.; Jackson, T. *Geophys. Res. Lett.* **1988**, *15*, 639-642.
21. Morton, J.; Barnes, J., Schueler, B., Mauersberger, K. *J. Geophys. Res.* **1990**, *95*, 901-907.
22. Bhattacharya, S. K.; Thiemens, M. H. *Geophys. Res. Lett.* **1988**, *15*, 9-13.
23. Yang, J.; Epstein, S. *Geochim. Cosmochim. Acta.* **1987**, *51*, 2019-2024.
24. Troe, J.; *Ann. Rev. Phys. Chem.* **1978**, *29*, 223-250.
25. Troe, J.; *J. Chem. Phys.* **1977**, *66*, 4758-4775.
26. Schroeder, J.; Troe, J. *J. Ann. Rev. Phys. Chem.* **1987**, *38*, 163-190.
27. Croce de Cobos, A. E.; Troe, J. *Int. J. Chem. Kinet.* **1984**, *16*, 1519-1529.
28. Hippler, H.; Rahn, R., Troe, J. *J. Chem. Phys.* **1990**, *93*, 6560-6569.
29. Thiemens, M. H.; Jackson, T. *Geophys. Res. Lett.* **1990**, *17*, 717-719.
30. Morton, J.; Schueler, B., Mauersberger, K. *Chem. Phys. Lett.* **1989**, *154*, 143-145.
31. Anderson, S. M.; Morton, J., Mauersberger, K. *Chem. Phys. Lett.* **1989**, *156*, 175-180.
32. Bates, D. R.; *J. Chem. Phys.* **1990**, *93*, 8739-8744.
33. Bates, D. R.; *J. Chem. Phys.* **1990**, *93*, 2158.
33a Bates, D. R.; *Planet. Space Sci.* **1991**, *39*, 945-950.
34. Valentini, J. J.; Garrity, D. P., Phillips, D. L., Nieh, J.-C., Tabor, K. D., *J. Chem. Phys.* **1987**, *86*, 6745-6756.
35. Valentini, J. J.; *J. Chem. Phys.* **1987**, *86*, 6757-6765.
36. Bates, D. R.; *Geophys. Res. Lett.* **1988**, *15*, 13-16.
37. Anderson, S. M.; Morton, J., Mauersberger, K. *Geophys. Res. Lett.* **1987**, *14*, 1258-1261.
38. Bhattacharya, S. K.; Thiemens, M. H. *Z. Naturforsch.* **1989**, *44a*, 435-444.
39. Bhattacharya, S. K.; Thiemens, M. H. *Z. Naturforsch.* **1989**, *44a*, 811-813.
40. (Kondratiev, V.N. In IUPAC CONGRESS, edited in reference (41)).
41. Slanger, T. G.; Black, G. *J. Chem. Phys.* **1970**, *53*, 3722-3725.
42. Bottinga, Y.; Craig, H. *Earth Planet. Sci. Lett.* **1969**, *5*, 285-295.
43. Bains-Sahota, S. K.; Thiemens, M. H. *Anal. Chem.* **1988**, *60*, 1084-1088.
44. Bains-Sahota, S. K.; Thiemens, M. H. *J. Chem. Phys.* **1989**, *90*, 6099-6109.
45. Wen, J.; Thiemens, M. H. *Chem. Phys. Lett.* **1990**, *172*, 416-420.
46. Wen, J.; Thiemens, M. H. *J. Geophys. Res.* **1991**, *96*, 10,911-10,921.
47. Yung, Y. L.; DeMone, W. B., Pinto, J. P. *Geophys. Res. Lett.* **1991**, *18*, 13-16.

RECEIVED October 3, 1991

Chapter 10

Heavy Ozone Anomaly
Evidence for a Mysterious Mechanism

S. M. Anderson[1], K. Mauersberger[2], J. Morton[2], and B. Schueler[2,3]

[1]Department of Physics, Augsburg College, Minneapolis, MN 55454
[2]School of Physics and Astronomy, University of Minnesota, Minneapolis, MN 55455

> The unexpectedly large abundances of isotopically heavy ozone observed in a variety of environments indicates that our understanding of how ozone is formed is incomplete and may involve significant participation of one or more low-lying metastable electronic states. A summary of the experimental observations of ozone isotopomers in the stratosphere and in laboratory studies is presented. It is argued that these data are inconsistent with the standard picture of recombination reactions, and an alternative symmetry-selective mechanism which involves multiple electronic states is proposed. Recent progress in theoretical and experimental determinations of the energies of candidate states indicates that the $^3B_2$ state, and to a lesser extent the $^3A_2$, may be involved.

During the past two decades the stratospheric ozone layer has been subject to intensive research. The equilibrium between formation and destruction of ozone may be altered by man-made chemicals which become activated in the stratosphere to participate in catalytic destruction processes. Since ozone is the only constituent which absorbs significant amounts of solar radiation between the visible and 240 nm, a decrease in stratospheric ozone will result in an increase of UV radiation at the Earth's surface and consequent damage to biological systems. Its long term stability has been and is continuously monitored, and atmospheric photochemical models which include established chemical and dynamic processes are used to predict future changes in ozone.

Formation of ozone is generally assumed to follow the well-known Chapman reactions:

and
$$O_2 + h\nu \rightarrow O + O \qquad (R1)$$
$$O + O_2 + M \rightarrow O_3 + M \qquad (R2)$$

---

[3]Current address: Department of Diagnostic Radiology, Mayo Clinic, Rochester, MN 55905

where M is a third body molecule to stabilize ozone. Photochemistry models assume that ozone is formed only in its ground electronic state. Loss processes include

$$O_3 + h\nu \rightarrow O + O_2 \tag{R3}$$

$$O_3 + O \rightarrow 2 O_2 \tag{R4}$$

and catalytic cycles where Cl, NO and OH convert O_3 and O into 2 O_2.

Ozone is a fragile molecule with a dissociation energy of only 1.1 eV. For comparison, dissociation of O_2 requires 5.1 eV, and a similar triatomic molecule such as NO_2 requires 3.1 eV. Throughout the atmosphere ozone is always present as a trace gas; its mixing ratio does not exceed 10 ppmv in the stratosphere and 50 ppbv in the free troposphere. Though it can form many isotopomers due to the existence of three oxygen-atom isotopes (^{16}O, ^{17}O and ^{18}O), the natural abundance of those atoms makes ozone molecules of mass numbers 48, 49 and 50 the only ones of atmospheric significance. Statistical distribution should lead to the following isotope ratios:

$$48/50 = 163 \text{ and } 48/49 = 900$$

Whenever the two heavy isotopomers are formed, one-third of the molecules should be of symmetric molecular structure ($^{16}O^{18}O^{16}O$) and two-thirds asymmetric (e.g. $^{18}O^{16}O^{16}O$). Recognition of these two forms is important in ozone isotope studies and will be discussed below in more detail.

A measurement *(1)* of the two most abundant isotopes of ozone at mass 48 and 50 has led to speculation that the formation mechanism stated above may not be entirely correct. A balloon-borne mass spectrometer experiment measured ozone of mass 48 and 50 over the altitude range of 38 to 22 km and revealed the astonishing result that the heavy isotope was as much as 40% enriched relative to a statistical distribution at 32 km. Theoretical analysis *(2, 3)* predicted a small depletion in the heavy isotope rather than an enhancement. More atmospheric measurements have since become available, and numerous laboratory studies have been performed to elucidate this unusual enhancement in heavy ozone. A variety of experimental techniques have been employed, including emission and absorption spectroscopy as well as mass spectrometry and the collection of stratospheric ozone samples. The ozone formation process has been clearly identified as the step where the isotope fractionation occurs.

This paper provides a review of stratospheric ozone measurements and laboratory studies with the goal of identifying the mechanism responsible for the heavy isotope enrichment. Although a number of explanations have been proposed, it appears that none of them is able to explain all of the detailed results now available, and a more radical approach to the problem seems required.

A Brief Historical Account of the Isotope Enrichment

The first paper on a possible isotope enrichment in heavy ozone was published by Cicerone and McCrumb *(4)* who suggested that $^{34}O_2$ may be preferentially dissociated in the upper atmosphere by selective absorption in the Schumann-Runge bands when compared to $^{32}O_2$, resulting in an increased production of ^{18}O. This, in turn, would result in an enhancement of $^{50}O_3$. In 1981, Mauersberger *(1)* published the first heavy ozone measurement between 22 and 38 km, showing an unexpected enhancement of about 40% near 32 km, decreasing both toward higher and lower altitudes. An analysis of the ozone formation process in the stratosphere *(2)* questioned the proposed source of the enhancement *(4)* since an increased production of ^{18}O would be rapidly diluted by the exchange reaction

$$^{18}O + {}^{32}O_2 \Leftrightarrow {}^{16}O + {}^{34}O_2 \tag{R5}$$

Kaye and Strobel *(2)* concluded that a paradox must exist since the mass spectrometer measurement *(1)* appeared to be of good quality while their calculation predicted a depletion in the heavy isotopes. A detailed analysis *(5)* of the absorption cross-section in the Schumann-Runge bands resulted in a production rate of ^{18}O an order of magnitude less than originally proposed by Cicerone and McCrumb *(4)*, and thus eliminated conclusively an enhancement of ^{18}O as the source of heavy ozone found during the balloon flight.

Another indication of an unusual isotope effect in ozone was published by Heidenreich and Thiemens *(6)* who produced a mass-independent enhancement in $^{49}O_3$ and $^{50}O_3$ of about 3% in a laboratory O_2-discharge experiment, considerably less than measured in the stratosphere, yet clearly different from statistically expected values. The experiment was performed at low temperatures, an important fact which determines the magnitude of the isotope effect as will be shown later.

Summary of Stratospheric Ozone Isotope Measurements

During the last ten years numerous ozone isotope measurements have been published, involving very different experimental techniques. The initial mass spectrometer balloon-borne measurements have been complemented by spectroscopic investigations and by ozone sample collection. In this section an attempt will be made to summarize these measurements; details can be found in the papers cited. It is fascinating to note that every technique employed to measure heavy ozone has detected a substantial and variable isotopic enhancement.

In situ mass spectrometer measurements. During a number of balloon descents, seven isotope ratio profiles have been obtained. They are explained and discussed in detail by Mauersberger *(7)* ; Figure 1 shows four representative profiles. All flight data have been analyzed in a consistent, uniform way to derive background corrections for intensities measured at mass 50. This procedure is justified since essentially the same equipment was used in each of the flights. The isotope enhancement shown in Figure 1 is defined in the usual way.

$$\text{Enh. (\%)} = \{(50/48)_{sample} - (50/48)_{stand.}\} / (50/48)_{stand.} \times 100$$

The standard is the expected statistical distribution of natural ^{16}O and ^{18}O in the atmosphere, although the observed effect is so large that virtually any standard would suffice. Most pronounced is the variability in the enhancement. There are flights with large high altitude enhancements and others which show "only" 10-20%. The standard isotope ratio, corresponding to an enhancement of zero, was only found in one of the flights at lower altitude. There appears to be no correlation of heavy isotope enrichment with the time of the day, season or even altitude, although upper stratospheric enrichments (>30 km) are more frequent.

Ozone Sample Collection. The *in situ* mass spectrometer measurements do not provide sufficient accuracy to measure the less abundant smaller isotopomers of mass 49, although high altitude data indicated that it was also enhanced. A unique ozone collection system has been developed to improve the accuracy and precision of the measurement of mass 50, and to add measurements of 49. Details about the collection procedure and the subsequent laboratory data analysis are provided by Schueler et al. *(8)*. Table I shows the isotope enhancements in 49 and 50 obtained from three balloon flights. The gondola had three sample collectors on board; during the second flight only two samples were obtained due to balloon problems. After the third flight, two of

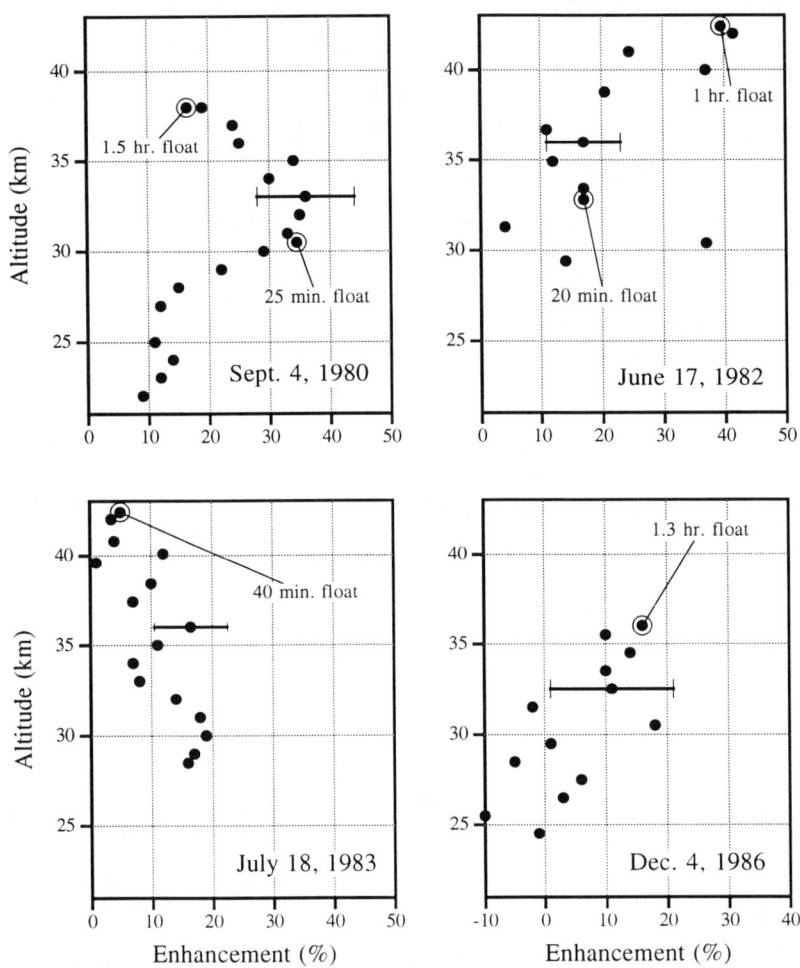

Figure 1. Enrichment of $^{50}O_3$ measured with a mass spectrometer during four balloon flights.

the containers developed leaks and analysis of the ozone sample was not possible. As discussed below, the 8 to 9% enhancements found in the last two collector flights agree well with laboratory studies which predict such enhancements when the temperature dependence is taken into account.

Table I. Flight Parameters and Results

Flight	Date Location	Altitude Range (km)	$^{49}O_3$	±	$^{50}O_3$	±
I	25 Aug 1988 Palestine, TX	35.5-32.3	11.2	1.2	16.1	0.9
		32.0-29.0	10.7	1.0	14.3	0.6
		28.7-26.2	8.7	1.0	12.1	0.6
II	31 July 1989 Palestine, TX	35.2-31.8	9.2	1.7	8.9	1.2
		31.1-26.8	8.1	1.8	8.1	1.5
III	24 Oct 1989 Fort Sumner, NM	32.0-29.1	8.3	0.8	9.1	0.5

Optical Measurements. Infrared absorption and emission spectroscopy have a clear advantage over mass spectrometer investigation since a separation of symmetric and asymmetric molecules contributing to $^{50}O_3$ is possible. The first absorption data were published by Rinsland et al. *(9)* using ground-based observations to obtain column density isotope enhancements. For three measurements the asymmetric molecules showed an enhancement of 11% ± 11%, and 5% ± 5% for the symmetric molecules. The large error bars make it difficult to detect any symmetry-dependence. Somewhat unusual results were published by Abbas et al. *(10)*, who reported analysis of far infrared emission spectra which led to a higher enhancement of the symmetric molecules (60% at 33 km) than found for the asymmetric (20% at 33 km). The enhancement showed considerable variability with altitude.

The most recent data were reported by Goldman et al. *(11)*, who used a Fourier-transform IR absorption spectrometer on board a balloon to determine column densities of ozone and its isotopomers above 37 km. For two flights the following enhancements were found:

	$^{16}O^{18}O^{16}O$	$^{18}O^{16}O^{16}O$
Nov. 18, '87, Fort Sumner, NM	20 ± 14 %	40 ± 18 %
June 6, '88, Palestine, TX	16 ± 8 %	25 ± 12 %

Because column density ratios are reported, local enhancements could be substantially larger.

Summary of Laboratory Ozone Isotope Measurements

The discovery of an unusual enrichment in stratospheric heavy ozone has resulted in an extensive effort to simulate and understand the effect in dedicated laboratory experiments. The first laboratory studies of ozone isotope fractionation were published by Heidenreich and Thiemens *(6)* who reported a mass independent enrichment of approximately 3% in both $^{49}O_3$ and $^{50}O_3$. Since this time, ozone has been produced using a variety of techniques including O_2 dissociation by UV absorption, electric discharge, microwave discharge and others *(12, 13)*.

In a novel experiment reported by Morton et al. *(14)*, ozone was formed from O and O_2 in their ground states ("Visible Light Experiment") to investigate whether excited states may play a role in producing the isotope enhancement. Results of this experiment are shown in Figure 2. Near room temperature, the enrichment is approximately 12.5% for $^{50}O_3$ and 10.5% for $^{49}O_3$. The enrichment decreases toward higher pressures, a result also found when other methods of O production were used *(12-14)*. By 10 atmospheres the isotope effect has decreased dramatically to about 3% *(15)*. Figure 3 shows the change in the enhancement when the temperature of the vessel in which ozone was formed was varied over a wide range. The early laboratory results of 3% enhancement *(6)* were obtained when ozone was formed in a liquid N_2 cooled container.

The laboratory data shown in Figures 2 and 3 would predict a stratospheric enrichment in heavy ozone (near temperatures of - 35°C) of 8-9%, slightly larger for $^{50}O_3$ then for $^{49}O_3$. While such enhancements have been observed it is not unusual, as shown in Figure 1, to find the enrichments higher than this by factors of 5 or more.

Pathways to Symmetry-Selective Isotope Effects in Ozone

Within our current picture of how ozone is formed we can understand neither the direction nor magnitude of the isotope effects discussed above. Kaye *(3)* has carried out extensive theoretical calculations based on established recombination theory and found that a small depletion of the heavier isotopomers would be expected. Several alternative explanations have been put forward. It has been proposed *(6)* that since the asymmetric ozone molecules can exist in every rotational state while the symmetric ones can only exist only in every other one, the asymmetric molecules possess twice the density of states and a concommitantly longer $O-O_2$ collision complex lifetime. Bates *(16)* has suggested that the collision complex may be so short-lived that energy never becomes completely randomized, and the complex 'remembers' which bond originally belonged to the O_2. As an alternative to conventional theory, Valentini *(17)* has suggested that the striking isotope effects observed in the $O_2(^1\Delta)$ photoproduct of ozone photolysis could account for the ozone observations, provided that enough of these excited molecules were present.

None of these explanations accounts adequately for the observations presented previously. The minor modifications of recombination theory are inconsistent with the pressure and temperature dependence, and the $O_2(^1\Delta)$ mechanism cannot explain the results of experiments designed to preclude the presence of these metastables *(14)*.

The inability to understand these isotope effects in the context of conventional theories of recombination reactions suggests that Reaction R2 is not as simple as one might have thought, although we are not the first to propose this. In a study of the pressure-and third body-dependence of its rate coefficient, Cobos and Troe *(18)* concluded that the standard picture of recombination may not apply to this process. Its rate coefficient is exceedingly small, factors of 200 and 20 below that for O + NO + M at low and high pressures, respectively, and even lower than that for a typical atom-atom recombination reaction. Locker et al. *(19)* have pointed out that time-resolved experiments in which both O and O_3 are monitored indicate that the O-atoms are consumed faster than the ozone appears. Based on detailed study of the kinetics they suggest that a second electronic surface must be involved in the recombination process, and that as much as 60% of the recombination occurs into this state.

It seems to us that a similar mechanism, as indicated in Figure 4, could account for the isotope effects which have been observed, provided that some type of symmetry-selective relaxation of the metastable to the ground state occurs. It therefore becomes crucial to understand the nature of the states involved.

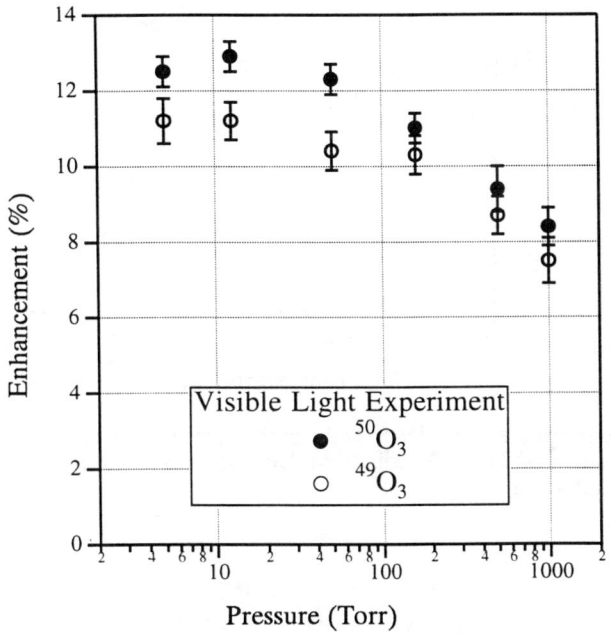

Figure 2. Pressure dependence of enhancement in $^{49}O_3$ and $^{50}O_3$ when ozone was formed from ground state O and O_2 at a gas temperature of 320 K.

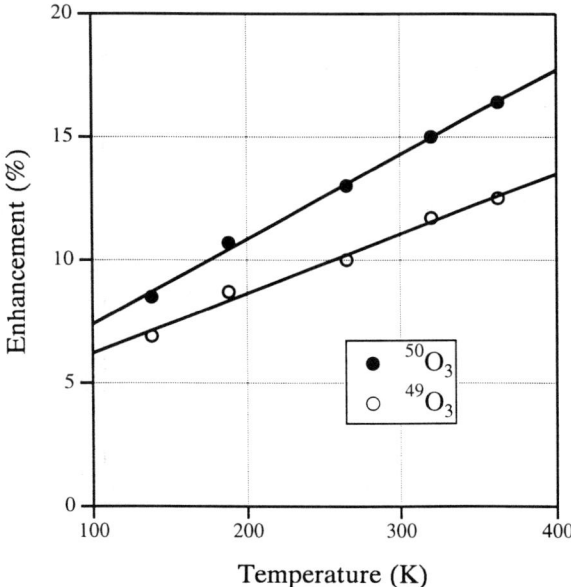

Figure 3. Temperature dependence of the enrichment in $^{49}O_3$ and $^{50}O_3$ measured at an O_2 pressure of 50 Torr. Contributions due to the isotope exchange reaction *(2)* have been subtracted.

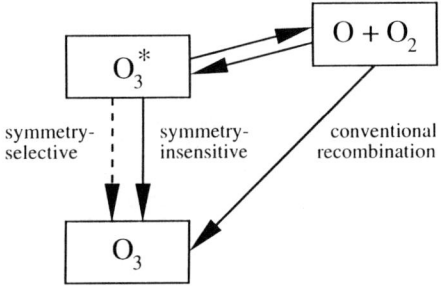

Figure 4. Schematic representation of the role of a low-lying electronic state of ozone in $O + O_2$ recombination. Isotope effects could come about from symmetry-selective relaxation to the ground state.

Relevant Electronic States

Due to the atmospheric importance of ozone there has been considerable effort expended to understand its electronic structure *(20)*. Several groups have performed ab-initio quantum calculations to determine electronic energies and the corresponding geometries *(21-23)*, and the results have been critical for the interpretation of ozone's optical absorption spectrum. For the present purposes, however, we need to know which electronic configurations correspond to bound states, i.e., their adiabatic electronic energies, and of these which are accessible from thermal, ground state O and O_2. The triplet nature of the reagents and ozone's small binding energy places considerable demands on calculations designed to answer this question.

To date there is a clear consensus about which states are likely candidates for bound systems, namely the $^1A_1$, $^3B_2$, $^3A_2$, and perhaps the $^1A_2$ C_{2v} states as well as a D_{3h} ring state. It is also clear that the first two of these are in fact bound, and that there are no barriers to association for either one. Whether or not the $^3A_2$, $^1A_2$ or ring states are bound is still an open question theoretically. Given that the most recent global calculation of ozone electronic surfaces is nearly 15 years old, this may not be the case for long.

Experimental determination of accurate adiabatic electronic energies also presents a challenge. The diffuse character of ozone's absorption bands from the ultraviolet into the near IR (including the Hartley, Huggins, Chappuis and Wulf bands) precludes applying the full power of modern laser techniques to the problem, although some headway has been made in the case of the quasi-bound Huggins system *(24)*. Nonetheless, it is possible to obtain some valuable information even at relatively low resolution from changes in the vibrational structure of the absorption bands under isotopic substitution.

To show this, we write the energy, T, of a given vibronic level as

$$T = T_e + \sum_i (v_i + 1/2)\omega_i + \sum_i \sum_k (v_i + 1/2)(v_k + 1/2) x_{ik} \tag{1}$$

where ω and x are the harmonic and anharmonic constants associated with the normal-mode vibrations *(25)*. Transitions originating in the lowest vibrational level of the ground electronic state (denoted by ") will have frequencies given by

$$v = T_e' - T_e'' + \sum_i [v_i'\omega_i' + 1/2(\omega_i' - \omega_i'')] + \sum_i \sum_k [(v_i' + 1/2)(v_k' + 1/2)x_{ik}' - 1/4 x_{ik}''] . \tag{2}$$

Since the vibrational constants for the fully-substituted molecule are *(26)*

$$\omega_i^* = \rho \omega_i, \qquad x_{ik}^* = \rho^2 x_{ik}, \qquad \text{with } \rho^2 = m/m^*, \tag{3}$$

one can rewrite equation (2) for the labeled molecule and obtain an expression for the isotope shifts of the vibronic bands. For small vibrational excitations we substitute fundamental for harmonic frequencies and drop the quadratic terms to obtain a pleasingly simple relationship between these shifts and the vibrational energies:

$$\Delta v \equiv v^* - v \cong (\rho-1) \left((1/2 \sum_i v_i' + \sum_i v_i' v_i') - 1/2 \sum_i v_i'' \right) \tag{4}$$

which is just

$$\Delta v \cong (\rho-1)(E_v' - E_o''). \tag{5}$$

The isotope shift of a given vibronic band is seen to be proportional to the difference between the vibrational energies in the upper and lower states; see Figure 5a for a pictorial summary. Within a vibrational progression one expects the isotope shifts

(a)

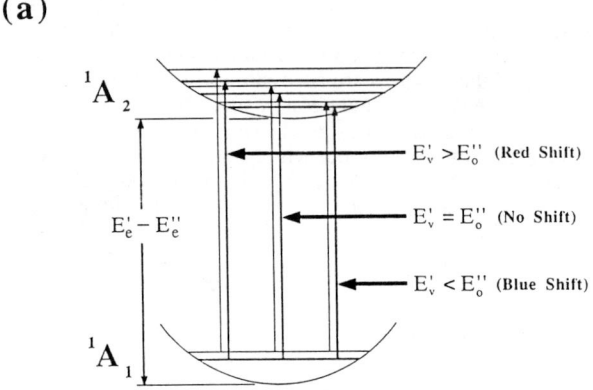

(b)

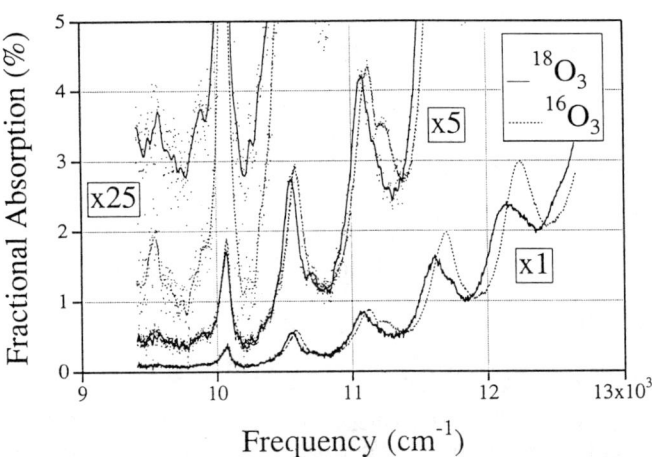

Figure 5. Determination of adiabatic electronic energies from vibronic isotope shifts. a) Illustration of the effect of heavy-atom substitution (heavy lines) on the vibrational manifolds of the ground and excited electronic states. For the upper state, the well represents motion along bound degrees of freedom only. b) Absorption spectrum of $^{16}O_3$ and $^{18}O_3$ near 1μ (300K, ≈100 Torr, 46 cm path). Note pattern of isotope shifts decreasing towards the adiabatic electronic energy of 9990 cm^{-1}.

to decrease with transition energy towards the band origin. In particular, note that the isotope shift vanishes when the vibrational energies are equal, and that the photon energy corresponding to this 'shiftless' transitition is the energy difference between the bottoms of the wells on the two electronic potential surfaces.

Figure 5b shows our absorption spectra of normal and fully ^{18}O-substituted ozone in the Wulf region *(27)*. The dominant peaks clearly demonstrate this pattern of decreasing isotope shifts, with the band near 10080 cm^{-1} moving slightly to the red while the very weak one near 9550 cm^{-1} moves slightly to the blue when the atomic masses are increased. The adiabatic energy determined from the spectrum was 9990±70cm^{-1}, or 1.24±0.01 eV.

This technique was first used by Katayama *(28, 29)* to find the correct origin of the Huggins bands and has been exploited recently by us *(30, 27)* to estimate adiabatic energies for the states associated with the Chappuis and Wulf bands, which have historically been associated with the $^1B_1$ and $^1A_2$ states, respectively. Preliminary results from very recent ab-initio calculations indicate that these assignments may not be entirely correct, however. They suggest that the $^1B_1$ and $^1A_2$ states interact to produce the Chappuis system *(31, 32)* and that the $^3A_2$ is more likely responsible for the Wulf bands *(32)*. These results are presented with the theoretical energies in Table II. If the assignment is correct, the experimental result for the Wulf bands together with the energy ordering from the theoretical calculations suggests that the $^3A_2$ state is not quite bound, while the $^3B_2$ state very likely is. This state could have a significant impact on ozone formation due to its high multiplicity compared to the singlet ground state *(33)*. Moreover, the metastable character of these states could provide multiple access to symmetry-controlled relaxation opportunities, if they exist, and could therefore be responsible for the observed isotope effects described here.

Table II. Summary of Electronic States of Ozone
Theoretical vertical energies from refs. 23 and 22; adiabatic energies from ref. 22. Experimental vertical energies from ref. 20, adiabatic energies from refs. 27-30.

State	Vertical Energy			Adiabatic Energy		Assignment
	Ref. 22	Ref. 21	Expt.	Ref. 22	Expt.	
$^3B_2$	1.20	1.60	-	0.92	-	-
$^3A_2$	1.44	2.09	-	1.35	1.24±0.01	Wulf Bands
$^1A_2$	1.72	2.34	1.6	1.66	-	-
$^3B_1$	1.59	2.01	-	1.74	-	-
$^1B_1$	1.95	2.41	2.1	2.06	2.03±0.04	Chappuis Bands
2 $^3B_2$	3.27	4.71	-	2.92	-	-
2 $^1A_1$	3.60	4.58	-	1.20	-	(ring state?)
$^1B_2$	4.97	6.12	4.9	5.54	3.41	Hartley, Huggins Bands

Conclusion

The preponderance of experimental evidence on isotope effects in ozone formation points to very strange behavior in the recombination reaction between O and O_2. The large enrichments observed in the heavy isotopomers apparently comes about through a mechanism outside the conventional picture of recombination processes on a single potential energy surface. No satisfactory explanation has been found. One very attractive possibility is the participation of the low-lying $^3B_2$ and perhaps the $^3A_2$ metastable states. Combined with an appropriate, but as yet, unknown symmetry-selective relaxation process, such states could provide the explanation of the anomalous isotope effects. More work, both theoretical and experimental, will be needed to characterize these states to determine whether this hypothesis is correct and to assess the implications for the stratosphere.

Literature Cited

1. Mauersberger, K., *Geophys. Res. Lett.* 1981, **8**, 935
2. Kaye, J.A.; Strobel, D.F., *J. Geophys. Res.* 1983, **88**, 8447
3. Kaye, J.A., *J. Geophys. Res.* 1986, **91**, 7865.
4. Cicerone, R.J.; McCrumb, J.L., *Geophys. Res. Lett.* 1980, **7**, 251.
5. Blake, A.J.; Gibson, S.T.; McCoy, D.G.,*J. Geophys. Res.* 1984, **89**, 7277.
6. Heidenreich, J.E.; Thiemens, M.H., *J. Chem. Phys.* 1986, **84**, 2129.
7. Mauersberger, K., to be published.
8. Schueler, B.A., Morton, J.; Mauersberger, K., *Geophys. Res. Lett.* 1990, **17**, 1295.
9. Rinsland, C.P.; Devi, V.M.; Flaud, J.-M.; Camy-Peyret, C. Smith, M.A.; Stokes, G.M., *J. Geophys. Res.* 1985, **90**, 10719.
10. Abbas, M.M.; Guo, J.; Carli, B.; Mencaraglia, F.; Carlotte, M.; Nolt, I.G., *J. Geophys. Res.* 1987, **92**, 13231.
11. Goldman, A.F.; Murcray, J.; Murcray, D.G.; Kosters, J.J.; Rinsland, C.P.; Flaud, J.M.; Camy-Peyret, C.; Barbe, A., *J. Geophys. Res.* 1989, **94**, 8467.
12. Thiemens, M.H.; Jackson, T., *Geophys. Res. Lett.* 1987, **14**, 624.
13. Thiemens, M.H.; Jackson, T., *Geophys. Res. Lett.* 1988, **15**, 639.
14. Morton, J.; Barnes, J.; Schueler, B.; Mauersberger, K., *J. Geophys. Res.* 1990, **95**, 901.
15. Thiemens, M.H.; Jackson, T., *Geophys. Res. Lett.* 1990, **17**, 717.
16. Bates, D.R., *Geophys. Res. Lett.* 1988, **15**, 13.
17. Valentini, J.J., *J. Chem. Phys.* 1987, **12**, 6757.
18. Croce De Cobos, A.E.; Troe, J., *Int. J. Chem. Kinet.* 1984, **16**, 1519.
19. Locker, J.R.; Joens, J.A.; Bair, E.J., *J. Photochem.* 1987, **36**, 235.
20. Steinfeld, J.I.; Adler-Golden, S.M.; Gallagher, J.W., *J. Phys. Chem.* Ref. Data 1987, **16**, 911.
21. Hay, P.J.; Goddard III, W.A.,*Chem. Phys. Lett.* , 1972, **14**, 46.
22. Hay, P.J.; Dunning, Jr., W.A., *J. Chem. Phys.*, 1977, **67**, 2290.
23. Thunemann, K.-H.; Peyerimhoff, S.G.; Buenker, R.J., *J. Mol. Spectrosc.* 1978, **70** ,432.
24. Sinha, A.; Imre, D.; Goble, Jr., J.H.; Kinsey, J.L., *J. Chem. Phys.* , 1986, **84**, 6108.
25. Herzberg, G., *Molecular Spectra and Molecular Structure, III. Electrronic Spectra and Electronic Structure of Polyatomic Molecules*, Van Nostrand Reinhold: New York, NY, 1966; p. 142 ff.
26. Barbe, A.; Secroun, C.; Jouve, J.P., *J. Mol. Spectrosc.* 1974, **49**, 171.
27. Anderson, S.M.; Morton, J.; Mauersberger, K., *J. Chem. Phys.* , 1990, **93**, 3826.
28. Katayama, D.H., *J. Chem. Phys.* , 1979, **71**, 815.
29. Katayama, D.H., *J. Chem. Phys.* , 1986, **85**, 6809.
30. Anderson, S.M.; Maeder, J.; Mauersberger, K., *J. Chem. Phys.*, 1991, **94**, 6351.
31. Banichevich, A.; Peyerimhoff, S.D.; Grein, F., *Chem. Phys. Lett.*, 1990, **173**, 1.
32. Braunstein, M.; Hay, P.J.; Martin, R.L.; Pack, R.T., *J. Chem. Phys.*, submitted May, 1991.
33. Smith, I.W.M., *Int. J. Chem. Kinet.*, 1984, **16**, 423.

RECEIVED October 3, 1991

Chapter 11

Isotopic Study of the Mechanism of Ozone Formation

N. Wessel Larsen, T. Pedersen, and J. Sehested

Department of Chemistry, University of Copenhagen, H. C. Ørsted Institute, 5, Universitetsparken, DK–2100 Copenhagen, Denmark

Thompson and Jacox [1] reported, in a recent study using isotopic oxygen species, that apparently ozone formation in a matrix took place in a straightforward end-on manner. However rather surprisingly an isotopic study of the formation of ozone in the gas phase has not previously been undertaken. The most obvious reason for *not* undertaking an investigation would be to tacitly assume that the incoming oxygen atom had no choice but to attach itself at one of the ends of the oxygen molecule. Indeed our original motive for undertaking this investigation was to find out whether a symmetrically triangular (D_{3h}) form of ozone might be involved in the transition state. (We call it "cyclic ozone" following Wright [3]).

Cyclic ozone has a fairly long history in chemistry, see Wright [3], but it has never been observed. Quantum chemical calculations invariably seem to come up with either an excited state with D_{3h} symmetry or with a dip in the higher energy ranges of the ground state surface with the same symmetry. See Murrell et. al. [4] and Burton [5].

To study the formation of ozone by means of isotopes would seem to be a good way of collecting evidence for the involvement of cyclic ozone in ozone formation. During the formation of ozone O and O_2 must come down from the dissociation limit on the potential surface (or perhaps slightly above), and hence have a chance of falling into the D_{3h} region, provided it exists, but also assuming that it lies below the dissociation limit.

If the mechanism of ozone formation is purely "end-on", then we expect only one isotopically substituted version of ozone in the reaction:

$$O + QQ + M \rightarrow OQQ + M \qquad (1)$$

Q denotes ^{18}O while O denotes ^{16}O. If on the other hand the reaction passes via cyclic ozone, then we expect a mixture of products:

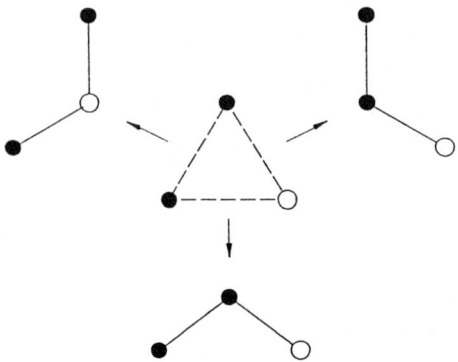

Figure 1: The passage from the reactants to the ground state of ozone via the hypothetical, cyclic intermediate. Supposing an equal propability for the breaking of each bond this channel will lead to a 2:1 ratio of the asymmetrical to the symmetrical isotopomer.

$$O + QQ + M \rightarrow aOQQ + bQOQ + M \qquad (2)$$

where the "isotopomer ratio" r_{52} = a/b (52 refers to the common mass of the ozone isotopomers) depends on the propability of the reacting species falling into the dip on the potential surface. In particular a ratio close to 2 is expected if all ozone formation takes place via cyclic ozone, assuming that the three bonds open up with equal propability to form ground state ozone, see Figure 1.

A third channel, which might be referred to by the self-explanatory term "insertion", is also conceivable:

$$O + QQ + M \rightarrow QOQ + M \qquad (3)$$

Bates [6] has proposed a socalled "flip-over" mechanism, which essentially amounts to the same thing as our cyclic mechanism. The only difference is whether the transition state is a minimum (our assumption) or just a shallow region on the potential surface, where the molecule is floppy.

In Table 1 we show the very first results we obtained for the isotopomer yields. (The experimental details are reviewed in the Appendix.) As is evident, the isotopomer ratios are invariably close to 2 and rather insensitive to the nitrogen partial pressure, i. e. to the rate of ozone formation.

Somewhat to our surprise we found that $^{54}O_3$ was present among the products and that there was more of $^{50}O_3$ than was to be expected from the natural abundance of these isotopomers in the ozone. We realized that a wellknown exchange reaction [7] was taking place:

$$O + QQ \rightarrow OQ + Q \qquad (4)$$

This reaction is in fact about 300 times faster than ozone formation (depending on pressure and temperature). Only when we started to do kinetic

Table 1: Experimental abundances in pct. (percentages uncertain by .05 except for OOO which is uncertain by .1) and isotopomer ratios (uncertain by .2) after UV-photolysis for 10 s of O_3 mixed with Q_2 at room temperature and varying partial pressures of nitrogen.

	Nitrogen partial pressure/Torr				
Species or Ratios	0	64	134	256	476
OOO	99.4	96.7	97.6	96.4	93.3
OOQ	1.03	1.39	1.37	1.60	1.53
OQO	0.49	0.64	0.60	0.71	0.65
r_{50}	2.10	2.17	2.28	2.25	2.35
OQQ	0.75	1.18	1.50	1.97	2.07
QOQ	0.38	0.55	0.70	0.84	0.91
r_{52}	1.97	2.15	2.14	2.35	2.27
QQQ	0.62	0.84	1.72	1.93	2.41

simulations did we realize that the isotopomer ratio r_{52} would end up being close to 2 also for the end-on or insertion mechanisms, when this process is taken into account.

In the following section we analyse the situation in more detail and try to explain the strategy that we used in order find out which channel is actually followed.

A strategy aimed to influence the isotopomer ratios

The ratios r_{50} and r_{52} between the two isotopomers of $^{50}O_3$ and of $^{52}O_3$ may be stated in terms of the concentrations of the isotopomers as follows:

$$r_{50} = \frac{[OOQ]}{[OQO]} \tag{5}$$

$$r_{52} = \frac{[OQQ]}{[QOQ]} \tag{6}$$

If the channel via cyclic ozone prevails, then the two ratios can only deviate from 2 to the extent that isotope effects make the breaking of one chemical bond more preferable than the breaking of a symmetrically non-equivalent bond, see Figure 1. If such is the case, then we would still anticipate *a constant ratio, independent of the experimental conditions*. (To simplify the following discussion we shall assume that the ratios will be exactly statistical, i.e. equal to 2 for this mechanism).

If therefore, by the right choice of experimental conditions, r_{50} and r_{52} can be caused to deviate significantly from the value 2, then this is evidence for either end-on addition (both ratios becoming larger than 2) or insertion (both ratios becoming smaller than 2).

If the isotopomer ratios cannot be influenced, then it is more difficult to reach a definite conclusion. It can either mean that scrambling processes dominate, thus masking the end-on or insertion mechanisms, or that the mechanism is via cyclic ozone.

In the following we shall assume that at any stage during the photolysis, the free atoms will become scrambled. This means that the ratio between the atomic concentrations will attain the same value as the ratio between amounts of the oxygen nuclides present:

$$\frac{[O]}{[Q]} = \frac{2[OO] + [OQ]}{2[QQ] + [OQ]} \tag{7}$$

This relation follows from the approximate steady-state conditions for the atoms:

$$k[O][QQ] + \frac{1}{2}k[O][OQ] = k[Q][OO] + \frac{1}{2}k[Q][OQ] \tag{8}$$

here k is the rate constant for the exchange reaction Eqn. (4) (left to right) while $\frac{1}{2}k$ is the rate constant for the opposite reaction (right to left, neglecting isotope effects).

There are two approximations involved in the steady-state condition Eqn. (8): the neglect of ozone formation, and the neglect of ozone destruction, according to the process:

$$O + O_3 \rightarrow 2\ O_2 \tag{9}$$

Both approximations are very good indeed, since these processes occur on much larger time scales. Ozone formation is some three hundred times slower than exchange, and ozone destruction, is even slower.

Obviously, if the dioxygen molecules also become scrambled, in the sense that the nuclides are statistically distributed among the molecular dioxygen species as well, then the ozone isotopomer ratios will both become equal to 2 *irrespective of the mechanism*. We believe that this was the situation in our early experiments, because the concentration of O_2 (which was only present to the extent that it was produced by the photolysis) was insignificant at any time during the experiments, while OQ was formed at exactly the same rate as Q according to Eqn. (4).

It then occurred to us, that we might delay the molecular part of the scrambling by having not only Q_2 present, but by adding O_2 in comparable quantity. This would delay the formation of the statistical amount of OQ, since the formation of OQ from O_2 would now be a process of substantial bulk. We have simulated the time development of the molecular concentrations in Figure 2.

The strategy in our final experiments has therefore consisted in letting the ozone photolysis take place in a mixture of approximately equal and relatively high concentrations of O_2 and Q_2 (numbers are found in the appendix).

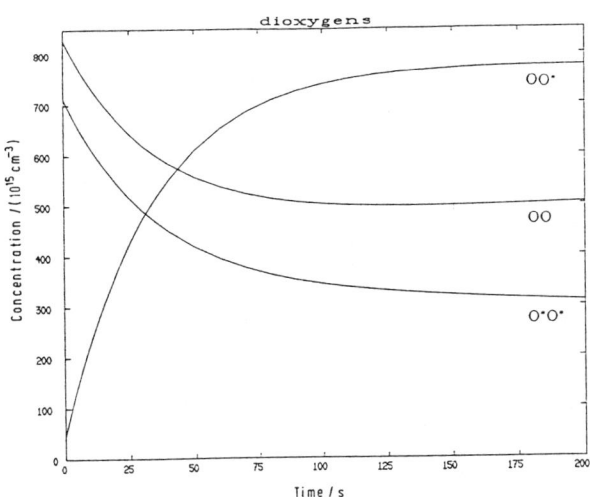

Figure 2: Simulation of the time development of the concentrations of the molecular dioxygen species. The measurements were performed at 120 s. It is seen that at long times [OQ] approaches its "scrambled value" (there is continuous formation of O_2 due to ozone photolysis), while there is a range (0–125 s) in which it has not yet caught up. This is the preferable range for obtaining isotopomer ratios different from 2.
Reprinted from ref. 2. Copyright 1991 Wiley.

The mechanism of ozone formation

The ratios observed in an experiment performed with visible light at room temperature are shown in Figure 3. As is evident from the figure, the ratios are both larger than 2, so that we may conclude that *the mechanism is dominated by the end-on channel*. The simulations were made under the assumption that there were no mass dependent isotope effects.

The error ranges (9-13 % for r_{50}, 13-30 % for r_{52}) do of course allow for contributions from other channels. In the figure we have made a simulation in which a contribution of 5 % insertion to the reaction has been assumed, 5 % is the limit, imposed by the error ranges, to which insertion might contribute to the reaction. However *alternatively we might have added 15 % of the channel via the cyclic intermediate*. (Since 1/3 becomes symmetrical).

In the discussion we mention a series of new, as yet unpublished, experiments in which the number of data points has been increased, and where the temperature has been varied. In Figure 4 we show that the resulting simulations corroborate the findings of [2].

Discussion

We have presented experimental evidence for a mechanism of ozone formation in the gas phase, that is predominantly end-on, [2]. This evidence has been corroborated by a new series of experimental results (as yet unpublished, see

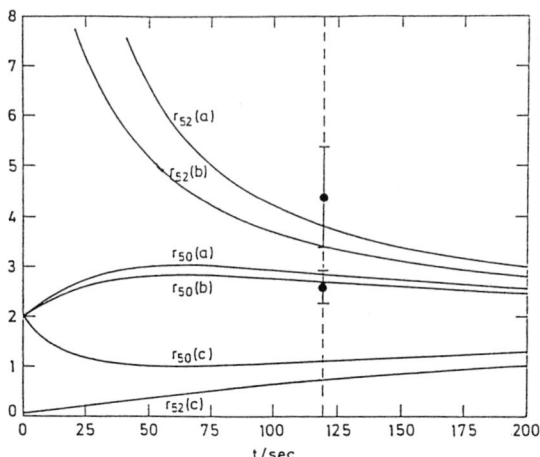

Figure 3: Simulations of the time development of the isotopomer ratios r_{50} and r_{52} for VIS-photolysis experiment assuming: (a) the end-on mechanism, (b) 95 % end-on plus 5 % insertion and (c) 100 % insertion. The dashed line indicates the time (120 s) where the measurements are performed. Also shown are the the experimental ratios with error bars.
Reprinted from ref. 2. Copyright 1991 Wiley.

below). It is important to emphasize that our conclusion is based on the fact that we have been able to cause the isotopomer ratios to increase significantly above the value 2, by the strategy discussed above. It does not rely, therefore, on absolute determinations of the isotopic abundances in the resulting mixtures.

It is worth noting that in our first experiments see Table 1, where we attempted to influence the reactions by manipulating the nitrogen pressure, we were unable to obtain results differing significantly from 2. At that stage we actually thought that we had evidence for the channel via the cyclic transition state.

As implied above (see also [2] for details) we rely on a calibration mixture for our measurements on the microwave spectrometer. This is a drawback, which we have not sofar been able to avoid, although an attempt has been made, which is explained in the second part of the appendix.

At the moment we attempt to come around this difficulty along two lines: Firstly we are trying to measure a larger set of microwave lines. Secondly we are working on obtaining the rotational spectrum of a calibration mixture, of isotopically substituted ozone species, in the far-IR range using our high-resolution Fourier Transform spectrometer (Bruker FS120HR). In both cases we aim at applying the absolute line-strength for concentration assessments. The calculation of absolute line strength, in the rotational part of the spectrum, requires knowledge about the *permanent* dipole moment and the partition functions. The permanent dipole moment for the parent molecule has been measured, while its components in the principal axis systems for the isotopically substituted species

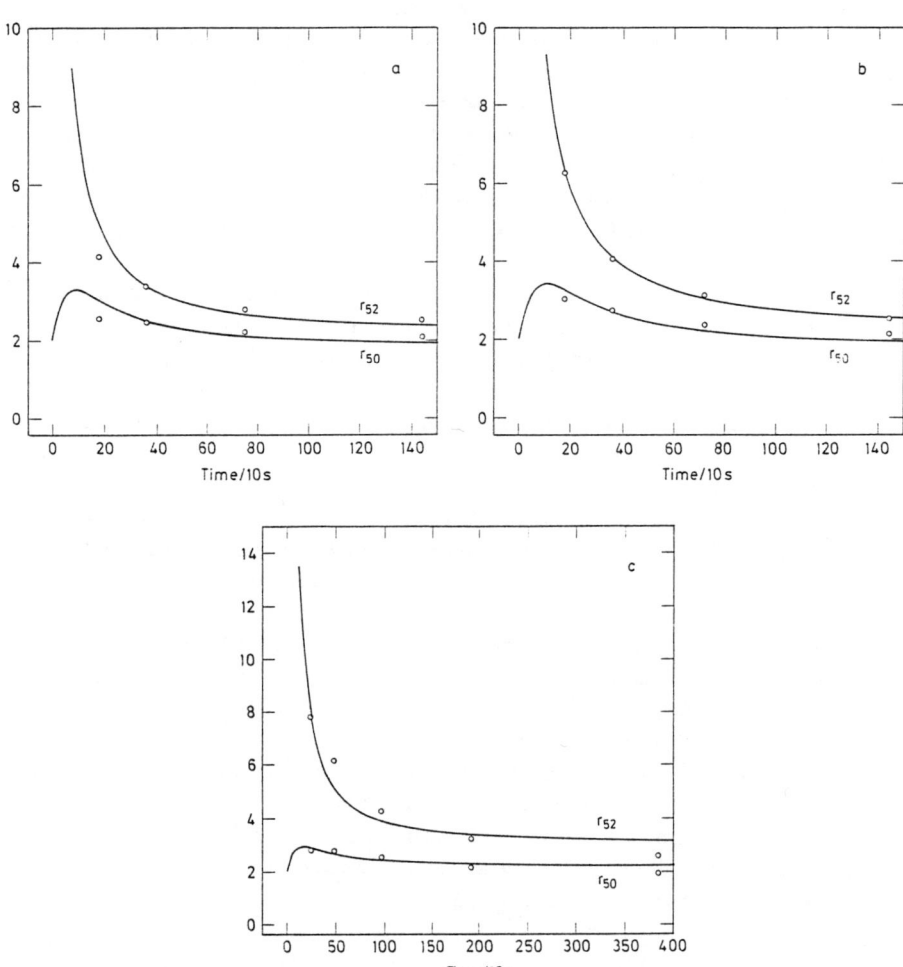

Figure 4: Simulations of the time development of the isotopomer ratios r_{50} and r_{52} for three, more recent VIS-photolysis experiments, run at different temperatures. The experimental set-up is different from that of [2], that is the reason why the time scale is different. The low temperature experiments b) and c) were simulated using the newly found temperature dependence for the scrambling process – see discussion and appendix. a) 10 °C ; b) –70 °C ; c) –130 °C

may be calculated by simple vector algebra. The partition functions for all the species may be obtained from the rotational constants and a general force field, data which are also available.

The reason why we put so much emphasis on the calibration procedure is, that we wish to contribute to the solution of a number of unsolved problems in the atmospheric behaviour of heavy ozone. These problems have arisen through the findings of Maursberger and collaborators [10], [11] and [12], Abbas et. al. [13], Goldberg et. al. [14] and Thiemens et. al. [15] of unexpected enhancements of 50-ozone in stratosphere as well as in the laboratory. Such enhancements were originally predicted by Cicerone and McCrumb [16] but Kaye and Strobel [17] and Kaye [18] questioned the predictions, pointing out that the exchange process Eqn. (4) would render any major enrichment impossible, leaving room for only very minor, mass dependent effects. We aim at doing laboratory studies of absolute isotopomer ratios. (As opposed to the relative values we are presently using). Such ratios are crucial in order to understand how the observed enhancements arise.

Even though we have not as yet accomplished reliable, absolute measurements of individual isotopomer abundances for 50- and 52-ozone, we have been able to improve our calibration (see appendix), so that we can at least compare enhancements obtained for total abundances with those of Morton, Barnes, Schueler and Maursberger [12]. Results of scrambling experiments at three different temperatures are shown in Table 2. (The scrambling experiments are described in the first part of the appendix.)

While Morton et al [12] use mass spectrometry to ascertain the reference composition of the original oxygen mixture, relative to which the enhancements are calculated, we have had to use the isotope composition of the collected ozone as the reference composition (using the same formula as Morton et al in [11] for the calculation of the enhancements). We believe that the uncertainty stemming from this source is small compared to that originating from the experiments. It should be kept in mind, that the enhancements are relatively more uncertain than the abundances, because they are ratios between two abundances divided by ratios between two standard abundances.

Table 2: Abundances and enhancements of isotopomers of ozone (both in %), obtained by scrambling using VIS-photolysis and measured by microwave spectroscopy applying a new calibration method, described in the appendix. Enhancement of OOO is zero by definition.

	Abundance / Enhancement		
Isotopomer	−70°C	20°C	70°C
OOO	14.1(.5)/0(0)	13.4(.5)/0(0)	14.1(.5)/0(0)
OOQ+OQO	41.0(1.7)/14(10)	42.0(1.8)/21(11)	42.6(1.8)/15(10)
OQQ+QOQ	34.4(1.6)/12(4)	34.2(1.6)/15(4)	37.3(1.6)/15(4)
QQQ	8.5(1.6)/−3(7)	8.7(1.6)/1(8)	9.0(1.6)/−5(7)

We can compare with Morton et al [12] for 50-ozone only. For this isotopomer we find from their Fig. 3 $-70°C : 8\%$, $20°C : 11\%$, $70°C : 14\%$. Within our large error bars – and their smaller – we seem to be in fair agreement with the enhancements previously found for 50-ozone produced by photolysis.

We would now like to discuss the low temperature experiments performed at 143 K in [2]. These presented us with a puzzling discrepancy between simulated isotopomer ratios and the corresponding experimental values. In one experiment we determined the values $r_{50} = 2.8(0.3)$ and $r_{52} = 7.3(2.1)$, while the simulated values were $r_{50} = 4.9$ and $r_{52} = 13.3$. We speculated that the reason for such a large discrepancy had to do with lack of knowledge of the temperature dependence of either the exchange process or the process of ozone formation.

The relation for the temperature dependence of the process of ozone formation was extended to low temperatures by Hippler, Rahn and Troe [19]. It was found to predict a value at 143 K, not deviating much from the value we had been using. So the focus had to be put on the exchange process, which was until then believed (but not measured) to be independent of temperature [9] at least for temperatures at or above room temperature.

We have examined this process at 143 K (see appendix for details) and found, very much to our surprise, that its temperature dependence is pronounced at lower temperatures, and that it mimicks that of ozone remarkably. This means, among other things, that *the rate of exchange increases with decreasing temperature*. With this relation for the rate of exchange, we are now able to simulate also the low temperature experiments reasonably well, as is evident from Figure 4. (The new measurements underlying the figure aimed at putting more data points – spanning a suitable time domain – onto the graphs. The data are as yet unpublished, proper publication awaits one of our – hopefully successful – new calibration procedures. The original calibration method has been used to obtain the results.)

This new finding is very interesting, since it points towards a relationship between ozone formation and exchange. If we consider the results in relation to the theoretical considerations also presented in the paper by Hippler et. al. [19], then it appears that our experiments take place in a pressure/temperature-regime, where the function of the third body is to form a Van der Waals-type complex *before* the final reaction takes place, rather than to act as an energy transport agent *after* the reaction has taken place. In the light of this theory one might suggest that the reaction sequences leading to exchange or ozone formation are the following:

$$Q + O_2 \rightleftharpoons Q \cdot O_2; \qquad k_{1\rightarrow}; \; k_{1\leftarrow} \qquad (10)$$

$$Q \cdot O_2 \rightarrow QO + O; \qquad k_2 \qquad (11)$$

$$Q \cdot O_2 + M \rightarrow QO_2 + M; \qquad k_3 \qquad (12)$$

Assuming that the the concentration of the Van der Waals complex reaches a steady-state we obtain the concentration of the complex:

$$[Q \cdot O_2] = \frac{k_{1\to}[Q][O_2]}{k_{1\leftarrow} + k_2 + k_3[M]} \approx \frac{k_{1\to}[Q][O_2]}{k_{1\leftarrow} + k_2} \qquad (13)$$

Hippler et. al. formulated their socalled "Radical Complex"(RC)-mechanism in terms of an unspecified M. What we suggest is that M be O_2. This suggestion is advanced partly because it explains the relationship we have established between exchange and ozone formation, partly because the biradical $O_2(^3\Sigma_g^-)$ is more likely to form a relatively long-lived Van der Waals complex with the atom $Q(^3P)$ than is a closed shell molecule like N_2.

The primary argument used by Hippler et. al. for the RC-mechanism was based on the temperature dependence for ozone formation. The theoretically predicted $T^{-6/2}$ dependence was sufficiently close to the experimentally determined $T^{-2.6}$-dependence to corroborate the mechanism, whereas a T^{-1}-dependence was required theoretically for the "Energy Transfer"-(ET)-mechanism.

If we assume, in accordance with Hippler et. al., that the temperature dependence of $k_{1\to}$ is $\sim T^{-5/2}$, those of $k_{1\leftarrow}$ and $k_2 \sim T^0$ and that of $k_3 \sim T^{-1/2}$, then the resulting temperature dependence of the exchange process will become $\sim T^{-5/2}$, while that of ozone formation becomes $\sim T^{-6/2}$, both in accordance with the experimental findings within their uncertainties. (We must also assume $k_3[M] \ll k_2$, i. e. the low pressure limit, in Eqn. (13) to obtain the correct kinetic reaction orders, and so must Hippler).

APPENDIX

Experimental

The experiments have been fully described elsewhere [2]. Here we shall therefore only give at brief outline.

The assessment of the concentrations is made by means of a microwave spectrometer, that allows the independent measurement of all the isotopomers. We calibrate using a supposedly scrambled mixture of ozone isotopomers.

The photolysis takes place in a quartz bulb using either UV-light from a high pressure mercury lamp (for 10 s) or visible light from a horticultural lamp (for 120 s, to achieve a similar degree of ozone photolysis).

The first experiments were performed at room temperature using varying nitrogen pressures, 10 Torr Q_2 and 10 Torr ozone, while the last experiments were performed with about 30 Torr of Q_2 and O_2 and 10 Torr of ozone. Nitrogen pressure was fixed – by adding nitrogen until a pressure of about 1 atm was reached. The latter conditions aimed at speeding up ozone formation as much as possible.

In the last experiments we also photolysed at 143 K by immersing the bulb into a pentane/pentane-ice bath prior to photolysis. The idea was again to speed up the ozone formation reaction, which has a negative temperature coefficient.

In our unpublished experiments we have used a different set-up. The main difference is in the size of the bulb and the way it is illuminated. To achieve the same degree of conversion as in [2] substantially longer exposure times were needed.

In the discussion we have made an attempt to compare some of our results with enhancements found by various authors. The experiments – one at $-70°C$, one at $20°C$ and one at $70°C$ – were made as follows. The oxygen/ozone mixtures, of the same composition isotopewise as in our other experiments, were photolysed with intense visible light for about an hour. The ozone was preconditioned in the sense that it had been prepared out of a small portion of the oxygen mixture and scrambled – using the Tesla coil as described in [2]. The idea was to have the ozone as close to its final composition as possible before the illumination started.

The simulations were made using conventional routines (NAG-library) and rate constants from the NASA compilation [8] and from Anderson et. al. [9] (exchange reaction). We took the temperature dependent expression for the rate of ozone formation from Hippler et. al. [19] to simulate our low-temperature experiments.

The exchange process was measured after illuminating a mixture of 2.7 Torr O_3, 33 Torr of a mixture of Q_2, OQ and Q_2 and finally 588 Torr N_2 for 2 minutes at 143 K. (In the new set-up). The dioxygen species were analysed mass spectrometrically before and after the illumination: ($O_2\%$, $OQ\%$, $Q_2\%$)= (48.9%, 5.6%, 45.4%) (before) and (32.8%, 37.4%, 29.7%) (after). The (preliminary) scrambling rate constant determined was $k_{exch}(143\ K)=3.5\ 10^{-11}$ cm^{-3} s^{-1}, which is *12 times the room temperature value.* In order to be able to simulate also the -70 °C experiments we have actually found a functional form of $k_{exch}(T)$: $k_{exch}(T) = 3.8\ 10^{-12}\ (\frac{T}{300\ K})^{-2.9}$ cm^{-6} s^{-1}, but we must emphasize that this expression is even more preliminary.

While we have no doubt about the qualitative behaviour of $k_{exch}(T)$, the expression needs further experimental confirmation, which we are presently working at obtaining. We bring it here in order to point out its close similarity to the corresponding expression for the (three-body) rate constant for ozone formation [8]: $k_{ozoneform.}(T) = 6.0\ 10^{-34}\ (\frac{T}{300\ K})^{-2.3}$ cm^{-9} s^{-1}. The thing to notice is the value of the exponent, which, within the quoted uncertainty (0.5), is the same as the one we have found for the scrambling rate constant. (Hippler et. al. [19] found $5.5\ 10^{-34}\ (\frac{T}{300\ K})^{-2.6}$ cm^{-9} s^{-1} for the interval 100–400 K with nitrogen as third body).

A new attempt to calibrate

We want here to describe a new but only partly succesful attempt to calibrate the microwave spectrometer measurements (not in our first paper [2]). The basic idea is the following. While the signal intensity of a given rotational line is proportional to the partial pressure of substance present (if care is taken to avoid saturation), then the proportionality constant is a function, not only of factors intrinsic to the molecule to which it belongs (its "theoretical line strength" τ, that depends on the quantum numbers implied, the rotational constants of the molecule and on temperature), but also of factors having to do with the spectrometer. If the latter factors were independent of the wavelength of the microwave at which the absortion takes place, then this would not be a serious complication, since the intensity ratios would still be reliable. This unfortunately is not the case and this is what causes all the trouble – an example is shown in Table 3.

However, since proportionality with partial pressure can be taken for granted, then we can still assign a definite constant to each absorption line, which is a measure of its intensity *on our particular instrument*. We shall call it the "specific line intensity" σ.

The attempt we have made, to determine specific line intensities, consists in using 26 different isotopomer mixtures to collect information on σ_l, using a least squares procedure. (We have used one absorption line for each isotopomer, therefore $l = 1, 2, \ldots 6$, also designates the 6 isotopomers, see the first 2 columns of Table 3.)

The equations we solve are the following:

$$Y_k = \sum_{l=1}^{6} a_{k,l} \sigma_l; \quad k = 1, 2, \ldots 26 \quad (14)$$

Y_k is the total percentage present in mixture "k" and hence equal to 100% for all k, $a_{k,l}$ is the intensity of the measured signal belonging to isotopomer "l". The six terms on the right hand side of each equation are therefore the percentages present of the six isotopomers. The results obtained are found in Table 3

The correlation matrix of the fit reveals strong correlation between the pairs (σ_2, σ_3) and (σ_4, σ_5): |correlation coefficients| ≥ 0.995. This is the main reason

Table 3: The 6 rotational lines used to assess the percentages of the pertinent isotopomers of ozone. The line assignments Q_l are the asymmetric top quantum numbers $J_{K_{-1}, K_{+1}}$ for the lower and upper state respectively. Frequencies ν_l are in MHz. The theoretical line strength τ_l are in units of $(10^{-20}$ cm^2 MHz) and the specific line intensities σ_l, in units of (lineheigth/%), have both been given for comparison. The uncertainties on σ_l are given in parantheses. Note that the ratio τ_1/τ_6 differs appreciably from σ_1/σ_6. In the absence of effects due to the spectrometer, the ratios should be equal.

l	Isotop.	Q_l	ν_l	τ_l	σ_l
1	OOO	$19_{2,18} \to 18_{3,15}$	23859.67	0.2149	6.94(.13)
2	OOQ	$17_{2,15} \to 16_{3,14}$	20076.27	0.2466	-10.1(10)
3	OQO	$23_{3,21} \to 22_{4,18}$	26040.29	0.2936	28.6(10)
4	OQQ	$18_{2,17} \to 17_{3,14}$	24932.74	0.1694	25.9(8)
5	QOQ	$21_{2,20} \to 20_{3,17}$	19270.82	0.3990	-14.17(15)
6	QQQ	$19_{2,18} \to 18_{3,15}$	21022.42	0.4115	7.00(.15)

for the unacceptably large uncertainties – or even wrong signs in two cases – of these four σ's. The reason for the strong correlation is to be found in the limited variability of the implied four abundances. The fitting procedure is a weighted least squares method, (the equations have been given weigths reciprocal to the uncertainty of Y_k derived from those of the measured $a_{k,l}$-values. The relative uncertainties are assumed to be 6% for all k for $a_{k,1}$ and 2% for $a_{k,l\geq 2}$). So the fit cannot be characterised by a total rms, but the relative rms is close to 1 with our uncertainty estimates, as it should be.

The obvious implication is that this calibration method cannot be used to determine the individual abundances of the symmetric/asymmetric forms of 50- and 52-ozone. However, it may still be used to determine the total abundances of the isotopomers 48, 50, 52 and 54. This is in fact done when we calculate enhancements in the discussion.

In conclusion the new method does not allow us to improve on our previous results, but it does make a comparison with other methods feasible, when only the total abundances are concerned, this is done in the discussion.

Literature Cited

1. Thompson, W.E. and Jacox, M.E., J. Chem. Phys. **91**, 3826-3837, (1989).

2. Wessel Larsen, N., Pedersen, T., and Sehested, J., Int. J. Chem. Kin **23**, 331-343 (1991).

3. Wright, J.S., Canad. J. Chem. **51**, 139-146 (1973).

4. Murrell, J.N., Carter, S., Farantos, S.C., Huxley, P. and Varandas, A.J.C., "Molecular Potential Functions", John Wiley & Sons Ltd., 1984.

5. Burton, P.G., J. Chem. Phys. **71**, 961-972, (1979).

6. Bates, D.R., J. Chem. Phys. **93**, 8739-8744 (1990).

7. Ogg, R.A., Jr. and Sutphen, W.T., Discuss. Faraday Soc. **17**, 47-54 (1954).

8. "Chemical Kinetics and Photochemical Data for Use in Stratospheric Modelling", Evaluation no. 9, JPL-publication 90-1. NASA Jet Propulsion Laboratory, 1990.

9. Anderson, S.M., Klein, F.S. and Kaufman, F., J. Chem. Phys. **83**, 1648-55 (1985).

10. Mauersberger, K., Geophys. Res. Lett **8**, 935-937 (1981).

11. Morton, J., Schueler, B. and Mauersberger, K., Chem. Phys. Lett. **154**, 143-145 (1989).

12. Morton, J., Barnes, J., Schueler, B. and Mauersberger, K., J. Geophys. Res. **95**, 901-907 (1990).

13. Abbas, M.M., Guo, J., Carli, B., Mencaraglia, F., Carlotti, M. and Nolt, I.G., J. Geophys. Res. **92**, 231-239 (1987).

14. Goldman, A., Murcray, F.J., Murcray, D.G., Kosters, J.J., Rinsland, C.P., Flaud, C.P., Camy-Peyret, C. and Barbe, A., J. Geophys. Res. **94**, 8467-8473 (1989).

15. Heidenreich, J.E. III, Thiemens, M.H., J. Chem. Phys. **84**, 2129-2136 (1985).

16. Cicerone, R.J. and McCrumb, J.L., Geophys. Res. Lett. **7**, 251-254 (1980).

17. Kaye, J. and Strobel, D.F., J. Geophys. Res. **88**, 8447-8452 (1983).

18. Kaye, J., J. Geophys. Res. **91**, 7864-7874 (1986).

19. Hippler, H., Rahn, R., Troe, J., J. Chem. Phys. **93**, 6560-6569 (1990).

RECEIVED May 11, 1992

Chapter 12

Negative-Ion Formation by Rydberg Electron Transfer
Isotope-Dependent Rate Constants

Howard S. Carman, Jr., Cornelius E. Klots, and Robert N. Compton

Chemical Physics Section, Health and Safety Research Division, Oak Ridge National Laboratory, Oak Ridge, TN 37831−6125

The formation of negative ions during collisions of rubidium atoms in selected ns and nd Rydberg states with carbon disulfide molecules has been studied for a range of effective principal quantum numbers ($10 \le n^* \le 25$). For a narrow range of n^* near $n^* = 17$, rate constants for CS_2^- formation are found to depend upon the isotopic composition of the molecule, producing a negative ion isotope ratio (mass 78 to mass 76, amu) up to 10.5 times larger than the natural abundance ratio of CS_2 isotopes in the reagent. The isotope ratio is found to depend strongly upon the initial quantum state of the Rydberg atom and perhaps upon the collision energy and CS_2 temperature.

In recent years there has been rapid growth in the study of the properties and collision dynamics of atoms in highly excited (Rydberg) states (*1*). Collisions involving these "Rydberg atoms" are important in many processes which occur in high-energy environments such as the interstellar medium, plasmas, and combustion flames. In addition to their importance in such processes, Rydberg atoms have also been recently used as tools in the laboratory to probe low-energy electron-molecule interactions and, in particular, low energy electron attachment to molecules (*2*). In a recent publication (*3*) we reported an unusually large isotope dependence of rate constants for CS_2^- ion formation during collisions between Cs (ns, nd) Rydberg atoms and CS_2 molecules. For a narrow range of effective principal quantum number near $n^* = 17$ we found that rate constants for formation of $^{32}S^{12}C^{34}S^-$ ions were up to 4.5 times greater than those for $^{32}S^{12}C^{32}S^-$ formation. Preliminary measurements using jet-cooled CS_2 molecules suggested that the ratio of rate constants for the two isotopes was also dependent upon the CS_2 temperature (*3*). The origin of the isotope dependence, however, remains unexplained. In this paper we present the results of a recent study of the isotope dependence of rate constants for CS_2^- formation during collisions of Rb (ns, nd) Rydberg atoms with CS_2 molecules.

Rydberg Electron Transfer (RET). Several unique properties of Rydberg atoms (summarized in Table I) make them quite useful as tools for probing low-energy electron-molecule interactions. When one electron in an atom is excited to an orbital of large principal quantum number, n, its average distance from the nucleus and inner electrons (together referred to as the "core") is much larger than the range of interaction between a charged particle and a neutral molecule.

Table I. Properties of Hydrogen-like Rydberg Atoms

Property	n-dependence	n=1	n=25	n=50
Bohr radius r_n (Å)	$n^2 a_o$	0.53	331	1325
rms velocity of electron v_{rms} (cm s^{-1})	v_o/n	2.2×10^8	8.8×10^6	4.4×10^6
binding energy E_n (eV)	R/n^2	13.6	2.2×10^{-2}	5.4×10^{-3}

Therefore, during a collision between a highly excited Rydberg atom and a neutral molecule, the excited electron and the core behave as independent particles. The interaction between the Rydberg electron and a molecule is then essentially that of a free electron interacting with a molecule. This idea forms the basis for the "free electron model" which has been used to theoretically model Rydberg atom-molecule collisions (4,5).

One advantage of using Rydberg atoms to probe electron-molecule interactions is that the root-mean-square velocity, v_{rms}, of the Rydberg electron decreases as n increases (see Table I). Thus at large n, it is possible to produce electrons with subthermal velocities, which is not feasible with alternative methods. Since the binding energy of the electron to the core also decreases rapidly as n increases ($\propto 1/n^2$), it is easily removed from the atom during collisions and may attach to a colliding molecule, providing a method for studying very low-energy electron attachment processes.

In general, for Rydberg electron transfer (RET) reactions of the type

$$Ry(n,l) + AB \rightarrow Ry^+ + AB^- \qquad (1)$$
$$\rightarrow Ry^+ + A + B^- \quad ,$$

it has been shown (2) that, for large values of n ($n \geq 25$), rate constants for negative ion production agree well with rate constants for free electron attachment and are in accord with the "free electron model." However, for lower values of n, the Rydberg electron and ion core no longer behave as independent particles and the "free electron model" is no longer applicable. In these cases, interactions between the nascent negative and positive (Rydberg core) ions formed by RET become important and may greatly affect the rates for ion production (6-10). In particular, for thermal collision energies, the strong Coulombic attraction between the nascent positive and negative ions (which becomes more important at lower n) may prevent the ions from separating, resulting in a rapid decrease in the rate constants for ion production as n decreases (6-10). More interesting perhaps are several recent studies which have shown that such interactions may also lead to stabilization of negative ions which otherwise have very short autodetachment or dissociation lifetimes, including CS_2 (3,10,11), O_2 (10,12), CH_3NO_2 (13), HI (13), and DI (13). The CS_2 molecule is especially interesting in this regard.

Carbon Disulfide. The electron affinity of CS_2 is known to be positive and the CS_2^- ion has been observed by many workers using a variety of techniques (14-20). Using photodetachment electron spectroscopy, Oakes and Ellison (21) recently determined the electron affinity of CS_2 to be 0.895 ± 0.020 eV. The neutral CS_2 molecule has the electronic configuration ... $(5\sigma_g)^2(4\sigma_u)^2(6\sigma_g)^2(5\sigma_u)^2(2\pi_u)^4(2\pi_g)^4$ and is linear in its $^1\Sigma_g^+$ ground state ($D_{\infty h}$ symmetry). The lowest unoccupied orbital of CS_2 is a doubly degenerate π_u orbital in the linear molecule. However, due to vibronic coupling (the Renner-Teller effect), the degeneracy of the π_u orbital is lifted as the molecule bends and it evolves into two nondegenerate orbitals (a_1 and b_1) for the bent configuration (C_{2v} symmetry) (22). The a_1 orbital is lowered in energy relative to the π_u orbital (23) and therefore the ground state of CS_2^- is expected to be bent with a bond angle of ~132° (24). A negative ion produced by capture of an electron into the π_u orbital will therefore be very short-lived (on the order of a bending vibrational period) unless energy is somehow removed from the ion to stabilize it in a bent configuration. Measurements of electron attachment rate constants for CS_2 in the presence of various buffer gases (25) are consistent with this notion. The measured attachment rates are well-described in terms of a two-step three-body (Bloch-Bradbury) attachment mechanism (25),

$$e^- + CS_2 \rightarrow CS_2^{-*}$$
$$CS_2^{-*} + M \rightarrow CS_2^- + M + energy ,$$
(2)

where a third body (buffer gas, M) collisionally stabilizes the short-lived temporary negative ion.

Several groups have previously studied Rydberg electron transfer to carbon disulfide (3,10,11). During collisions of K (nd) Rydberg atoms (10 < n < 20) with CS_2, Kalamarides et al. (11) observed production of both long-lived CS_2^- ions and free electrons. These authors suggested that free electron production resulted from Rydberg electron capture into short-lived CS_2^- states which underwent rapid autodetachment, as discussed above. The observation of long-lived CS_2^- ions suggested that, for the intermediate values of n studied, a fraction of the nascent CS_2^- ions may be stabilized by energy transfer with the Rydberg core before autodetachment occurs. Since the average Bohr radius of the Rydberg electron is proportional to n^2 (see Table I) the nascent ions are formed in closer proximity at lower n and stabilization is expected to become more efficient. In this case the Rydberg core acts as a "built-in" third body during the collision. In accord with this hypothesis, the Rice group (11) found that rate constants for production of stable CS_2^- ions increased as n was decreased from 18 to 12. However, a significant fraction of the long-lived CS_2^- ions were found to undergo field-induced detachment in moderate (1-3 kV cm^{-1}) electric fields (11). These results suggested that perhaps more than one mechanism for production of long-lived ions was important and that ions in different electronic or vibrational states may be produced (11). Harth et al. (10) measured rate constants for both Ne$^+$ and CS_2^- formation during collisions between Ne (ns, nd) Rydberg atoms and CS_2. For large values of the effective principal quantum number ($n^* \geq 25$) they found that the rate constants for formation of long-lived CS_2^- ions were much smaller than those for Ne$^+$ formation, indicating that CS_2 was indeed capturing the Rydberg electron but that the resulting negative ion was very short-lived. However as n^* was decreased, the rate constant for CS_2^- production increased rapidly, becoming comparable to that for Ne$^+$ formation for $n^* < 25$. These results provided further evidence that the nascent negative ions could be stabilized by energy transfer with the Rydberg core.

Although these previous studies have provided some insight into the dynamics of CS_2^- formation during RET, the isotope dependence observed for production of CS_2^- is still not understood. Since it is clear from these previous studies that energy transfer between the negative ion and the Rydberg core are necessary for production of long-lived CS_2^- ions, we have studied the isotope dependence of CS_2^- formation rates using rubidium Rydberg atoms, which have a core configuration different from that of cesium.

Experimental

The experimental apparatus has been described in detail previously (3,6) and is only briefly described here. Rubidium atoms in a collimated effusive beam (~240°C) were excited to ns and nd Rydberg levels in a field-free region (defined by two parallel mesh grids) by resonant two-photon absorption using the outputs of two independently tunable nitrogen-pumped dye lasers (Molectron). The first laser was fixed at the $5S \rightarrow 5P_{3/2}$ transition (780.2 nm) of atomic Rb while the second laser was tuned to a specific $5P_{3/2} \rightarrow nS$ or nD transition. The excited Rb beam was crossed at 90° with a beam of CS_2 containing naturally occurring isotopic abundances. Two different methods were used to produce the CS_2 beam. A glass capillary array was used to produce an effusive beam (cw) of CS_2 at room temperature. Alternatively, a pulsed nozzle expansion of neat CS_2 was collimated to produce a pulsed beam of jet-cooled CS_2. When using the pulsed nozzle (Lasertechnics), the time delay between the opening of the nozzle and the firing of the lasers could be varied in order to sample different regions of the gas pulse. In addition, the pulsed nozzle could be heated up to a temperature of ~75°C. After the lasers were fired, collisions were allowed to occur for a specified time (typically 0.5 - 10 μs) after which a voltage pulse was applied to one of the grids to accelerate the positive or negative ions present in the collision zone into the drift tube of a 1.5 meter time-of-flight mass spectrometer (TOFMS). At the exit of the TOFMS the ions were further accelerated into a dual microchannelplate multiplier and detected. The output signal from the multiplier was averaged over several hundred laser pulses and recorded using a 200 MHz transient digitizer (DSP Technology) controlled by a laboratory computer (Compaq 286 DeskPro) via a CAMAC interface. Negative ion isotope ratios were determined by integrating each peak in the mass spectrum and determining the ratios of the peak areas.

Results

Typical negative ion mass spectra obtained following collisions of Rb (nd) Rydberg atoms with a room temperature effusive beam of CS_2 are shown in Figure 1. As was found previously for Cs Rydberg atoms (3), the ratio of the intensity for mass 78 ions to that for mass 76 ions is dependent upon the initial state of the Rydberg atom. The concentrations of the various CS_2 isotopes in the reagent gas are in the same proportion as their natural abundances. (This was

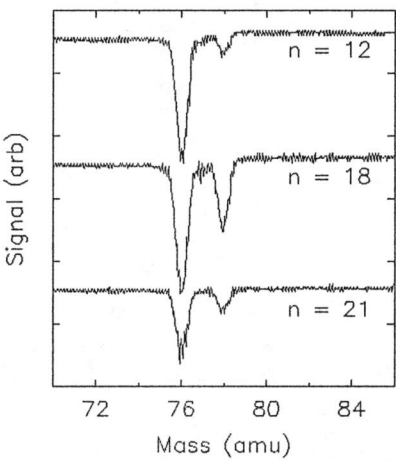

Figure 1. Negative ion mass spectra obtained for several nd states of Rb with a room temperature effusive beam of CS_2. The plots have been shifted vertically for clarity.

verified by an independent mass spectral analysis of the reagent sample). Therefore, ratios of negative ion intensities which differ from the natural abundance ratio indicate that the rate constants for CS_2^- formation are different for CS_2 molecules of different isotopic composition. Figure 2 shows the negative ion intensity ratio, R = I(mass 78)/I(mass 76) as a function of the effective principal quantum number, n^*, for both the ns and nd states of Rb. For $10 < n^* < 23$, R deviates significantly from the value expected based upon the natural abundances of isotopes with these masses (R = 0.089). For $n^* \approx 17$, the rate constant for production of mass 78 ions is ~4.8 times larger than that for production of mass 76 ions. The results shown in Figure 2 for Rb are nearly identical with our previous results for Cs (3), which were also obtained for a room-temperature effusive beam of CS_2.

As we mentioned in our previous report (3), the negative ion isotope ratios observed when CS_2 was expanded in a nozzle jet depended strongly upon the jet expansion conditions, suggesting that the isotope effect may depend upon either the internal temperature of the CS_2 or upon the collision energy (or both). We have previously demonstrated (6) that rate constants for negative ion formation during RET are sensitive to the collision energy for $Cs(n,l)$ - SF_6 collisions. Our nozzle jet experiments with CS_2 were complicated, however, by the presence of CS_2 clusters in the jet which also capture Rydberg electrons to produce negative ions (26). Figures 3 and 4 show negative ion mass spectra obtained for the 18D state of Rb with a nozzle jet expansion of neat CS_2 and with different delay times, t_d, between the opening of the nozzle and the firing

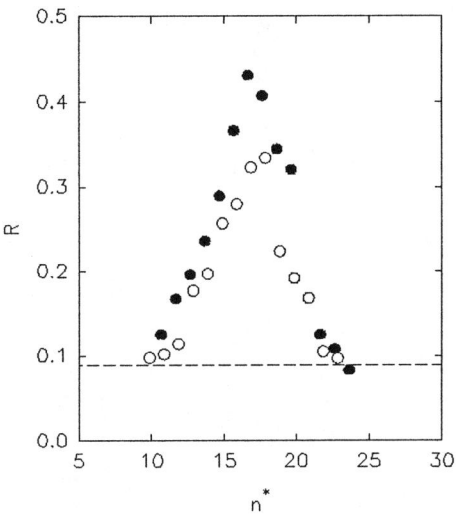

Figure 2. Negative ion intensity ratio, R, vs. n^* for the ns (○) and nd (●) states of Rb and an effusive beam of CS_2. The dashed line shows the ratio expected based upon natural abundances.

of the lasers used to produce the Rydberg atoms. By varying t_d it was possible to sample regions of the jet expansion which had undergone differing amounts of collisional relaxation and cooling. Shorter delay times (Figure 4) sample the leading edge of the jet pulse where fewer collisions (and therefore less cooling and clustering) have occurred during the expansion. As can be seen in the

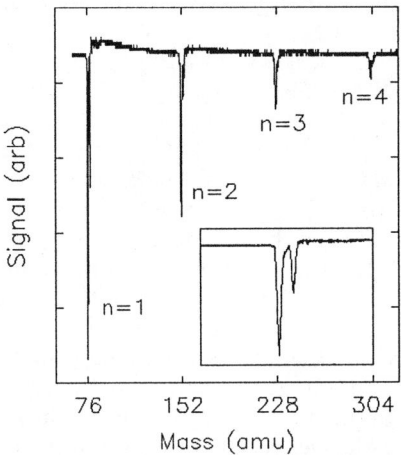

Figure 3. Negative ion mass spectrum obtained for the 18D state of Rb and a jet expansion of CS_2 (t_d = 350 μs). $(CS_2)_n^-$ ions (n=1-4) are evident. Inset shows an expanded view near mass 76 and mass 78 peaks.

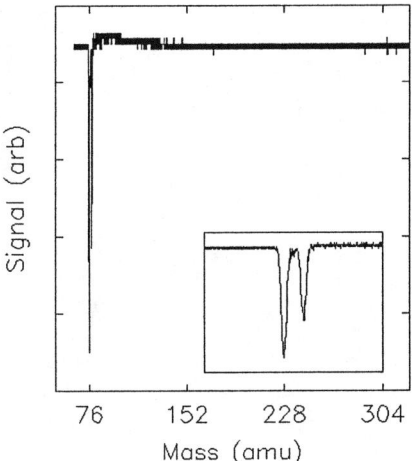

Figure 4. Negative ion mass spectrum obtained for the 18D state of Rb and a jet expansion of CS_2 (t_d = 210 μs). Inset shows an expanded view near mass 76 and mass 78 peaks.

figures, the number of cluster ions observed depended strongly upon which region of the jet was being probed. It is interesting to note that the isotope ratio, R, also depended upon t_d: R = 0.68 and 0.47 for t_d = 210 μs and 350 μs, respectively. However, the fact that dissociative attachment to clusters of CS_2 may also lead to CS_2^- formation precludes any definitive correlation of R with the internal temperature of CS_2. We cannot distinguish between CS_2^- ions which have been formed by dissociative attachment to clusters and those formed directly by attachment to CS_2.

Negative ion mass spectra were also obtained for different temperatures of the pulsed nozzle, T_n, while keeping t_d fixed (~250 μs) such that no cluster ions were observed. Under these conditions, R was found to increase linearly from 0.55 at T_n = 25°C to 0.88 at T_n = 70°C for the 18D state of Rb. Figure 5 shows a plot of R vs. n^* for T_n = 70°C. Qualitatively, the results are similar to those obtained for a room temperature effusive beam of CS_2. However, the maximum deviation of R from the natural abundance ratio is much larger for the 70°C jet source than for the room temperature effusive source, reaching a maximum of 0.94 for $n^* \approx 18$. The rate constant for production of mass 78 ions is thus ~10.5 times larger than that for production of mass 76 ions under these conditions. Comparison of Figures 2 and 5 suggests that, indeed, R is dependent upon either the collision energy or the internal energy of CS_2. Unfortunately, direct measurements of the velocity distributions or the rotational and vibrational state populations of molecules in the jet were not possible at the time of these measurements. Such measurements will be necessary before any definitive conclusions can be reached about the effects of collision energy or CS_2 internal energies on the negative ion isotope ratios. Another test would be to vary the temperature of the effusive source, which produces a beam of molecules with

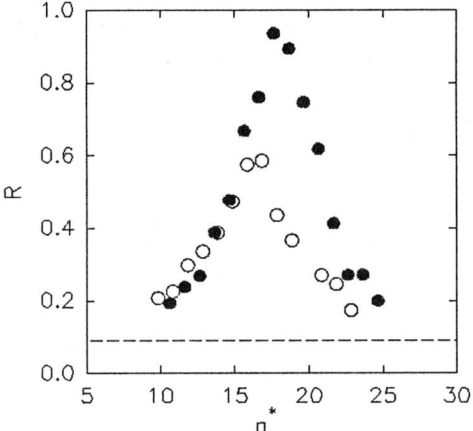

Figure 5. R vs. n^* for the ns (○) and nd(●) states of Rb with CS_2 expanded from a 70°C pulsed nozzle (t_d=250μs). The dashed line shows the ratio expected based upon natural abundances.

well-characterized (Boltzmann) distributions of rotational, vibrational, and translational energies. However such a study was not possible with the effusive source used in these experiments.

Discussion

The fact that stable CS_2^- ions can be produced by Rydberg electron capture but not by free electron capture is easily rationalized by invoking energy transfer between the nascent (short-lived) negative ion and the Rydberg core. The origin of the isotope dependence for the rates of CS_2^- formation, however, remains a mystery. Although there are twenty stable isotopes of CS_2, our sensitivity limits us to observations of the two most abundant isotopes: $^{32}S^{12}C^{32}S$ (89.25% natural abundance) at mass 76 and $^{32}S^{12}C^{34}S$ (7.93% natural abundance) at mass 78. Two additional mass 78 isotopes exist ($^{33}S^{12}C^{33}S$, <0.01% and $^{32}S^{13}C^{33}S$, 0.02%) but we assume, based upon the natural abundances, that $^{32}S^{12}C^{34}S$ is the only significant contributor to the signal at mass 78.

Isotope effects for *nonionizing* processes involving Rydberg atoms have been previously reported (27,28). Gallagher et al. (27) found that cross sections for quenching of the ns states of Na during collisions with CH_4 and CD_4 were strongly dependent upon n and that the n-dependence was different for the two molecules. The quenching cross sections were greatly enhanced by resonant electronic-to-vibrational energy transfer when the energy difference between two vibrational levels of the molecule was nearly equal to the energy difference between two electronic levels of Na. Since the vibrational level spacings for CH_4 and CD_4 are different due to the isotopic substitution, the resonance energies for the two molecules are matched by different $nl \rightarrow n'l'$ transitions of Na. Similar results were reported by Petitjean et al. (28) for Rb(ns, nd) + NH_3, ND_3 collisions. While these studies demonstrated isotope effects for deexcitation (quenching) of Rydberg states, it is feasible that such effects may also occur for excitation (including ionization) of Rydberg atoms, whereby molecular rotational or vibrational energy is transferred to electronic energy of the atom. However, Kalamarides et al. (11), based upon measurements of n-changing cross sections for K(nd)-CS_2 collisions, concluded that transfer of molecular vibrational or rotational energy was insignificant for these collisions. Nevertheless it is quite interesting to note that, for $n^* \approx 17$ (where we observed the largest isotope effect), the electron binding energy for Rb (~380 cm^{-1}) is very nearly equal to one quantum of energy for the bending vibration for CS_2: $\omega_2 = 395.99$ cm^{-1} and 395.08 cm^{-1} for $^{32}S^{12}C^{32}S$ and $^{32}S^{12}C^{34}S$, respectively (29). The total cross section for CS_2^- formation also peaks near $n^* \approx 17$ (3, 10). Since the ground state of CS_2 is linear and the ground state of CS_2^- is bent, is seems likely that the bending vibration may play an important role in the electron capture process. However, due to the fact that the bending vibrational frequencies of the two isotopes are nearly equal, it seems unlikely that a resonant energy transfer mechanism involving the bending mode would produce the large isotope effect we observe.

While chemical isotope effects in general are related to the mass of a particle through its effect on the Hamiltonian for the system (kinetic isotope effect), several examples are known of isotope effects which arise due to parity or symmetry constraints and are not directly correlated with the mass difference of the isotopes. Several examples of such isotope effects are discussed elsewhere in this volume. Valentini (*30*) has recently shown that such parity and symmetry constraints may result in isotope-dependent selection rules for nonadiabatic transitions involving linear molecules which contain atoms with isotopes of zero nuclear spin located in two equivalent positions (e.g., O_2, CO_2, CS_2). Specifically, he showed that when such molecules posses a Σ electronic state which is coupled via a nonadiabatic transition (curve crossing) with a state of nonzero electronic angular momentum (e.g., Π or Δ), the nonadiabatic transition rate will be twice as great for non-symmetric molecules (different isotopes in equivalent positions) than for symmetric molecules (identical isotopes in equivalent positions). The basis for this isotope dependence lies in the selection rules associated with the parity of the rotational wave functions of the molecule (*30*). For the Σ state of the symmetric molecule, half of the rotational levels are missing due to the fact that their wave functions do not possess the required symmetry with respect to interchange of the identical boson nuclei. In contrast, all rotational states are allowed for the Σ state of the non-symmetric molecule and for the non-Σ states of both the symmetric and the non-symmetric molecules. The dynamical effects of these symmetry-related selection rules were elegantly demonstrated by Valentini *et al.* (*30,31*) for the photodissociation of ozone. Similar selection rules may apply to collisions of Rydberg atoms with CS_2.

For the $^1\Sigma^+$ ground state of CS_2, only even-J rotational states are allowed for the $^{32}S^{12}C^{32}S$ isotope whereas all J values are allowed for $^{32}S^{12}C^{34}S$. In its linear configuration, the ground state of the CS_2^- ion has Π_u electronic symmetry with all J values allowed for both $^{32}S^{12}C^{32}S^-$ and $^{32}S^{12}C^{34}S^-$. During a collision with a Rydberg atom, the rates for nonadiabatic transitions between states correlating with the neutral and ionic states of the CS_2 molecule may therefore be isotope-dependent. However, the Rydberg atom-CS_2 system is quite complex, with a number of intersecting potential energy surfaces. As a result, the system is difficult to model, especially when the vibrational motion of the molecule is considered. It is therefore not clear at present if symmetry-related dynamical constraints can account for the large isotope dependence observed for the rates of CS_2^- formation.

More than likely, several different factors are contributing to the isotope effect. As can be seen in Figures 2 and 5 and in Figure 1 of reference (*3*), the values of the negative ion isotope ratio, R, are consistently smaller for the *ns* states of the alkali atom compared to the *nd* states. This observation suggests that the symmetry of the Rydberg atom states must also be considered and that conservation of angular momentum for the total system may present dynamical constraints which are isotope-dependent. Further experimental and theoretical studies are necessary to provide more insight into the dynamics of these interesting reactions. Experiments with isotopically enriched CS_2 (especially of the symmetric and non-symmetric heavier isotopes) are currently being planned in our group. Also planned are experiments using stimulated emission pumping to prepare vibrationally and rotationally state-selected CS_2 molecules. Methods for determining the translational energy and state populations of CS_2 in the jet beam are currently being explored.

Acknowledgments

This research was sponsored by the Office of Health and Environmental Research, U.S. Department of Energy under contract DE-AC05-84OR21400 with Martin Marietta Energy Systems, Inc. Useful discussions with Professors R.F. Curl, Jr. and J.J. Valentini are gratefully acknowledged.

Literature Cited

(1) *Rydberg States of Atoms and Molecules*; Stebbings, R.F.; Dunning, F.B., Eds.; Cambridge University Press: New York, NY, 1983.
(2) Dunning, F.B. *J. Phys. Chem.* 1987, *91*, 2244 and references therein.
(3) Carman, Jr., H.S.; Klots, C.E.; Compton, R.N. *J. Chem. Phys.* 1990, *92*, 5751.
(4) Matsuzawa, M. in Reference (*1*), p. 267.
(5) Hickman, A.P.; Olson, R.E.; Pascale, J. in Reference (*1*), p. 187.
(6) Carman, Jr.,H.S.; Klots, C.E.; Compton, R.N. *J. Chem. Phys.* 1989, *90*, 2580.
(7) Zollars, B.G.; Walter, C.W.; Lu, F.; Johnson, C.B.; Smith, K.A.; Dunning, F.B. *J. Chem. Phys.* 1986, *84*, 5589.
(8) Zheng, Z.; Smith, K.A.; Dunning, F.B. *J. Chem. Phys.* 1988, *89*, 6295.
(9) Beterov, I.M.; Vasilenko, G.L.; Riabtsev, I.I.; Smirnov, B.M.; Fateyev, N.V. *Z. Phys. D* 1987, *6*,55.
(10) Harth, K.; Ruf, M.-W.; Hotop, H. *Z. Phys. D* 1989, *14*, 149.
(11) Kalamarides, A.; Walter, C.W.; Smith, K.A.; Dunning, F.B. *J. Chem. Phys. 1988, 89*, 7226.
(12) Walter, C.W.; Zollars, B.G.; Johnson, C.B.; Smith, K.A.; Dunning, F.B. *Phys. Rev. A* 1986, *34*, 4431.
(13) Carman, Jr., H.S., Klots, C.E., Compton, R.N. unpublished data.
(14) Compton, R.N.; Reinhardt, P.W.; Cooper, C.D. *J. Chem. Phys.* 1975, *63*,3821.
(15) Tang, S.Y.; Rothe, E.W.; Reck, G.P. *J. Chem. Phys.* 1974, *61*, 2592.
(16) Caldwell, G.; Kebarle, P. *J. Chem. Phys.* 1984, *80*, 577.
(17) Hughes, B.M.; Lifshitz, C.L.; Tiernan, T.O. *J. Chem. Phys.* 1973, *59*, 3162.
(18) MacNeil, K.A.C.; Thynne, J.C.J. *J. Phys. Chem.* 1969, *73*, 2960.
(19) Dillard, J.G.; Franklin, J.L. *J. Chem. Phys.* 1968, *48*, 2349.
(20) Kraus, K; Muller-Duysing, W.; Nuert, H.Z. *Z. Naturforsch. A* 1961, *16*, 1385.

(21) Oakes, J.M.; Ellison, G.B. *Tetrahedron* **1986**, *42*, 6263.
(22) Herzberg, G. *Molecular Spectra and Molecular Structure*; D. Van Nostrand Company, Inc.: Princeton, NJ, 1966; Vol. III, pp 23-37.
(23) Walsh, A.D. *J. Chem. Soc.* **1953**, 2266.
(24) Benz, A.; Leisin, O.; Morgner, H.; Seiberle, H.; Stegmaier, J. *Z. Phys. A* **1985**, *320*, 11.
(25) Wang, W.C.; Lee, L.C. *J. Chem. Phys.* **1986**, *84*, 2675.
(26) Kondow, T.; Mitsuka, K. *J. Chem. Phys.* **1985**, *83*, 2612.
(27) Gallagher, T.F.; Ruff, G.A.; Safinya, K.A. *Phys. Rev. A* **1980**, *22*, 843.
(28) Petitjean, L.; Gounand, F.; Fournier, P.R. *Phys. Rev. A* **1986**, *33*, 1372.
(29) Jolma, K.; Kauppinen, J. *J. Mol. Spectrosc.* **1980**, *82*, 214.
(30) Valentini, J.J. *J. Chem. Phys.* **1987**, *86*, 6757.
(31) Valentini, J.J.; Gerrity, D.P.; Phillips, D.L.; Nieh, J.-C.; Tabor, K.D. *J. Chem. Phys.* **1987**, *86*, 6745.

RECEIVED September 4, 1991

Experimental Studies of Isotope Effects in Reactions of Ionic Systems

Chapter 13

Isotope Effects in the Reactions of Atomic Ions with H_2, D_2, and HD

Peter B. Armentrout

Department of Chemistry, University of Utah, Salt Lake City, UT 84112

> Reactions of various atomic ions with H_2, D_2, and HD have been studied as a function of kinetic energy by using guided ion beam mass spectrometry. For exothermic reactions, the dependence on translational and rotational energy and the effect of angular momentum conservation are illustrated. For endothermic reactions, the observed behavior falls into several distinct groups (statistical, direct and impulsive) that can be used to characterize the potential energy surfaces for the reactions. The characteristic behavior of each of these groups is illustrated and then used to understand more complex reaction systems.

Of all the systems where isotope effects might be observed, the simplest is that of atomic species with H_2, D_2, and HD, reactions 1-4.

$$A^+ + H_2 \longrightarrow AH^+ + H \tag{1}$$

$$A^+ + D_2 \longrightarrow AD^+ + D \tag{2}$$

$$A^+ + HD \begin{cases} \longrightarrow AH^+ + D & (3) \\ \longrightarrow AD^+ + H & (4) \end{cases}$$

In our laboratory, such reactions for the atomic ions of 44 different elements have now been studied (1). A wide variety of different types of reactivity are displayed by these systems, but several unifying themes are found. Among these are the observation that the inter-molecular and intramolecular isotope effects fall into several distinct categories. Here, we illustrate such behavior and review its origins.

A unique aspect of these studies is that the reactions are studied over a broad range of kinetic energies. The kinetic energy dependence provides a more complete evaluation of the origins of the isotope effects observed. It also allows the characterization of isotope effects for endothermic reactions, processes that have not been studied in as much detail as reactions accessible at thermal energies.

We do not discuss charge transfer or dissociative charge transfer reactions of A^+ with H_2, D_2, or HD, processes that can compete directly with reactions 1-4. For most elements, however, these reactions occur only at high energies (since the ionization energy of H_2 exceeds that of A) and consequently are not influential in the reaction dynamics of reactions 1-4. This is also true in borderline cases such as A = Ar and N where we have studied these charge transfer channels (2,3). In only a few systems (A = He, Ne, and F) are the charge transfer processes strongly exothermic. In these cases, a complete understanding of the interactions of A^+ with dihydrogen should include a consideration of the charge transfer processes (4).

Experimental Method

The experimental technique used in our laboratory to examine the reactions of atomic ions with hydrogen is guided ion beam tandem mass spectrometry, as detailed elsewhere (2). In this instrument, ions are formed in one of several available sources that enable the populations of different electronic states of the atomic ion to be manipulated. In all cases, the results discussed below correspond to a single electronic state, usually the ground state, of the atomic ion. These ions are extracted from the source and focused into a 60° magnetic sector for mass analysis. The mass-selected beam is decelerated to a kinetic energy that can vary from ~0.05 eV to over 500 eV and is focused into an rf octopole ion beam guide (5) that passes through a collision cell containing the neutral reactant. The collision zone is designed so that reactions occur over a well-defined path length and at a pressure low enough that all products are the result of *single* ion-neutral encounters, as verified by pressure dependence studies. The octopole helps ensure efficient collection of both product and reactant ions by containing them until they are extracted and focused into a quadrupole mass filter. After mass analysis, ions are detected by using a secondary electron scintillation ion detector (6) and counted by using standard pulse counting electronics.

The absolute intensities of the reactant and product ions as a function of the ion kinetic energy in the laboratory frame are converted to absolute reaction cross sections as a function of the kinetic energy in the center-of-mass frame, $\sigma(E)$, as described previously (2). Conversion of the laboratory ion energy to the center-of-mass frame energy involves a simple mass factor (except at very low energies where truncation of the ion beam must be accounted for) (2). The absolute zero of energy is determined by a retarding potential analysis that is facilitated by the use of the octopole beam guide.

Exothermic Reactions

Intermolecular Isotope Effects. For most atomic ions, we find that the total cross sections for reactions 1, 2 and the sum of reactions 3 and 4 are very similar, although small differences can be observed. For instance, when $A^+ = O^+(^4S)$, the cross section for reaction 1 is 19% larger than that for reaction 2 and 12% larger than that for the sum of reactions 3 and 4 (7). While these differences do fall within our absolute experimental error of ±20%, they are reproducible and fall outside of our estimated relative error of ±5%. Such intermolecular isotope effects have not been explained.

A much more severe and interesting exception to the norm is the case where $A^+ = Kr^+(^2P_{3/2})$ (8). Here, the cross section for reaction 1 is 43% larger than that for reaction 2 and 50% *smaller* than that for the sum of reactions 3 and 4. The origins of this unusual result are not well characterized although we speculate on several possibilities elsewhere (8).

Intramolecular Isotope Effects. Effect of Translational Energy. When reactions 3 and 4 are exothermic, the competition between them usually shows a fairly strong dependence on translational energy. At low kinetic energies, the systems yield about 50 ± 10% of the AH^+ product which then gradually increases with energy. The results for $A^+ = O^+(^4S)$ shown in Figure 1 are typical (7). At higher energies, the behavior can be understood in terms of the models developed for endothermic reactions, as described below.

As discussed in detail previously (7,9-12), the explanation for the intramolecular isotope effect observed at low energy lies in the fact that the attractive ion-induced dipole interaction exerts a torque on the HD molecule since the center of polarizability (which is at the center of the molecule) is displaced from the center of mass (which is nearer the D atom). This force rotates the H atom toward the incoming O^+ ion, and thus H is more likely to be abstracted in the reaction. Detailed calculations by Dateo and Clary (12) verify this, as can be seen in Figure 1 by the good agreement between their theoretical and our experimental results. (Any deviations are within the experimental error in the energy scale.) Their results show that as the kinetic energy increases, the maximum impact parameter that can lead to a reactive collision decreases; and as the impact parameter decreases, the torque exerted on the HD molecule increases, thereby enhancing the probability that hydrogen is abstracted.

Orbital Angular Momentum Conservation. Above about 0.3 eV, the theoretical predictions no longer describe the experimental work, Figure 1. In addition, at about this same energy, the reaction cross sections for reactions 1-4 cease to follow the predictions of the Langevin-Gioumousis-Stevenson (LGS) model for ion-molecule reactions (13), namely $\sigma_{LGS} = \pi e(2\alpha/E)^{1/2}$, where e is the charge on the electron, α is the polarizability of the neutral reactant, and E is the kinetic energy of the reactants. We have shown that these deviations can be explained in terms of the conservation of orbital angular momentum (which couples the entrance and exit channels) (7). Such an exit channel effect was not included in the theoretical calculations.

To see the origins of this effect, we consider the general case where reactants with relative velocity v and reduced mass μ evolve to products with similar quantities denoted by primes. The orbital angular momentum of the reactants is $L = \mu v b = (2\mu E)^{1/2} b$ and that for the products is $L' = (2\mu' E')^{1/2} b'$, where b is the impact parameter. For hyperthermal kinetic energies, the rotational angular momentum of the reactants, J, is small compared with L. As a first approximation, we further assume that the rotational angular momentum of the products, J', is also small, and consequently, angular momentum conservation requires that $L \approx L'$. This leads to the relationship that $b \approx b'(\mu'E'/\mu E)^{1/2}$, and thus to an expression for the reaction cross section, $\sigma_X = \pi b^2 \approx \pi b'^2(\mu'E'/\mu E)$, or upon applying the LGS criterion for reaction in the exit channel, $\sigma_X = \pi e(2\alpha'/E')^{1/2}(\mu'E'/\mu E)$.

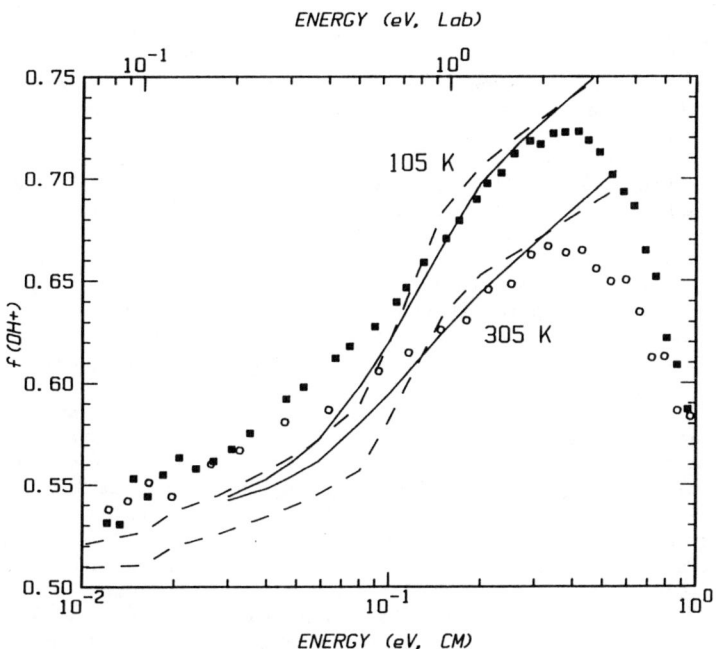

Figure 1. Fraction of OH^+ formed in reactions 3 and 4 ($f(OH^+) = \sigma(3)/[\sigma(3) + \sigma(4)]$) with $A^+ = O^+(^4S)$ as a function of kinetic energy in the lab frame (upper axis) and center-of-mass frame (lower axis) for HD temperatures of 305 K (open circles) and 105 K (closed squares). The dashed lines are calculated values from ref. 12 and the solid lines are the calculated values convoluted over the experimental energy distribution. Reproduced with permission from ref. 16. Copyright 1990. Elsevier Science Publishers

We now need to address whether the reaction cross section is limited by σ_X or by σ_{LGS}. Comparison of these two expressions shows that σ_X is the smaller of the two as long as $(\alpha'E'/\alpha E)^{1/2}(\mu'/\mu) < 1$. For reactions 1-4, μ' (which is approximately the mass of the neutral product, H or D) is necessarily less than μ (which is approximately the mass of the neutral reactant, H_2, D_2 or HD), and $\alpha'(H,D) \approx 0.67$ Å^3 (14) is slightly less than $\alpha(H_2,D_2,HD) \approx 0.80$ Å^3 (15), such that these terms favor a reaction cross section limited by σ_X. Physically, this result indicates that because the products of reactions 1-4 have smaller polarizabilities and reduced masses than the reactants, it is difficult for the products to conserve orbital angular momentum. The only way that this restriction can be overcome is if the (E'/E) term in the inequality is large enough to overcome the factors introduced by the polarizabilities and reduced masses. This is possible for exothermic reactions since the maximum value of E' is simply $E + \Delta H$, where ΔH is the exothermicity, and thus, $(E'/E) = 1 + \Delta H/E$.

At this point, the results of this analysis are most readily seen by considering a specific system, such as the reaction of $O^+ + H_2$ which is exothermic by 0.54 eV. Here, $\mu = 1.79$ amu and $\mu' = 0.95$ amu, such that $(\alpha'/\alpha)^{1/2}(\mu'/\mu) = 0.49$. Substituting these various quantities into the inequality above, we find that the cross section is limited by σ_{LGS} when $E < 0.17$ eV and by σ_X above this energy. For reactions 2-4, similar limits ranging from 0.12 to 0.43 eV are obtained (7). These predictions are in qualitative agreement with the observations that above these energies the cross sections for reactions 1-4 fall below σ_{LGS} (7) and the theory of Dateo and Clary no longer accurately models the branching ratio, Figure 1.

Effect of Rotational Energy. One interesting observation that we have made recently is a direct comparison of the effects of rotational and translational energy on the branching ratio in reactions 3 and 4 when $A^+ = O^+(^4S)$ (16). In our experiments, the rotational energy of the HD reactant is changed by altering its equilibrium temperature. Figure 1 shows that higher rotational energy results in a decrease in the fraction of OH^+ produced, exactly the opposite effect that increased kinetic energy has. As demonstrated by the calculations of Dateo and Clary, this is because the orientation effect discussed above is stronger for lower rotational states, and thus stronger at lower HD temperatures.

Endothermic Reactions

While fewer studies of the isotope effects of endothermic reactions have been made, our studies of reactions 1-4 have actually concentrated on such processes, which are simpler to understand in many respects than exothermic reactions. Effects due to zero point energy differences are readily observed and we find that the isotope effects for many atomic ions that undergo endothermic reactions 1-4 fall into one of three broad categories: statistical, direct, or impulsive behavior. These are outlined and illustrated below.

Zero Point Energy Effects. In all cases that we have studied where reactions 1-4 are endothermic, it is possible to measure the consequences of the different endothermicities of reactions 1-4. These differences arise due to the zero point energy differences between H_2,

D_2, HD, AH^+ and AD^+. These differences are such that $D°(D_2) = D°(H_2)$ + 77 meV and $D°(HD) = D°(H_2)$ + 35 meV. For most A^+, $D°(AD^+) \approx D°(AH^+)$ + 45 ± 10 meV. An example where the effect that this has on the reaction cross section is noticeable is the near-thermoneutral reaction of $N^+(^3P)$. In this system, reaction 1 is believed to be endothermic by 18 ± 2 meV at 0 K (17,18). Reactions 2-4 are therefore endothermic by 46, 54, and 4 meV, respectively. Our results for these four reactions are shown in Figure 2 (3). These cross sections do not exhibit explicit thresholds at the calculated endothermicities because there are also contributions to the energy from rotation of the hydrogen molecules and the translational energy distributions of both reactants. However, it can be seen that the two reactions for which the endothermicities are relatively low have cross sections that vary approximately as $E^{-1/2}$ (and hence are nearly linear on this log-log plot), while the two reactions that have higher endothermicities have cross sections that increase less rapidly as the kinetic energy is decreased. (For a more detailed characterization of these reaction cross sections, see reference 3.) This behavior is a direct consequence of the different endothermicities of these four reactions.

Statistical Behavior. A good example of one type of isotopic behavior is provided by Si^+ (and other group 14 atomic ions as well). Figure 3 shows results for reaction of ground state $Si^+(^2P)$ with D_2 and HD. The reaction is known to be endothermic by about 1.25 eV, consistent with the observed threshold (19). (For reactions 1-4, sharp features in the cross sections like the threshold are broadened by the kinetic energy distributions of both reactants. The effects of this broadening can be accounted for explicitly (2), thus revealing a true threshold in excellent agreement with literature thermochemistry.) Cross sections for reactions 1 and 2 have similar energy dependences and absolute magnitudes (within ~20%). Overall, the intermolecular isotope effects are small. In the HD system, the branching ratio between reactions 3 and 4 is nearly 1:1 at lower energies, but reaction 3 clearly dominates at the higher energies where the cross section declines, Figure 3. Table I lists other atomic ions that exhibit such statistical behavior.

Low Energies. The observation that reactions 3 and 4 have similar cross sections at low kinetic energies is the effect expected for a statistically behaved system, i.e. if all degrees of freedom are in equilibrium. This can be seen by examining the density of states for the products of reactions 3 versus 4. The following arguments presume that the mass of A greatly exceeds the mass of H and D, a reasonable approximation for all but the lightest elements. For vibrations, the density of states in the classical limit is $1/\hbar\omega$. Since $\omega \propto 1/m^{1/2}$, where m is the reduced mass of the diatomic product [$m(AD^+) \approx 2$ and $m(AH^+) \approx 1$], this favors AD^+ by a factor of $2^{1/2}$. The classical density of rotational states is $1/hcB$ where $B \propto 1/m$ such that AD^+ is favored by a factor of 2. The density of translational states is proportional to $\mu^{3/2}$, where μ is the reduced mass of the reactant or product channel [$\mu(AD^+ + H) \approx 1$ while $\mu(AH^+ + D) \approx 2$]. This favors AH^+ by a factor of $2^{3/2}$. Overall, these factors cancel such that the classical statistical isotope effect is about 1:1 formation of AH^+ and AD^+. While this simple treatment ignores quantum effects, it does capture the essence of a statistically behaved system as verified by more detailed calculations using classical phase space theory (20).

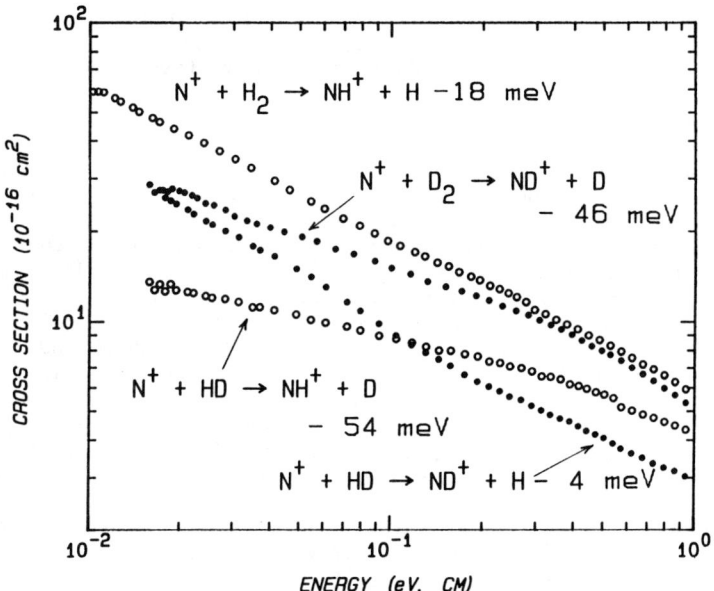

Figure 2. Cross sections for reactions 1-4 with $A^+ = N^+(^3P)$ as a function of the ion kinetic energy in the center-of-mass frame.

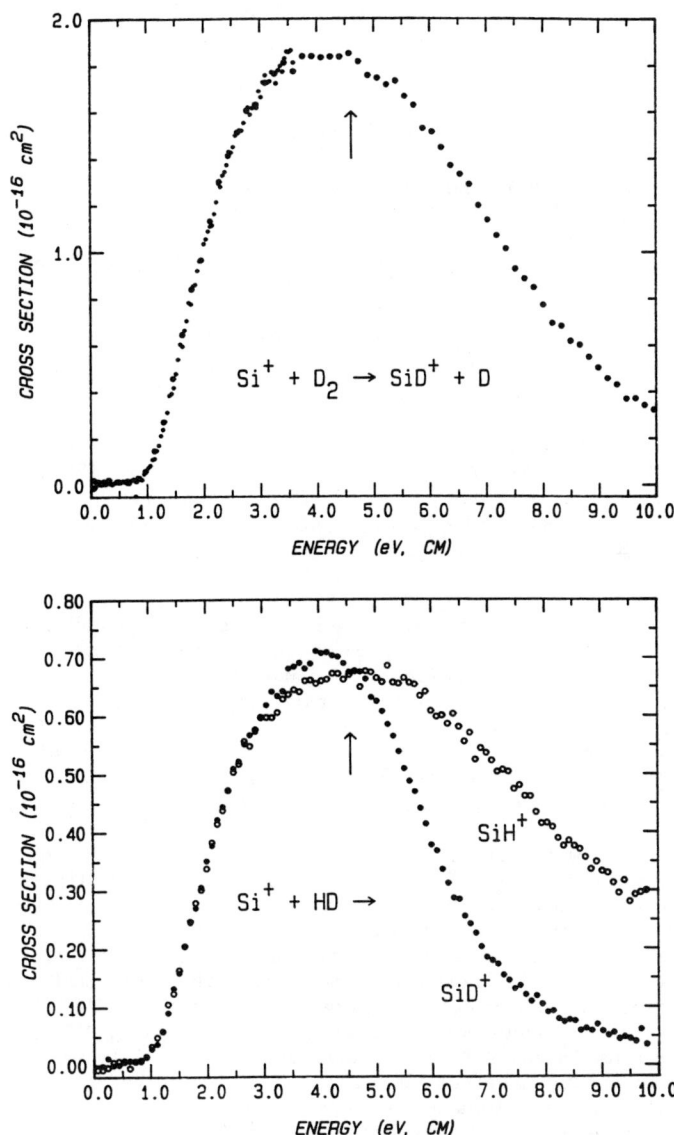

Figure 3. Cross sections for reactions 2-4 with $A^+ = Si^+(^2P)$ as a function of the ion kinetic energy in the center-of-mass frame. Arrows indicate the bond dissociation energies of D_2 and HD at ~4.5 eV.

Table I. Characteristic HD Intramolecular Isotope Effects

Category[a]	Reactant Ions
Exothermic	$O^+(^4S)$,[b] $F^+(^3P)$,[c] $Ar^+(^2P)$,[d] $Kr^+(^2P)$[e]
Statistical	$C^+(^2P)$,[f] $Si^+(^2P)$,[g] $Ge^+(^2P)$,[h] $Sn^+(^2P)$,[h] $Ti^+(^4F)$,[i] $V^+(^5D)$[j]
Direct	$Ca^+(^2S)$,[k] $Ti^+(^2F)$,[i] $V^+(^3F)$,[j] $Cr^+(^4D,^4G)$,[l] $Mn^+(^5S,^5D)$[m] $Fe^+(^4F)$,[n] $Co^+(^3F)$,[o] $Ni^+(^2D)$,[o] $Cu^+(^1S)$[o]
Impulsive	$He^+(^2P)$,[p] $Ne^+(^2P)$,[p] $Xe^+(^2P)$[q] $Mn^+(^7S)$,[m] $Fe^+(^6D)$,[n] $Ni^+(^4F)$,[o] $Zn^+(^2S)$[k]
Mixed	$B^+(^1S)$,[h] $Al^+(^1S)$,[h] $N^+(^3P)$,[r] $P^+(^3P)$,[h] $S^+(^4S)$,[s] $Cr^+(^6S)$[l]

[a]See text for definitions. [b]Ref. 7. [c]Lin, K.-C.; Cotter, R. J.; Koski, W. S. *J. Chem. Phys.* 1974, *61*, 905. [d]Ref. 2. [e]Ref. 8. [f]Ref. 21. [g]Ref. 19. [h]Ref. 1. [i]Elkind, J. L.; Armentrout, P. B. *Int. J. Mass Spectrom. Ion Processes* 1988, *83*, 259. [j]Ref. 27. [k]Ref. 25. [l]Elkind, J. L.; Armentrout, P. B. *J. Chem. Phys.* 1987, *86*, 1868. [m]Elkind, J. L.; Armentrout, P. B. *J. Chem. Phys.* 1986, *84*, 4862. [n]Elkind, J. L.; Armentrout, P. B. *J. Phys. Chem.* 1986, *90*, 5736. [o]Elkind, J. L.; Armentrout, P. B. *J. Phys. Chem.* 1986, *90*, 6576. [p]Ervin, K. M.; Armentrout, P. B. *J. Chem. Phys.* 1987, *86*, 6240. [q]Ref. 30. [r]Ref. 3. [s]Ref. 32.

In general, we have observed statistical behavior when the generation of a stable AH_2^+ species is feasible; that is, when the reaction occurs via *insertion* of A^+ into the H_2 bond. The comparison between C^+ and Si^+ serves to illustrate the range of AH_2^+ stabilities that still allow statistical behavior. In the case of C^+, reactions 1-4 are endothermic by only ~0.4 eV and the CH_2^+ intermediate lies in a potential well that is very deep, 4.3 eV below the reactants (*21*). Such a situation clearly can lead to a long-lived AH_2^+ intermediate, as also demonstrated by crossed beam studies (*22*). In contrast, reactions 1-4 with Si^+ are endothermic by ~1.25 eV and the SiH_2^+ potential well is only ~0.8 eV deep (*23*). Such conditions may not lead to an intermediate that survives for a rotational period, but the statistical behavior shown in Figure 3 demonstrates that energy randomization occurs in such a system nevertheless.

Clearly, as the potential well for the intermediate becomes shallower or as the endothermicity increases, the ability for such randomization to occur must decrease and eventually disappear. The energy regions in which this shift from statistical to alternate types of behavior occurs have not been adequately characterized, although we have found that reactions of Ge^+ and Sn^+, endothermic by ~1.6 and ~2.3 eV, respectively, still exhibit statistical intramolecular isotope behavior (*24*).

High Energies. The cross sections for reactions 1-4 decline at higher energies due to dissociation of the product ion in process 5.

$$A^+ + H_2 \; (D_2, HD) \longrightarrow A^+ + H(D) + H(D) \qquad (5)$$

This dissociation can begin at the bond energy of the hydrogen molecule, $D°(H_2) = 4.5$ eV, and occurs when the product diatom has more internal energy than its bond dissociation energy. Differences in the density of states of the two product ions can again be used to explain why the cross section for SiD^+ decreases more rapidly than that for SiH^+. A statistical distribution of energies between translation and internal degrees of freedom places more energy in internal modes of SiD^+ than in SiH^+, and thus the probability of its dissociation is relatively higher at any given collision energy.

Direct Behavior. A different type of isotope effect is illustrated by the reactions of $Ca^+(^2S)$. Figure 4 shows results for reaction of this ion with H_2 and HD (25). While the thermochemistry of CaH^+ is not well-established, the observed reaction threshold is at an energy consistent with theoretical calculations (26). Overall, intermolecular isotope effects are small since cross sections for reactions 1 and 2 have similar energy dependences and absolute magnitudes (within ~20%). In contrast to a statistical 1:1 branching ratio between reactions 3 and 4, this system exhibits a branching ratio that is close to 4:1 until $D°(HD)$ where it increases further (consistent with the high energy behavior discussed above). Table I lists other atomic ions that show similar behavior, typically with process 3 being favored by a factor of 3-4. In general, such direct behavior is observed when generation of a stable AH_2^+ species is not feasible, but rather, the reaction occurs via *abstraction* of H from H_2 by A^+.

The analysis of statistical behavior showed that the internal density of states favors process 4. Thus, instead of assuming that the internal and translational degrees of freedom equilibrate fully, we assume that the reaction is no longer sensitive to the density of internal energy states. This would be true if the reaction were not long-lived, but *direct*. According to the analysis above, the translational density of states favors production of AH^+ by a factor of $2^{3/2} = 2.8$, consistent with the experimentally observed branching ratio. The reason that these reactions continue to be sensitive to the translational but not the internal density of states can be understood by again considering the consequences of angular momentum conservation.

Orbital Angular Momentum Conservation. An alternative way of thinking about the intramolecular isotope effect observed for direct reactions is in terms of orbital angular momentum conservation (27-29). As found above, these arguments lead to an expression for the reaction cross section, σ_X, that becomes the limiting cross section if $(\alpha'E'/\alpha E)^{1/2}(\mu'/\mu) < 1$. As noted above, α' and μ' are less than α and μ for reactions 1-4, and if the reactions are endothermic, then E' must be less than E. Thus, for endothermic reactions, the inequality is true at all kinetic energies, and therefore the entrance channel impact parameter that can lead to products is *always* limited by angular momentum conservation in the exit channel.

This result can now be used to predict the intramolecular isotope effect for reaction of A^+ with HD. The ratio of the cross sections for reactions 3 and 4 (indicated by single and double primes, respectively) is given by $\sigma(AH^+)/\sigma(AD^+) = \sigma_X'/\sigma_X'' = (\alpha'E'/\alpha''E'')^{1/2}(\mu'/\mu'')$. Since $\alpha'(H) \approx \alpha''(D)$ and $\mu' \approx 2\mu''$, $\sigma'/\sigma'' \approx 2(E'/E'')^{1/2}$. The ratio E'/E'' can

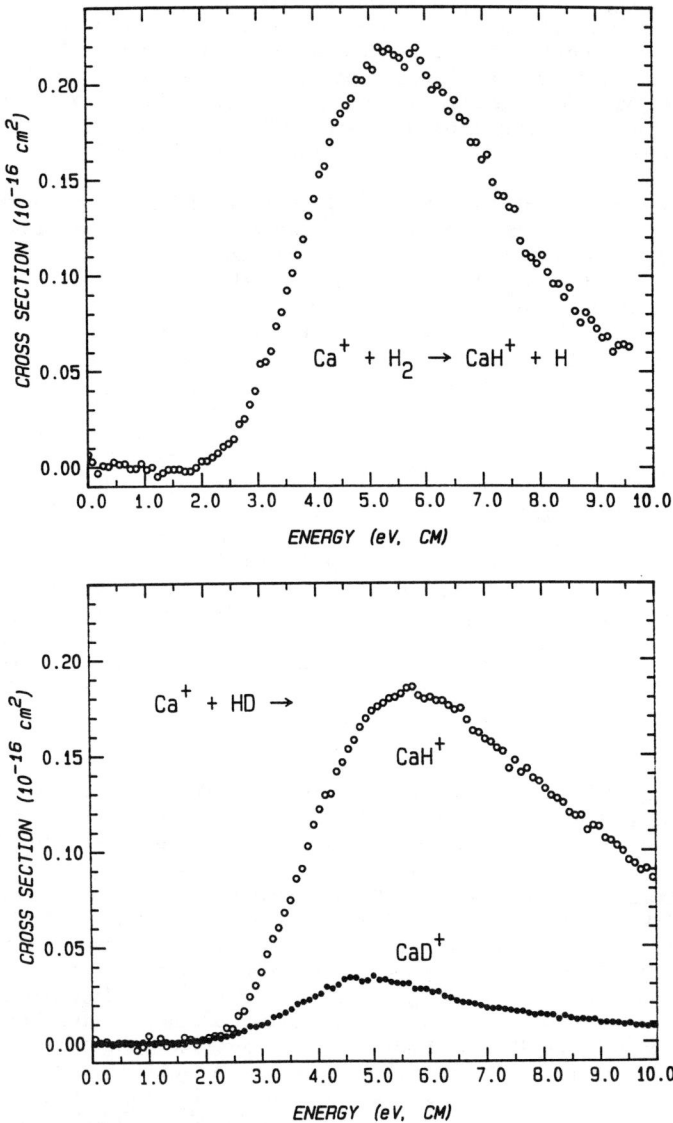

Figure 4. Cross sections for reactions 1, 3, and 4 with A^+ = $Ca^+(^2S)$ as a function of the ion kinetic energy in the center-of-mass frame. Adapted from ref. 25.

conceivably range from ~1 (if the energy release is the same for both channels) to ~2.8 (if energy of the products tracks with the density of translational states). (Experimentally, the observation that the CaD^+ cross section declines more rapidly at higher energies than the CaH^+ cross section directly shows that $E' > E''$.) Thus, the branching ratio in the threshold region is predicted to range between 2 and 3.4, again consistent with experimental observation.

Impulsive Behavior. A third class of intramolecular isotope effect is illustrated by the reaction of $Xe^+(^2P_{3/2})$ *(30)*. Results for reactions 1 and 2 have similar energy dependences and absolute magnitudes (within ~20%), but the results for reactions 3 and 4 are clearly different, Figure 5. The threshold and peak energies for reaction 4 are lower than those for reactions 1 and 2, while these energies are higher for reaction 3. The magnitude of the cross section for reaction 4 is larger than half those for reactions 1 and 2, while that for reaction 3 is quite small. Thus, in contrast to the results for statistical and direct behavior, there are strong intermolecular isotope effects. While this behavior is unusual, it is not an uncommon one for reactions 1-4 as shown in Table I. We find that this behavior is characteristic of ions that are expected to have largely repulsive interactions with H_2.

A simple model that explains this isotopic behavior is an *impulsive* model. In the limit of a hard-sphere type of reaction, the collision will be a "pairwise" one between A^+ and closest atom of H_2, D_2, or HD. Thus, the energy relevant to reactions 1-4 is *not* the center-of-mass (CM) energy (relevant when the interation is between A^+ and the entire hydrogen molecule) but a pairwise interaction energy. In the CM system, the energy available to cause chemical change is the relative kinetic energy between the incoming atom (having mass A) and the reactant molecule BC (having mass B + C). Hence, E(CM) = E(lab)· (B + C)/(A + B + C) where E(lab) is the ion energy in the laboratory frame (and the molecule is stationary in the lab frame). In a pairwise interaction, A is sensitive only to the potential between A and the atom B which is transferred in reactions 1-4. Hence, the pairwise energy for transfer of B from molecule BC, given by equation 6,

$$E(pair) = E(lab) \cdot B/(A + B) = E(CM)(A + B + C)B/(B + C)(A + B) \quad (6)$$

is *always* less than the energy available in the CM frame. For situations like reactions 1-4 where A >> B or C, E(pair) ≈ E(CM)·B/(B + C). For reactions where BC = H_2 or D_2, the mass factor B/(B + C) is 1/2; for BC = HD, it is 1/3; and for BC = DH, it is 2/3.

For the example of Xe^+, reactions 1-4 have endothermicities of ~0.9 eV *(30)*. Thus, in reaction with H_2 or D_2, the pairwise threshold and dissociation energies will occur at approximately twice the CM energies, 1.8 and 9.0 eV, respectively. It can be seen in Figure 5 that these energies are close to the observed threshold and somewhat above the maximum in the cross section for reaction 2. For reaction 4, the threshold and dissociation energies are predicted to shift up from the CM energies by 50%, to 1.3 and 6.8 eV, respectively, again in rough agreement with the experimental cross section, Figure 5. For reaction 3, the shift is now a factor of 3 such that the predicted threshold and dissociation energies are 2.7 and 13.3 eV, respectively. These energies describe the second feature in the cross section most closely, and thus we have attributed the small low energy feature to reaction

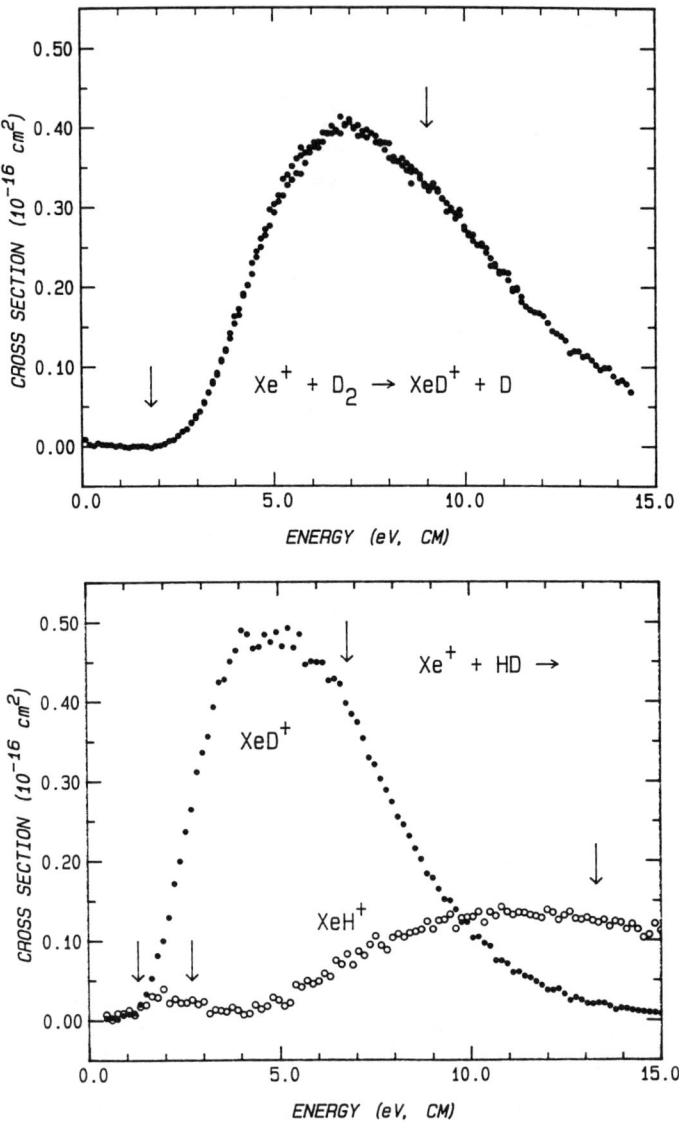

Figure 5. Cross sections for reactions 2-4 with $A^+ = Xe^+$ as a function of the ion kinetic energy in the center-of-mass frame. The reactivity shown is due primarily to $Xe^+(^2P_{3/2})$. Arrows indicate the thresholds and dissociation energies predicted by the pairwise model for all three reactions, see text. Adapted from ref. 30.

occuring via a non-impulsive pathway (30). While the absolute energy predictions of the impulsive model do not correspond directly with observation, the pairwise scheme readily explains the relative shifts in the thresholds and dissociation energies observed for the H_2, HD and D_2 systems, and suggests that the enhanced production of AD^+ in process 4 is due the much lower threshold for this reaction compared to that for reaction 3.

The pairwise energy frame may be familiar to some as the spectator stripping model (SSM) (31). The SSM is a highly specific example of a model which incorporates the pairwise energy concept. The difference is that the SSM assumes there is no momentum transfer to the product atom C, while the more general pairwise model allows such transfer. As a consequence, the SSM makes very specific predictions about the velocity and internal energy of the products, while the pairwise model allows for distributions of these quantities. The latter, not surprisingly, corresponds more closely to observation. As shown above, the extremely useful concept of a pairwise energy scale can easily be derived without the severe assumption made in the SSM.

Mixed Behavior. While the reactions of many atomic ions fall nicely into the categories described above, we have observed a number of systems that exhibit mixtures of these behaviors, Table I. An example that we have recently studied is the reactions of $S^+(^4S)$ (32). This species is isovalent with $O^+(^4S)$, but now reaction 1 is endothermic by 0.92 eV. As shown in Figure 6, the cross section for reaction 2 has two features (as does that for reaction 1). The first begins promptly at the thermodynamic threshold. When HD is the reactant, the first feature does not shift in energy and exhibits a statistical branching ratio between SH^+ and SD^+. The second, higher energy feature, however, does shift in energy and clearly favors formation of SD^+, results that correspond to an impulsive mechanism. The threshold and dissociation energies predicted by the impulsive model (calculated as for the Xe^+ example above) are indicated in Figure 6. It can be seen that the predictions match the shifts observed in the experimental cross sections reasonably well.

These results indicate that the reaction must pass through a stable SH_2^+ intermediate at low energy. However, this species has a $^2B_1$ ground state and therefore its formation from $S^+(^4S) + H_2(^1\Sigma_g^+)$ is spin-forbidden. This explains the relative inefficiency of this pathway. The second feature in the cross section can then be assigned to reaction proceeding along a spin-allowed but relatively repulsive potential energy surface. Ab initio calculations on this system confirm these qualitative ideas concerning the potential energy surfaces (32).

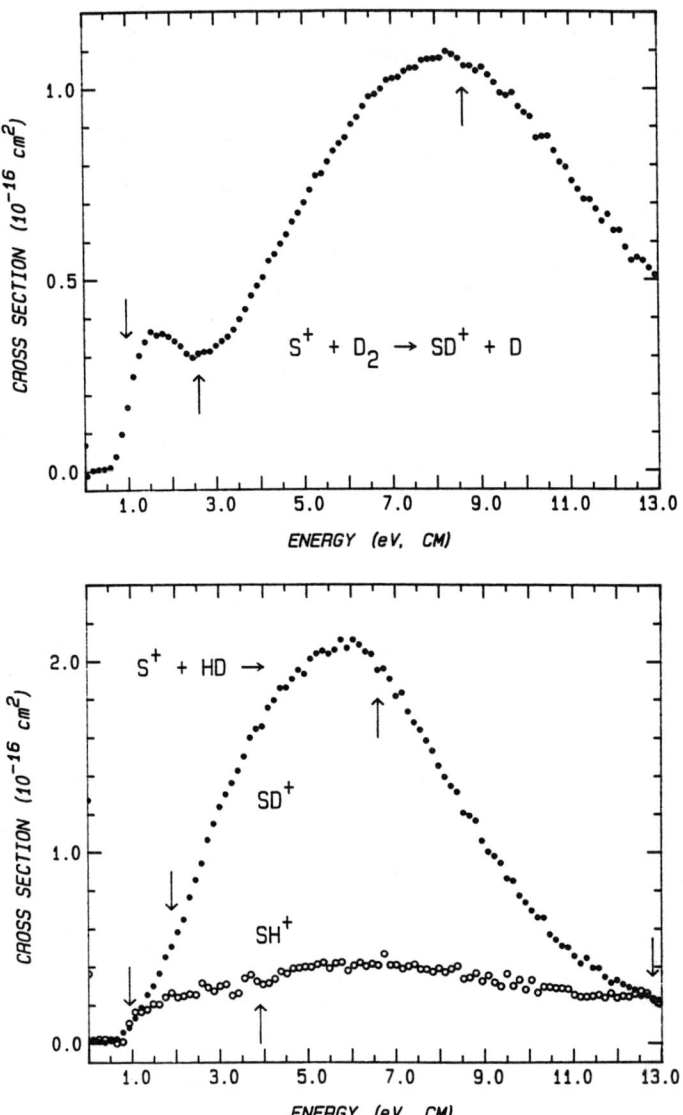

Figure 6. Cross sections for reactions 2-4 with $A^+ = S^+(^4S)$ as a function of the ion kinetic energy in the center-of-mass frame. Arrows indicate the thermodynamic thresholds at ~0.9 eV, and the thresholds and dissociation energies predicted by the pairwise model for all three reactions, see text. Adapted from ref. 32.

Acknowledgments. This work is supported by the National Science Foundation. I also thank my research collaborators for their contributions to the work discussed here.

Literature Cited

1. Armentrout, P. B. *Int. Rev. Phys. Chem.* **1990**, *9*, 115.
2. Ervin, K. M.; Armentrout, P. B. *J. Chem. Phys.* **1985**, *83*, 166.
3. Ervin, K. M.; Armentrout, P. B. *J. Chem. Phys.* **1987**, *86*, 2659.
4. See for example, Jones, E. G.; Wu, R. L. C.; Hughes, B. M.; Tiernan, T. O.; Hopper, D. G. *J. Chem. Phys.* **1980**, *73*, 5631.
5. Teloy, E.; Gerlich, D. *Chem. Phys.* **1974**, *4*, 417.
6. Daly, N. R. *Rev. Sci. Instrum.* **1959**, *31*, 264.
7. Burley, J. D.; Ervin, K. M.; Armentrout, P. B. *Int. J. Mass Spectrom. Ion Processes* **1987**, *80*, 153.
8. Ervin, K. M.; Armentrout, P. B. *J. Chem. Phys.* **1986**, *85*, 6380.
9. Light, J. C.; Chan, S. *J. Chem. Phys.* **1969**, *51*, 1008.
10. George, T. F.; Suplinskas, R. J. *J. Chem. Phys.* **1971**, *54*, 1046.
11. Hierl, P. M. *J. Chem. Phys.* **1977**, *67*, 4665.
12. Dateo, C. E.; Clary, D. C. *J. Chem. Soc. Faraday Trans 2*, **1989**, *85*, 1685.
13. Gioumousis G.; Stevenson, D. P. *J. Chem. Phys.* **1958**, *29*, 292.
14. Miller, T. M.; Bederson, B. *Adv. Atomic Molec. Phys.* **1977**, *13*, 1.
15. Rothe, E. R.; Bernstein, R. B. *J. Chem. Phys.* **1959**, *31*, 1619.
16. Sunderlin, L. S.; Armentrout, P. B. *Chem. Phys. Lett.* **1990**, *167*, 188.
17. Marquette, J. B.; Rebrion, C.; Rowe, B. R. *J. Chem. Phys.* **1988**, *89*, 2041.
18. Gerlich, D. *J. Chem. Phys.* **1989**, *90*, 3574.
19. Elkind, J. L.; Armentrout, P. B. *J. Phys. Chem.* **1984**, *88*, 5454.
20. Ervin, K. M.; Armentrout, P. B. *J. Chem. Phys.* **1986**, *84*, 6750.
21. Ervin, K. M.; Armentrout, P. B. *J. Chem. Phys.* **1986**, *84*, 6738.
22. Mahan, B. H.; Sloane, T. M. *J. Chem. Phys.* **1973**, *59*, 5661.
23. Boo, B. H.; Armentrout, P. B. *J. Am. Chem. Soc.* **1987**, *109*, 3549.
24. Elkind, J. L.; Ervin, K. M.; Armentrout, P. B. unpublished work.
25. Georgiadis, R.; Armentrout, P. B. *J. Phys. Chem.* **1988**, *92*, 7060.
26. Schilling, J. B.; Goddard, W. A.; Beauchamp, J. L. *J. Phys. Chem.* **1987**, *91*, 5616.
27. Elkind, J. L.; Armentrout, P. B. *J. Phys. Chem.* **1985**, *89*, 5626.
28. Aristov, N.; Armentrout, P. B. *J. Am. Chem. Soc.* **1986**, *108*, 1806.
29. Sunderlin, L.; Aristov, N.; Armentrout, P. B. *J. Am. Chem. Soc.* **1987**, *109*, 78.
30. Ervin, K. M.; Armentrout; P. B. *J. Chem. Phys.* **1989**, *90*, 118.
31. Henglein, A. In *Ion-Molecule Reactions in the Gas Phase*; Ausloos, P. J., Ed.; ACS: Washington, D.C., 1966; p 63.
32. Stowe, G. F.; Schultz, R. H.; Wight, C. A.; Armentrout, P. B. *Int. J. Mass Spectrom. Ion Processes* **1990**, *100*, 177.

RECEIVED October 3, 1991

Chapter 14

Non-Mass-Dependent Isotope Effects in the Formation of O_4^+
Evidence for a Symmetry Restriction

K. S. Griffith and Gregory I. Gellene

Department of Chemistry and Biochemistry, University of Notre Dame, Notre Dame, IN 46556

> Non-mass dependent isotopic enrichment of ^{17}O and ^{18}O has been observed mass spectrometrically in O_4^+ ions produced by termolecular association reactions of O_2^+ and O_2 where the O_2^+ was generated by electron ionization of O_2. The enhancement is strongly dependent on the energy of the ionizing electrons, decreasing from a near ten-fold enhancement at threshold for O_2^+ production to no enhancement above 40 eV. Additionally, O_2^+ generated near threshold was found to be significantly less efficient in producing O_4^+ than O_2^+ ions generated at higher energies. A permutation inversion symmetry analysis of the termolecular association reaction suggests that the results can be understood in terms of a symmetry restriction on the $O_2^+(^2\Pi_g)$ rotational states which can efficiently access the electronic ground state of O_4^+ upon collisions with O_2. The restriction is rooted in the Pauli principle and vanishes when the O_2^+ ion is isotopically heteronuclear. The possible relevance of the non-mass dependent isotope enhancement in O_4^+ to similar enhancements found in stratospheric and laboratory-produced ozone is discussed.

Motivated by relevance to interstellar and upper atmosphere chemistry, termolecular association reactions producing O_4^+ have been the subject of experimental and theoretical investigations for almost thirty years. In 1963, Curran (1) investigated the effect of O_2 pressure and ionizing electron energy on the rate of O_4^+ production and found the results to be consistent with the two step mechanism:

$$O_2 + e^- \rightarrow O_2^+ + 2e^- \tag{1a}$$
$$O_2^+ + 2O_2 \rightarrow O_4^+ + O_2 \tag{1b}$$

Surprisingly, the appearance potential of O_4^+ was found to be 16.9±0.1 eV, almost 5 eV greater than that of ground state O_2^+. On the basis of this observation Curran

0097–6156/92/0502–0210$06.00/0
© 1992 American Chemical Society

suggested a role for the metastable $a^4\Pi_u$ electronic state of O_2^+ in reaction (1b), although no consideration was given to the origin of the unusual requirement of electronic excitation for the production of a cluster ion. Unfortunately, no notice of the O_4^+ appearance potential seems to have been taken by other workers in subsequent kinetic studies of reaction (1b) leaving the detailed nature of the apparent electronic energy requirement unknown.

In 1973 Conway and Janik (2) using high pressure mass spectrometry determined the O_2--O_2^+ bond energy to be 0.457±0.005 eV with respect to $O_2(X^3\Sigma_g^-)$ and $O_2^+(X^2\Pi_g)$. Similar studies by Kebarle et. al. (3,4) and a photoionization study (5) confirmed this result. Until recently, this was the only experimental information available for O_4^+. However, within the last two years ESR and IR spectroscopic data have been obtained for matrix isolated O_4^+. The ESR spectrum provided clear evidence for the quartet nature of the ground state (6) and the IR spectrum was found to be compatible with a C_{2h} structure (7) in agreement with theoretical predictions (8). Taken together these results indicate a ^4B_u electronic ground state for O_4^+ (8).

Kebarle et. al. (3,4) also determined the termolecular association rate (k_{1b}) over the range of 90 - 340K and found an inverse temperature dependence described by $k_{1b} = C\ T^{-n}$ with n = 3.2. Böhringer et. al. (9) extended the kinetic measurements to 51K and observed a marked deviation from the T^{-n} proportionality below 80 K with k_{1b} leveling off and perhaps decreasing at the lowest temperatures investigated. This result prompted speculation about the existence of energetic barriers to ion-molecule association reactions (10,11) and incomplete energy randomization (12) as the origin of the maximum in k_{1b} as T is decreased. Interestingly, no deviation from the T^{-n} proportionality was observed with He as the third body (13) as in:

$$O_2^+ + O_2 + He \rightarrow O_4^+ + He \qquad (2)$$

Rowe et.al. (14) further extended the kinetic measurements to 20 K using O_2 as the third body, and although only lower limits to k_{1b} could be measured below 50 K, they found no evidence for a rate maximum at low temperature. In all of these kinetic studies, O_2^+ was produced by electron ionization (EI) under conditions where the $a^4\Pi_u$ state would be readily produced and thus these studies do not address the origin of the electronic energy requirement identified by Curran (1). However, it can be noted that the electronic energy of the $a^4\Pi_u$ state is almost ten times the cluster binding energy making unlikely the termolecular association reaction:

$$O_2^+(a^4\Pi_u) + 2O_2 \rightarrow O_4^+ + O_2 \qquad (3)$$

Very recently, Smith et. al. (15) have determined k_{1b} over the temperature range of 4 - 20 K using a free jet expansion technique. They found that all of the available kinetic data for reaction 1b could be well described by the expression $k_{1b} = C\ T^{-1.81}$ (4 K < T < 300 K) and confirmed the conclusions of Rowe et. al. (14) that there was no evidence for either a deviation from the inverse temperature rule or an energetic barrier to association. The work by Smith et. al. differed from the previous kinetic studies in that the O_2^+ ion was produced by a resonance-enhanced multiphoton ionization (REMPI) scheme which is known to produce only electronic ground state ions (*16*). Thus the results of Smith *et. al.* can be compatible with those of Curran only if REMPI and near threshold EI produce ion state populations which differ in some critical aspect affecting the kinetics of reaction 1b and this difference is preserved in "thermalizing" collisions of the ions.

The present work addresses this question by reexamining the threshold EI production of O_4^+ paying particular attention to the production of cluster ions containing a single ^{17}O or ^{18}O atom from O_2 molecules with a natural abundance isotopic distribution (0.037% and 0.204% respectively). The results confirm the apparent enhanced clustering reactivity for the $a^4\Pi_u$ ion and most significantly indicate a non-mass dependent kinetic isotope effect for ^{17}O and ^{18}O which approaches an order of magnitude near threshold for O_2^+ production (~12 eV). The results are interpreted in terms of a Pauli principle based symmetry restriction which allows only J-S = odd states of $^{32}O_2^+(^2\Pi_g)$ ion to cluster efficiently. A direct implication of the present work is that threshold EI of $O_2(^3\Sigma_g^-)$ has a strong propensity to produce J-S = even ions which are not readily interconverted to J-S = odd by collisions with O_2.

Experimental

O_2^+ ions were produced by EI of room temperature neat O_2 at an ion source pressure of 0.75 Torr. It is known that EI of O_2 produces vibrationally excited $X^2\Pi_g$ ions (*17*) as well as vibronically excited metastable $a^4\Pi_u$ ions. The efficiency of the latter process increases with ionizing electron energy and can account for as much as 50% of the ions at 80 eV (*18*). The O_2^+ ions have an ion source residence time of 0.3 - 0.4 µs in these experiments which allows for approximately 5-7 "thermalizing" collisions with O_2 assuming a Langevin collisional cross section. This is sufficient to ensure that more than 90% of $X^2\Pi_g$, v = 1, 2 and $a^4\Pi_u$ excited ions have been relaxed to the $X^2\Pi_g$, v = 0 state (*17,19*). Because of the efficient collisional relaxation of the $a^4\Pi_u$ state, termolecular association reactions of O_2^+ ions increasingly involve ions originally produced in the $a^4\Pi_u$ state as the electron energy is increased. Ions exiting the source were accelerated through 5 keV, magnetically mass resolved, and detected by a Channeltron electron multiplier operating in a single ion counting mode.

The ionizing electrons were produced from a hot Re filament located external to the ion source with total electron emission current regulated. The ionizing energy (E_e) was taken to be the nominal voltage difference between the filament and the source. Because the experimental emphasis was not on obtaining accurate appearance potential measurements, the actual electron energy distribution was not determined. Nevertheless, the distribution could not have been excessively wide because no ion signals could be observed above background signals (~0.05 s^{-1}) at E_e below the ionization threshold.

Results

The intensity of mass spectrometric counts corresponding to $^{32}O_2^+$ and $^{64}O_4^+$ (denoted I_{32} and I_{64} respectively) as a function E_e is shown in Figure 1. As E_e is increased above the O_2 threshold, I_{32} initially increases substantially faster than does I_{64} indicating that the simple kinetic description:

$$d[O_4^+]/dt = k_{1b}[O_2]^2[O_2^+] \quad (4)$$

is not valid at low E_e. This is a remarkable result considering that for $E_e < 16$ eV, no electronically excited states of O_2^+ can be produced and the vibrationally excited levels of the ground electronic state produced are efficiently relaxed to the $v = 0$ level under our ion source conditions (17). Further evidence that the ionizing conditions have a significant effect on clustering efficiency is provided by the pronounced increase in I_{64}/I_{32} at $E_e = 19\text{-}20$ eV in qualitative agreement with the results of Curran (1). As mentioned previously, the metastable $O_2^+(a^4\Pi_u)$ ions produced in the ion source are substantially relaxed to the ground state by collisions with O_2. Above $E_e = 20$ eV, I_{64}/I_{32} is largely constant and Equation 4 is valid. Thus Figure 1 indicates that under EI conditions, the rate of reaction 1b for ions arising from relaxation of the $a^4\Pi_u$ state is significantly greater than that for $O_2^+(^2\Pi_g)$ ions produced directly.

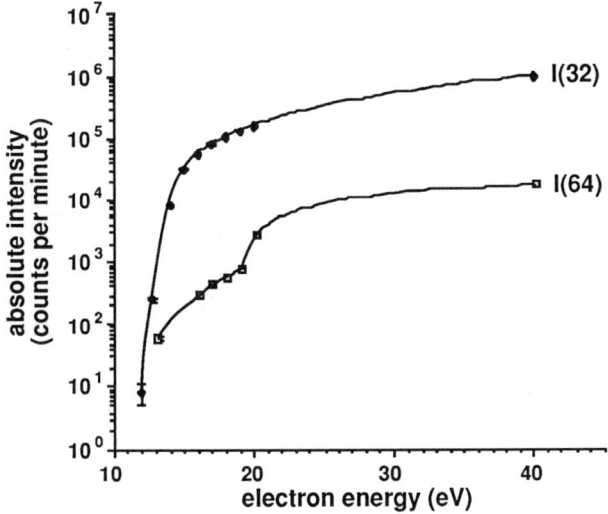

Figure 1. Intensity of O_2^+ (I(32)) and O_4^+ (I(64)) ion signal as a function of ionizing electron energy.

The intensities of O_4^+ ions containing one ^{17}O (I_{65}) or one ^{18}O (I_{66}) relative to the intensity of the all ^{16}O containing species (I_{64}) as a function of E_e is shown in Figure 2 where the dotted lines indicate the expected ratios based on natural isotopic abundance. Although the experiment does not have the mass resolution to distinguish an O_4^+ ion containing two ^{17}O atoms from one containing one ^{18}O, natural isotopic abundances predict that the contribution of the former to the m/e = 66 peak would be only 0.01%. At low E_e reaction 1b shows a strong kinetic isotope effect with the intensity of mixed isotopic O_4^+ ions approaching a ten fold enhancement relative to natural abundance expectations near threshold ($E_e \sim 12$ eV). Furthermore the enhancement is mass independent with the I_{65}/I_{66} ratio always observed to have its natural abundance value. As E_e is increased, the I_{65}/I_{64} and I_{66}/I_{64} ratios continually decrease and natural abundance values are observed for $E_e \geq 40$ eV. The most dramatic change in the I_{65}/I_{64} and I_{66}/I_{64} ratios occurs near the threshold $O_2^+(a^4\Pi_u)$ production ($E_e \sim 16$ eV) suggesting that the E_e dependance of the kinetic isotope effect in reaction 1b results from the increasing participation of O_2^+ originally produced in the $a^4\Pi_u$ as E_e is increased. Phenomenologically, the results indicate a strong kinetic isotope effect favoring mixed isotopic cluster ion formation when the $O_2^+(^2\Pi_g)$ is produced by EI directly and little or no kinetic isotope effect when the $O_2^+(^2\Pi_g)$ arises from relaxation of metastable $O_2^+(a^4\Pi_u)$.

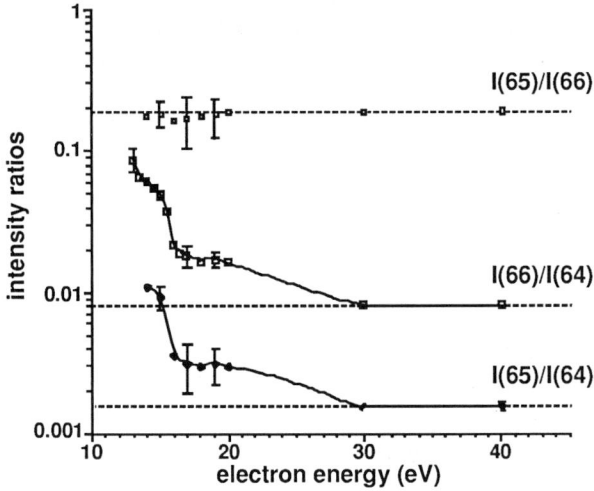

Figure 2. Intensity ratios of isotopically substituted O_4^+ ion signals as a function of ionizing electron energy. I(64) corresponds to O_4^+ ions containing four ^{16}O atoms whereas for I(65) and I(66) one ^{16}O atom is replaced with an ^{17}O and ^{18}O respectively.

Discussion

A fundamental understanding of the preceding results requires that the following three points be addressed: (1) A characteristic of the $O_2^+(^2\Pi_g)$ ions must be identified which can give rise to largely varying state specific values of k_{1b}. (2) The effect of this characteristic on k_{1b} must be strongly isotopic dependent. And (3) the population distribution of ions with regard to this characteristic must depend on the details of ion formation and must be largely preserved in a collision environment.

Termolecular association reactions are generally considered to consist of two bimolecular energy transfer reactions (20,21) which for reaction 1b can be written:

$$O_2^+(^2\Pi_g) + O_2(^3\Sigma_g^-) \underset{k_d}{\overset{k_a}{\rightleftarrows}} (O_4^+)^* \tag{5a}$$

$$(O_4^+)^* + O_2(^3\Sigma_g^-) \overset{k_s}{\rightarrow} (O_4^+) + O_2(^3\Sigma_g^-) \tag{5b}$$

where k_a, k_d, and k_s are association, dissociation, and stabilization rate constant respectively and $(O_4^+)^*$ represents a collision complex. Ordinarily all of the reactions are considered to occur on a single Born-Oppenheimer surface correlating with electronic ground state products. However there is no rigorous reason why this should be so. In particular the reactants of reaction 5a have a 24 fold electronic state degeneracy (3 for $O_2(^3\Sigma_g^-)$ times 4 for $O_2^+(^2\Pi_g)$ times 2 for ion/neutral exchange) of which the ^4B_u ground state of O_4^+ (in C_{2h} symmetry) accounts for only 4. Considering the weak binding of the O_4^+ ion (2-4), it is reasonable to assume that the remaining Born-Oppenheimer surfaces will have little or no binding and thus be characterized as repulsive with respect to the $O_2^+(^2\Pi_g) + O_2(^3\Sigma_g^-)$ asymptote. This situation for one dimensional cuts of the full Born-Oppenheimer surfaces is qualitatively depicted in Figure 3 where only a single repulsive surface is represented. If, for a particular set of

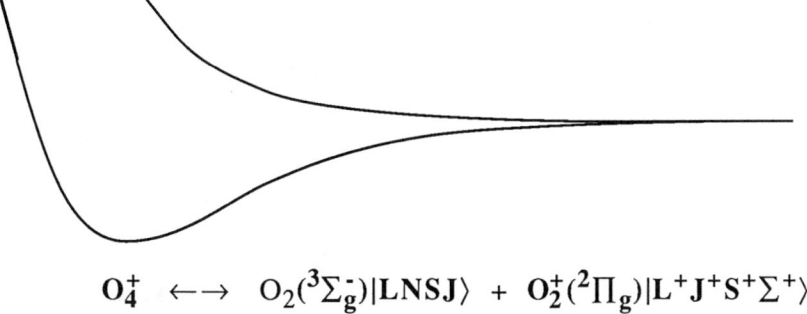

$$O_4^+ \longleftrightarrow O_2(^3\Sigma_g^-)|LNSJ\rangle + O_2^+(^2\Pi_g)|L^+J^+S^+\Sigma^+\rangle$$

Figure 3. Qualitative one dimensional potential energy surfaces illustrating the problem of correlating states of the separated diatoms with electronic states of the collision complex.

collision conditions, reaction 5a preferentially took place on one of the repulsive surfaces, then the $(O_4^+)^*$ collision complex would be much shorter lived and termolecular association would be inhibited relative to reaction 5a occurring on the binding surface. With this qualitative picture of the termolecular association process, the points raised at the beginning of the discussion can be address by a detailed consideration of the symmetry correlations of the states of the $O_2^+(^2\Pi_g) + O_2(^3\Sigma_g^-)$ super-molecule with those of the $(O_4^+)^*$ collision complex. These two limits will be referred to as O_2^+/O_2 and $(O_4^+)^*$ respectively. Because neither O_2^+/O_2 nor $(O_4^+)^*$ can be properly described by a point group, the symmetry analysis must be performed using a permutation-inversion (PI) symmetry group. However, the stabilized (O_4^+) complex can be described by the C_{2h} point group (7,8) and the correlation of PI symmetry species with those of C_{2h} will provide the connection needed to identify the ground ^4B_u state among the PI symmetry adapted electronic states.

The PI symmetry analysis can be divided into four parts. First is the development of the appropriate PI group which appears not to have been considered previously. Here only the rationale for its development will be given as the full details will be published elsewhere (22). Second, the rotational-vibrational-electronic-translational wavefunctions (Ψ_{rvet}) for O_2^+/O_2 and electronic wavefunctions for $(O_4^+)^*$ must be written in terms of an appropriate set of basis functions having well defined transformation properties under the operations of the PI group. Third, Pauli-allowed symmetry adapted Ψ_{rvet} are projected out of the set of basis functions. Finally, symmetry restrictions are determined by correlating the symmetry adapted Ψ_{rvet} with the electronic states of $(O_4^+)^*$.

PI Symmetry Analysis

The PI Group. The full permutation group of four identical nuclei (S_4) is of order 24 with 5 symmetry species. In the present application, however, it is more appropriate to employ a subgroup of S_4, denoted S_4', consisting of the energetically feasible permutations (23). It is clear that the 16 permutations of S_4 resulting in atom exchange between the diatoms is not feasible when applied to O_2^+/O_2. In addition, experimental and theoretical evidence indicates that $(O_4^+)^*$ consists of two weakly interacting diatomic subunits (7,8) between which atom exchange is also unlikely. Taking the permutation subgroup of "diatom preserving" permutations, S_4' is of order 8 with 5 symmetry species. A coordinate system suitable for use with S_4' is defined in Figure 4 and will be discussed in greater detail in the next section. In terms of the atomic labels in Figure 4, the permutation group S_4' contains the elements {E, (12), (34), (12)(34), (13)(24), (14)(23), (1324), (1423)}, where (ij), (ij)(kl), and (ijkl)

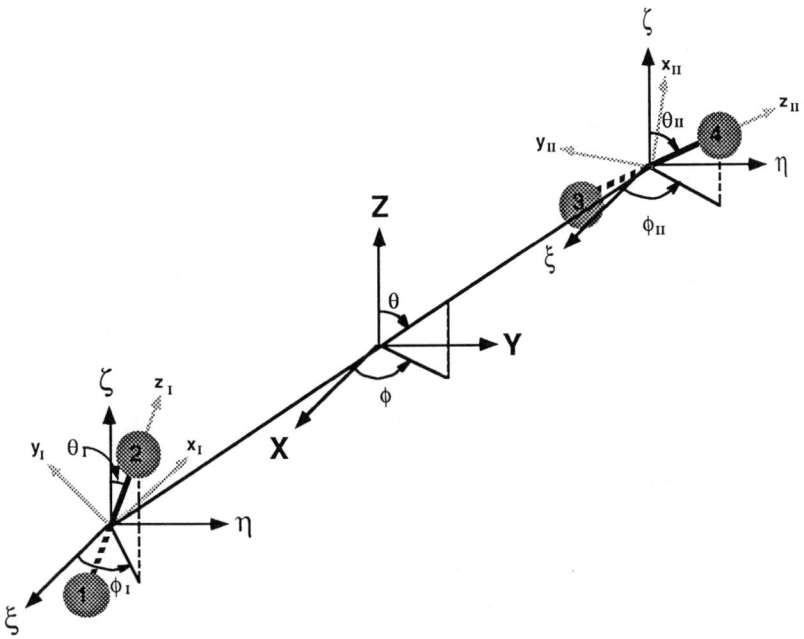

Figure 4. Coordinate system used to describe two diatoms moving with respect to each other.

denote permutation of the labeled nuclei (24). The PI group obtained by taking the direct product of S_4' and the inversion group, $\mathcal{E} = \{E, E^*\}$, is of order 16 with 10 symmetry species and will be denoted $G_{16}(A_2/A_2)$ (i.e. $G_{16}(A_2/A_2) = S_4' \otimes \mathcal{E}$).

$G_{16}(A_2/A_2)$ has been used previously for the symmetry classification of the rotational-vibrational-translational wavefunctions of a general A_2/A_2 collision system assuming that only totally symmetric singlet electronic states of A_2 are involved (25). For the present situation, an extension of that analysis to include diatomic electronic states of any symmetry or spin multiplicity is required. The spatial dependence of electronic functions is best described using a body fixed coordinate system. Although the diatoms have only two rotational degrees of freedom corresponding to the two Euler angles θ and ϕ, it is possible through the use of Hougen's isomorphic Hamiltonian (26) to introduce a third Euler angle χ_e as a dynamical variable with well defined symmetry properties in $G_{16}(A_2/A_2)$. In this way diatom electronic states of any spatial symmetry can be accommodated.

Dealing with the symmetry properties of electronic states of arbitrary multiplicity is not straightforward due to the well known problem (27) of specifying unique symmetry properties for wavefunctions having half-integral values for J (arising from half-integral total electron spin S). The problem is usually addressed by adding a fictitious operation of a rotation through 2π which is not considered the identity and requiring a rotation through 4π to return the system to itself (28). For the present situation, this approach must be extended because the coordinate system consists of two completely independent sets of coordinates, one for each diatom. A group of fictitious operations $\mathcal{R} = \{E, R_I, R_{II}, R=R_I R_{II}\}$, where the subscript indicates the particular diatom coordinate system affected by the operation is required. The PI symmetry group appropriate for classifying any electronic-vibrational-rotational-translational wavefunction of an A_2/A_2 collision system is the spin quadrupled group given by the direct product of $G_{16}(A_2/A_2)$ with $\mathcal{R}$ and is denoted $G_{16}{}^4(A_2/A_2)$ in analogy with the usual nomenclature used for spin doubled groups (i.e. $G_{16}{}^4(A_2/A_2) = S_4' \otimes \mathcal{E} \otimes \mathcal{R}$). The character table for $G_{16}{}^4(A_2/A_2)$, which is of order 64 and has 16 symmetry species, will be published elsewhere (22).

Basis Functions. Before considering Ψ_{rvet} for O_2^+/O_2 or $(O_4^+)^*$ the coordinate system of Figure 4 will be discussed. A space fixed (XYZ) Cartesian system is located with its origin at the center of mass (CM) of the whole system of four atoms and can therefore be taken to be at rest. At the CM of each of the diatoms is located a $(\xi\eta\zeta)$ Cartesian system which moves with the CM of its respective diatom but always remains parallel to the corresponding axes of the (XYZ) system. Also having its origin at the CM of each of the diatoms is a body fixed (xyz) Cartesian system which rotates with its respective diatom. The orientation of an (xyz) system relative to the $(\xi\eta\zeta)$ system with the same origin is given by Euler angles (θ,ϕ) with $\chi = 0°$. Taken together the dynamical angular variables $(\theta\phi\chi_e)$ will be abbreviated by $\hat{r}$. The coordinates associated with diatom I (atoms 1 and 2) are distinguished from the analogous coordinates associated with diatom II (atoms 3 and 4) by the subscripts I and II. For example, the internuclear separation of diatoms I and II is denoted r_I and r_{II} respectively. Unsubscripted coordinates involve both diatoms with r being the separation of CM of the two diatoms; θ the angle r makes with the Z axis; and ϕ the angle formed by the X axis and the projection of r in the XY plane.

Using these coordinates, zero order wavefunctions for O_2^+/O_2 can be written as a product of two diatomic wave functions and a relative translational wavefunction. The latter is composed of a spherical Bessel function ($\langle kr|j_L\rangle$) for the component of the translational wavefunction along r and a spherical harmonic ($\langle\theta\phi|LM\rangle$) for the rotational component. The diatomic wavefunctions can be written as products of normalized Slater determinants of $\pi_{g\pm}$ molecular orbitals, harmonic oscillator vibrational functions ($\langle r_i|v\rangle$ and $\langle r_j|v^+\rangle$ for the neutral and ion respectively), and rotational functions. Hund's case (b) rotational functions ($|\Lambda NSJ\rangle$) with $\Lambda = 0$ and $S = 1$ are used for $O_2(^3\Sigma_g^-)$ and Hund's case (a) rotational functions ($|\Lambda^+J^+S^+\Sigma^+\rangle$) with $\Lambda^+ = 1$ and $S^+ = 1/2$ are used for $O_2^+(^2\Pi_g)$ where the angular momentum quantum numbers have their usual meaning. Recalling that O_4^+ is known to be a quartet, zero order wavefunctions for the $S = 3/2$, $m_S = 3/2$ state of O_2^+/O_2 are written as:

$$\Psi_a^\pm = (1/\sqrt{2})(|\pi_{g+}^I\pi_g^I\pi_{g+}^{II}|\langle\hat{r}_{II}|\Lambda^+J^+S^+\Sigma^+\rangle \pm |\pi_{g+}^I\pi_g^I\pi_{g-}^{II}|\langle\hat{r}_{II}|-\Lambda^+J^+S^-\Sigma^+\rangle) \times$$
$$\langle\hat{r}_I|\Lambda NSJ\rangle\langle r_I|v\rangle\langle r_{II}|v^+\rangle\langle\theta\phi|LM\rangle\langle kr|j_L\rangle \qquad (6a)$$

$$\Psi_b^\pm = (1/\sqrt{2})(|\pi_{g+}^{II}\pi_g^{II}\pi_{g+}^I|\langle\hat{r}_I|\Lambda^+J^+S^+\Sigma^+\rangle \pm |\pi_{g+}^{II}\pi_g^{II}\pi_{g-}^I|\langle\hat{r}_I|-\Lambda^+J^+S^-\Sigma^+\rangle) \times$$
$$\langle\hat{r}_{II}|\Lambda NSJ\rangle\langle r_{II}|v\rangle\langle r_I|v^+\rangle\langle\theta\phi|LM\rangle\langle kr|j_L\rangle \qquad (6b)$$

In equations (6a) and (6b), the molecular orbitals each contain an α spin electron and are defined by the following linear combination of atomic $n = 2$ orbitals:

$$\pi_{g\pm}^k = (1/\sqrt{2})(P_{\pm 1}^a - P_{\pm 1}^b) \qquad (7)$$

where (a,b) denotes atomic centers (1,2) or (3,4) for $k = I$ or II respectively.

The vibrational, rotational, and translational wavefuctions used in equations (6a) and (6b) are not appropriate basis functions for the nuclear motion wavefunctions for $(O_4^+)^*$. However, considering the low binding energy of O_4^+, the diatomic molecular orbitals are taken to remain as good electronic basis functions. This is an essential assumption of the correlation scheme and allows zero order electronic wavefunctions (Φ) for $(O_4^+)^*$ correlating with (6a) and (6b) to be written as:

$$\Phi_a^\pm = (1/\sqrt{2})(|\pi_{g+}^I\pi_g^I\pi_{g+}^{II}| \pm |\pi_{g+}^I\pi_g^I\pi_{g-}^{II}|) \qquad (8a)$$

$$\Phi_b^\pm = (1/\sqrt{2})(|\pi_{g+}^{II}\pi_g^{II}\pi_{g+}^I| \pm |\pi_{g+}^{II}\pi_g^{II}\pi_{g-}^I|) \qquad (8b)$$

Pauli allowed Wavefunctions. Because ^{16}O is a boson ($I = 0$) the Pauli principle requires that an acceptable O_2^+/O_2 wavefunction must be symmetric with respect to all permutation operations although it may be either symmetric or antisymmetric with respect to inversion. A detailed consideration (22) of the effect of the operations of $G_{16}{}^4(A_2/A_2)$ on $\Psi_a^\pm$ and $\Psi_b^\pm$ shows that they are transformed into +/- themselves or +/- each other depending on the even or oddness of the angular momentum quantum numbers: N, (J+-S+), and L. The Pauli allowed O_2^+/O_2 wavefuctions for various combinations of the value of these quantum numbers is listed in the fourth column of Table I. Only odd values of N appear in Table I resulting from Pauli principle restrictions on isolated $O_2(^3\Sigma_g^-)$.

Table I. Correlation of Pauli allowed wavefunctions for $^{32}O_2^+/^{32}O_2 \to (^{64}O_4^+)^*$

N	J+-S+	L	$\Psi(O_2^+/O_2)$	$\Phi((O_4^+)^*)$	$\Gamma_\Phi(G_{16}(A_2/A_2))$	$\Gamma_\Phi(C_{2h})$
odd	even	even	$\Psi_a^- + \Psi_b^-$	$\Phi_a^- + \Phi_b^-$	E^-	$A_u \oplus B_g$
odd	even	odd	$\Psi_a^- - \Psi_b^-$	$\Phi_a^- - \Phi_b^-$		
odd	odd	even	$\Psi_a^+ + \Psi_b^+$	$\Phi_a^+ + \Phi_b^+$	A_2^+	A_g
odd	odd	odd	$\Psi_a^+ - \Psi_b^+$	$\Phi_a^+ - \Phi_b^+$	B_2^+	B_u

Electronic State Correlation. Electronic states of $(O_4^+)^*$ which correlate with each of the Pauli allowed wavefunctions of O_2^+/O_2 are given by the linear combinations of Slater determinants listed in fifth column of Table I. A consideration of the symmetry properties of these electronic wavefunctions (22) indicates that they transform as the representations of $G_{16}(A_2/A_2)$ listed in column six. The question of which of these electronic functions correlates with the electronic ground state of O_4^+, can be answered by correlating the representations of $G_{16}(A_2/A_2)$ with those of C_{2h} (i.e. the point group characterizing ground state O_4^+). The results of this correlation (22) are listed in the last column of Table I. Recalling that the ground electronic state of O_4^+ is B_u in C_{2h} symmetry, Table I indicates that O_2^+ ions in a state with J+-S+ = odd having an odd L collision with an O_2 molecule will access the ground electronic state of O_4^+ and lead to termolecular association more efficiently than any other type of collision. Although for any particular collision L can be asymptotically defined, it can not be experimentally controlled and collisions with odd and even values for L occur with equal probability. Alternatively, J+-S+ is a well defined quantum number for an O_2^+ ion, with odd and even values corresponding to e and f parity labels respectively (29). Further there is experimental and theoretical evidence for a strong propensity for conservation of e/f in J-changing collisions for some atom-diatom systems (30). Thus Table I indicates that $^{32}O_2^+(^2\Pi_g)$ ions can be thought of as existing in two different modifications having very different efficiencies for termolecular association with $^{32}O_2(^3\Sigma_g^-)$.

The preceding symmetry arguments indicate that depending on the extent of e/f production, $^{32}O_2^+(^2\Pi_g)$ ions produced by different ionizing methods may exhibit different rates of $^{64}O_4^+$ production by termolecular association. This effect may be partly responsible for the variation in k_{1b} reported by different workers (3,4,9,13-15). It remains to be shown, however, that the observed non-mass dependent kinetic isotope effects can also be understood in terms of these symmetry considerations. This question can be addressed by considering electronic state correlation in the subgroup of $G_{16}{}^4(A_2/A_2)$ which contains the permutation of nuclei on only one diatom as its sole permutation element. The subgroup which is denoted $G_4{}^4(A_2/AB)$ contains 16 elements and 10 symmetry species (22). The two cases where either the neutral or the ion is isotopically homonuclear must considered separately. The results of these symmetry analyses are summarized in Tables II and III.

Table II. Correlation of Pauli allowed wavefunctions for $^{33,34}O_2^+/^{32}O_2 \to (^{65,66}O_4^+)^*$

N	J+-S+	L	$\Psi(O_2^+/O_2)$	$\Phi((O_4^+)^*)$	$\Gamma_\Phi(G_4(A_2/AB))$	$\Gamma_\Phi(C_{2h})$
odd	even	even	$\Psi_a^+ + \Psi_b^+$	$\Phi_a^+ + \Phi_b^+$		
odd	even	odd	$\Psi_a^+ - \Psi_b^+$	$\Phi_a^+ - \Phi_b^+$	$E^- \oplus$	$A_u \oplus B_g \oplus$
odd	odd	even	$\Psi_a^- + \Psi_b^-$	$\Phi_a^- + \Phi_b^-$	$B_2^+ \oplus A_2^+$	$A_g \oplus B_u$
odd	odd	odd	$\Psi_a^- - \Psi_b^-$	$\Phi_a^- - \Phi_b^-$		

Table III. Correlation of Pauli allowed wavefunctions for $^{32}O_2^+/^{33,34}O_2 \to (^{65,66}O_4^+)^*$

N	J+-S+	L	$\Psi(O_2^+/O_2)$	$\Phi((O_4^+)^*)$	$\Gamma_\Phi(G_4(A_2/AB))$	$\Gamma_\Phi(C_{2h})$
even	even	even				
even	even	odd	$\Psi_a^- + \Psi_b^-$	$\Phi_a^- + \Phi_b^-$		
odd	even	even	$\Psi_a^- - \Psi_b^-$	$\Phi_a^- - \Phi_b^-$	E^-	$A_u \oplus B_g$
odd	even	odd				
even	odd	even				
even	odd	odd	$\Psi_a^+ + \Psi_b^+$	$\Phi_a^+ + \Phi_b^+$		
odd	odd	even	$\Psi_a^+ - \Psi_b^+$	$\Phi_a^+ - \Phi_b^+$	$B_2^+ \oplus A_2^+$	$A_g \oplus B_u$
odd	odd	odd				

The symmetry correlations in Table II indicate that if the O_2^+ ion is the heteroatomic diatom, then no symmetry restrictions occur and all combinations of N, J+-S+, and L for $^{33,34}O_2^+/^{32}O_2$ wavefunction correlate with the ground electronic state of $(^{65,66}O_4^+)^*$. Taken together Tables I and II provide a framework for understanding the major experimental observations in this study. In particular they show that a Pauli symmetry restriction can cause different clustering rates for the *e/f* parity levels of $O_2^+(^2\Pi_g)$ in the all ^{16}O system and the difference vanishes in a non-mass dependent way if the ion is isotopically heteronuclear. Interestingly, Table III indicates that if the neutral is heteroatomic, the symmetry restriction occurring in the all ^{16}O system essentially remains (i.e. only e parity levels of $O_2^+(^2\Pi_g)$ correlate with ground electronic state O_4^+). Unfortunately, the experiment is not sensitive to the origin of the unique oxygen isotope in the termolecular association so that this prediction could not be tested.

Effects of the Ionization Scheme

The dependance of the rate of reaction (1b) on ionizing conditions can be accounted for by the preceding PI symmetry analysis only if the e/f parity level ratio of the $O_2^+(^2\Pi_g)$ ions is strongly dependent on the mode of ion production. In particular, threshold EI must be highly selective for f level production relative to the e/f parity level ratio occurring with higher energy EI, REMPI, or schemes proceeding through the $a^4\Pi_u$ metastable state of O_2^+.

Currently there is very little information available concerning rotational transitions occurring in molecular EI. A detailed theoretical treatment is a formidable problem which has not been discussed in the literature. Muntz (*31*) and Coe *et. al.* (*32*), using a theoretical treatment based on the Born approximation (valid at high E_e) which included only molecular angular momentum terms, have predicted electric dipole selection rules. However, the validity of such predictions for threshold EI which involves the motion of three highly correlated particles (an ion plus two near zero kinetic energy electrons) is highly questionable (*33*). Indeed, experimental studies on N_2 (*34-36*) and CO (*37*) have shown that the selection rules begin to significantly breakdown for $E_e < 800$ eV. Although those experimental studies investigated E_e as low as 60 eV, the ion product fine structure states were not resolved and thus the results do not bear on the question of e/f parity level selectivity.

Little more can be said about e/f parity level selectivity for $O_2^+(^2\Pi_g)$ produced by collisional relaxation of EI produced $O_2^+(a^4\Pi_u)$. If the symmetry restriction derived in the previous section is accepted as the origin of unusual kinetic behavior of reaction 1b, then the results indicate that the production of e parity levels of $O_2^+(^2\Pi_g)$ is significant by this ionization scheme. However, whether the e parity levels are produced by ionization or collisional relaxation or both can not be determined from the present results.

By comparison to EI, questions of final state selectivity is much more tractable for photoionization and detail selection rules for diatomic molecules have recently been derived by Xie and Zare (*38*). In principle, these selection rules and the previous experimental study of termolecular association of REMPI produced O_2^+ (*15*) could be used to assess the validity of the symmetry restrictions derived in the previous section. Unfortunately, the state specific nature of the intermediate levels occurring in the REMPI process were not specified in that study, so that a detailed comparison is not possible.

Relationship To Heavy Ozone Production

Over the past ten years evidence for a non-mass dependent enhancement of $^{49}O_3$ and $^{50}O_3$ relative to $^{48}O_3$ in stratospheric ozone has been continually mounting (*39-41*).

The degree of enhancement appears to increase with altitude with maximum values of about 40% being reported. Analogous enhancements have also been reported for laboratory produced ozone (42-47) although the magnitudes have typically been somewhat less (~10%). Considering the termolecular association process:

$$O_2 + O \rightarrow (O_3)^* \qquad (9a)$$
$$(O_3)^* + M \rightarrow O_3 + M \qquad (9b)$$

theoretical treatments of these results have attempted to explain the reported isotope effects by incomplete energy randomization in $(O_3)^*$ (48) or non-adiabatic effects involving states of ozone correlating with $^3\Sigma_g^-$ and $^1\Delta$ states of O_2 (49). Reactions (9a) and (9b) are generally accepted to represent the most important route to atmospheric and laboratory ozone production (50), and thus previous modeling of the isotope effects have focused exclusively on neutral chemistry. In light of the large non-mass dependent isotopic enhancements observed for O_4^+, it is interesting to consider the possibility of ion-molecule chemistry transferring the enhancement to ozone. In particular the reaction:

$$O_4^+ + O \rightarrow O_3 + O_2^+ \qquad (10)$$

which could effect the isotopic enhancement transfer in a single step, proceeds efficiently at approximately half the Langevin collision rate (51). Additionally, the known switching reaction (52):

$$O_4^+ + O_3 \rightleftharpoons O_5^+ + O_2 \qquad (11)$$

may also transfer isotopic enhancement from O_4^+ to ozone depending on the structure of O_5^+.

At this point it is difficult to assess the importance of O_4^+ ion-molecule chemistry in the non-mass dependent production of stratospheric heavy ozone which will depend on altitude dependent ion concentrations, vertical mass transport rates, and other considerations of atmospheric chemistry modeling (53,54) in addition to individual chemical reaction rates. Further, some laboratory observed enhancements have occurred under conditions where ions could not have been produced (45-47), indicating an enhancement mechanism involving neutral chemistry must also exist. Nevertheless, the present results indicate that the possible contribution of reactions such as (10) and (11) to the non-mass dependent enhancement of stratospheric (and some laboratory produced) ozone is worth more detailed consideration.

Acknowledgments

K. S. G. thanks the Henry Luce Foundation for a graduate fellowship. Acknowledgment is made to the donors of the Petroleum Research Fund administered by the American Chemical Society (PRF# 22721-AC5) for support of this work.

Literature Cited

1. Curran, R. K. *J. Chem. Phys.* **1963**, *38*, 2974.
2. Conway, D. C.; Janik, G. S. *J. Chem. Phys.* **1970**, *53*, 1859.
3. Durden, D. A.; Kebarle, P.; Good, A. *J. Chem. Phys.* **1969**, *50*, 805.
4. Payzant, J. D.; Cunningham, A. J.; Kebarle, P. *J. Chem. Phys.* **1973**, *59*, 5615.
5. Linn, S. H.; Ono, Y.; Ng, C. Y. *J. Chem. Phys.* **1981**, *74*, 3348.
6. Knight, L. B. Jr.; Cobranchi, S. T.; Petty, J. *J. Chem. Phys.* **1989**, *91*, 4423.
7. Thompson, W. E.; Jacox, M. E.; *J. Chem. Phys.* **1989**, *91*, 3826.
8. Conway, D. C. *J. Chem. Phys.* **1969**, *50*, 3864.
9. Böhringer, H.; Arnold, F.; *J. Chem. Phys.* **1982**, *77*, 5534.
10. Mickens, R. E. *J. Chem. Phys.* **1983**, *79*, 1102.
11. Ferguson, E. E. *Chem. Phys. Lett.* **1983**, *101*, 141.
12. Bates, D. R. *J. Chem. Phys.* **1984**, *81*, 298.
13. Böhringer, H.; Arnold, F.; Smith, D.; Adams, N. G.; *Int. J. Mass Spectrom. Ion Phys.* **1983**, *52*, 25.
14. Rowe, B. G.; Dupeyrat, G.; Marquette, J. B.; Gaucherel, P. *J. Chem. Phys.* **1984**, *80*, 4915.
15. Randeniya, L. K.; Zeng, X. K.; Smith, R. S.; Smith, M. A. *J. Phys. Chem.* **1989**, *93*, 8031.
16. Sur, A.; Ramana, C. V.; Colson, S. D.; *J. Chem. Phys.* **1985**, *83*, 904.
17. Böhringer, H.; Durup-Ferguson, M.; Fahey, D. W.; Fehsenfeld, F. C.; Ferguson, E. E. *J. Chem. Phys.* **1983**, *79*, 4201
18. Cosby, P. C.; Moran, T. F.; Hornstein, J. V.; Flannery, M. R. *Chem. Phys Lett.* **1974**, *24*, 431.
19. Lindinger, W.; Albritton, D. L.; McFarland, M.; Fehsenfeld, F. C.; Schmeltekopf, A. L.; Ferguson, E. E. *J. Chem. Phys.* **1975**, *62*, 4101.
20. Bates, D. R. *J. Phys. B* **1979**, *12*, 4135.
21. Herbst, E. *J. Chem. Phys.* **1979**, *70*, 2201.
22. Gellene G. I. submitted for publication.
23. Longuet-Higgins, H. C. *Mol. Phys.* **1963**, *6*, 445.
24. Bunker, P. R. *Molecular Symmetry and Spectroscopy;* Academic Press Inc.: London, 1979; pp 5-19.
25. Metropoulos, A. *Chem. Phys. Lett.* **1981**, *83*, 357.
26. Hougen, J. T. *J. Chem. Phys.* **1962**, *36*, 519.
27. Herzberg, G. *Molecular Spectra and Molecular Structure;* Van Nostrand Reinhold Co.: New York, NY, 1966; Vol. 3, p 15.
28. Bethe, H. *Ann. Physik.* **1929**, *3*, 133.
29. Brown, J. M.; Hougen, J. T.; Huber, K.-P.; Johns, J. W. C.; Kopp, I.; Lefebvre-Brion, H.; Merer, A. J.; Ramsay, D. A.; Rostas, J.; Zare, R. N. *J. Mol. Spectrosc.* **1975**, *55*, 500.
30. Pouilly, B.; Alexander, M. H. *J. Chem. Phys.* **1988**, *88*, 3581.
31. Muntz, E. P. *Phys. Fluids* **1962**, *5*, 80.
32. Coe, D.; Robben, F.; Talbot, L.; Cattolica, R. *Phys. Fluids* **1980**, *23*, 706.
33. Read, F. H. In *Electron Impact Ionization;* Märk, T. D.; Dunn, G. H., Eds.; Springer-Verlag: New York, NY, 1985, pp 42-88.
34. Hernandez, S. P.; Dagdigian, P. J.; Doering, J. P. *Chem. Phys. Lett.* **1982**, *91*, 409.

35. Hernandez, S. P.; Dagdigian, P. J.; Doering, J. P. *J. Chem. Phys.* **1982**, *77*, 6021.
36. Helvajian, H.; Dekoven, B. M.; Baronavski, A. P. *Chem. Phys.* **1984**, *90*, 175.
37. Dagdigian, P. J.; Doering, J. P. *J. Chem. Phys.* **1983**, *78*, 1846.
38. Xie, J.; Zare, R. N. *J. Chem. Phys.* **1990**, *93*, 3033.
39. Mauersberger, K. *Geophys. Res. Lett.* **1981**, *8*, 935.
40. Mauersberger, K. *Geophys. Res. Lett.* **1987**, *14*, 80.
41. Abbas, M. M.; Guo, J.; Carli, B.; Mencaraglia, F.; Carlotti, M.; Nolt, I. G. *J. Geophys.Res.* **1987**, *92*, 13231.
42. Heidenreich, J. E. III; Thiemens, M. H. *J. Chem. Phys.* **1983**, *78*, 892.
43. Heidenreich, J. E. III; Thiemens, M. H. *J. Chem. Phys.* **1986**, *84*, 2129.
44. Morton, J.; Schueler, B.; Mauersberger, K. *Chem. Phys. Lett.* **1989**, *154*, 143.
45. Thiemens, M. H.; Jackson, T. *Geophys. Res. Lett.* **1987**, *14*, 624.
46. Thiemens, M. H.; Jackson, T. *Geophys. Res. Lett.* **1988**, *15*, 639.
47. Morton, J.; Barnes, J.; Schueler, B.; Mauersberger, K. *J. Geophys. Res.* **1990**, *95*, 901.
48. Bates, D. R. *Geophys. Res. Lett.* **1988**, *15*, 13.
49. Valentini, J. J. *J. Chem. Phys.* **1987**, *86*, 6757.
50. Crutzen, P. J. *J. Geophys. Res.* **1971**, *76*, 7311.
51. Fehsenfeld F. C.; Ferguson, E. E. *Radio Sci.* **1972**, *7*, 113.
52. Dotan, I.; Davidson, J. A.; Fehsenfeld, F. C.; Albritton, D. L. *J. Geophys. Res.* **1978**, *83*, 4036.
53. Arnold F.; Viggiano, A. A. *Planet Space Sci.* **1982**, *30*, 1295.
54. Pfeilsticker, K.; Arnold, F. *Planet Space Sci.* **1989**, *37*, 315.

RECEIVED October 3, 1991

Chapter 15

Temperature, Kinetic Energy, and Rotational Temperature Effects in Four Reactions Involving Isotopes

A. A. Viggiano[1], Robert A. Morris[1], Jane M. van Doren[1], John F. Paulson[1], H. H. Michels[2], R. H. Hobbs[2], and Christopher E. Dateo[3]

[1]Phillips Laboratory, Geophysics Directorate, Ionospheric Effects Division (LID), Hanscom Air Force Base, MA 01731-5000
[2]United Technologies Research Center, East Hartford, CT 06108
[3]Department of Chemistry, University of California, Santa Barbara, CA 93106

Data on four reactions involving isotopes taken in a variable temperature-selected ion flow drift tube are presented. A study of the reaction of O^- with N_2O indicates that the reaction proceeds preferentially by bonding of the O^- to the central nitrogen in N_2O. The preference for O^- attack at the central nitrogen over attack at the terminal nitrogen decreases at higher temperatures. In the atom abstraction reaction of O^+ with HD, OH^+ is formed more efficiently than is OD^+ at low temperatures and moderate energy. The branching fraction favoring OH^+ production is also sensitive to the rotational temperature of the HD. The results for this reaction are consistent with a model of the reaction based on the long range part of the ion-neutral potential. Rate constants for the reactions of O^- with H_2, D_2, and HD vary with mass of the hydrogen molecule as predicted from the collision rate constant dependence. These reactions proceed by two channels: hydrogen abstraction and associative electron detachment. The rate constant for the minor hydrogen abstraction channel increases with increasing kinetic energy. The efficiency of the abstraction reaction of O^- with D_2 is significantly smaller than that in the reaction with H_2. More OD^- as compared with OH^- is produced in the reaction of HD. The results are explained by a direct two step mechanism. A large preference for OH^- over OD^- formation in the atom abstraction reaction of O^- with CH_2D_2 is observed. An *ab initio* potential for the reaction path reveals a barrier and indicates that zero point energy effects play a major role in the observed isotope effects. The four reactions studied in this paper, when taken together, show that isotopically labelled reactants can be important in elucidating widely different reaction mechanisms.

NOTE: The Phillips Laboratory was formerly the Air Force Geophysics Laboratory.

Studies involving isotopically labeled reactants can yield important insights into reaction mechanisms. In particular, the extent of isotope incorporation into the products formed can shed light on the structure of the prinicipal complex(es)/intermediate(s) involved in the reaction. In addition, kinetic isotope effects may provide insight into the reaction mechanism by providing information about the reaction coordinate potential energy curve. Frequently, large kinetic isotope effects are seen only in reactions involving hydrogen and deuterium. However, several examples in this volume show that significant kinetic isotope effects can occur in reactions involving heavier isotopes. The temperature, kinetic energy, and rotational energy dependence of the rate constant and branching ratios can provide further information on the reaction coordinate potential and mechanism.

The selected ion flow tube is a versatile system for studying reactions involving isotopes. The reactant ion is formed in a remote ion source, mass selected, and injected into the reaction flow tube. This "sifting" prevents the neutral precursor of the reactant ion from entering the reaction region and eliminates unwanted isotopic mixtures of the reactant ion at different masses. Many isotopically labeled reactions have been studied using such systems.[1] In addition to these advantages, the present selected ion flow tube can be heated or cooled over a wide range, and the kinetic energy of the reactant ions can be varied by application of a drift field.[2] This apparatus is designated VT-SIFDT for variable temperature-selected ion flow drift tube. From studies of rate constants or branching ratios as a function of kinetic energy at several temperatures, information on the influences of rotational and vibrational energy on a variety of reactions can be obtained.[2-5]

In this paper we report results on four reactions involving isotopically labeled reagents. The goal in all of these studies was to elucidate the reaction mechanism. An important aspect of these studies was the variation of the isotopic effects with temperature, kinetic energy, and rotational energy.

Experimental

The measurements were made using the Phillips Laboratory (formerly the Geophysics Laboratory) variable temperature-selected ion flow drift tube apparatus.[2] Instruments of this type have been the subject of review,[6] and only those aspects important to the present study will be discussed in detail. Ions are created by electron impact in a moderate pressure ion source (~ 0.1 torr). The ions are extracted from the source and mass selected in a quadrupole mass filter. Ions of the desired mass are injected into a meter-long flow tube through a small orifice, 2 mm in diameter. A buffer gas, helium unless otherwise noted, transports the ions along the length of the flow tube. The pressure in the flow tube is ~0.5 torr. The buffer gas is added through a Venturi inlet surrounding the ion injection orifice and thus aids in injecting the ions at low energy. A drift tube, consisting of 60 electrically insulated rings connected by resistors, is positioned inside the flow tube. A voltage can be applied to the resistance chain in order to produce a uniform electric field inside the flow tube for studies of energy dependence. The bulk of the gas in the flow tube is pumped by a Roots type blower. The flow tube is terminated by a truncated nose cone. A small fraction of the ions in the flow tube is sampled through a 0.2 mm hole in the nose cone, mass analyzed in a second quadrupole mass spectrometer, and detected by a channel electron multiplier.

Neutral reactant gas is added through one of two inlets. The inlets are rings with a series of holes pointing upstream.[2] The area inside the ring is equal to the area between the ring and the tube wall to aid in quickly distributing the reactant gas throughout the cross sectional area of the flow tube. Rate constants are measured by monitoring the decay of the primary ion signal as a function of added neutral flow. This is done at each of the two neutral inlets, and an end correction is determined from those data. Ion flight times are measured by applying an electrical retarding pulse to two of the drift tube rings consecutively and measuring the two arrival time spectra of the ions. The ion velocity and therefore the reaction time are determined from these data and from knowledge of the relevant distances. Pressure is monitored by a capacitance manometer. Flow rates of the buffer and of the reactant gas are controlled and measured by MKS flow controllers. The rate constant is determined, using data obtained at each inlet, from the slope of the line obtained by plotting the logarithm of the reactant ion signal decay versus neutral reactant flow, and from the values of the pressure, temperature, flow rates of the reactant and buffer, and ion velocity. The final rate constant incorporates the end correction. The entire flow tube can be heated or cooled over the range 85 to 550 K.

Branching fractions are measured by monitoring the signals from the ionic products and determining the fraction of the total products that each product represents at each neutral flow rate. These fractions are plotted versus neutral flow rate, and the resulting curves are extrapolated to zero neutral flow to account for secondary reactions. The branching fractions reported are the extrapolated values. No correction for mass discrimination was made in the studies reported here because the different product ions in each reaction have very small mass differences.

The average kinetic energy in the ion-neutral center-of-mass system, $\langle KE_{cm} \rangle$, in the drift tube is derived from the Wannier formula[7] as

$$\langle KE_{cm} \rangle = \frac{(m_i + m_b)m_n}{2(m_i + m_n)} v_d^2 + \frac{3}{2} kT \tag{1}$$

where m_i, m_b, and m_n are the masses of the reactant ion, buffer gas, and reactant neutral, respectively; v_d is the ion drift velocity; and T is the temperature. The first term in the formula is the energy supplied by the drift field, and the second term is the thermal energy. This formula is an excellent approximation of the ion energy at low ion energies.[8,9] At energies approaching 1 eV the formula is still good to within ±10%. The neutral reactant temperature under the conditions of the present experiments is the same as that of the buffer gas since the neutral gas enters the flow tube through inlets at the same temperature as the flow tube and suffers many collisions with the walls of the inlets before entering the flow tube. The ions used in the present studies are all monatomic, and therefore the drift tube can influence only the ion translational energy (no electronic excitation is expected for the drift tube energies employed here). The distribution of ion energies in a helium buffer is very nearly Maxwellian at an effective temperature, T_{eff}, such that $\langle KE_{cm} \rangle = 3/2kT_{eff}$.[8-11] By "nearly Maxwellian", it is meant that the distribution of ion energies is such that the rate constant measured in the drift tube is indistinguishable from that which would be measured for a Maxwellian distribution (within the 15% experimental precision).

It has recently been shown in our laboratory that dependences of rate constants or branching ratios on the internal temperature of the reactant neutrals can be derived for a variety of ion-molecule reactions by measuring rate constants or branching ratios as a function of $\langle KE_{cm} \rangle$ at several temperatures.[2-5,9,12,13] If the ions are monatomic, as in the present studies, comparing the rate constants or branching ratios at a particular $\langle KE_{cm} \rangle$ but at different temperatures yields the dependence of the rate constant on the internal temperature of the reactant neutral. If the neutral has no low frequency vibrational modes, this dependence is a rotational temperature dependence.

O^+ was formed from CO by dissociative electron impact ionization. This source produced 98% of the O^+ ions in the ground ^4S state. The fraction of excited state O^+ was monitored by allowing the O^+ ions to react with CO in the flow tube. CO reacts with the excited metastable states of O^+ and does not react with the ground state.[1] CO_2 and O_2 were tested as alternative sources of O^+ but produced more metastable ions. The 2% fraction of O^+ in the excited state remained constant over the course of these experiments. $^{16}O^-$ was generated by dissociative electron attachment to N_2O, while $^{18}O^-$ was generated by dissociative attachment to $^{18}O_2$.

All reactant gases were obtained commercially and used without further purification. The HD was ≥98.7% isotopically pure, and no corrections for incomplete labeling were made to the data. According to the manufacturers the HD may be purer than this but the absolute purity could not be measured with greater accuracy. The $^{14}N^{15}NO$ was of 99% isotopic purity, and the product distributions were corrected for the slight $^{14}N_2O$ impurity. The CH_2D_2 was 98% isotopically pure, and no corrections to the data were made since it is not certain which isotopic form of CH_4 the impurities are.

The rate constant for the reaction of O^- with unlabeled N_2O was measured in a helium buffer as a function of $\langle KE_{cm} \rangle$ at several temperatures. The reactions with labeled N_2O were studied in an argon buffer. For the reaction of the unlabeled species we were interested in the dependence of the rate constant only on the internal temperature of N_2O. This required the use of a helium buffer so that the distribution of kinetic energies at a particular $\langle KE_{cm} \rangle$ would be essentially Maxwellian and thus approximately independent of temperature.[8-11] The goal of the isotopic labeling studies of N_2O was to determine the reaction mechanism by monitoring the distribution of isotopes in the NO^- products. This required an argon buffer so that no detachment of the NO^- product ions would occur. Helium is known to detach the weakly bonded electron[14] from NO^-, but the electron remains attached to NO^- in an argon buffer.[12] In argon, the product ion signals balanced the loss of primary ion signal, indicating that electron detachment from the product ions was minimal. The requirement of an argon buffer (and the cost of $^{14}N^{15}NO$) prevented us from studying the reaction of the completely labeled reactants using the drift tube because the ion kinetic energy distributions produced in an argon buffer are known to be non-Maxwellian.[11,15,16] Such non-Maxwellian distributions would confuse any attempt to derive internal temperature dependences in the branching fraction for the labeled systems.

O^+ reacts with water to produce H_2O^+. This ion has the same mass as that of the OD^+ product ion in the reaction of O^+ with HD. Since we expected small effects on the branching ratio of this reaction, the flow tube had to be essentially free of water vapor. In order to accomplish this goal several steps were taken. Ultra high purity helium was used (99.9999%) and passed through a molecular sieve trap cooled to liquid nitrogen temperature. The flow tube was repeatedly baked at 550 K over the course of several weeks, and the helium inlets were also heated over the same time frame. The system was not exposed to air during the course of this baking. The baking reduced the amount of conversion of O^+ to H_2O^+ at 300 K from over 20% at the start of the process to an undetectable amount. The maximum water vapor concentration in the flow tube is estimated to be 10^8 cm^{-3}, or a mixing ratio of

~10^{-8}. At 509 K, about 1% of the O^+ was converted to H_2O^+. The branching fractions observed were corrected for this small mass 18 impurity. The water impurity at 509 K leads to an uncertainty of 2.5 percentage points in the branching fractions (expressed as percentages) derived at 509 K. The uncertainty at other temperatures is 1.5 percentage points. All of the branching fraction data were extrapolated to zero HD flow rate since OH^+ and OD^+ react with HD to form various H_2O^+ isotopomers which further react to form H_3O^+ in all of the isomeric forms. Reactions involving OH^- and OD^- products were simpler since no secondary reactions occurred under the conditions of our experiments.[17]

Results and Discussion

a) The Reaction of O^- With N_2O. The reaction of unlabeled reactants produces NO^- and NO,

$$O^- + N_2O \rightarrow NO^- + NO. \quad (2)$$

The reaction is exothermic by 0.14 eV.[18] The goal of using isotopically labeled reagents to study this reaction was to determine the fraction of interactions in which the O^- attacks the central nitrogen versus the terminal nitrogen in N_2O. Our results for this reaction, for both labeled and unlabeled reactants, have been published previously[19] and will be summarized below.

The rate constant for this reaction was studied at four temperatures over the range 143 K to 515 K.[19] The rate constant was found to decrease with increasing temperature or energy from a value of 2.9×10^{-10} cm^3 s^{-1} at 143 K (0.019 eV) to a minimum of 8-9 $\times 10^{-11}$ cm^3 s^{-1} at approximately 0.5 eV. Above 0.5 eV, the rate constant increases to a value of 1.1×10^{-10} cm^3 s^{-1} at 0.92 eV. However, at a particular kinetic energy, no temperature dependence of the rate constant was found. This implies that there is no dependence on the internal temperature of the N_2O over this temperature range, indicating that neither rotational energy nor excitation of the N_2O bending mode, the lowest energy vibrational mode, has a large effect on reactivity between 143 and 515 K.

Several aspects of the reaction have been investigated using isotopic labeling. Van Doren et al.[20] have shown that the oxygen atoms exchange with a rate constant of 1.7×10^{-10} cm^3 s^{-1},

$$^{18}O^- + N_2{}^{16}O \rightarrow {}^{16}O^- + N_2{}^{18}O. \quad (3)$$

The same oxygen exchange rate was measured for the reaction of totally labeled reactants, i.e., labeling of both nitrogen and oxygen.[19,21] At 143 K, the rate constant for O^- exchange was found to be 1.4×10^{-10} cm^3 s^{-1}, indicating that the rate constant for this channel has a small positive temperature dependence. Interestingly, the efficiency for the sum of the two channels, reactions 2 and 3, is approximately temperature independent over this temperature range. Therefore, as the efficiency for NO^- production decreases with increasing temperature, the efficiency for O^- exchange appears to increase by the same amount. The positive temperature dependence for oxygen exchange may result from a small activation energy (presumably from the breaking of the $O - N_2$ bond). Alternatively, it may be due to the fact that the contribution from the reactive channel producing NO⁻ decreases, leaving a larger fraction of "unreactive collisions" which can result in O⁻ exchange.

In one set of experiments, we studied this reaction using unlabeled oxygen and labeled nitrogen.[19]

$$O^- + {}^{14}N{}^{15}NO \rightarrow {}^{14}NO^- + {}^{15}NO \qquad (4)$$
$$\rightarrow {}^{15}NO^- + {}^{14}NO.$$

In that study, we observed equal production of $^{14}NO^-$ and $^{15}NO^-$ at both 143 K and 298 K. This is consistent with the findings of Barlow and Bierbaum.[21] At higher energies (about 1 eV), Paulson[22] found that there was a preference for the O⁻ to pick up the terminal nitrogen in the ratio of 1.25:1. This suggests that at higher energies there may be a change in the mechanism from one involving a long lived complex/intermediate to a direct process such as atom stripping. The involvement of a direct mechanism at high energy could explain the increase in the rate constant at energies above 0.5 eV and the change in the energy dependence of the rate constant from negative to positive.

We have also studied the reaction using both labeled nitrogen and oxygen atoms,

$$^{18}O^- + {}^{14}N{}^{15}N{}^{16}O \rightarrow {}^{14}N{}^{16}O^- + {}^{15}N{}^{18}O \qquad (5)$$
$$\rightarrow {}^{15}N{}^{16}O^- + {}^{14}N{}^{18}O$$
$$\rightarrow {}^{14}N{}^{18}O^- + {}^{15}N{}^{16}O$$
$$\rightarrow {}^{15}N{}^{18}O^- + {}^{14}N{}^{16}O$$
$$\rightarrow {}^{16}O^- + {}^{14}N{}^{15}N{}^{18}O.$$

Figure 1 shows a plot of the percentage of the total NO⁻ production contributed by each of the four possible NO⁻ isotopic products as a function of temperature. The data sort themselves into two groups. Masses 30 and 33 are produced on the order of 15% each and masses 31 and 32 on the order of 35% each. This is again consistent with the findings of Barlow and Bierbaum.[21] Increasing temperature leads to decreases in the percentages of masses 30 and 33, and increases in the percentages of masses 31 and 32. The negative temperature dependence of mass 33 and the positive dependence of the mass 32 are outside our estimated uncertainty. The uncertainies in the fractions of masses 30 and 31 are larger due to secondary chemistry and the temperature dependences are within the uncertainty. However, it is expected that the fractions of masses 30 and 33 are equal, as are those of masses 31 and 32. The apparent difference in the dependences of masses 31 and 32 probably results in the uncertainty in determining the fraction of mass 31 which is affected by secondary chemistry.

As stated previously, the main goal of the labeling studies was to determine whether the reaction producing NO⁻ proceeds by O⁻ reaction at the central or terminal nitrogen. If the reaction were to proceed exclusively by terminal attack without migration of O⁻, one would expect equal abundances of masses 31 and 32 with no production of masses 30 and 33. This expectation assumes that it is equally probable that the charge remains on either NO fragment. This assumption is supported by the data obtained using labeled nitrogen and unlabeled oxygen, reaction 4, where equal fractions of the two possible NO⁻ products were observed. Attack exclusively at the central nitrogen should produce all four isotopically labeled NO⁻ species with equal

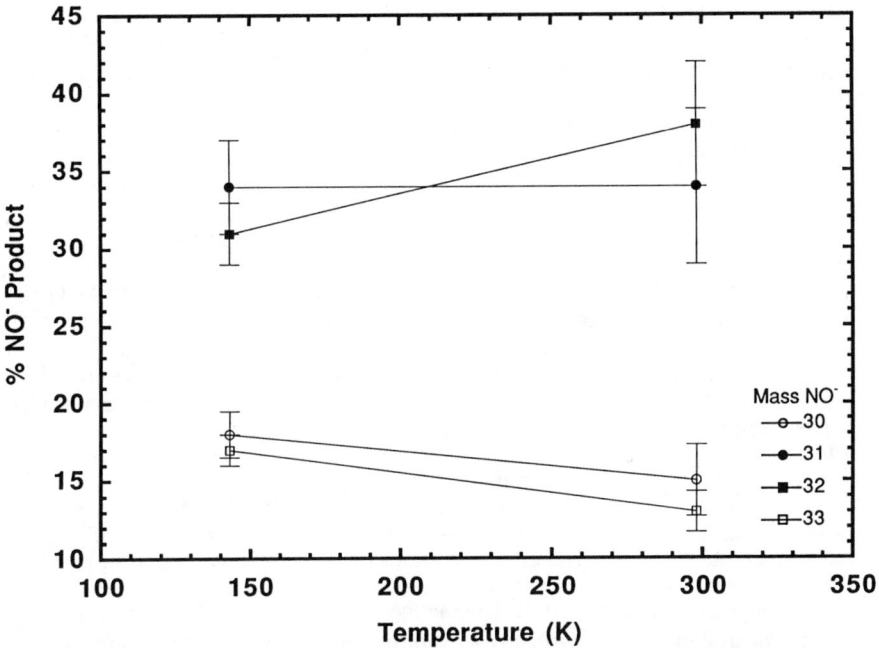

Figure 1. Percent contribution to the total NO⁻ production by the individual NO⁻ isotopes in the reaction of $^{18}O^-$ with $^{14}N^{15}N^{16}O$ as a function of temperature. Open circles, solid circles, solid squares, and open squares represent $^{14}N^{16}O^-$, $^{15}N^{16}O^-$, $^{14}N^{18}O^-$, and $^{15}N^{18}O^-$, respectively.

probability since the stable form of the complex formed by central attack is of C_{2v} symmetry.[23-25]

Assuming the above probabilities for forming the various isotopically labeled NO^- products for the two positions of attack on N_2O by O^-, we can derive the fractions of the reaction that proceed through reaction at the central and terminal nitrogen. The fraction of the total NO^- produced that originates from the central nitrogen is twice the sum of the mass 30 and 33 fractions, which are equal within experimental uncertainty. Thus we find that in 70% of the reactive collisions at 143 K the O^- reactant bonds to the central nitrogen. At 298 K, bonding to the central nitrogen proceeds in 56% of the reactive collisions. Multiplying these fractions by the rate constants for the reaction of the unlabeled reactants, reaction 2, yields rate constants of 2.0×10^{-10} cm^3 s^{-1} at 143 K and 1.1×10^{-10} cm^3 s^{-1} at 298 K for reaction at the central nitrogen. The rate constants for the unlabeled reagents are used since they are known the most accurately. Within our experimental uncertainty isotopic substitution does not change the rate constant for production of NO^-, i.e., there is no kinetic isotope effect in this reaction, as expected.[19] Using the same procedure, we find the rate constants for reaction at the terminal nitrogen to be 0.9×10^{-10} cm^3 s^{-1} at both temperatures. The rate constant for the central nitrogen reaction channel has a negative temperature dependence of $T^{-0.8}$, while the rate constant for the terminal nitrogen channel is independent of temperature. The negative temperature dependence of the reaction rate constant for NO^- formation, reaction 2, between 143 and 298 K arises from the temperature dependence of the reaction at the central nitrogen.

The difference in the temperature dependences of these two reactive channels forming NO^- may arise from increased rotational energy of the reactant N_2O with increasing temperature. As the molecule rotates faster it is harder for the O^- to approach the central nitrogen before encountering one of the end atoms. This increases the relative number of encounters in which the O^- hits the terminal nitrogen on first approach. Alternatively, the reaction may be initiated by terminal attack followed by migration of one of the O atoms to the central nitrogen. Increasing temperature would decrease the amount of migration, either through a decrease in the complex lifetime or because the migration involves a tight transition state which is known to lead to a negative temperature dependence.[26-28] Other mechanisms are also conceivable, and an answer to the exact nature of the competition between central and terminal attack awaits detailed calculations.

The fact that more than half of the reactive collisions proceed via reaction of the O^- at the central nitrogen is probably a result of the central nitrogen having a partial positive charge, while the terminal nitrogen is approximately neutral.[29] In spite of this, the abundance of reaction at the central nitrogen was surprising since it has been demonstrated that most reactions of anions with N_2O appear to proceed via reaction of the anion at the terminal nitrogen in N_2O.[30,31] The difference between the reaction of O^- with N_2O and the other systems studied may simply be due to steric factors. O^- is a small negative ion that can easily approach the central nitrogen. The other anions studied in reaction with N_2O are molecular, and the approach to the central nitrogen may be sterically hindered.

b) The Reaction of O^+ with HD. The reaction of O^+ with HD proceeds by hydrogen abstraction,

$$O^+ + HD \rightarrow OH^+ + D \qquad (6)$$
$$\rightarrow OD^+ + H.$$

The reaction is exothermic by approximately 0.5 eV.[18] We have measured the rate constant and product branching fractions as a function of $\langle KE_{cm} \rangle$ at temperatures of 93, 300, and 509 K. The rate constant was found to be 1.2×10^{-9} cm^3 s^{-1} independent of temperature or energy within a rather large experimental uncertainty (±40%). The large uncertainty results from taking only one set of data at each inlet port, resulting in a larger than normal error in the end correction. In addition, each run included recording of the signals from the ionic products in order to determine the branching fractions. The consequent increased data collection time results in greater scatter in the rate constant data due to drift in the reactant ion signal over time. More runs were not taken due to the cost of HD and the fact that our interest was mainly in the branching ratio, which does not depend on the end correction.

In Figure 2, the fraction of OH$^+$ produced in the reaction is plotted as a function of $\langle KE_{cm} \rangle$. Circles, squares and diamonds represent data taken at 93, 300, and 509 K, respectively. At each temperature, the fraction of OH$^+$ increases with increasing kinetic energy. At a particular $\langle KE_{cm} \rangle$, the OH$^+$ fraction decreases with increasing temperature. The difference between the fractions measured at 93 K and 300 K is larger than that between 300 K and 509 K. The latter difference for each point is smaller than our estimated error limits. The difference is probably real since every 509 K point is lower than the corresponding 300 K point at the same $\langle KE_{cm} \rangle$ and the error limits refer mostly to random error. A difference between 300 K and 509 K is also consistent with a difference between 93 and 300 K. As explained in the experimental section, the observation of different values for the branching fraction at the same $\langle KE_{cm} \rangle$ but different temperatures indicates that the branching fraction depends on the rotational temperature of the HD. Our observations indicate that kinetic and rotational energy have opposite effects on the branching fraction.

The observed energy dependences for this reaction are explained theoretically by the work of Dateo and Clary,[32] who assumed that the reactivity (both rate constant and branching fraction) is dominated by the long range part of the potential. This assumption seems justified in the present case because the reaction is very exothermic, proceeds on every collision, and has no competing channels. The form of the potential assumed by Dateo and Clary is

$$V(R,\theta) = -\frac{q\alpha}{2R^4} + q^2\left(\frac{\Theta}{R^3} - \frac{\alpha_2}{2R^4}\right)P_2\cos(\theta) \qquad (7)$$

where q is the ionic charge, α is the polarizability of HD, R is the length of a line L connecting the center of the O$^+$ ion to the center-of-mass of the HD, θ is the angle between the line L and the HD bond axis, Θ is the quadrupole moment of HD, and α_2 is the anisotropic polarizability of HD. The potential is the sum of the ion-induced dipole potential term, the ion-quadrupole potential term, and a term that allows the polarizability to be anisotropic. This potential was used in a rotationally adiabatic capture theory. The rate constant calculated using this theory is determined by the rate of passage over the centrifugal barrier, while the branching fraction is determined by the end of the HD molecule which points at the O$^+$ at the centrifugal barrier. Dateo and Clary assumed that the reaction proceeds rapidly once over the centrifugal barrier, i.e., there is no re-crossing of the barrier, and that the product distribution is frozen at that point.

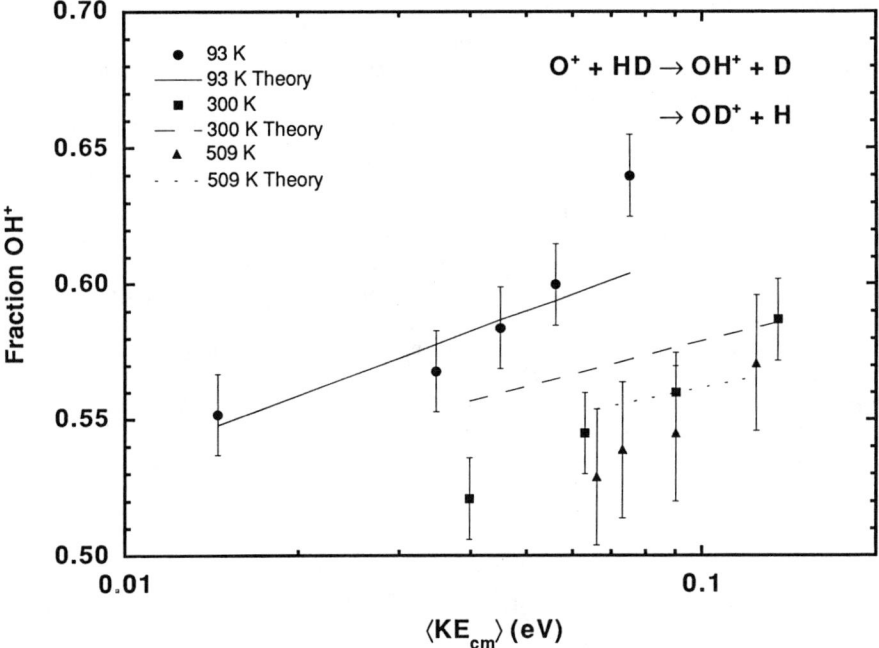

Figure 2. Branching fraction of OH^+ produced in the reaction of O^+ with HD as a function of $<KE_{cm}>$. Circles, squares, and triangles represent data taken at 93 K, 300 K, and 509 K, respectively. Theoretical curves[32] are given as solid, long dashed, and short dashed lines for 93 K, 300 K, and 509 K, respectively. (Reproduced with permission from ref. 38. Copyright 1992 American Institute of Physics)

The results of the theoretical calculations are shown in Figure 2 as a solid line, a large dashed line, and a small dashed line for temperatures of 93 K, 300 K, and 509 K, respectively. The theoretical results are integrated over Boltzmann distributions of rotations at the three temperatures and Boltzmann distributions of kinetic energies such that the average kinetic energy is given by $\langle KE_{cm} \rangle$. As stated earlier, drift tube measurements made in a helium buffer are adequately represented by Boltzmann distributions.[8-11] The low temperature data and theory are in very good agreement except at the highest kinetic energy. The branching fractions measured at higher temperature are smaller than the values predicted by theory by about 2 percentage points. This indicates that the theory slightly underestimates the rotational and kinetic energy dependences observed. The trends with energy, however, are reproduced very well.

The observed kinetic and rotational energy dependences may be understood in terms of simple physical ideas. The preference for OH^+ results from the fact that the center-of-mass (nearer to D than H) and the center of polarizability (the center of the HD bond) are different. This difference creates a torque on the HD molecule that tends to orient the H side of the molecule toward the O^+. The decreasing fraction of OH^+ formed in the reaction with increasing temperature at a given $\langle KE_{cm} \rangle$, i.e., the rotational energy dependence, arises from rotational averaging of the anisotropic potential as the HD rotates faster. As expected from this argument, the largest enhancement for OH^+ production should occur for low rotational quantum numbers. In fact, the theory of Dateo and Clary predicts the largest effect for $J = 0$ and smaller effects for large J (see Figure 5 in Dateo and Clary[32]). The increase in the fraction of OH^+ at higher kinetic energies may be understood as follows. At higher kinetic energies, the average impact parameter that leads to reaction is smaller. This constrains the centrifugal barrier to smaller radii where the anisotropy of the potential is larger, resulting in more alignment.

This reaction has also been studied by Burley et al.[33] in a guided ion beam mass spectrometer and more recently by Sunderlin and Armentrout in a modified version of the same apparatus with a variable temperature reaction cell.[34] The trends observed with ion kinetic energy and neutral gas temperature were the same as those observed in the present study. However, the magnitude of the temperature dependence at a given ion kinetic energy was smaller in value than that reported in this work. The difference in these observations arises from different ion velocity distributions in the two experiments. As mentioned above, the ion velocity distributions in our experiment are Maxwellian while those in the beam instrument are not. Support for this explanation comes from the theoretical predictions for the two experiments which incorporate the appropriate velocity distributions into the calculations and are in good agreement with the data.[34] As observed, the temperature dependence at a given ion kinetic energy is predicted to be smaller in the beam apparatus compared with that in the variable temperature drift tube.

The agreement between the predicted rotational energy dependence, the beam data, and the data obtained using our apparatus further validates the derivation of internal energy dependences from the observation of a temperature dependence of a kinetic parameter at a given $\langle KE_{cm} \rangle$. This is the first time that we have been able to compare data of this sort to other experimental or theoretical data.

c) **The Reaction of O^- with H_2, D_2 and HD.** The reaction of O^- with H_2 (and HD and D_2) proceeds by two channels, atom abstraction and associative detachment,

$$O^- + H_2 \rightarrow OH^- + H \quad (8)$$
$$\rightarrow H_2O + e.$$

Our results for the reactions of O^- with H_2 and D_2 have been published elsewhere along with measurements on the effects of hydration of the O^- reactant on the reaction.[35] Figure 3 shows the rate constants for the reactions of O^- with H_2 and D_2 at three temperatures and with HD at 300 K. The rate constants are large and decrease slightly with increasing $\langle KE_{cm} \rangle$ or temperature. At a given $\langle KE_{cm} \rangle$, no temperature dependence of the rate constant is observed in the reactions of H_2 and D_2 over the range 176 K to 490 K, indicating that the reaction is not sensitive to the rotational temperature of the reactant neutral in this range, as expected for efficient reactions. The rate constants for the isotopes of hydrogen are in the order $k(H_2) > k(HD) > k(D_2)$, i.e., the rate coefficient decreases as the mass of the hydrogen isotope increases. The differences in the rate constants can be totally accounted for by the differences in the collision rate constants, which are inversely related to the square root of the reduced mass of the ion-neutral system. All of the reactions proceed with the same efficiency, on the order of 50%.

We have also measured the branching fractions of the two product channels. Figure 4 plots the percentage of the H/D atom abstraction channel forming the ionic products OH^- or OD^- versus $\langle KE_{cm} \rangle$. At low energies, only a few percent of ionic product is formed. This percent increases rapidly with increasing energy or temperature. The atom abstraction channel is more efficient in the reaction with H_2 than in the reaction with D_2. The efficiency of this channel for the reaction with HD was found to be approximately the same as measured for the reaction with H_2. The fact that the branching fraction for reaction with HD is similar to that for H_2 may be a real effect or may arise from a small systematic error due to the fact that the HD experiments were performed over a year after the H_2 and D_2 experiments, which were conducted in the same week. No rotational energy dependence is observed for the branching fractions in either of the reactions with H_2 or D_2, within our experimental uncertainty.

In addition to the total branching fraction to form ionic products, the relative abundances of OD^- and OH^- in the reaction of O^- with HD is of interest. We found that OD^- is formed preferentially in this reaction. The branching ratio of OD^- to OH^- was found to be 1.5:1, approximately independent of $\langle KE_{cm} \rangle$ over the entire energy range explored, 0.038 to 0.11 eV. This preference for formation of OD^- seemed surprising at first considering that OH^- is formed more frequently than is OD^- in the reactions with only one isotope. The preference for OD^- would be consistent with the fact that OD^- formation is about 0.04 eV more exothermic than OH^- production. However, this is inconsistent with preliminary results that indicate that at low temperature the preference for OD^- production is less. If thermodynamics controlled the branching, a larger, not smaller, preference for OD^- would be expected at low temperature. As will be shown below, the preference for OD^- can be readily explained.

The room temperature reactions of O^- with H_2 and D_2 were studied in a drift tube at the NOAA laboratories by McFarland et al.,[36] and our results are in very good agreement with theirs. McFarland et al. postulated a simple mechanism to explain their results. This mechanism also explains the HD data. They postulated a direct mechanism that proceeds in two steps. The first step is atom abstraction,

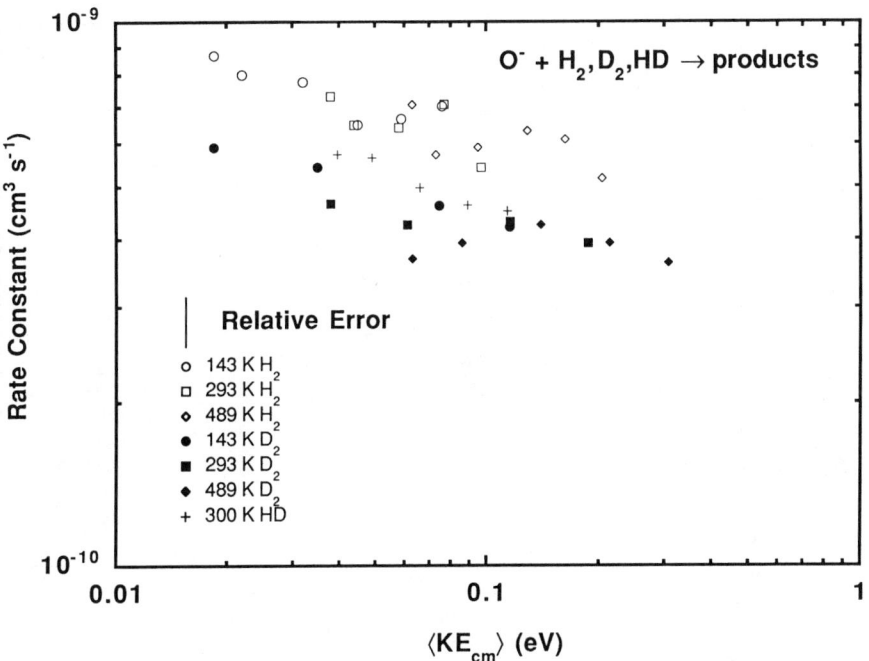

Figure 3. Rate constants for the reactions of O⁻ with H_2, D_2, and HD as a function of $\langle KE_{cm}\rangle$. Data for H_2 are represented by open circles, squares, and diamonds for temperatures of 143 K, 293 K, and 489 K, respectively. Data for D_2 are represented by solid circles, squares, and diamonds for temperatures of 143 K, 293 K, and 489 K, respectively. Data for HD are represented by plusses.

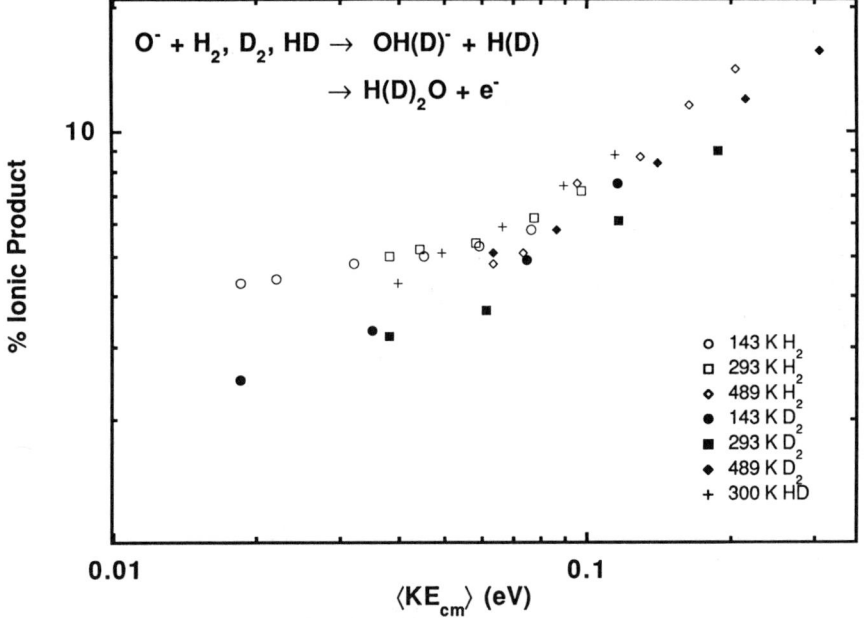

Figure 4. Branching percentages for the channel to produce the total ionic product (OH⁻ or OD⁻) in the reactions of O⁻ with H_2, D_2, and HD as a function of $\langle KE_{cm} \rangle$. Data for H_2 are represented by open circles, squares, and diamonds for temperatures of 143 K, 293 K, and 489 K, respectively. Data for D_2 are represented by solid circles, squares, and diamonds for temperatures of 143 K, 293 K, and 489 K, respectively. Data for HD are represented by plusses.

$$O^- + H_2 \rightarrow \{OH^- \cdots H\}. \tag{9}$$

The H atom can then quickly separate from the OH^- leaving an ionic product,

$$\{OH^- \cdots H\} \rightarrow OH^- + H \tag{10}$$

or in the lifetime of the collision the OH^- can associatively detach,

$$\{OH^- \cdots H\} \rightarrow H_2O + e^-. \tag{11}$$

This simple mechanism seems to explain all of the features of the data. The associative detachment reaction (9 and 11) is known from separate experiments to be efficient.[1] This efficiency explains the predominance of this channel at low energy. As the energy is raised, the time that the OH^- and H spend near each other decreases, and the contribution by the OH^- channel increases. The overall reaction is relatively efficient, and rotational energy should have little influence. The relative branching fractions for the H_2 and D_2 reactions are explained by the ease with which the H or D atom can escape the initially formed $\{OH^- \cdots H\}$ complex. D atoms move more slowly than H atoms and will not escape from the complex as quickly as the H atom. As a result, the complex containing a D atom will have more time to pass through the critical configuration(s) for associative detachment. Therefore, one expects more associative detachment and thus less ionic product in the D_2 reaction compared with the H_2 reaction.

This prediction is borne out in the data. In the reaction with HD, two possible complexes may be formed initially. Either the D atom is transferred or the H atom is transferred. When the H atom is transferred to the O^-, a D atom is left behind. This complex, $\{OH^- \cdots D\}$ will result in more associative detachment than the other complex for the reasons discussed above, and less OH^- will be formed. Assuming that the two complexes are formed with equal probability, one predicts that OD^- will be formed more efficiently than OH^-, again in agreement with the observation. A rotational energy dependence similar to that observed in the reaction of O^+ with HD may occur in the atom abstraction step. However, such a dependence favors formation of OH^- over OD^-, a preference which is not observed. Therefore, any rotational energy dependence in the atom abstraction step is overwhelmed by the preference for formation of OD^- in the second step. We plan to examine the reaction for a small rotational energy dependence in this branching ratio in the future. In summary, the mechanism for this reaction, postulated almost twenty years ago, continues to explain all of the very detailed data observed for the isotopes of hydrogen.

d) **The Reaction of O^- with CH_2D_2.** The reaction of O^- with CH_4 produces OH^- with a rate constant on the order of 10^{-10} cm^3 s^{-1}.[3,4] At low energies the rate constant shows small negative dependences on both temperature and energy. Above approximately 0.1 eV the rate constant increases with increasing energy. In order to learn more about the mechanism of this reaction, we have begun a study involving isotopically labeled CH_4. Preliminary results (final results were not obtained in time to be included in this volume) indicate that substituting D for H decreases the rate constant substantially. For CD_4, the rate constant is on the order of a factor of ten lower than that for CH_4.

We have studied the branching fraction for the reaction of O^- with CH_2D_2, which produces both OH^- and OD^-. The results are shown in Figure 5. Plotted is the percent of the OH^- product formed as a function of $\langle KE_{cm} \rangle$. At low energies and temperatures, 90% of the total product is OH^-. This can be compared with the statistical (by number of H's or D's) prediction of 50%. Increasing temperature or energy decreases the OH^- branching fraction. Even at $\langle KE_{cm} \rangle$ approaching 0.5 eV the OH^- product still dominates, constituting 60% of the products formed. The data also indicate that the branching fraction does not depend on the rotational energy or low frequency vibrations of the CH_2D_2.

The strong preference for formation of OH^- over OD^- and the decrease in the rate constant with deuterium substitution can be explained by a barrier in the potential energy surface. Figure 6 shows a calculated ab initio potential energy curve for the reaction coordinate of the reaction of O^- with CH_4. The calculations were carried out at the HF/6-31++G(d,p) level of theory.[37] This curve represents the minimum energy pathway along the ground $\{^2A_1\}$ surface for the reaction of O^- with CH_4 leading to OH^- and CH_3 products. A second reaction pathway along the first excited $\{^2E\}$ surface leads to OH and CH_3^- products. This latter reaction pathway was not energetically accessible for the range of collision energies in our studies. Figure 6 indicates that a small barrier to the ground state reaction is found with a classical barrier height of approximately 0.02 eV, essentially at the zero of energy within the uncertainty of the calculations. The classical reaction pathway illustrated in Figure 6 must be modified to account for quantum effects of rotation and vibration. The quantum corrected barrier will be slightly smaller than the classical barrier. Correlated energy calculations along this reaction pathway, which are currently in progress, should reduce the uncertainty in locating the barrier height to less than 0.08 eV.

In Table 1, we list the thermochemistry for reactions of O^- with isotopically substituted CH_4. We find that the reaction exothermicity is greatest for O^- with CH_4 and least for O^- with CD_4. The strong preference observed in the branching fraction and the decrease observed in the rate constant for abstraction with D substitution can be explained by zero point energy effects. The overall exothermicity[37] for the reaction of O^- with CH_4 is 0.258 eV, while that for the reaction of O^- with CD_4 is 0.205 eV, i.e., the reaction is 0.053 eV less exothermic for the perdeuterated form. Similar differences are found for the two channels in the reaction of O^- with CH_2D_2, since the bond strengths for the C-H and C-D in CH_2D_2 are similar to those in the pure isotopes. This exothermicity difference shows that the difference in zero point energies makes it slightly harder to break the C-D bond than the C-H bond. This will be reflected in the quantum corrected barrier height. For reactions involving D abstraction, the barrier will be slightly higher than for reactions involving H abstraction. The increased barrier height will in turn slow the rate of D abstraction compared to the rate of H abstraction. This is reflected in the data in both the branching fraction and the rate constants. We note that Figure 6 indicates a broad barrier along the minimum energy reaction coordinate. This suggests that quantum tunneling effects are negligible for this system.

Conclusions

We have presented data on four reactions in which isotopic labeling is used to help determine the reaction mechanism. In the reaction of O^- with N_2O, the labeling of the

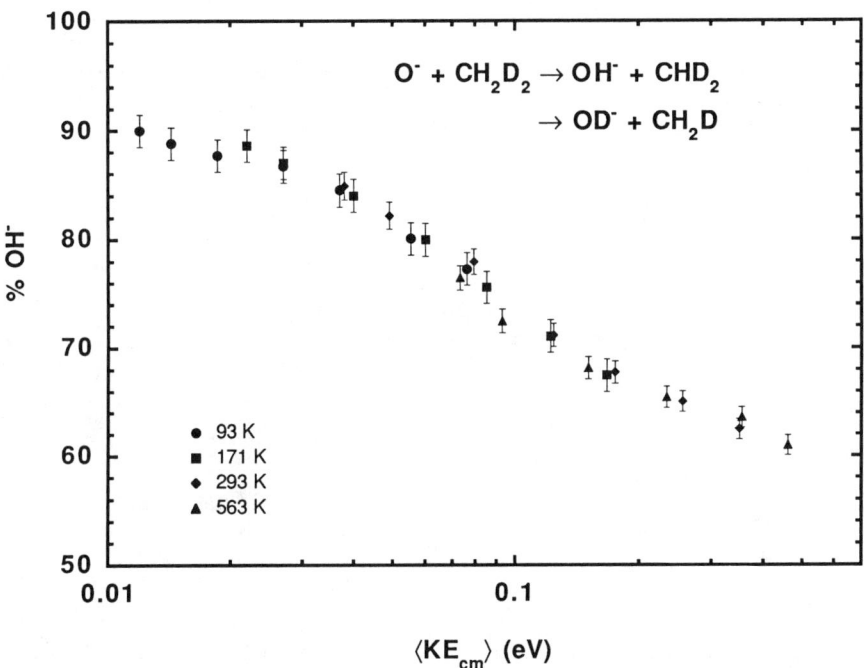

Figure 5. Branching percentage of OH⁻ produced in the reaction of O⁻ with CH_2D_2 as a function of $\langle KE_{cm}\rangle$. Circles, squares, diamonds, and triangles represent data for temperatures of 93 K, 171 K, 293 K, and 563 K, respectively.

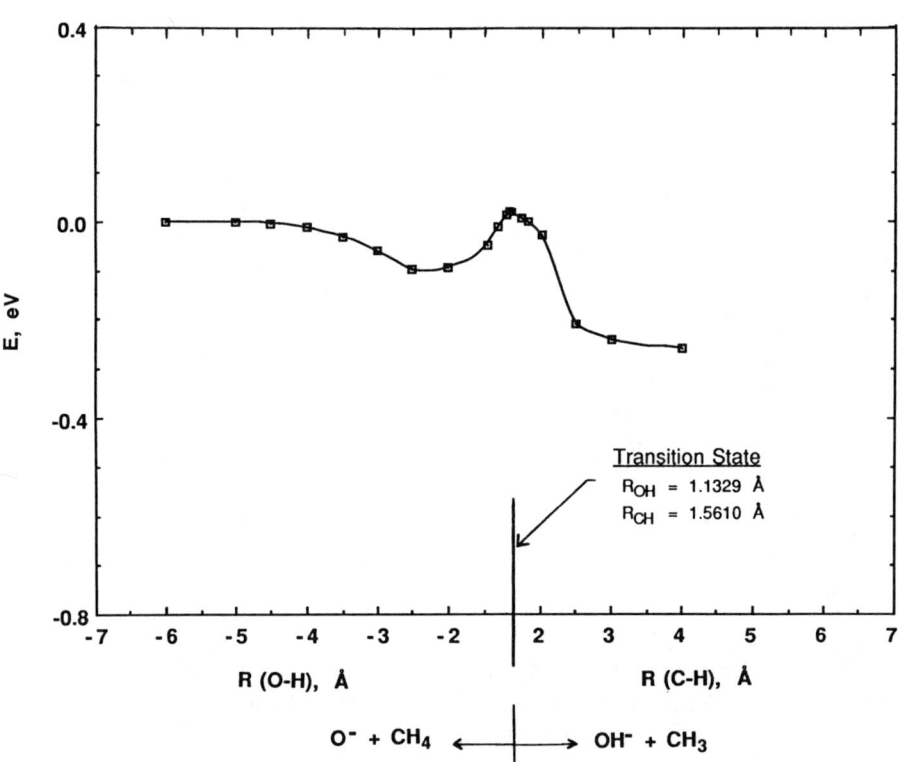

Figure 6. Reaction path diagram for the reaction of O⁻ with CH₄.

Table 1. Thermochemistry of the Reactions of O^- with CH_4, CD_4, and CD_2H_2.[37]

Reaction	$-\Delta H_R$ (eV)		
	96 K	298 K	500 K
$O^- + CH_4 \rightarrow OH^- + CH_3$	0.284	0.258	0.232
$O^- + CD_4 \rightarrow OD^- + CD_3$	0.232	0.205	0.188
$O^- + CD_2H_2 \rightarrow OH^- + CD_2H$	0.263	0.235	0.212
$O^- + CD_2H_2 \rightarrow OD^- + CDH_2$	0.257	0.231	0.210

reactants helped to determine the relative preference for attack of the O^- at the central and terminal nitrogens. The study did not yield unambiguous results but did indicate that a large fraction of the reactive encounters occurs by attack at the central nitrogen, in contrast with studies involving other anions reacting with N_2O.[30,31]

The reaction of O^+ with HD shows a large preference for OH^+ production at low temperatures and moderate energies. A model based on entrance channel effects seems to explain the data.[32] The branching fraction is determined by such effects as the speed of the HD rotations, the torque arising from the different positions of the center-of-mass and center-of-polarizability, and the location of the centrifugal barrier. The confirmation of a rotational energy dependence in this reaction helps to validate the technique in which variable temperature drift tube studies are used to derive information on internal energy effects in ion-molecule reactions.

The reactions of O^- with H_2, D_2, and HD are explained by a two-step mechanism. The rate constants vary with the mass of the hydrogen molecule, as predicted by the mass-dependent collision rate, while the efficiency of the reaction remains constant. Isotopic effects in the branching fractions are most easily explained by the relative velocity of the departing H or D atom. The slower D atom allows more time for the complex to assume the critical configuration(s) necessary for associative detachment.

Finally, in the reaction of O^- with CH_4 and its deuterated analogs, zero point energy effects play a major role in both the observed rate constant and product branching fractions. The greater exothermicity of the reactions involving H abstraction leads to the rate constant for CH_4 being larger than that for CD_4 and also leads to a branching fraction for the reaction with CH_2D_2 which greatly favors OH^- production.

The mechanisms for all four of these reactions have little in common. Taken together, these four studies show the utility of using isotopically labeled reactants to elucidate a variety of reaction mechanisms. One has only to look at other chapters in this volume to see more such examples.

Acknowledgements

This research was sponsored in part by the United States Air Force under Contract No. F19628-86-C0224 and F49620-89-C-0019. The United States Government is authorized to reproduce and distribute reprints for governmental purposes notwithstanding any copyright notation herein.

Literature Cited

1. Ikezoe, Y.; Matsuoka, S.; Takebe, M.; Viggiano, A. A. *Gas Phase Ion-Molecule Reaction Rate Constants Through 1986.*; Maruzen Company, Ltd.: Tokyo, 1987.
2. Viggiano, A. A.; Morris, R. A.; Dale, F.; Paulson, J. F.; Giles, K.; Smith, D.; Su, T. *J. Chem. Phys.* **1990**, *93*, 1149.
3. Viggiano, A. A.; Morris, R. A.; Paulson, J. F. *J. Chem. Phys.* **1988**, *89*, 4848.
4. Viggiano, A. A.; Morris, R. A.; Paulson, J. F. *J. Chem. Phys.* **1989**, *90*, 6811.
5. Viggiano, A. A.; Van Doren, J. M.; Morris, R. A.; Paulson, J. F. *J. Chem. Phys.* **1990**, *93*, 4761.
6. Smith, D.; Adams, N. G. *Adv. At. Molec. Phys.* **1988**, *24*, 1.
7. Wannier, G. H. *Bell. Syst. Tech. J.* **1953**, *32*, 170.
8. Viehland, L. A.; Robson, R. E. *Int. J. Mass Spectrom. Ion Processes* **1989**, *90*, 167.
9. Viehland, L. A.; Viggiano, A. A.; Mason, E. A. *J. Chem. Phys.* **1991**, submitted.
10. Dressler, R. A.; Beijers, J. P. M.; Meyer, H.; Penn, S. M.; Bierbaum, V. M.; Leone, S. R. *J. Chem. Phys.* **1988**, *89*, 4707.
11. Dressler, R. A.; Meyer, H.; Langford, A. O.; Bierbaum, V. M.; Leone, S. R. *J. Chem. Phys.* **1987**, *87*, 5578.
12. Viggiano, A. A.; Morris, R. A.; Paulson, J. F. *J. Phys. Chem.* **1990**, *94*, 3286.
13. Su, T.; Morris, R. A.; Viggiano, A. A.; Paulson, J. F. *J. Phys. Chem.* **1990**, *94*, 8426.
14. Travers, M. J.; Cowles, D. C.; Ellison, G. B. *Chem. Phys. Lett.* **1989**, *164*, 449.
15. Albritton, D. L.; Dotan, I.; Lindinger, W.; McFarland, M. *J. Chem. Phys.* **1977**, *66*, 410.
16. Albritton, D. L. In *Kinetics of Ion-Molecule Reactions*; Ausloos, P., Ed.; Plenum Press: New York, 1979.
17. Grabowski, J. J.; DePuy, C. H.; Bierbaum, V. M. *J. Am. Chem. Soc.* **1983**, *105*, 2565.
18. Lias, S. G.; Bartmess, J. E.; Liebman, J. F.; Holmes, J. L.; Levin, R. D.; Mallard, W. G. *J. Phys. Chem. Ref. Data* **1988**, *17, Supplement 1*, 1.
19. Morris, R. A.; Viggiano, A. A.; Paulson, J. F. *J. Chem. Phys.* **1990**, *92*, 3448.
20. Van Doren, J. M.; Barlow, S. E.; DePuy, C. H.; Bierbaum, V. M. *J. Am. Chem. Soc.* **1987**, *109*, 4412.
21. Barlow, S. E.; Bierbaum, V. M. *J. Chem. Phys.* **1990**, *92*, 3442.
22. Paulson, J. F. *Adv. Chem. Ser.* **1966**, *58*, 28.
23. Posey, L. A.; Johnson, M. A. *J. Chem. Phys.* **1988**, *88*, 5383.
24. Hacaloglu, J.; Suzer, S.; Andrews, L. *J. Phys. Chem.* **1990**, *94*, 1759.
25. Jacox, M. E. *J. Chem. Phys.* **1990**, *93*, 7622.
26. Magnera, T. F.; Kebarle, P. In *Ionic Proc. Gas Phase*; Almoster Ferreira, M.A., Eds.; D. Reidel Publishing: Boston, 1984.
27. Troe, J. *Int. J. Mass Spectrom. Ion Processes* **1987**, *80*, 17.
28. Troe, J. In *Advances in Chemical Physics. State-selected and State-to-state Ion-molecule Reaction Dynamics: Theoretical Aspects, Part 2;* Baer, M., Ng, C. Y., Eds.; John Wiley: New York, in press.
29. Deakyne, C. A., private communication, 1989.

30. Kass, S. R.; Filley, J.; Van Doren, J. M.; DePuy, C. H. *J. Am. Chem. Soc.* **1986**, *108*, 2849.
31. Bierbaum, V. M.; DePuy, C. H.; Shapiro, R. H. *J. Am. Chem. Soc.* **1977**, *99*, 5800.
32. Dateo, C. E.; Clary, D. C. *J. Chem. Soc. Faraday Trans. II* **1989**, *85*, 1685.
33. Burley, J. D.; Erwin, K. M.; Armentrout, P. B. *Int. J. Mass Spectrom. Ion Processes* **1987**, *80*, 153.
34. Sunderlin, L. S.; Armentrout, P. B. *Chem. Phys. Lett.* **1990**, *167*, 188.
35. Viggiano, A. A.; Morris, R. A.; Deakyne, C. A.; Dale, F.; Paulson, J. F. *J. Phys. Chem.* **1991**, *95*, 3644.
36. McFarland, M.; Albritton, D. L.; Fehsenfeld, F. C.; Ferguson, E. E.; Schmeltekopf, A. L. *J. Chem. Phys.* **1973**, *59*, 6629.
37. Frisch, M. J.; Head-Gordon, M.; Trucks, G. W.; Foresman, J. B.; Schlegel, H. B.; Raghavachari, K.; Robb, M. A.; Binkley, J. S.; Gonzalez, C.; DeFrees, D. J.; Fox, D. J.; Whiteside, R. A.; Seeger, R.; Melius, C. F.; Baker, J.; Martin, R. L.; Kahn, L. R.; Stewart, J. J. P.; Topiol, S.; Pople, J. A. *Gaussian 90* (Carnegie - Mellon Quantum Chemistry Publishing Unit, Pittsburgh, PA, 1990).
38. Viggiano, A. A.; Van Doren, J. M.; Morris, Williamson, J.S., Mundis, P. L.; R. A.; Paulson, J. F. *J. Chem. Phys.* **in press**.

RECEIVED October 16, 1991

Chapter 16

Isotope-Exchange Reactions Within Gas-Phase Protonated Cluster Ions

Susan T. Graul[1], Mark D. Brickhouse[2], and Robert R. Squires

Department of Chemistry, Purdue University, West Lafayette, IN 47907

> The process of H/D isotope exchange within gas-phase cluster ions has been examined for a variety of small proton-bound cluster ions. Product ion distributions from low-energy collision-induced dissociation (CID) of partially deuterated water cluster ions reveal enrichment of deuterium in the neutral water products due to an equilibrium isotope effect wherein deuterium atoms preferentially occupy peripheral sites in the cluster. By comparison, product ion distributions from CID of partially deuterated ammonia cluster ions are controlled by both equilibrium and kinetic isotope effects, compounded by incomplete exchange within the clusters. Isotope exchange reactions and labeling studies have provided information about structures of cluster ions and mechanisms of bimolecular ion–molecule reactions. A summary of these studies is presented.

Neutral clusters and cluster ions have been the subject of intensive study in the past decade (*1-11*). Our interest in cluster ions derives in part from their potential role as models for the intermediates of bimolecular ion-molecule reactions (*12*). The study of cluster ions also provides a means to bridge the gap between the properties and reactivities of isolated gas-phase ions and their fully solvated counterpart species in solution (*13-17*). Several years ago, we initiated a program for studying structures of protonated cluster ions and the mechanisms of their formation and reactions. We were particularly interested in small cluster ions containing one or more molecules of typical solvents such as water, methanol, and ammonia loosely bound to a core ion via hydrogen bonds or electrostatic interactions (i.e., ion-dipole or ion-induced dipole). A significant aspect of our cluster ion research has involved the study of kinetic and thermodynamic isotope effects in H/D exchange reactions of cluster ions in order to reveal details of their structural features and the mechanisms of their fragmentation reactions.

---

[1]Current address: Department of Chemistry, University of California, Santa Barbara, CA 93106
[2]Current address: W. R. Grace, 7379 Route 32, Columbia, MD 21044

The experimental techniques employed to characterize the cluster ions include the use of isotope labeling and H/D isotope exchange reactions. We analyze the product distributions from thermal energy (~300 K) isotope exchange reactions and the collision energy dependence of the product distributions from low-energy collision-induced dissociation. In this chapter, we describe studies of isotope exchange reactions of a variety of protonated cluster ions composed of small inorganic and organic molecules such as H_2O, NH_3, CH_3CN, and CH_3OH. Hydrogen/deuterium isotope exchange occurs within these clusters by means of reversible deuteron and proton transfers between the component molecules. In bimolecular reactions of these clusters with labeled reactants, we can observe the competition between H/D exchange and molecular (solvent) exchange.

Experimental Approach

These experiments were carried out in a flowing afterglow – triple quadrupole apparatus (*18,19*). The flowing afterglow section of this instrument is a 1-meter long flow reactor in which the cluster ions are generated by termolecular association reactions and thermalized to 300 K by collisions with the helium buffer gas (0.30–0.50 torr). Reactant species can be added to the flowing gases via a series of inlets along the length of the flow tube, allowing for spatial and temporal separation of ion formation and reaction. The ions are sampled at a 1-mm orifice at the terminus of the flow tube and focused into a triple quadrupole mass spectrometer. The reactant ion of interest is mass-selected at the first quadrupole and injected into the second quadrupole, which is enclosed in shrouding to permit maintenance of a local pressure of 0.01-0.5 mtorr of reactant or collision gas. The second quadrupole thus functions as a collision and/or reaction chamber for collision-induced dissociation (CID) and activated bimolecular reactions. Product ions from reactions or fragmentations are mass-analyzed at the third quadrupole.

Isotope Fractionation within Protonated Cluster Ions

In this section, we will describe the results of several studies of the deuterium isotope distribution in partially deuterated cluster ions. We refer specifically to the distribution of deuterium among different sites <u>within an isolated cluster ion</u>, and not between that cluster ion and other neutral or ionic species present in the reaction or collision region. The isotope distributions are probed by means of collision-induced dissociation of the cluster ions. The product distributions we discuss herein correspond to those products formed by collision-induced dissociation (CID) that result from loss of a <u>single</u> deuterated or undeuterated molecule from a precursor cluster with a given number of deuterium atoms. For example, CID of the d_2-water dimer cluster ion yields three sets of products (equation 1), and it is the significance of the relative amounts of these products that we will discuss.

$$H_3D_2O_2^+ \xrightarrow{CID} \begin{cases} H_3O^+ + D_2O \\ H_2DO^+ + HOD \\ HD_2O^+ + H_2O \end{cases} \quad (1)$$

Water Clusters. Protonated water clusters $(H_2O)_nH^+$ are readily formed in the flow tube by means of bimolecular ion–molecule reactions and termolecular association reactions that occur when water vapor is added into the ion source (equations 2-4; M = a third body in a stabilizing thermal energy collision).

$$H_2O^+ + H_2O \longrightarrow H_3O^+ + OH \qquad (2)$$

$$H_3O^+ + H_2O \rightleftharpoons \left[(H_2O)_2H^+\right]^* \xrightarrow{M} (H_2O)_2H^+ \qquad (3)$$

$$(H_2O)_nH^+ + H_2O \rightleftharpoons \left[(H_2O)_{n+1}H^+\right]^* \xrightarrow{M} (H_2O)_{n+1}H^+ \qquad (4)$$

Dimer, trimer and tetramer clusters (that is, $(H_2O)_{2-4}H^+$) can be formed in abundance. However, the pentameric cluster is detected at only a small fraction of the abundance of the tetramer, and larger clusters are not observed at all. The low abundance of the larger cluster ions is probably a consequence of thermal dissociation (20). The heat capacity of these clusters increases with cluster size, because the addition of each water ligand creates a new internal rotor and several low frequency vibrations that can store a significant amount of energy at room temperature. Under the pressure and temperature conditions of the flow tube, cluster formation is controlled by free energy. The ΔG of association decreases with increasing cluster size. At the transition from tetramer to pentamer, the ΔG of association is only 5.6 kcal/mol (21), which is probably comparable to the internal energy of the pentamer at 300 K. These factors preclude the production of clusters larger than the pentamer in our experiments.

When $(H_2O)_nH^+$ clusters are allowed to react with D_2O in the flow tube, rapid deuterium exchange ensues (22-25). The H/D exchange product distributions observed at low extent of reaction reveal that incorporation of a single deuterium occurs faster than incorporation of two, indicating that exchange occurs primarily by proton/deuteron transfer rather than solvent switching. Measurements by Adams, Smith and Henchman of the kinetics of isotope exchange in the bimolecular reactions of H_3O^+ with D_2O (or D_3O^+ with H_2O) (equation 5) showed a nearly statistical distribution of hydrogen and deuterium in the products (25).

$$H_3O^+ + D_2O \rightleftharpoons \left(H_3D_2O_2^+\right) \begin{array}{c} \nearrow H_2DO^+ + HOD \\ \searrow HD_2O^+ + H_2O \end{array} \qquad (5)$$

This indicates that the exchange is essentially complete within the lifetime of the collision complex. However, close inspection of the data reveals small deviations in the experimental distributions from those predicted for statistical scrambling. One plausible reason for these differences is equilibrium isotope effects associated with the differences in zero-point energies between the deuterated and undeuterated species. The proton affinity of H_2O is slightly higher than that of D_2O (25). This favors enrichment of deuterium in the neutral reaction product.

A useful model for the intermediate species in the bimolecular proton transfer reaction of H_3O^+ with H_2O is the protonated water dimer $(H_2O)_2H^+$. Inspection of the favored structures for the dimer, trimer and tetramer water cluster, as shown in Figure 1(26-28), reveals two different types of sites occupied by hydrogens: bridging sites associated with the ion core and peripheral sites associated with the solvent shell (29). For the partially deuterated water clusters, there are, in principle, multiple isotopomeric structures wherein the deuterium atoms occupy bridging or peripheral sites. For example, the isotopomers of the d_2-water dimer cluster are shown in Figure 2.

Based on the rapidity of the H/D exchange observed by Adams et al. for the relatively short-lived collision complexes formed in the bimolecular reaction of H_3O^+ with D_2O (25), we expect that in the the stabilized partially deuterated water clusters, isotope scrambling will also be complete. It seems a natural question to ask whether the distribution of deuterium between the different types of sites in the "completely scrambled" clusters is perturbed from a statistical distribution as a result of thermodynamic (equilibrium) isotope effects. Do isotope effects lead to a measurable preference for deuterium occupation of a particular site - bridging or peripheral - over the other?

To address this question, we carried out low-energy collision-induced dissociation (CID) of partially deuterated water clusters and carefully analyzed the product distributions for loss of neutral water as H_2O, HOD, or D_2O (*30*). The distributions were monitored over a range of collision energies from about 0.5 to 6 eV in the center of mass frame of reference. For these low-energy collisions, the average internal energy deposited in the clusters should scale approximately linearly with collision energy (*31,32*). Consistent with this prediction, the yield of CID products increases with collision energy, and the fraction of products due to loss of two or three neutral water molecules increases.

The observed product distributions were close to those predicted by simple statistics, but showed a reproducible and experimentally significant preference for loss of deuterated neutral water (Figures 3-5). The distributions were reproducible within 1–2% (which is the experimental error for these measurements) over a wide range of pressure conditions and collision energies. This indicates that the relative rates of dissociation for the different isotopomers and different sets of isotopically substituted products do not change measurably in the energy range of 0.5 to 6 eV. The lack of energy dependence suggests that we are observing a thermodynamic isotope effect. (Kinetic isotope effects generally are dependent on internal energy.) The neutral water product of CID is removed from the "solvent shell" or periphery of the cluster ion. Thus, the preferential loss of deuterated neutral water indicates that the peripheral sites of the cluster ions are enriched in deuterium.

In bulk water, deuterium isotope fractionation is a well-known phenomenon manifested by enrichment of deuterium in neutral water and depletion in solvated hydronium ions (*33-36*). The equilibrium constant for the exchange reaction 6 in bulk water is only 0.69±0.02 at 25°C rather than 1.0 as predicted for no isotope effect (*37-40*).

$$\frac{1}{3}H_3O^+ + \frac{1}{2}D_2O \rightleftharpoons \frac{1}{3}D_3O^+ + \frac{1}{2}H_2O \qquad (6)$$

This fractionation effect has been attributed to vibrational zero-point energy differences and to entropy effects of hydration of the ions (*41-46*). Fractionation has also been observed for H_3O^+ in solution with acetonitrile (*47*), and recent studies have shown that isotope fractionation occurs in the gas phase as well, with an equilibrium constant for gas-phase reaction 6 comparable to that in acetonitrile solution (*22,23*). Our CID results provide an indication that this fractionation effect is further reproduced within gas-phase cluster ions, which represent the reaction intermediates of the bimolecular gas-phase reactions. The enhanced loss of deuterated water from CID of the partially deuterated water clusters points to enrichment of the peripheral sites of the cluster ions, which are comparable to the neutral water component in the binary equilibrium in bulk solution the gas phase.

Ammonia Clusters. From the relative rates of proton or deuteron transfer for reactions 7 and 8, Adams et al. concluded that the proton affinity of ND_3 is higher than that of NH_3 (*25*).

$$NH_4^+ + ND_3 \longrightarrow NH_3D^+, NH_2D_2^+, NHD_3^+ \qquad (7)$$

$$ND_4^+ + NH_3 \longrightarrow NHD_3^+, NH_2D_2^+, NH_3D^+ \qquad (8)$$

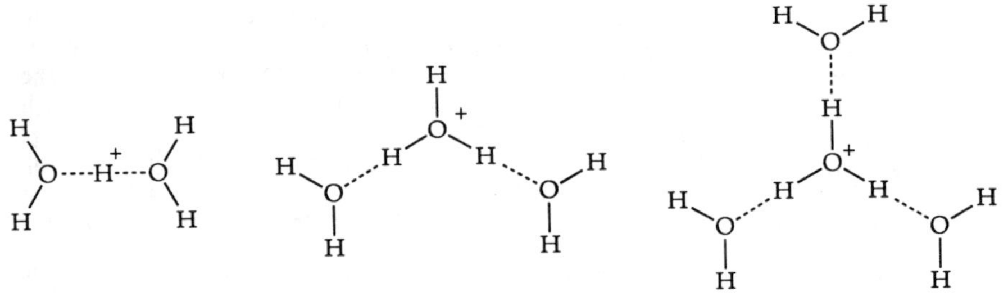

Figure 1. The protonated water dimer ion has a symmetrical hydrogen bond. The trimer and tetramer have core H_3O^+ ions with H_2O molecules in the solvent shell, bound by hydrogen bonding.

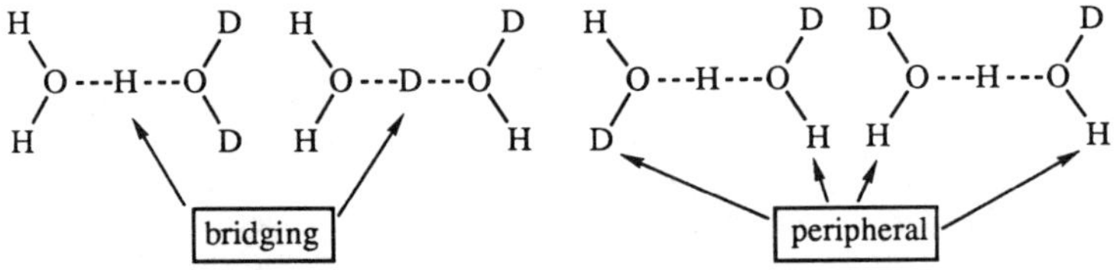

Figure 2. Isotopomeric structures for the dideuterated (d_2) water dimer cluster ion. Bridging and peripheral sites are indicated.

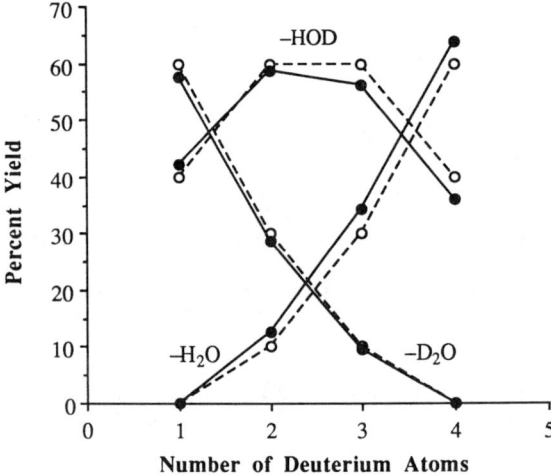

Figure 3. Water loss from the dimer cluster ion $D_nH_{5-n}O_2^+$. The statistical distributions are shown by open circles and dashed lines and the experimental distributions by closed circles and solid lines. The connecting lines are intended only to aid the eye in detecting trends in the deviations.

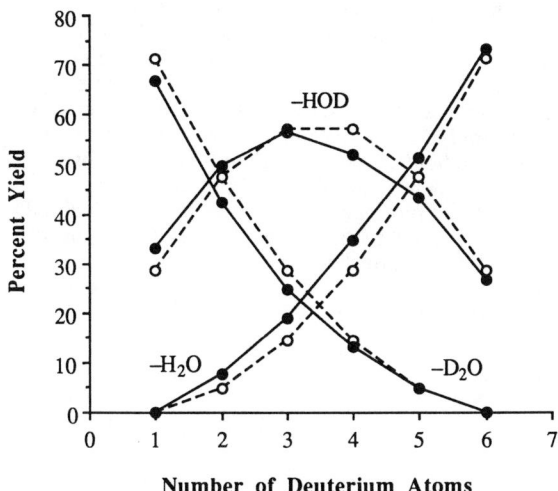

Figure 4. Water loss from the trimer cluster ion $D_nH_{7-n}O_3^+$. Symbols are as described in Figure 3.

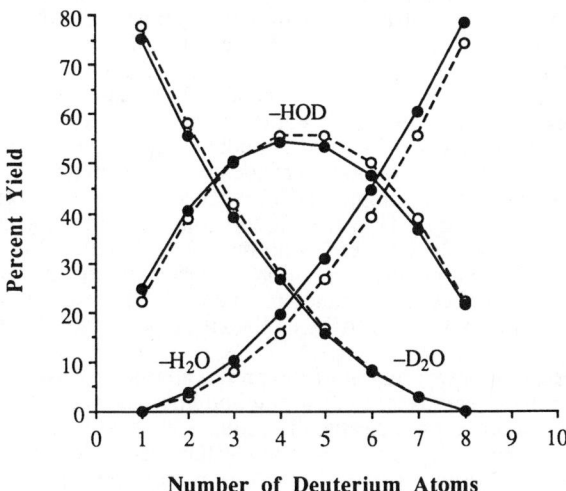

Figure 5. Water loss from the tetramer cluster ion $D_nH_{9-n}O_4^+$. Symbols are as described in Figure 3.

This ordering is the reverse of the relative proton affinities of H_2O and D_2O, wherein the deuterated species has the lower proton affinity. Thus, in contrast to the water clusters, one might predict that equilibrium isotope effects within partially deuterated ammonia clusters will result in deuterium enrichment in the <u>bridging</u> sites, leading to preferential loss of undeuterated ammonia upon CID. In view of the different isotope effects expected for the ammonia system compared with water clusters, it is of interest to examine the product distributions for CID of partially deuterated ammonia clusters.

Qualitative observations of the H/D exchange reactions of $(NH_3)_nH^+$ with ND_3 in the flow tube revealed a fundamental difference between this system and the water clusters. The rate at which a given $(NH_3)_nH^+$ cluster ion incorporated three deuterium atoms in reaction with ND_3 was comparable to the rate of incorporation of one deuterium atom. This indicates that the lifetimes of the collision complexes are not long enough to allow complete scrambling in a single collision. This is consistent with incomplete exchange in the bimolecular reactions of NH_4^+ with ND_3 and ND_4^+ with NH_3 (25). Thus, in this system, the rate of exchange of ammonia <u>molecules</u> competes with the rate of proton/deuteron exchange. This competition is depicted in equation 9, in which equation 9a corresponds to exchange of an ammonia molecule and equation 9b to the proton/deuteron exchange.

$$(NH_3)_nH^+ + ND_3 \longrightarrow (NH_3)_{n-1}(ND_3)H^+ + NH_3 \quad (9a)$$
$$(NH_3)_nH^+ + ND_3 \longrightarrow (NH_3)_{n-1}(NH_2D)H^+ + NHD_2 \quad (9b)$$

The product distributions from CID of the partially deuterated ammonia clusters were within 10% of the distributions predicted for complete scrambling of the isotopes. The observed deviations from the statistical distributions reflected a preference for loss of neutral NH_3 or ND_3 relative to NH_2D or NHD_2 (Figures 6-8). Such a trend is not readily explained in terms of equilibrium isotope effects. These product distributions are probably a result of a complex interplay of equilibrium isotope effects and relatively slow H/D exchange within the cluster ions. Whereas the former effect should result in migration of deuterium to the ion core, and consequently enrichment of deuterium in the <u>ionic</u> products of CID, the latter effect would retard the H/D exchanges necessary to attain equilibrium. The observed CID product distributions for the partially deuterated ammonia clusters were independent of the cluster residence time in the flow tube for residence times between 2 and 7 msec. This indicates that the apparently incomplete H/D exchange within the clusters is not a result of insufficient reaction time. However, if collisional stabilization of the ammonia cluster ions in the flow tube competes with the rate of internal H/D scrambling, we might expect to observe a greater population of cluster structures with intact NH_3 or ND_3 molecules than predicted for a random isotope distribution. These structures are essentially "frozen in" by the stabilizing collisions of the adduct ions with the helium buffer gas.

An alternative intrepretation of the preferential loss of NH_3 and ND_3 is that H/D exchange <u>is</u> in fact complete, and that equilibrium isotope effects favor localization of hydrogen or deuterium on the same molecule. This could result if the zero-point energies for the partially deuterated NH_2D and NHD_2 species do not fall between those of NH_3 and ND_3. We consider this situation improbable, but cannot rule it out. Accurate vibrational frequencies are required to calculate zero-point energies, but these are not available for NH_2D or NHD_2.

In further contrast to the water system, the product distributions from CID of the ammonia clusters were dependent on collision energy, indicating a kinetic effect. Changes in the product distributions as the collision energy was decreased to 0.5 eV seemed to indicate a trend toward enrichment of deuterium in the ionic products (as

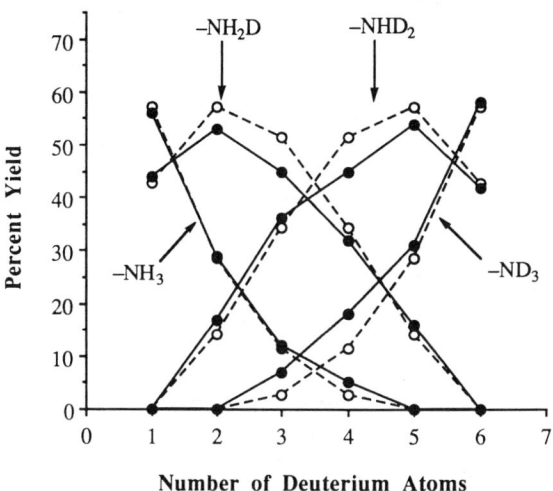

Figure 6. Ammonia loss from the dimer cluster ion $D_nH_{7-n}N_2^+$. Symbols are as described in Figure 3.

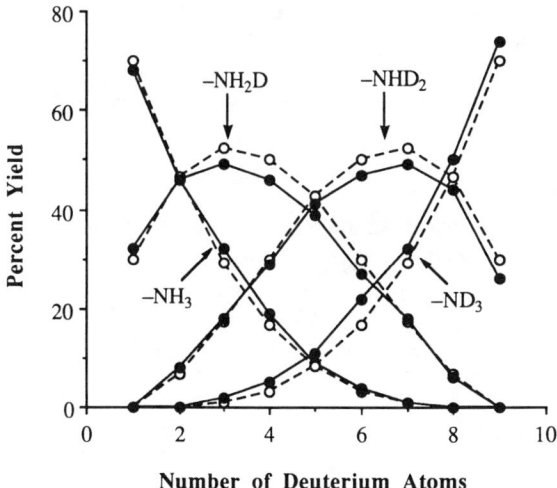

Figure 7. Ammonia loss from the trimer cluster ion $D_nH_{10-n}N_3^+$. Symbols are as described in Figure 3.

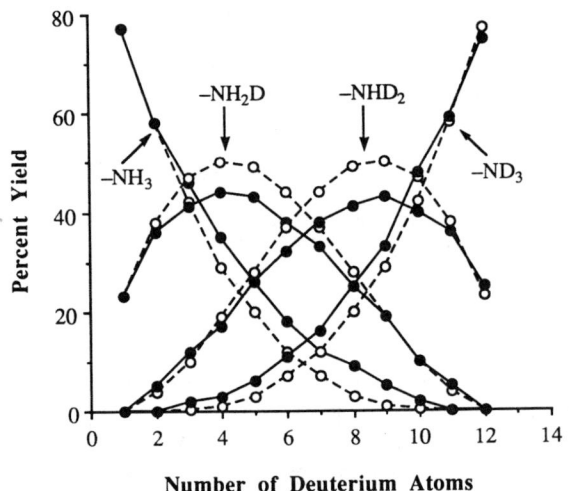

Figure 8. Ammonia loss from the tetramer cluster ion $D_nH_{13-n}N_4^+$. Symbols are as described in Figure 3.

would be predicted from the equilibrium isotope effect). At higher energies, the product distributions reverted to the situation noted above, with enhanced loss of both NH_3 and ND_3 relative to statistics. The origin of collision energy dependence in CID product distributions can be found in the kinetics for different dissociation processes. At low collision energies, less energy is deposited in the cluster, and the lifetime of the excited species may be long enough to permit rearrangement reactions (such as H/D exchange) prior to dissociation. At the higher collision energies used, more energy is deposited such that the lifetime of the excited species may be too short to permit rearrangement. Under these circumstances, simple cluster bond cleavage is expected to dominate. (We note that CID employing very high energy collisions (>100 eV) can result in electronic excitation, and, for many ions, rearrangement reactions can be observed. Under the conditions used in this study however, the mechanism of excitation is most probably vibrational/rotational excitation, and the above generalizations concerning the energy dependence of CID product distributions should hold.)

These ammonia cluster ion CID results provide an interesting contrast to the water cluster ion results. In both systems, we are able to make conclusions about the distribution of deuterium and hydrogen within the clusters. Whereas clear evidence for isotope fractionation is found for the water cluster ions, the results for the ammonia cluster ions instead seem to indicate that H/D exchange is incomplete. It seems clear that both kinetic and equilibrium isotope effects, as well as non-equilibrium isotope distributions in the reactant ions, are involved in determining the ammonia cluster ion CID product distributions.

Methanol Clusters. Deuterium enrichment in the neutral products of CID has also been observed for methanol clusters (48). The focus of this study was elucidation of the mechanism of the displacement reaction 10 (49,50).

$$CH_3OH_2^+ + CH_3OH \longrightarrow (CH_3)_2OH^+ + H_2O \qquad (10)$$

We examined the mechanism of this reaction by using $CH_3{}^{18}OH_2^+$ reactant ions and natural isotope abundance CH_3OH to track the fate of the ^{18}O label, and found a 2:1 preference for loss of ^{18}O in the neutral water product, suggesting a backside displacement mechanism (equation 11).

$$\begin{matrix} H \\ \diagdown \\ O: \curvearrowright CH_3 \text{---}{}^{18}OH_2^+ \\ \diagup \\ CH_3 \end{matrix} \longrightarrow (CH_3)_2OH^+ + H_2{}^{18}O \qquad (11)$$

The bimolecular reaction of $CH_3{}^{18}OH_2^+$ with CH_3OD yielded the same 2:1 ratio for distribution of the ^{18}O label between the protonated dimethyl ether and water products, but a nearly statistical distribution of hydrogen and deuterium (equation 12).

$$CH_3{}^{18}OH_2^+ + CH_3OD \longrightarrow \begin{cases} \left. \begin{array}{l} (CH_3)_2OH^+ + H^{18}OD \\ (CH_3)_2OD^+ + H_2{}^{18}O \end{array} \right\} 2 \\ \left. \begin{array}{l} (CH_3)_2{}^{18}OH^+ + HOD \\ (CH_3)_2{}^{18}OD^+ + H_2O \end{array} \right\} 1 \end{cases} \qquad (12)$$

We examined the product distribution from CID of the singly deuterated methanol dimer cluster ion, and found that it agreed within 5% with the statistical H/D distributions for both direct cleavage and displacement products (equation 13).

$$d_1-(CH_3OL)_2L^+ \quad L = H,D \begin{cases} a \begin{cases} CH_3OHD^+ + CH_3OH \\ CH_3OH_2^+ + CH_3OD \end{cases} \\ b \begin{cases} (CH_3)_2OD^+ + H_2O \\ (CH_3)_2OH^+ + HOD \end{cases} \end{cases} \quad (13)$$

The branching between cluster bond cleavage channel (a) and displacement channel (b) in equation 13 was dependent on collision energy, but the relative quantities of H- and D-labeled products within each channel were independent of collision energy. For the d_1-dimer, about twice as much CH_3OHD^+ was formed as $CH_3OH_2^+$, and about twice as much $(CH_3)_2OH^+$ was formed as $(CH_3)_2OD^+$. These near-statistical product ratios were reversed for the d_2-dimer, as expected. However, small deviations from a statistical distribution of H/D isotopes between the products in each channel were observed, and these corresponded to deuterium enrichment of 2-5% in the neutral products. As with the water cluster ions, these results may be due to an isotope effect that favors deuterium fractionation into the neutral products. The results suggested that H/D scrambling in this reaction is facile and moreover can continue subsequent to displacement, leading to a near-statistical distribution of hydrogen and deuterium but a nonstatistical distribution of the oxygen label.

These studies demonstrate that low-energy collision-induced dissociation can be a sensitive probe of the distribution of hydrogen and deuterium within proton/deuteron-bound clusters. We have seen that equilibrium isotope effects associated with isotopomeric forms of cluster ions can control the distribution of hydrogen and deuterium *within* cluster ions. Such effects are analogous to the equilibrium isotope effects associated with differing zero-point energies of the separated products and reactants. It is intriguing to speculate that the isotope effects noted for water and methanol clusters may be general for many alcohol cluster ions (perhaps both positive and negative ions), in which proton transfers are facile (*51-53*). Similarly, other amine clusters might be characterized by relatively slow proton exchanges, in analogy with the ammonia system. These experiments provide a means to address such questions.

Isotope Exchange Reactions within Protonated Cluster Ions

Water/Ammonia Cluster Ions. An interesting extension of our studies of isotope fractionation in partially deuterated water and ammonia clusters is the mixed ammonia–water system. When $(NH_3)_nH^+$ clusters were allowed to react with H_2O in the flow tube, solvent exchange reactions were not observed, consistent with the endothermicity of the reaction for small clusters (*54,55*). However, when $(NH_3)_nH^+$ clusters were allowed to react with D_2O in the flow tube, H/D exchange into the clusters <u>was</u> observed, clearly indicating that D_2O molecules are incorporated at least into the transient intermediates of the exchange reactions. It is probable that the reason that $(H_2O)(NH_3)_nH^+$ cluster ions were not observed in the reaction of $(NH_3)_nH^+$ with H_2O is not that such species are not formed, but that they undergo rapid reactions with NH_3 to switch out H_2O. In order to form mixed clusters of NH_3 and H_2O, a large excess of H_2O must be present to drive the equilibrium shown in equation 14 toward the mixed clusters.

$$(NH_3)_nH^+ + H_2O \rightleftharpoons (NH_3)_{n-1}(H_2O)H^+ + NH_3 \quad (14)$$

For the reaction of $(NH_3)_nH^+$ with D_2O, it cannot be determined simply from the masses of the exchange products whether D_2O is incorporated into the stabilized partially deuterated clusters (as opposed to transient reaction intermediates). For example, the nominal masses of a d_n–ammonia dimer and a d_{n-1}–ammonia/water dimer are equal, and these clusters are not resolved in a quadrupole mass spectrometer. However, CID can be used to distinguish these species. Shown in Figure 9 is the product distribution for loss of 17, 18, 19, and 20 amu from the trimer cluster ions formed in the reaction of $(NH_3)_nH^+$ with D_2O. These neutral losses correspond to NH_3 (17 amu), NH_2D or H_2O (18 amu), NHD_2 or HOD (19 amu), and ND_3 or D_2O (20 amu). In Figure 9, we indicate the mass of the precursor cluster ions rather than the number of deuterium atoms in the clusters because of the possibility of D_2O-uptake by the ammonia clusters. For example, a cluster of mass 56 could be d_4-(ammonia)$_3$H$^+$ and/or d_3-(ammonia)$_2$(water)H$^+$. For comparison, the experimental distributions are compared to the predicted products from CID of pure ammonia cluster ions with statistical distributions of hydrogen and deuterium. It is clear that the CID results for the products of the $(NH_3)_nH^+/D_2O$ reactions differ significantly from those observed for the partially deuterated ammonia cluster ions discussed above and shown in Figures 6-8. The intensity of the CID product ion from loss of 20 amu is greatly enhanced relative to the statistical prediction, and furthermore is considerably more prominent than the corresponding product ion from CID of the partially deuterated ammonia cluster ions discussed above.

When the $(NH_3)_2(H_2O)H^+$ cluster is subjected to low-energy collision-induced dissociation, the product resulting from single ligand loss is almost exclusively $(NH_3)_2H^+$. By analogy, the the enhanced loss of 20 amu indicates that D_2O has been incorporated into the clusters. Moreover, the data shows that H/D exchange is far from complete. Thus, we can conclude that in the partially deuterated mixed clusters formed by reaction of $(NH_3)_nH^+$ with D_2O, the deuterium atoms remain largely on the water molecules. This, in turn, indicates that collisional stabilization of the nascent mixed ammonia/water clusters in the flow tube competes with H/D exchange.

The fact that loss of 18 and 19 amu is also observed for CID of the mixed NH_3/D_2O clusters indicates that H/D exchange does occur, albeit slowly. It has been demonstrated that nominally endothermic proton transfers can occur within collision complexes formed in bimolecular ion–molecule reactions, where the energy necessary to drive the reaction is provided by electrostatic interactions (56-58). Proton transfer is 40 kcal/mol endothermic for reaction 15 (59), but the energy provided by the interaction between NH_4^+ and D_2O is only half as large (54), and thus inadequate to drive exchange.

$$NH_4^+ + D_2O \longrightarrow D_2OH^+ + NH_3 \quad (15)$$

However, proton transfer between ammonia and water moieties within larger clusters is less endothermic (29), and can be driven by the combination of the solvation energy associated with the additional ligands and the electrostatic interaction energy between the $(NH_3)_nH^+$ and D_2O reactants. As noted above, the observation of deuterated product ions from CID of the mixed trimer indicated that H/D exchange had indeed occurred. The H/D exchange is far from complete, and the deuterium atoms generally remain on the water moiety within the cluster. This is almost certainly a kinetic effect associated with the nominally endothermic proton/deuteron transfer from the ammonium ion core to water within the clusters, which is a required step in isotope scrambling.

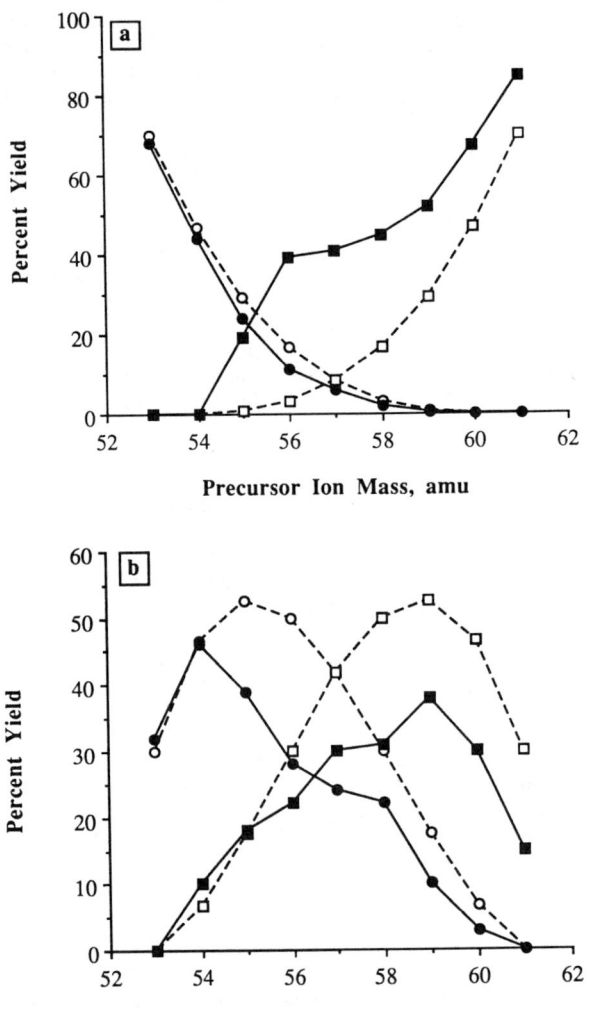

Figure 9. Neutral ligand loss from partially deuterated mixed clusters of ammonia and water. Shown for the purpose of comparison are the are statistical distributions predicted for loss of ammonia from partially deuterated ammonia cluster ions (open symbols and dashed lines). Closed symbols are experimental distributions. (a) Circles correspond to loss of 17 amu (NH_3) and squares to loss of 20 amu (D_2O or ND_3). (b) Circles correspond to loss of 18 amu (NH_2D or H_2O) and squares to loss of 19 amu (NHD_2 or HOD).

Acetonitrile/Water Clusters. Deuterium exchange reactions were utilized in a study of the structures of acetonitrile–water clusters $(CH_3CN)_n(H_2O)_mH^+$ (60). In this study, results of energy-resolved CID provided support for theoretically predicted cluster structures, which contain a core H_3O^+ ion for n + m = 3,4 (Figure 10) (55). These structures are favored by the improved hydrogen-bonding capability of an H_3O^+ ion in comparison to CH_3CNH^+. Alternative structures with core CH_3CNH^+ ions require hydrogen bonding through the methyl group (Figure 6), which is probably a relatively weak interaction (61-63). Reaction of D_2O with $(CH_3CN)_3(H_2O)H^+$ in the flow tube resulted in only three rapid deuterium exchanges. Collisional activation of the $(CH_3CN)_3(H_2OD^+)$, $(CH_3CN)_3(HOD_2^+)$, and $(CH_3CN)_3(D_3O^+)$ clusters formed by this reaction resulted in loss of acetonitrile and water in sequential desolvation steps. Acetonitrile was always lost as CH_3CN whereas the neutral water products were H_2O, HOD, or D_2O depending on the deuterium content of the cluster. These results are consistent with no isotope exchange occurring into the methyl group of the acetonitrile ligands.

Conclusions

Through the use of the technique of low-energy collision-induced dissociation, we have found evidence for an equilibrium isotope effect that causes preferential migration of deuterium to the periphery of partially deuterated gas-phase water cluster ions. This fractionation process is reflected in enrichment of deuterium in the neutral products of dissociation. A reversal of this isotope effect is expected for ammonia clusters, but CID of these species reflects additional kinetic effects, not observed for water clusters, that prevent complete equilibration of the cluster isotopomers and lead to energy-dependent CID product distributions. Fractionation in the methanol system parallels the water system, in that the neutral products of CID are deuterium-enriched. In contrast to the facile exchange in each of these systems, H/D exchange within mixed cluster ions of water and ammonia is inefficient, such that the deuterium label remains predominantly on the same moiety in which it first entered the cluster ion. For mixed acetonitrile/water clusters, H/D exchange involves only the hydroxyl sites of the water ligands; exchange into the methyl group of acetonitrile is not observed.

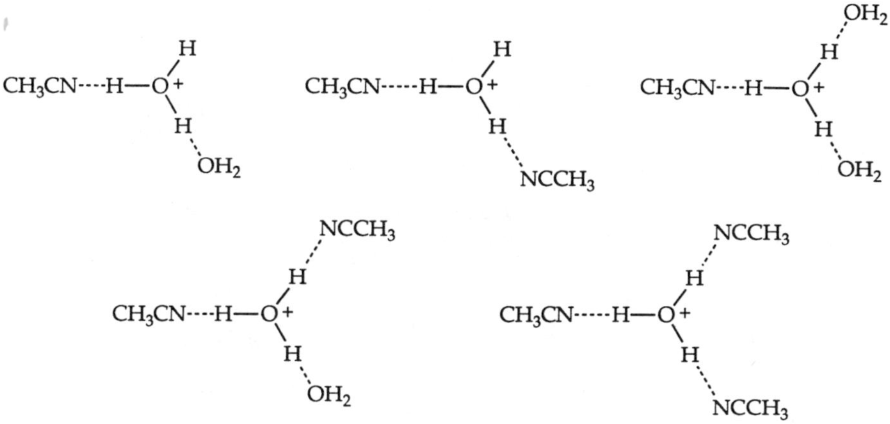

Figure 10. Although acetonitrile is a stronger base than water, the core ion in many mixed acetonitrile–water clusters is H_3O^+, which can form networks of strong hydrogen bonds.

Acknowledgment

This work was supported by the National Science Foundation (Grant CHE-8815502).

Literature Cited

1. Märk, T. D.; Castleman, A. W., Jr. *Adv. At. Mol. Phys.* **1985**, *20*, 65.
2. Castleman, A. W., Jr.; Keesee, R. G. *Annu. Rev. Phys. Chem.* **1986**, *37*, 825.
3. Castleman, A. W., Jr.; Keesee, R. G. *Z. Phys. D: At., Mol. Clusters* **1986**, *3*, 167.
4. Sattler, K. *Ultramicroscopy* **1986**, *20*, 55.
5. Castleman, A. W., Jr.; Keesee, R. G. *Acc. Chem. Res.* **1986**, *19*, 413.
6. Colton, R. J.; Kidwell, D. A.; Ross, M. M. In *Mass Spectrometric Analysis of Large Molecules*, McNeal, C. J., Ed.; Wiley: Chichester, U. K. 1986.
7. Brutschy, B.; Bisler, P.; Rühl, E.; Baumgärtel, H. *Z. Phys. D: At., Mol. Clusters* **1987**, *5*, 217.
8. DeHeer, W. A.; Knight, W. D.; Chou, M. Y.; Cohen. M. L. *Solid State Phys.* **1987**, *40*, 93.
9. Campana, J. E. *Mass Spectrom. Rev.* **1987**, *6*, 395.
10. King, B. V.; Tsong, I. S.; Lin, S. H. *Int. J. Mass Spectrom. Ion Proc.* **1987**, *79*, 1.
11. Kondow, T. *J. Phys. Chem.* **1987**, *91*, 1307.
12. Castleman, A. W., Jr.; Keesee, R. G. *Chem. Rev.* **1986**, *86*, 589.
13. Mackay, G. I.; Bohme, D. K. *J. Am. Chem. Soc.* **1978**, *100*, 327.
14. Bohme, D. K.; Mackay, G. I. *J. Am. Chem. Soc.* **1981**, *103*, 978.
15. Bohme, D. K.; Rakshit, A. B.; Mackay, G. I. *J. Am. Chem. Soc.* **1982**, *104*, 1100.
16. Hierl, P. M.; Ahrens, A. F.; Henchman, M.; Viggiano, A. A.; Paulson, J. F.; Clary, D. C. *J. Am. Chem. Soc.* **1986**, *108*, 3140, 3142.
17. Freiser, B. S.; Beauchamp, J. L. *J. Am. Chem. Soc.* **1977**, *99*, 3214.
18. Graul, S. T.; Squires, R. R. *Mass Spectrom. Rev.* **1988**, *7*, 263.
19. Squires, R. R.; Lane, K. R.; Lee, R. E.; Wright, L. G.; Wood, K. V.; Cooks, R. G. *Int. J. Mass Spectrom. Ion Proc.* **1985**, *65*, 185.
20. Viggiano, A. A.; Dale, F.; Paulson, J. F. *J. Chem. Phys.* **1988**, *88*, 2469.
21. Meot-Ner, M. *J. Am. Chem. Soc.* **1986**, *108*, 6189.
22. Larson, J. W.; McMahon, T. B. *J. Am. Chem. Soc.* **1986**, *108*, 1719.
23. Larson, J. W.; McMahon, T. B. *J. Am. Chem. Soc.* **1988**, *110*, 1087.
24. Smith, D.; Adams, N. G.; Henchman, M. J. *J. Chem. Phys.* **1980**, *72*, 4951.
25. Adams, N. G.; Smith, D.; Henchman, M. J. *Int. J. Mass Spectrom. Ion Proc.* **1982**, *42*, 11.
26. Newton, M. D.; Ehrenson, S. *J. Am. Chem. Soc.* **1971**, *93*, 4971.
27. Scheiner, S. *J. Phys. Chem.* **1982**, *86*, 376.
28. Del Bene, J. E.; Frisch, J. J.; Pople, J. A. *J. Phys. Chem.* **1985**, *89*, 3669.
29. Meot-Ner, M.; Speller, C. V. *J. Phys. Chem.* **1986**, *90*, 6616.
30. Graul, S. T.; Squires, R. R. *J. Am. Chem. Soc.* **1990**, *112*, 631.
31. Kenttämaa, H. I.; Cooks, R. G. *Int. J. Mass Spectrom. Ion Proc.* **1985**, *64*, 79.
32. Wysocki, V. H.; Kenttämaa, H. I.; Cooks, R. G. *Int. J. Mass Spectrom. Ion Proc.* **1987**, *75*, 181.
33. Purlee, E. L. *J. Am. Chem. Soc.* **1959**, *81*, 263.
34. Kreevoy, M. M. In *Isotopes in Organic Chemistry*; Buncel, E.; Lee, C. C., Eds.; Elsevier: Amsterdam, **1976**, Vol. 2.
35. Melander, L.; Saunders, W. H., Jr. In *Reaction Rates of Isotopic Molecules*; Wiley-Interscience: New York, 1980.

36. Kresge, A. J.; More O'Ferrall, R. A.; Powell, M. F. In *Isotopes in Organic Chemistry*; Buncel, E., Lee, C. C. Eds.; Elsevier: Amsterdam, 1987, Vol. 7.
37. Kresge, A. J.; Allred, A. L. *J. Am. Chem. Soc.* **1963**, *85*, 1541.
38. Gold, V. *Proc. Chem. Soc., London* **1963**, 14.
39. Heinzinger, K.; Weston, R. E. *J. Phys. Chem.* **1964**, *68*, 744.
40. Salomaa, R.; Aalto, V. *Acta Chem. Scan.* **1966**, *20*, 2035.
41. Swain, C. G.; Bader, R. F. *Tetrahedron* **1960**, *10*, 182.
42. Bunton, C. A.; Shiner, V. J., Jr. *J. Am. Chem. Soc.* **1961**, *83*, 42.
43. Halevi, A. E.; Long, F. A.; Paul, M. A. *J. Am. Chem. Soc.* **1961**, *83*, 305.
44. Gold, V.; Grist, S. *J. Chem. Soc. B* **1971**, 1665.
45. Gold, V.; Grist, S. *J. Chem. Soc., Perkin Trans. 2* **1972**, 89.
46. Taylor, C. E.; Tomlinson, C. *J. Chem. Soc., Faraday Trans. 2* **1974**, *70*, 1132.
47. Kurz, J. L.; Myers, M. T.; Ratcliff, K. M. *J. Am. Chem. Soc.* **1984**, *106*, 5631.
48. Graul, S. T.; Squires, R. R. *Int. J. Mass Spectrom. Ion Proc.* **1987**, *81*, 183.
49. Kleingeld, J. C.; Nibbering, N. M. M. *Org. Mass Spectrom.* **1982**, *17*, 136.
50. Bass, L. M.; Cates, R. D.; Jarrold, M. F.; Kirchner, N. J.; Bowers, M. T. *J. Am. Chem. Soc.* **1983**, *105*, 7024.
51. Wolfe, S.; Hoz, S.; Kim, C.-K.; Yang, K. *J. Am. Chem. Soc.* **1990**, *112*, 4186.
52. Szczesniak, M. M.; Scheiner, S. *J. Am. Chem. Soc.* **1982**, *77*, 4586.
53. Roos, B. O.; Kraemer, W. P.; Diercksen, G. H. F. *Theor. Chim. Acta* **1976**, *42*, 77.
54. Payzant, J. D.; Cunningham, A. J.; Kebarle, P. *Can. J. Chem.* **1973**, *51*, 3242.
55. Deakyne, C. A.; Meot-Ner, M.; Campbell, C. L.; Hughes, M. G.; Murphy, S. P. *J. Chem. Phys.* **1986**, *84*, 4958.
56. Squires, R. R.; Bierbaum, V. M.; Grabowski, J. J.; DePuy, C. H. *J. Am. Chem. Soc.* **1983**, *105*, 5185.
57. Grabowski, J. J.; DePuy, C. H.; Bierbaum, V. M. *J. Am. Chem. Soc.* **1985**, *107*, 7384.
58. Grabowski, J. J.; DePuy, C. H.; Bierbaum, V. M. *J. Am. Chem. Soc.* **1988**, *105*, 2565.
59. Lias, S. G.; Bartmess, J. E.; Liebman, J. F.; Holmes, J. L.; Levin, R. D.; Mallard, W. G. *J. Phys. Chem. Ref. Data*, **1988**, *17*, Suppl. 1.
60. Graul, S. T.; Squires, R. R. *Int. J. Mass Spectrom. Ion Proc.* **1989**, *94*, 41.
61. Kebarle, P. *Annu. Rev. Phys. Chem.* **1977**, *28*, 445.
62. Meot-Ner, M.; *J. Am. Chem. Soc.* **1984**, *106*, 1257.
63. Meot-Ner, M.; Deakyne, C. A. *J. Am. Chem. Soc.* **1985**, *107*, 469.

RECEIVED January 6, 1992

ISOTOPE EFFECTS IN PHOTODISSOCIATION PROCESSES

Chapter 17

Mechanisms for Isotopic Enrichment in Photochemical Reactions

Moshe Shapiro

Chemical Physics Department, Weizmann Institute of Science, Rehovot, 76100 Israel

We review the natural processes by which isotopic enrichment occurs in photodissociation processes. Special emphasis is given to the role of vibrational excitation in promoting isotopic selectivity. Examples include the HI/DI photodissociation, the role of vibrational excitation in CH_3I and the competition between the breakup of the HO and DO bonds in the B-continuum photodissociation of HOD,

$$H + O - D \leftarrow H - O - D \rightarrow H - O + D.$$

It is shown that excitation of either the OD or OH stretch modes results in a dramatic *inhibitory* isotopic effect.

The issue of isotopic enrichment while fragmenting a a polyatomic molecule is closely related to the issue of *mode specificity*. If one is able to preferentially dissociate one mode then one is also able to distinguish between different isotopes. In photodissociation, mode specificity can be achieved either by a process of *pre-selection* i.e., by dissociating an initially excited molecule, or by *post-selection* in which one excites to a mode that is inclined to dissociate to the desired product. (In recent years, an alternative route has also emerged in which one attempts to use the light's phase-coherence to control dissociation processes.(1,2) This method falls under the category of pre-selection for scenarios(1) involving the dissociation of a linear combination of excited states, or post-selection for scenarios(2) where one dissociates the same state via two different paths).

Before treating a polyatomic molecule it is worthwhile to examine diatomic molecules. For ground state diatomic molecules, the main isotope effect may be termed an "edge effect", because it is observed in the wings of the absorption line. Basically, the ground vibrational wavefunction of the lighter isotope extends more in coordinate space than the heavier isotope. This effect is mainly due to the ability of the lighter isotopic molecule to tunnel more deeply into the non-classical regions. To a lesser extent the effect is also felt in the classically allowed region: Given the same set of quantum numbers, the energy of the lighter isotope is higher, hence the volume occupied in coordinate space is larger. As a result, when the molecule is photodissociated, it follows from the reflection behavior(3) (which is substantiated by exact calculations, (cf. Ref. 4), that the absorption cross-section of the lighter isotope extends further to the red and further to the blue than that of the heavier isotope.

In the wings of the absorption we therefore expect to see preferential dissociation of the lighter isotope, whereas in the center of the absorption line the heavier isotope is preferred. The light-to-heavy isotopic ratio increases dramatically in the far wings, where the effect increases exponentially because of the fall-off of the absorption cross-section itself. In contrast, the isotopic ratio in the center (which favors the heavier isotope) is not very dramatic.

An example of this type of effect is given in Figs. 1,2 (taken from Ref. 5) and Figs. 3,4, where the absorption of HI is compared with that of DI. In Fig. 1,2 we show the partial absorptions due to the individual (decoupled) excited electronic states for both HI and DI. In Figs. 3,4 we display the HI/DI branching ratio and total absorption for the *fully coupled* case. (Details of the potential curves and the coupling between them are given in Ref. 5). Figure 3 depicts the photodissociation branching ratio from the ground state and Fig. 4 does the same for the first excited state.

Clearly, for both ground and first excited vibrational states, the absorption spectrum associated with the lighter isotope is much broader in frequency, resulting in a large isotope effect at the edges. For the first excited state the absorption has a pronounced minimum, corresponding to the node in the wavefunction. Because this node occurs near the equilibrium separation (which is identical for both isotopic molecules) and this gets mapped, by the reflection approximation, to the same frequency in the absorption line, there is very little isotopic discrimination in the line center, even for the excited state.

An additional isotopic effect exists if dissociation is not direct and relatively long-lived resonances (predissociations) appear in the spectrum. In this case it may be possible to tune directly to the resonance of one isotopic molecule in preference to another. This is an example of a "post-excitation" effect which is termed thus because it depends on properties of the continuum wavefunction to which the molecule has been excited. In contrast, the threshold effect discussed above may be classified under "pre-excitation" because it mainly depends on the shape of the wavefunction before the excitation act.

When a diatomic is (vibrationally) excited, the main isotope effect is also an "edge effect". The extension of the excited wavefunction in coordinate space exceeds that of the ground wavefunction. Accordingly, the absorption spectrum spreads out in frequency space, both to the red and to the blue. The red-shift of the onset of absorption due to this effect exceeds by far the shift caused by the mere increase in the initial energy. This enables one to achieve isotopic

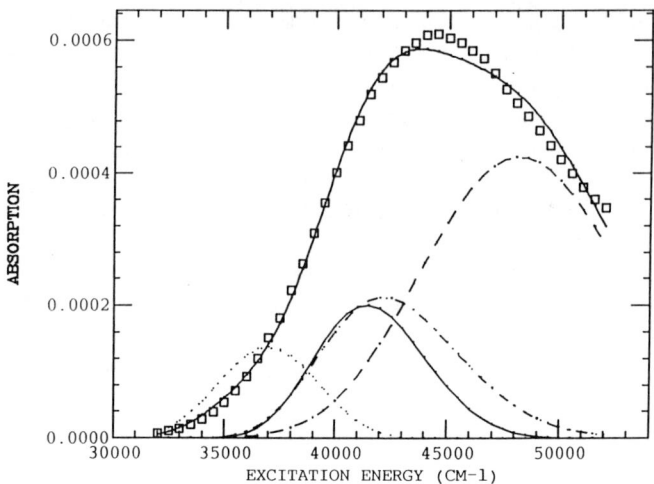

Figure 1. Decoupled partial and total photo-absorption cross-sections for HI. The total cross-section, marked as ___, is compared to the experimental points of ref. 19, marked as □. Partial cross-sections due to the individual electronic states are marked as: (....) $^3\Pi_1$; (_.._..) $^3\Pi_0$; (___) $^3\Sigma_1$; (_._.) $^1\Pi_1$ (Reproduced with permission from ref. 5. Copyright 1988 American Institute of Physics)

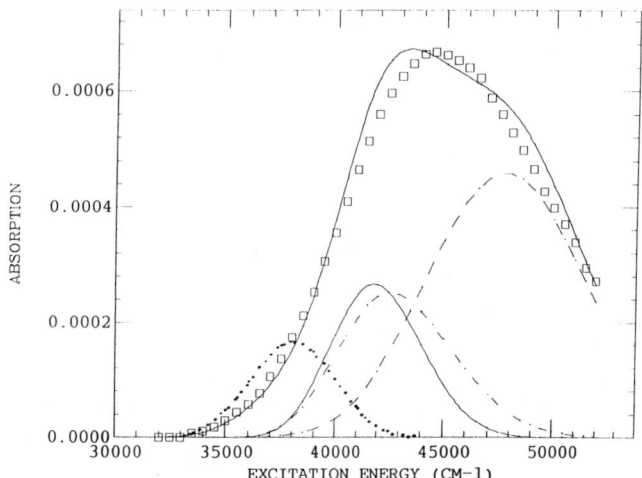

Figure 2. Decoupled partial and total photo-absorption cross-sections for DI. The total cross-section, marked as ___, is compared to the experimental points of ref. 19, marked as □. Partial cross-sections due to the individual electronic states are marked as: (....) $^3\Pi_1$; (_.._..) $^3\Pi_0$; (___) $^3\Sigma_1$; (_._.) $^1\Pi_1$ (Reproduced with permission from ref. 5. Copyright 1988 American Institute of Physics)

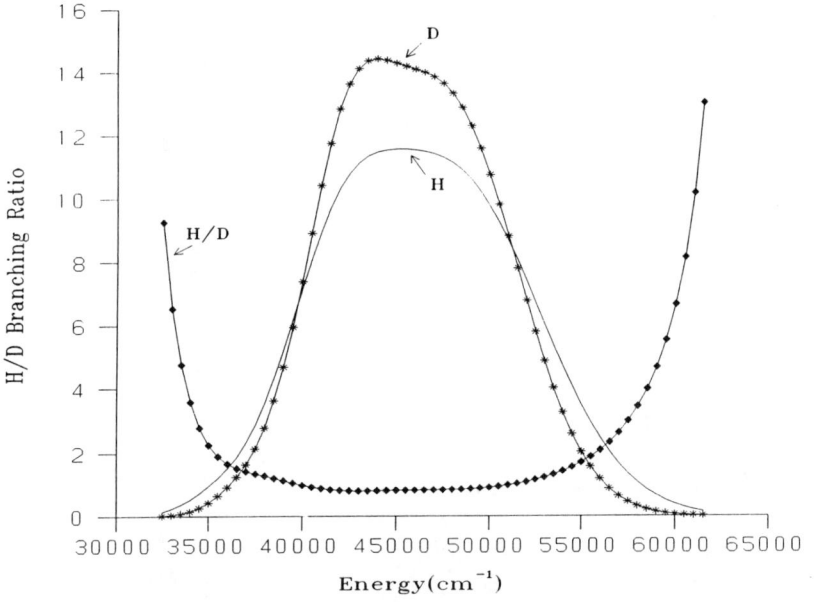

Figure 3. Branching ratio and total absorption of HI and DI from the ground ($v = 0$) vibrational state.
────── HI absorption cross-section (in arbitrary units)
-*-*-*-* DI absorption cross-section (in arbitrary units)
—◆—◆ H/D branching ratio.

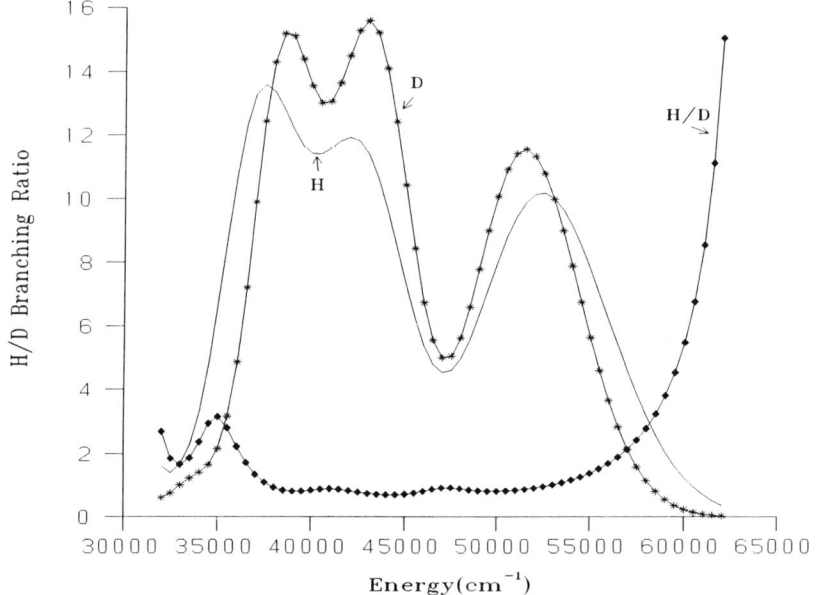

Figure 4. Branching ratio and total absorption of HI and DI from the ground ($v = 1$) vibrational state.
———— HI absorption cross-section (in arbitrary units)
-*-*-*-* DI absorption cross-section (in arbitrary units)
—◆—◆ H/D branching ratio.

17. SHAPIRO   *Mechanisms for Isotopic Enrichment*

selectivity by first frequency-tuning to excite one isotope and then dissociating exclusively the excited molecules.

The increased extension of a wavefunction upon excitation is also at the heart of many polyatomic selective bond breaking processes, discussed in more detail below. All the processes studied to date(6-12), ultimately rely on edge type effects, and they amount to *enhancing* the dissociation of one species while leaving the competing process virtually the same. Below we show that an alternative effect in which one *inhibits* the dissociation of one bond while leaving another unchanged also exists and leads to an even greater selectivity.

In isotopically pure polyatomic molecule the isotope effect results from the competition between *different* molecules. This case has a lot in common with diatomic molecules because the effect depends on the global aspects of tunneling and they also apply to polyatomic molecules. An isotopically *impure* polyatomic molecule is more complicated, because in that case the isotopic effect depends on an *intramolecular* competition between the dissociation rates of different bonds. This effect involves the full complexity of intramolecular dynamics.

We follow the evolution of the effect from diatomic molecules to polyatomic molecules by first looking at the pre-excitation process itself. In Figs. 5 and 6 we display the "photofragmentation maps"(13,14), of CH_3I. Photofragmentation maps are continuous interpolations of the (discrete) set of partial cross-sections, plotted as a function of the total energy and the fragment internal energy. The maps shown in Figs. 5 and 6 are for CH_3I in its ground and first excited vibrational state. The internal energy is that of the CH_3 umbrella-like vibration.

Figs. 5 and 6 clearly demonstrate the generalization to the polyatomic domain of the diatomic reflection principle(3). The ground-state map, shown in Fig. 5, mimics the shape of the ground state nuclear probability-density as a function of the C-I bond and the C-H_3(c.m.) distance. The first excited state map mirrors the image of the nuclear density of that initial state because it has a node running almost perpendicular to the total energy axis.

As in the diatomic case, the absorption cross-section of the excited state clearly extends over a wider frequency range. The excitation of the C-I stretch results in enhanced photodissociation probabilities at the wings of the absorption. Both the inhibitory effect of the node and the enhancement due to the frequency spread are not expected to be mode specific: They inhibit or enhance the act of photon-absorption but they will not in general change the final quantum state.

It can be shown(14) that excitation of another mode, e.g., the C-H_3 umbrella motion results in a map whose node runs *parallel* to the total energy axis. It is therefore impossible to find a wavelength which inhibits the entire photon absorption but the excitation is expected to be *state specific*, as some final quantum states will be turned off.

The pre-excitation patterns discussed here result in isotopic selectivity for molecules, such as HOD, containing two isotopes. This molecule has been studied both theoretically(6,7,8) and experimentally(9-12). Most of these studies

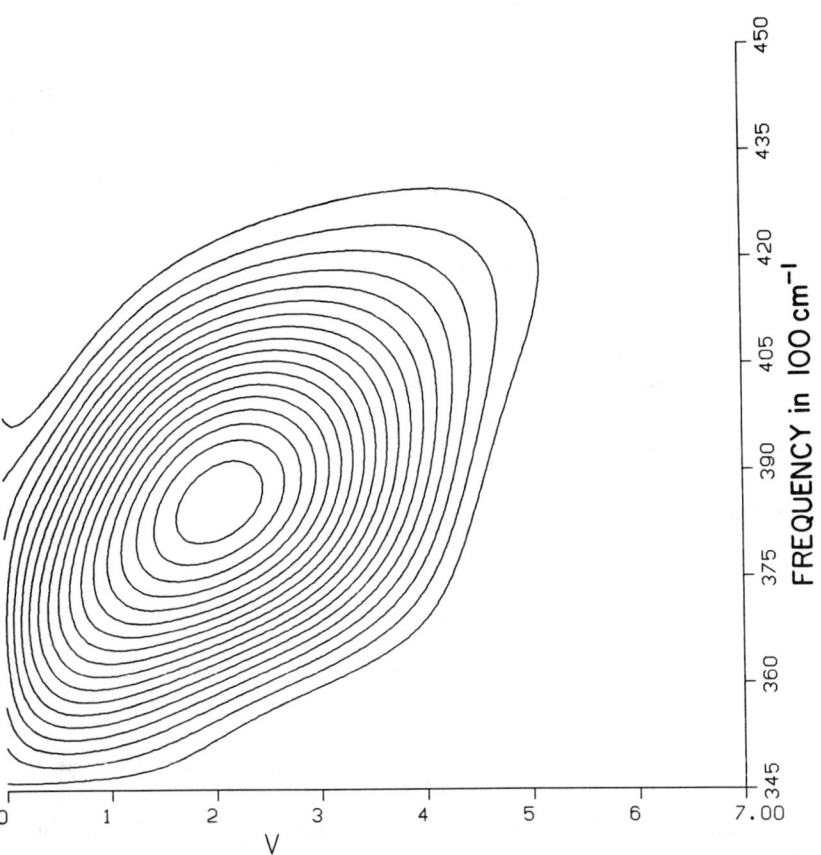

Figure 5. Photofragmentation map for the photodissociation of the ground vibrational state of CH_3I, via the parallel component of the transition dipole moment: $CH_3I \rightarrow CH_3(\tilde{v}) + I^{*2}P_{1/2}$). (Reproduced with permission from ref. 14. Copyright 1986 American Institute of Physics)

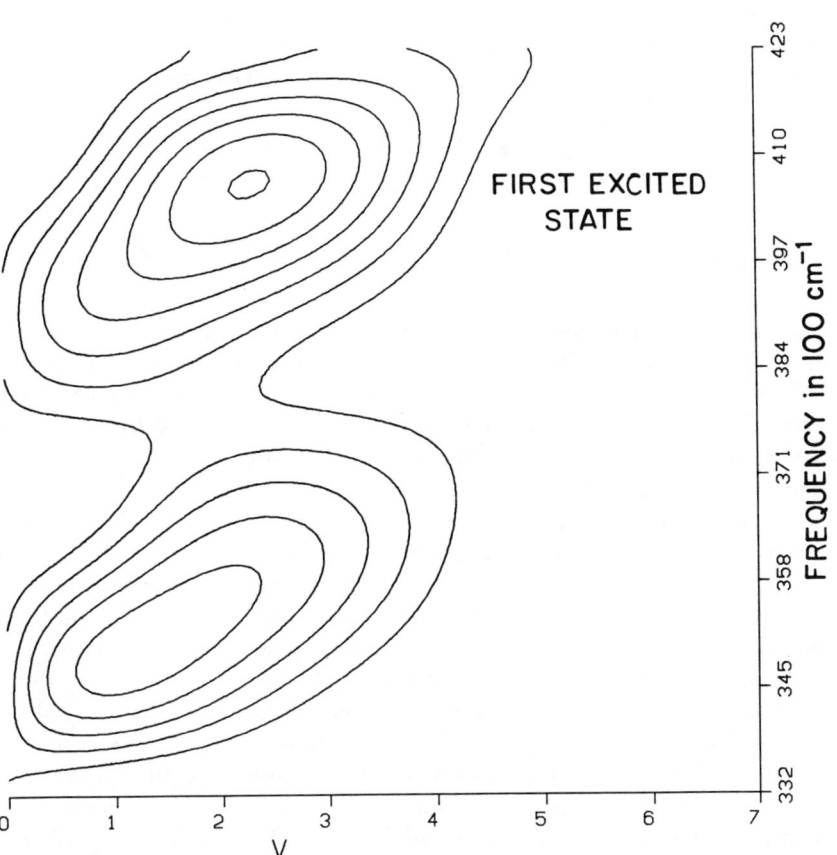

Figure 6. The same as Fig. 5 for dissociation from the first excited vibrational state ($v = 1$ of the C-I stretch) of CH_3I. (Reproduced with permission from ref. 14. Copyright 1986 American Institute of Physics)

concentrated on the A-system dissociation of HOD. It was shown that due to the local nature of the OH and OD bonds, an excitation of a stretch mode results in elongation of either an OH or an OD bond. As discussed above, such elongations result in a strong increase of the absorption at the edges. More importantly, because of the local nature of the excitation, the HOD molecule can be thought of as a direct product of an OH and OD molecules. Thus, in general, the bond that has been elongated is the one to be dissociated. However, as pointed out by Imre and Zhang(7) and by Bar et al.,(12) the effect is more pronounced for OH($v = 1$) than for OD($v = 1$). In order to see the effect for OD at threshold energies the OD bond must be elongated more than the OH bond (e.g., by exciting HOD to the (4,0,0) level(7,11)). This is due to the lesser penetration of the continuum wavefunction to the D+OH part of configuration space, as compared to the H+OD part, at threshold energies.

The effect of vibrational excitation on the HOD B-system was also studied(6), but so far only for excitations in the bend mode. In the B-Continuum most of the action is in the bending mode. This is illustrated in Fig. 7 - taken from Ref. 6- where the potential surface of the B($^1A'$) state of H_2O is plotted as a function of the R - the H-OH(c.m.) distance and $\gamma \equiv \cos^{-1}(\mathbf{R} \cdot \mathbf{r}_{OH})$. Clearly the molecule feels a strong torque towards the collinear configuration as the molecule falls apart, hence the prominent role of the bending mode in the dissociation.

In Fig. 8 (taken from Ref. 6) we show the calculated absorption spectrum and the D+OH($A^2\Sigma$)/H+OD($A^2\Sigma$) fragmentation ratio for ground state HOD. Because of the existence of resonances, which were postulated(6) to be linked to the existence of long-lived trajectories, (a schematic illustration of such trajectories whose existence was later verified(15), is shown in Fig. 7), a strong frequency-dependent isotope effect exists. Because both channels are disconnected in the threshold region (the barrier for the D + OH→H + OD exchange reaction in the B-state is estimated to be ≈1-2eV(16)), each arrangement gives rise to a separate set of resonances. These resonances vary both in width and in position, hence a strong "post-excitation" isotope effect can be found.

The calculations presented here are for the $J = 0 \to J = 1$ case. For room temperatures and even for supersonically cooled beams many more initial rotational states must be considered. When this is done, the resonance structure is expected to become less distinct and we essentially lose our ability to selectively dissociate one bond in preference to another by mere frequency tuning. It is of interest to see if one can find a more *global* effect which does not depend on the resonance structure.

Such effect is demonstrated in Fig. 9 where the photodissociation from the excited (1,0,0) state of HOD (basically the OD oscillator in its first excited state) is shown. The calculations were done using the artificial channel method(17) with a logarithmic-derivative propagator(18). Clearly the excitation of the OD stretch has resulted in an enormous isotope effect which favours the D+OH arrangement by two orders of magnitude! A close look at the absolute value of the cross-section for the D+OH production reveals that in fact it has not

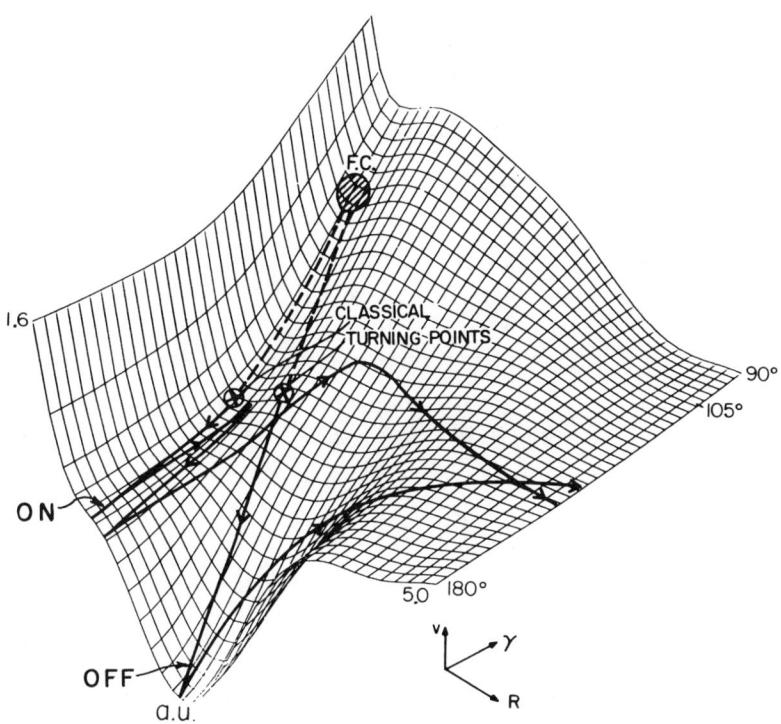

Figure 7. The B($^1A'$) excited surface of water as a function of R, the H-OH(c.m.) distance, and γ, the angle between R and r_{OH}. Also shown are two *schematic* trajectories both involving concerted bend *and* stretch motions, one displaying a direct dissociation route and one a compound process. (Reproduced with permission from ref. 6. Copyright 1982 American Institute of Physics)

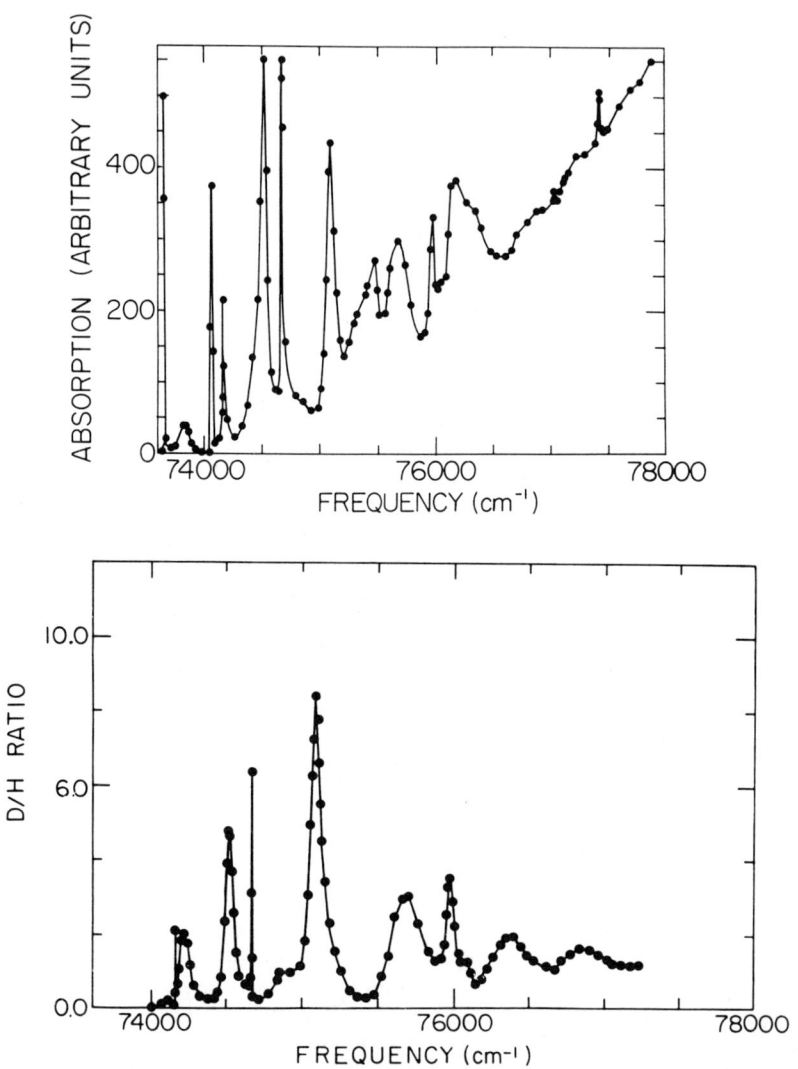

Figure 8. Calculated absorption spectrum and D + OH vs. H + OD isotopic ratio for the threshold region of absorption of HOD in the B-continuum. (Reproduced with permission from ref. 6. Copyright 1982 American Institute of Physics)

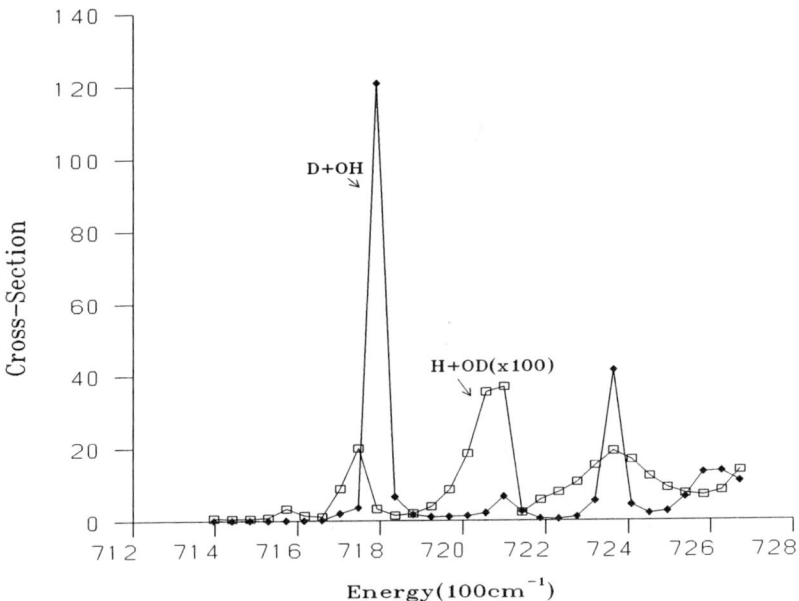

Figure 9. Calculated absorption spectrum and D + OH vs. H + OD isotopic ratio for the HOD(1,0,0) state. (Reproduced with permission from ref. 6. Copyright 1982 American Institute of Physics)

increased much as compared to photodissociation of HOD from its ground (0,0,0) state. What the excitation has done is to *inhibit* the H+OD channel.

When we photodissociate an HOD molecule in which the OH bond is excited, (this is achieved by looking at the (0,0,1) state), we see the opposite effect: As shown in Fig. 10, this time the D+OH channel is inhibited. The yield of the D+OH channel is two-orders of magnitude smaller than that of the H+OD channel.

In Fig. 11 we illustrate the underlying mechanism for this inhibitory effect for the case in which the O-D oscillator is excited. Shown are schematic diagonal (averaged over the vibrational (O-D) coordinate) channel potentials as a function of the reaction (O-H) coordinate, for the ground and excited surfaces. Also shown are the components of the (bound and dissociative) wavefunctions in each channel.

As shown in Fig. 11, there is good overlap along the reaction-coordinate between the bound and continuum wavefunction components. Hence the origin of the inhibitory effect is not in (OH) reaction coordinate but in the (OD) vibrational coordinate: In the B-state, the OD bond characteristics (the equilibrium position and the vibrational frequency) are very similar to those of the ground (X) state of HOD or free OD. As a result, neither the dissociation process nor the optical transition can couple different vibrational channels. Excitation of

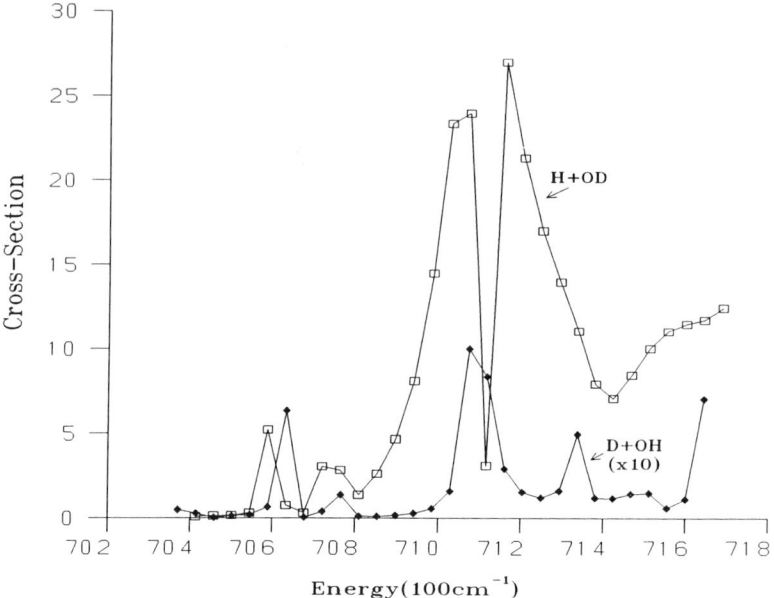

Figure 10. Calculated absorption spectrum and D + OH vs. H + OD isotopic ratio for the HOD(0,0,1) state. (Reproduced with permission from ref. 6. Copyright 1982 American Institute of Physics)

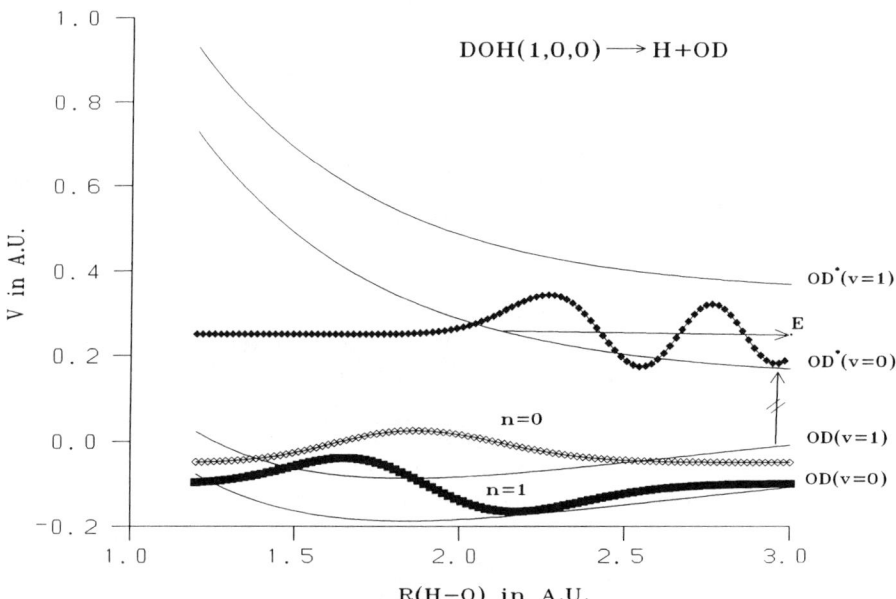

Figure 11. Schematic illustration of the diagonal channel-potentials and vibrational wavefunctions for the $v = 0$ and $v = 1$ states of OD.

the OD bond to $v = 1$ puts the molecule in a subspace which is essentially decoupled from the H + OD($v = 0$) channel. The molecule would dissociate to the H + OD($v = 1$) channel, but this channel is closed for $E_{av} < 2770 \text{cm}^{-1}$ where E_{av}, the *available* energy, $= h\nu - E_{th}$ with E_{th} being the *threshold* energy for dissociation in the B-state (to H+OD($A^2\Sigma$)). Thus the excitation completely turns-off the H + OD channel, leaving the D + OH channel as the only possibility.

As shown in Fig. 10, the reverse applies when we dissociate the (0,0,1) state, which is essentially a pure OH($v = 1$) local mode: The D + OH channel is inhibited, leaving the H + OD channel as the only possibility.

The effects predicted here for the B-state photodissociation of HOD are not really "edge-effects". It is true that the effects shown in Figs. 9 and 10 would disappear with the opening of the $v = 1$ channel, but we can easily maintain the effect even at higher energies by considering pre-excitation to the $v = 2$ channel etc... The inhibitory effect discussed here, besides being more dramatic then the enhancement effect of the A-state(7, 8,11) can therefore be made to operate at energies where the photo-absorption cross-section is at its peak.

Extension of our calculations to higher energies for higher pre-excitation states and the use of coherent control techniques to modify the H/D branching ratio will be reported shortly.

Acknowledgment

Figures 3 and 4 were produced with the help of I. Levy. This work was supported by a grant from the U.S. Israel Binational Science Foundation, grant no. 89/00397/1.

Literature Cited

1. Brumer P.; Shapiro M. *Chem. Phys. Lett.* **1986**, 126, 541.
   Seideman T.; Shapiro M. ; Brumer P. *J. Chem. Phys.* **1989**, 90 ,7132

2. Shapiro M. ; Hepburn J.W.; Brumer P. *Chem. Phys. Lett.* **1988**, 149 ,451.

3. Herzberg G. "Molecular Spectra and Molecular Structure. I. Spectra of Diatomic Molecules", Van Nostrand Reinhold, New York, N.Y. 1950.

4. Levy I.; Shapiro M. ; Brumer P. *J. Chem. Phys.* **1990**, 93 ,2493.

5. Levy I.; Shapiro M. *J. Chem. Phys.* **1988**, 89 ,2900.

6. Segev E.; Shapiro M. *J. Chem. Phys.* **1982**, 77 ,5601.

7. Zhang J.; Imre D.G. *Chem. Phys. Lett.* **1988**, 149 ,233.
   Imre D.G.; Zhang J. *Chem. Phys.* **1989**, 139 ,89.

8. Engel V.; Schinke R. *J. Chem. Phys.* **1988**, 88 ,6831
   Engel V.; Schinke R.; Staemmler V. *J. Chem. Phys.* **1988**, 88 ,129

9. Zittel P.I.; Little D.D. *J. Chem. Phys.* **1979**, 71, 713.
   Zittel P.I.; Lange V.I., *J. Photochem. Photobiol.* **1991**, 56, 149.

10. Shafer N.; Satyapal S.; Bersohn R. *J. Chem. Phys.* **1989**, 90, 6807.

11. Vander Wal R.L.; Scott J.L.; Crim F.F. *J. Chem. Phys.* **1990**, 92, 803.

12. Bar I.; Cohen Y.; David D.; Rosenwaks S.; Valentini J.J.
    J. Chem. Phys. **1990**, 93, 2146

13. Shapiro M. *Chem. Phys. Lett.* **1981**, 81, 521.

14. Shapiro M. *J. Phys. Chem.* **1986**, 90, 3644.

15. Weide K.; Kuehl K.; Schinke R. *J. Chem. Phys.* **1989**, 91, 3999.

16. Flouquet F.; Horsley J.A. *J. Chem. Phys.* **1974**, 60, 3767.

17. Shapiro M. *J. Chem. Phys.* **1972**, 56, 2582.

18. Johnson B.R. *J. Chem. Phys.* **1977**, 67, 4086

19. Oglivie J.F. *Trans. Faraday Soc.* **1971**, 67, 2205.

RECEIVED October 7, 1991

Chapter 18

State-Selected Dissociation of trans-HONO($\tilde{A}$) and DONO($\tilde{A}$)

Effect of Intramolecular Vibrational Dynamics on Fragmentation Rate, Stereochemistry, and Product Energy Distribution

R. Vasudev, S. J. Wategaonkar[1], S. W. Novicki[2], and J. H. Shan

Department of Chemistry, Rutgers—The State University of New Jersey, New Brunswick, NJ 08903

> The effect of coupling between vibrations on state-to-state photodissociation dynamics is probed through experiments on fragmentation of the $\tilde{A}^1A''$ state of *trans* HONO and its isotopic cousin DONO. Specific N=O stretching vibrational levels are excited in each case and the dissociation dynamics is probed through laser excitation, polarization and sub-Doppler spectroscopy of the ejected OH (OD) fragment. The optically excited -N=O stretching vibration ν_2 is coupled to the central O-N coordinate ν_4 (the recoil coordinate) and thus influences the fragmentation rate and OH (OD) recoil energy. In *trans* DONO($\tilde{A}$), the ν_2 vibration is accidentally in resonance with the in-plane DON angle-bending vibration ν_3. Thus, ν_2 in DONO is coupled to ν_3, in addition to ν_4. This coupling influences the rotational energy and anisotropy of the ejected OD fragment because the bending vibration ν_3 evolves into in-plane fragment rotation. Thus, the OD fragment is more rotationally excited and its motion is more frisbee-like than OH ejected by HONO.

An issue of basic interest in state-to-state chemistry is whether different states (especially vibrational states) of a reactant can influence various aspects of reactivity such as rates, product state distributions and anisotropies in a state-specific manner. It is easy to visualize such effects in a photodissociation process. Vibrational state-dependence of dissociation rates might be expected if the excited vibration is coupled to the reaction coordinate. The photoselected vibration may also leave a discernible fingerprint on other aspects of reactivity such as the energy content or anisotropic motion of a product

[1]Current address: Chemistry Department, Kansas State University, Manhattan, KS 66506
[2]On leave from the U.S. Air Force Academy, Colorado Springs, CO 80840

if it is coupled to a parent motion that evolves into a degree of freedom of the fragment. Photodissociation experiments on weakly bound states of molecules with four or more atoms are especially attractive in this regard because of the possibility of probing the consequences of V→T and V→V type energy transfer processes induced by coupling between vibrations in the parent molecule.

In this chapter, we address the above issues through dissociation of the $\tilde{A}^1A''$ state of trans HONO and its isotopic cousin DONO (1-6). This is an especially attractive system because the $\tilde{A}^1A''$-$\tilde{X}^1A'$ transition is accompanied by considerable vibrational structure, thus offering a wide selection of initial conditions for state-selected photodissociation experiments. In addition, the optically excited vibration in HONO($\tilde{A}$) and DONO($\tilde{A}$) is coupled to other vibrations that might be expected to leave a significant fingerprint on the fragmentation rate, energy distribution and motional anisotropy of the OH and OD photofragments. To orient the reader, we list the normal vibrational modes of HONO in Table I.

Table I. The normal vibrations of HONO.

Mode	Principal nuclear motion
ν_1	O-H stretch
ν_2	N=O stretch
ν_3	NOH bend
ν_4	O-N stretch*
ν_5	ONO bend
ν_6	O-H torsion

*The fragmentation coordinate

The HONO $\tilde{A}^1A''$-$\tilde{X}^1A'$ transition involves excitation of a nonbonded electron on the terminal oxygen to the lowest NO π^* orbital, and is accompanied by an extension of the terminal N=O bond (7-9). The $\tilde{A}$-$\tilde{X}$ absorption is thus accompanied by a progression in the N=O stretching vibration ν_2. The excited state is located well above the dissociation threshold (~2.1 eV) of the central O-N bond and is very short-lived, lasting from ≲1 to ~4-5 ν_2 vibrational periods. It is interesting that although the $\tilde{A}$-$\tilde{X}$ excitation is localized on the N=O chromophore, a large fraction of the available energy appears in the form of fragment recoil (i.e., along the O-N coordinate, ν_4), suggesting a coupling between ν_2 (the N=O coordinate) and ν_4. This is also corroborated by the facts that (a) the N=O vibrational frequency in HONO($\tilde{A}$) is very different from that in the free NO($X^2\Pi$) radical, and (b) widths of the HONO $\tilde{A}$-$\tilde{X}$ vibronic features decrease with increasing ν_2' content. These observations were previously interpreted by us (4) in terms of an excited state potential energy surface (PES) in which, in addition to the ν_2-ν_4 coupling, there is a bottleneck in the exit channel. The PES shown in Fig. 1 is similar to that proposed previously (4), and is analytically generated to mimic the general shape of the bottleneck in the approximate ab initio surface calculated recently by Hennig et al. (9). A consequence of the

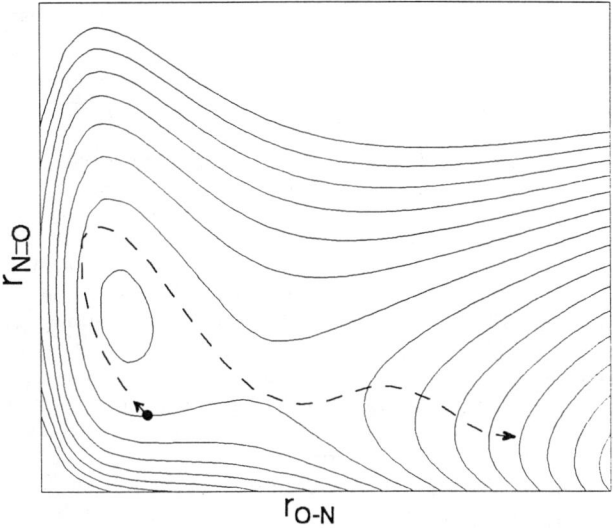

Figure 1. Contour diagram of the $\tilde{A}$ state potential energy surface showing the coupling between the terminal N=O and central O-N coordinates. The predominant change accompanying the $\tilde{A}$-$\tilde{X}$ transition is a lengthening of the N=O bond. Thus, trajectories on the $\tilde{A}$ state PES start near the inner turning point along the N=O coordinate. Trajectories with a low ν_2 content are quickly guided by the surface topology from the FC region towards the exit-channel, as depicted. By contrast, trajectories with a large initial kinetic energy along the N=O coordinate are temporarily trapped in the FC region due to the exit-channel bottleneck. Thus, the dephasing and fragmentation times decrease with increasing ν_2 excitation.

bottleneck in the PES is that although trajectories with a small initial kinetic energy along the N=O coordinate (i.e., with low ν_2 excitation) are guided promptly through the exit channel, as depicted in Figure 1, trajectories with a higher ν_2 excitation are temporarily trapped in the Franck-Condon (FC) region. Thus, the number of recurrences through the FC region increases, the resonance width decreases, and the fragmentation rate decreases with increasing ν_2 excitation, as observed (4,5). Similar effects have been observed in the case of *trans* DONO. Another consequence of the $\tilde{A}$ state PES topology depicted in Fig. 1 is that the partitioning of available energy into the recoil of OH (OD) and NO products ($\nu_2 \rightarrow \nu_4$ "V→T" energy transfer) and vibration of the NO fragment is a strong function of the parent ν_2 content. We have recently probed this effect through high-resolution sub-Doppler spectroscopy of OH ejected by 0^0, 2^1, 2^2, 2^3 and 2^4 states of HONO (5). These experiments furnish information on energy transfer effects and fragmentation time-scales. Our results, together with those of Dixon and Riley (10) on energy distribution in NO ejected by HONO (2^2), provide tests for the recent trajectory calculations of Hennig et al. (9). We find that the extent of V→T energy transfer is not so efficient as suggested by the calculations. We are also in considerable disagreement with the calculated excited-state dephasing and fragmentation time-scales. For example, we find that fragmentation from the 0^0 state is ultrafast, lasting ≲1 N=O vibrational period. Trajectory calculations (9), however, yield a lifetime that is orders of magnitude of longer due to an exit-channel barrier that must be tunnelled through for dissociation. It appears that the barrier has been overestimated in the approximate (two-dimensional) *ab initio* PES generated by Hennig et al. Details of this work are described elsewhere (5).

A major difference in the dissociation dynamics of *trans* HONO($\tilde{A}$) and DONO($\tilde{A}$) arises from the fact that in the latter the optically excited ν_2 vibration is accidentally in resonance with the in-plane DON angle-bending vibration ν_3, as shown by the spectroscopic work of King and Moule (7). The ν_2-ν_3 coupling, depicted schematically in Fig. 2(a), might be expected to leave a significant fingerprint on the OD photofragment's properties because the parent ν_3 DON bending motion is the major contributor to the product rotation (1-3,6). In particular, we expect to see increased OD fragment rotational energy when successive members of the ν_2' progression (2^n) are excited. In addition, since the ν_3 oscillation is localized in the parent DONO plane, the increased OD rotation should also be localized in this plane. Furthermore, since the half-filled 2pπ orbital in the OD fragment is initially associated with the central O-N bond in the parent, this orbital in the product should also be localized in the DONO plane. In other words, the half-filled 2pπ orbital in OD should be preferentially in the plane of product rotation and, in addition, this effect should become more pronounced when successive members of the ν_2' manifold are excited. By contrast, ν_2-ν_3 coupling in HONO is absent/negligible, as shown schematically in Figure 2(b), and thus the ν_2 dependence of OH rotational energy and anisotropy should be small.

The aim of the work summarized here is to probe the above differences in the photochemistry of HONO and DONO through measurements on the rotational energy and correlations among the vectorial (directional) properties of the ejected OH and OD fragments. The vectorial parameters of interest are depicted in Figure 3. As mentioned previously, the parent $\tilde{A}$-$\tilde{X}$ excitation involves excitation of a nonbonding electron on the terminal oxygen to the lowest NO π^* orbital, so that the transition moment $\vec{\mu}$ is perpendicular to the HONO(DONO) plane. Polarized excitation thus photoselects molecules preferentially aligned with the molecular frame perpendicular to the photolysis polarization $\hat{\epsilon}_p$, as shown in Figure 3. The central O-N bond is homolytically dissociated, and the OH(OD) fragment's half-filled $2p\pi$ orbital and recoil velocity vector $\vec{v}$ are localized in the parent HONO(DONO) plane (Plane I). The fragment's plane of nuclear rotation (plane II) makes a small angle with respect to plane I, as depicted in Figure 3. Of particular significance is the alignment of the fragment rotational angular momentum $\vec{J}$ ($\perp$ plane II) with respect to $\vec{\mu}$, $\vec{v}$ and the half-filled $2p\pi$ orbital because this is expected to be sensitive to v_2-v_3 coupling in DONO, as described previously.

Experimental

In the present work, the rotational energy content of the OH (OD) photofragment, alignment of $\vec{J}$ (i.e., $\vec{\mu},\vec{J}$ correlation), and alignment of the half-filled $2p\pi$ orbital are probed by polarized broadband (~1 cm^{-1}) laser excitation spectroscopy of the $A^2\Sigma^+$-$X^2\Pi$ transition. The correlations among $\vec{\mu}$, $\vec{J}$ and $\vec{v}$ are probed by sub-Doppler polarization spectroscopy. The experimental setup is shown in Figure 4. Polarized photodissociation and probe beams are counterpropagated through a reaction chamber. The former is generated either by a dye laser pumped by the third harmonic of a Nd:YAG laser, or by frequency-doubling the output of the dye laser excited by the second harmonic of the pump laser. The beam (FWHM~1 cm^{-1}) for probing the product state distributions and rotational alignment is generated by doubling the output of a second dye laser, also pumped by the Nd:YAG laser. In sub-Doppler spectroscopy experiments, the probe is pumped by a XeCl laser, and the bandwidth is narrowed with an intracavity pressure-scanned etalon. The fragment fluorescence is filtered from undesired photons, imaged on to a photomultiplier, and processed by a boxcar averager and a microcomputer. HONO (DONO) is prepared as described previously (1-3). The experiments are performed either in a supersonic jet (3,4) where rotational temperatures of ~3-6K are attained, or under low pressure (15-30 mTorr) flow conditions (5,6).

Photofragment Rotational Alignment and Energy Distribution

The major finding of this work is that OD ejected by specific vibrational states of DONO exhibits higher rotational anisotropy and excitation than OH ejected by the corresponding states of HONO (3,6). We describe the fragment rotational alignment in terms of the quadrupole alignment parameter (11) defined classically as $\mathcal{A}_0^{(2)}=(4/5)\langle P_2(\cos\theta)\rangle$,

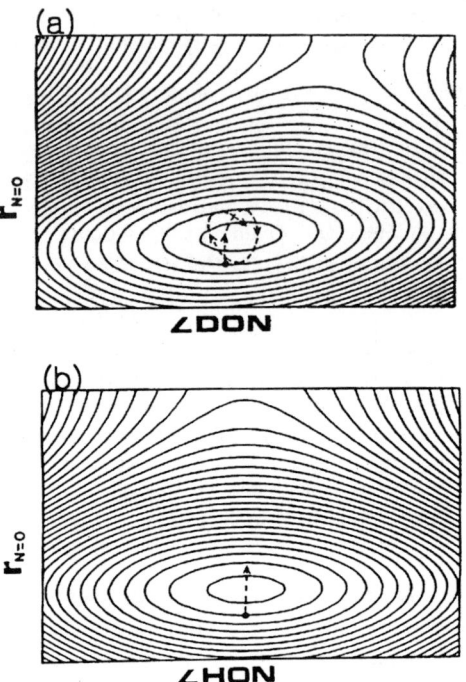

Figure 2. Qualitative difference between the $\tilde{A}$ state PES's of (a) *trans* DONO and (b) *trans* HONO. A possible dynamical consequence of the difference is illustrated by the trajectories. In DONO, because of ν_2-ν_3 coupling, motion along the N=O coordinate is accompanied by changes in the DON angle, so that it is possible to generate significant angular momentum in the DO moiety during the excited-state lifetime. In HONO, ν_2-ν_3 coupling is small/absent, and thus the motion is predominantly along the N=O coordinate (as shown), plus a small zero-point oscillation (not shown) in the HON angle. Reproduced with permission from ref. 3. Copyright 1989, by American Institute of Physics.

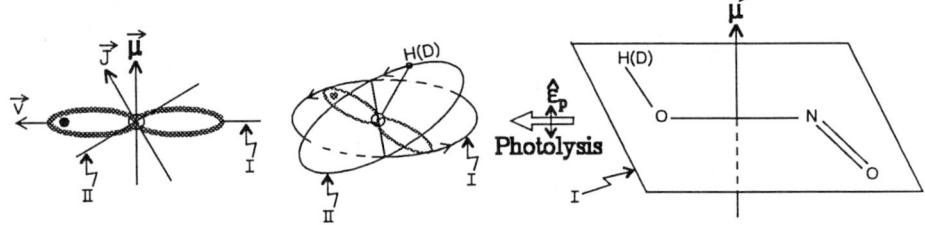

Figure 3. The vectorial (directional) parameters of interest.

where θ is the angle between the rotational angular momentum vector $\vec{J}$ and the photolysis polarization $\hat{\varepsilon}_p$, as shown in Figure 5(a). The parameter $\mathcal{A}_0^{(2)}$ lies between the limits of +4/5 and -2/5 corresponding to the extreme cases $\vec{J} \| \hat{\varepsilon}_p$ and $\vec{J} \perp \hat{\varepsilon}_p$, respectively, depicted in Figure 5(b).

We probe the product alignment through the dependence of rotational line intensities in the fragment excitation spectrum on the polarization directions $\hat{\varepsilon}_p$ and $\hat{\varepsilon}_a$ of the photolysis and analyzing probe lasers. The polarization effects are demonstrated in Figures 6(a) and (b) for DONO 2_0^3 photodissociation. Figure 6(a) shows part of the excitation spectrum of the ejected OD, recorded with $\hat{\varepsilon}_p \perp \hat{\varepsilon}_a$. The effect of rotating $\hat{\varepsilon}_p$ by 90° ($\hat{\varepsilon}_p \| \hat{\varepsilon}_a$) is shown in Figure 6(b) for a small portion of the fragment spectrum. For example, in comparison with the polarization insensitive $R_2(1)$ transition, the $P_1(4)$ line undergoes a ~40% reduction in intensity due to the above change in the experimental polarization geometry.

We use the line-intensities in the fragment excitation spectra recorded with $\hat{\varepsilon}_p \perp \hat{\varepsilon}_a$ and $\hat{\varepsilon}_p \| \hat{\varepsilon}_a$ to determine the fragment alignment $\mathcal{A}_0^{(2)}$, as described elsewhere (1,3,12). Briefly, the intensity I of a rotational line in a photofragment excitation spectrum is given by

$$I \sim \mathcal{P}(J)B[a_0 + a_1 \mathcal{A}_0^{(2)}] \qquad (1)$$

where $\mathcal{P}(J)$ is the fragment population in the state J, B is the absorption transition probability, and the parameters a_0 and a_1 depend upon the polarizations ($\hat{\varepsilon}_p, \hat{\varepsilon}_a$) and propagation directions of the photolysis and probe lasers, the direction in which the fragment fluorescence is monitored, and the fragment rotational quantum numbers in the absorption-emission cycle. The factors a_0 and a_1 are readily obtained from standard algorithms for calculating angular momentum coupling factors. The B factors for OH and OD used here are those of Crosley et al. (13). Thus, from the intensities of a rotational line in the fragment spectra recorded with two different polarization geometries, the alignment $\mathcal{A}_0^{(2)}$ is readily determined (1,3). The results for DONO photodissociation (3) are shown in Figure 7. The highest observed $\mathcal{A}_0^{(2)}$ is ~+0.6, which is quite close to the limiting value of +0.8 shown in Figure 5(b). We thus conclude that the OD fragment's plane of rotation has a strong preference to be in the initial DONO plane. By contrast, as shown in Figure 8, for OH ejected by HONO (3) the highest $\mathcal{A}_0^{(2)}$ is ~+0.3, corresponding to only a moderate preference for the plane of fragment rotation to be in the HONO plane. Unfortunately, the error bars associated with the calculated $\mathcal{A}_0^{(2)}$ for DONO photodissociation (Figure 7) are too large and variations in fragment alignment with the parent v_2 content are thus not easily discernible.

The influence of v_2 on the OD fragment's properties is more easily discernible in the product rotational energy distribution. To illustrate this effect, we show in Figure 6(c) the R_1 branch in the excitation

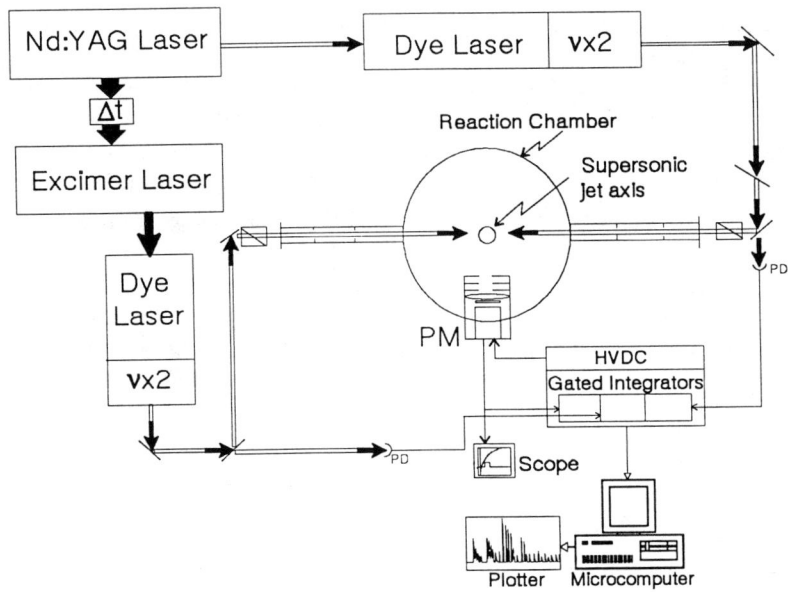

Figure 4. The experimental setup.

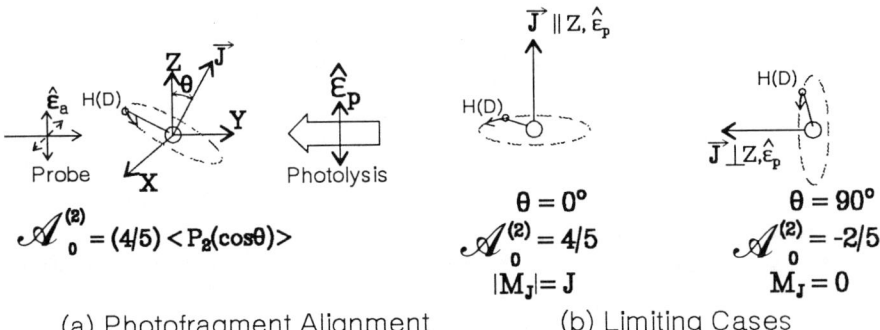

(a) Photofragment Alignment   (b) Limiting Cases

Figure 5. (a) The photofragment rotational alignment, defined in terms of the parameter $\mathcal{A}_0^{(2)}=(4/5)\langle P_2(\cos\theta)\rangle$, where θ is the angle between the photolysis polarization direction $\hat{\varepsilon}_p$ ($\|Z$ axis) and the fragment rotational angular momentum $\vec{J}$. (b) Two limiting cases of fragment alignment, corresponding to $\vec{J}\|Z$ ($\mathcal{A}_0^{(2)}=+4/5$) and $\vec{J}\perp Z$ ($\mathcal{A}_0^{(2)}=-2/5$).

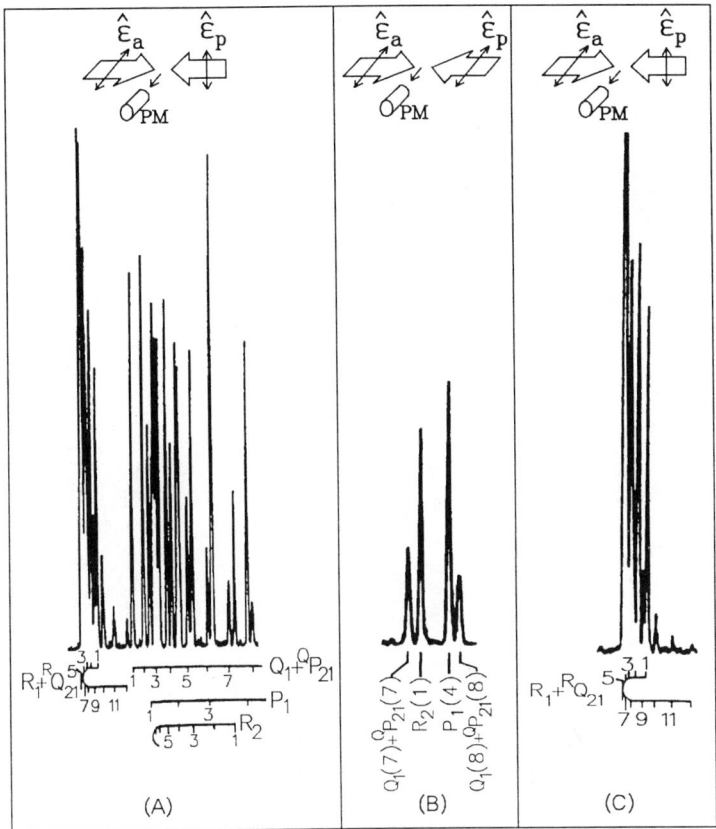

Figure 6. (a) Part of the A–X (1,0) excitation spectrum of OD ejected by *trans* DONO($\tilde{A}$,2^3). The experimental photolysis/probe/fluorescence-detection geometry is shown on top. The fragment's (1,1) undispersed fluorescence is monitored by a photomultiplier (PM) that senses all polarizations equally. (b) A small part of the fragment excitation spectrum recorded with $\hat{\varepsilon}_p \perp \hat{\varepsilon}_a$. (c) A small segment of the excitation spectrum of OD ejected by DONO($\tilde{A}$,2^1). The experimental geometry is the same as that in (a). Reproduced with permission from ref. 3. Copyright 1989, by the American Institute of Physics.

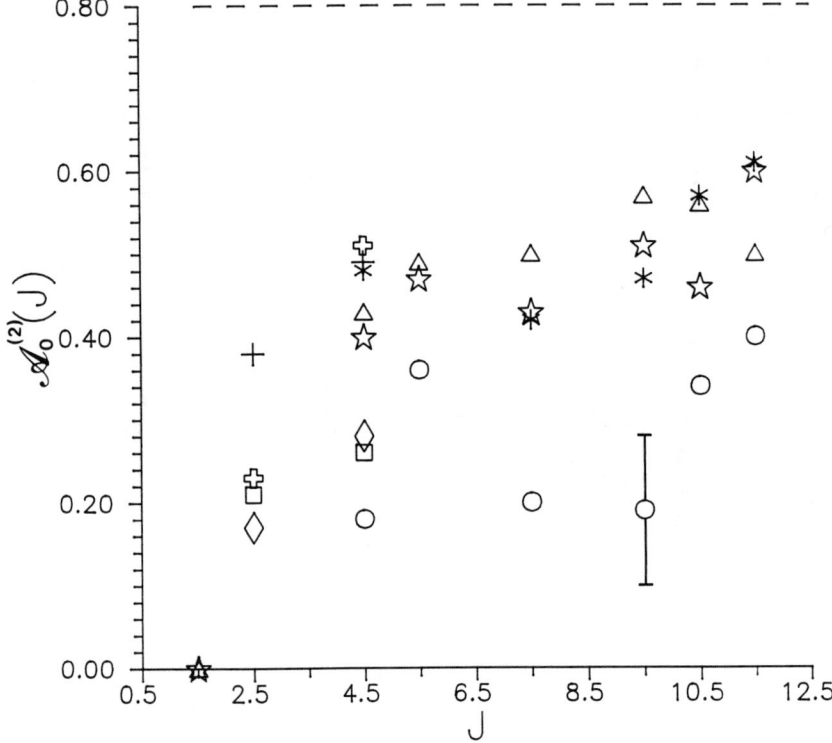

Figure 7. The alignment $\mathcal{A}_0^{(2)}(J)$ of OD generated in the $\Pi(A')$ Λ-doublet levels of the F_1 and F_2 spin-orbit components, originating from various vibrational levels $(0^0, 2^n)$ of *trans* DONO($\tilde{A}$). The symbols used are o:0^0,F_1; □:0^0,F_2; △:2^1,F_1; ◊:2^1,F_2; ☆:2^2,F_1; +:2^2,F_2; *:2^3,F_1; ⁑:2^3,F_2. The broken line shows the limiting value of $\mathcal{A}_0^{(2)}$ attained when $\vec{\mu} \parallel \vec{J}$. Reproduced with permission from Ref. 3. Copyright 1989, by American Institute of Physics.

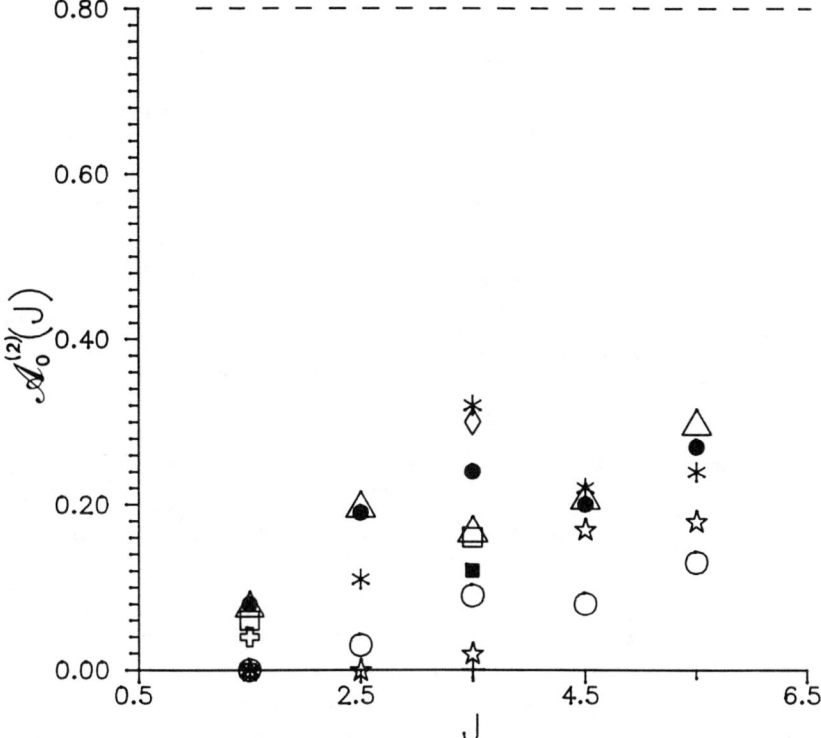

Figure 8. The alignment $\mathcal{A}_0^{(2)}(J)$ of OH ejected by *trans* HONO($\tilde{A}$). The symbols used for the dissociation of the 0^0, 2^1, 2^2 and 2^3 states are the same as those in Figure 7. The symbols used for fragmentation of the 2^4 state are: ●:F_1 and ■:F_2. Reproduced with permission from Ref. 3. Copyright 1989, by American Institute of Physics.

spectrum of OD ejected by the DONO 2^1 state. Comparison with Figure 6(a) clearly demonstrates that the rotational state distribution is shifted considerably to lower energies compared to the case of 2^3 fragmentation. The deduced ν_2-dependence of OD rotational excitation, described in detail elsewhere (3), is quite large. For example, the "rotational temperatures" of OD [$X^2\Pi(A'),F_1$] ejected by the 0^0, 2^1, 2^2, and 2^3 states of DONO are 366±8, 432±10, 565±12 and 656±22K, respectively. By contrast, the "rotational temperatures" of OH ejected by the corresponding states of HONO are 260±14, 348±25, 385±26 and 386±10K, showing a much milder ν_2 dependence (3).

Alignment of the OH (OD) fragment's half-filled 2pπ orbital

Another incisive probe of the photodissociation dynamics is the alignment of the OH (OD) fragment's half-filled 2pπ orbital with respect to the plane of nuclear rotation. As mentioned previously, since HONO($\tilde{A}$) dissociation involves homolytic fission of the central O-N bond, the half-filled 2pπ orbital in the OH product should be preferentially localized in the initial HONO plane. In addition, since the fragment's plane of rotation shows a preference to be in the HONO plane, as described above, the product's half-filled 2pπ orbital should be preferentially localized *in* the plane of rotation. Such orbital alignments are readily deduced from nonequilibrium populations in the so-called Λ-doublet fine-structure states of the fragment (14).

In the classical ("high J") description of OH($X^2\Pi$), the Λ-doublet states Π(A') and Π(A") correspond to the half-filled 2pπ orbital being in the plane of nuclear rotation and perpendicular to it, respectively, as shown in Figure 9. The populations in these states are readily obtained through rotational line intensities in the fragment excitation spectrum. We find that in "high J" OH and OD photofragments, the Π(A') states are more highly populated than the Π(A") states, so that the half-filled 2pπ orbital has a preference to be *in* the plane of fragment rotation. However, in the case of the OH product the Π(A')/Π(A") population ratio is not a strong function of the parent ν_2 content, whereas in OD it is ν_2 dependent (3). For example, for OH generated in J=5.5 ($X^2\Pi,F_1$) from the 0^0, 2^1, 2^2 and 2^3 states of HONO the above population ratio is 3.0±0.5, 3.5±0.5, 3.7±0.5 and 3.7±0.5, respectively. On the other hand, for the OD fragment in J=11.5 the corresponding ratios are 4.0±1.0, 5.7±0.5, 7.5±0.5 and 8.7±0.5, showing a very strong ν_2 dependence.

Photofragment $\vec{v},\vec{J}$ Correlation

The correlation between the fragment recoil velocity $\vec{v}$ and rotational angular momentum $\vec{J}$ also provides a direct probe of the HONO (DONO) dissociation dynamics because it directly reflects the contributions of in-plane and out-of-plane forces on the H (D) atom during O-N bond rupture. Dominance of the in-plane torque would result in a frisbee-type motion of the OH (OD) product, with the velocity vector $\vec{v}$ being in the plane of rotation (i.e., $\vec{v} \perp \vec{J}$). On the other hand, dominance of out-of-plane forces would generate a propeller-type motion ($\vec{v} \parallel \vec{J}$), as in the case of HONO($\tilde{B}^1A'$) dissociation (15). Such motional anisotropies can be probed by sub-Doppler polarization spectroscopy of the product because the $\vec{v},\vec{J}$ correlation results in a fragment rotational alignment with respect to the photolysis polarization $\hat{\varepsilon}_p$ (i.e., in the space-fixed frame) that is a function of the recoil direction and hence the Doppler shift (12,16,17). The dependence of product polarization on the recoil direction is depicted in Figure 10 for the limiting cases $\vec{v} \perp \vec{J}$ and $\vec{v} \parallel \vec{J}$.

In our Doppler spectroscopy setup, the probe's propagation direction (the Y axis in Figure 10) is perpendicular to the photolysis polarization direction (the Z axis). Thus, the experiments probe the Y-component (v_Y) of the fragment recoil, shown in Figure 10(a). The wings of the v_Y distribution are due to fragments moving along the Y axis. The center has a dominant contribution from products recoiling along the X axis, and a small contribution from fragments moving along the Z axis. From Figures 10(b) and (c), it is clear that the rotational alignments (polarizations) of fragments contributing to the wings and center of the v_Y distribution are different for frisbee-type and propeller-type fragment motions. Thus, the Doppler line-shape recorded with a high-resolution probe laser will be a function of the polarization directions $\hat{\varepsilon}_p$ and $\hat{\varepsilon}_a$. This provides us a means for quantifying the correlation between the fragment recoil direction $\vec{v}$ and rotational angular momentum $\vec{J}$ (12,16,17).

Polarization dependence of the fragment Doppler profiles is illustrated in Figure 11 where we show high-resolution spectra of the $P_2(5)$, $^PQ_{12}(5)$ transitions of OH generated by HONO 2_0^3 photolysis. A detailed analysis of the line-shapes allows us to quantify the fragment's vector correlations. Each profile observed with a given photolysis and probe polarization geometry is first fit to the line-shape function

$$S(\Delta\bar{\nu}_D) = (1/\Delta\bar{\nu}_D) [1 - \tfrac{1}{2} \beta_{eff} \, P_2(\Delta\bar{\nu}_0/\Delta\bar{\nu}_D)], \qquad (2)$$

convoluted with a Gaussian profile (FWHM=0.106 cm^{-1}) due to isotropic parent thermal (300K) motion and the probe laser bandwidth. In equation (2), $\Delta\bar{\nu}_0$ is the displacement from the line-center $\bar{\nu}_0$, $\Delta\bar{\nu}_D = \bar{\nu}_0 v/c$, and β_{eff} is the effective anisotropy parameter (12,17,18). The β_{eff} value for each $\hat{\varepsilon}_p, \hat{\varepsilon}_a$ polarization geometry is related to the bipolar moments β_0^k that describe correlations among $\vec{\mu}$, $\vec{v}$ and $\vec{J}$:

$$\beta_{eff} = -2[b_2 \beta_0^2(20) + b_3 \beta_0^0(22) + b_4 \beta_0^2(22)] / [b_0 + b_1 \beta_0^2(02)] \qquad (3)$$

The moments $\beta_0^2(20)$, $\beta_0^0(22)$ and $\beta_0^2(02)$ are measures of correlations between $\vec{\mu}$ and $\vec{v}$, $\vec{v}$ and $\vec{J}$, and $\vec{\mu}$ and $\vec{J}$ respectively (12). The moment

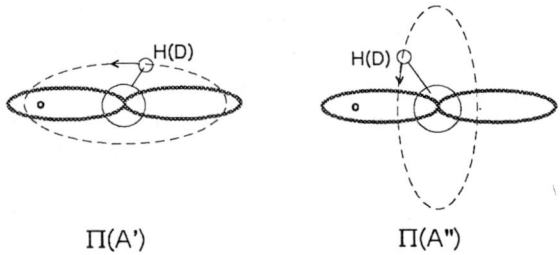

Figure 9. The $\Pi(A')$ and $\Pi(A'')$ Λ-doublets (high J limiting cases).

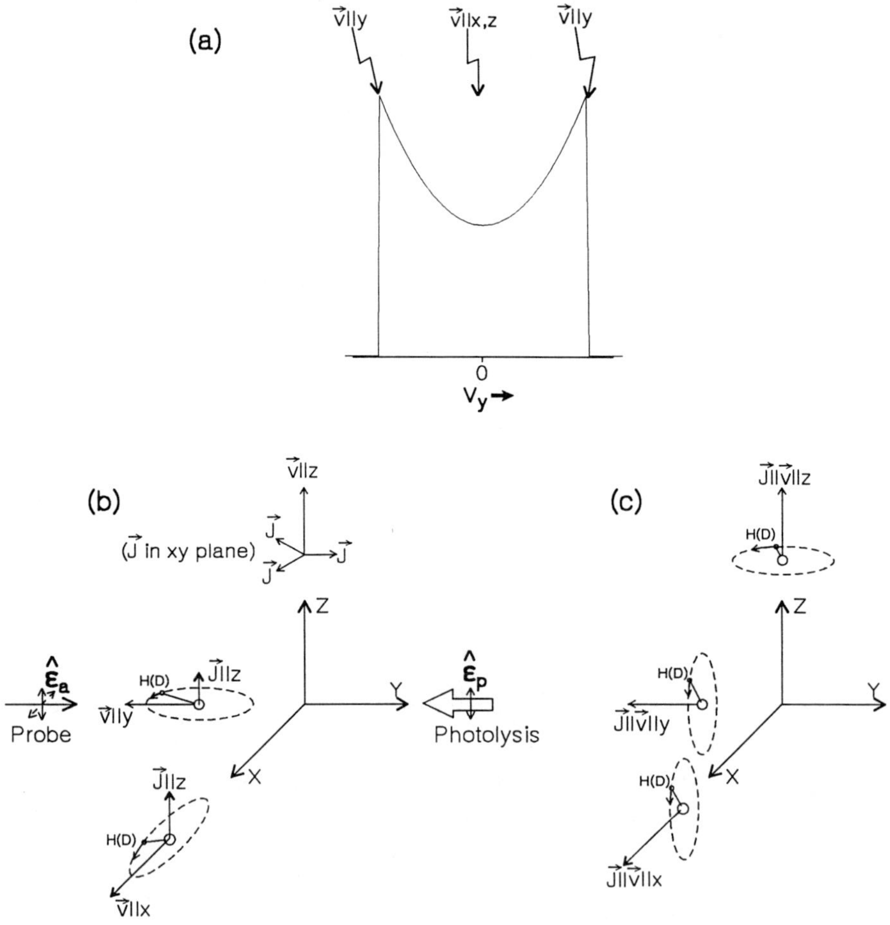

Figure 10. (a) The distribution of the Y-component (v_y) of the fragment recoil velocity $\vec{v}$. (b) Preferred directions of $\vec{J}$ for fragments recoiling along the X,Y and Z directions with a frisbee-type motion ($\vec{v} \perp \vec{J}$). (c) Preferred directions of $\vec{J}$ for fragments with a propeller-type motion ($\vec{v} \parallel \vec{J}$).

$\beta_0^2(22)$ describes the triple correlation $\langle\vec{\mu}.\vec{v}.\vec{J}\rangle$. The factors b_0-b_4 are readily calculated from angular momentum coupling factors and OH (OD) $A^2\Sigma^+$-$X^2\Pi$ Einstein A and B factors (13). The moment $\beta_0^2(02)=(5/4)\mathcal{A}_0^{(2)}$ (the fragment "bulk" rotational alignment) is obtained from the polarization dependence of integrated line-intensities. The deduced $\beta_0^2(02)$ is then used in equation (3) along with the experimental β_{eff} to determine the rest of the bipolar moments. Of particular interest here is the $\langle\vec{v}.\vec{J}\rangle$ correlation parameter $\beta_0^0(22)$ which has a value of -1/2 in the limit of a perfect frisbee-type fragment motion and +1 for a propeller-type motion.

Table II summarizes the results of vector analysis for HONO photodissociation shown in Figure 11. We find that $\langle\vec{v}.\vec{J}\rangle$ correlation $\beta_0^0(22)=-0.08\pm0.01$, suggesting a moderate dominance of frisbee-type motion. A similar analysis shows that OD ejected by DONO is more anisotropic. For example, in DONO (2^3) dissociation, analysis of the OD $P_2(7), ^PQ_{12}(7)$ transitions yields $\beta_0^0(22)=-0.15\pm0.02$, which is closer to the limiting value of -0.50. We thus conclude that the OD fragment has a higher tendency to mimic a frisbee-type motion than the OH product. A detailed analysis of vector-correlations in HONO and DONO dissociation will be described elsewhere (6).

Fragmentation Dynamics

As mentioned previously, the goal of the present work is to explore if coupling between vibrations of a reactant can leave a discernible fingerprint on the vectorial and scalar properties of a product. This work was, in fact, prompted by the spectrosopic work of King and Moule (7) who showed that the optically excited ν_2 vibration in DONO($\tilde{A}$) is coupled to the in-plane DON angle-bending vibration ν_3. As mentioned previously, this coupling is absent/small in the case of trans HONO. Thus, the OH fragment's rotation is uninfluenced by the ν_2 content and originates mainly from the zero-point angle-bending oscillations in the parent. There are two such angular vibrations in HONO: (a) the in-plane HON bend ν_3, and (b) out-of-plane torsion ν_6. Since ν_3 has a higher frequency than ν_6, the high-energy wing of the fragment rotational energy distribution has a larger contribution from the in-plane HON bending oscillation. Consequently, the fragment shows a moderate in-plane alignment and nonequilibrium $\Pi(A')/\Pi(A'')$ Λ-doublet populations. In DONO, however, since the optically excited ν_2 vibration is coupled to ν_3, some of the energy initially channelled into the ν_2 oscillation can be transferred to the DON angle-bending motion, as depicted in Figure 2(a). Consequently, increasing the ν_2 content results in an increased OD fragment rotation. In addition, since ν_3 is confined to the DONO plane, the increased rotation is also localized in this plane, thus resulting in higher rotational alignment and $\Pi(A')/\Pi(A'')$ population ratio.

We have recently obtained further incisive evidence in support of the above conclusions (6). In HONO($\tilde{A}$), the only significant coupling is that between ν_2 and the recoil coordinate ν_4 (shown in Figure 1). Thus,

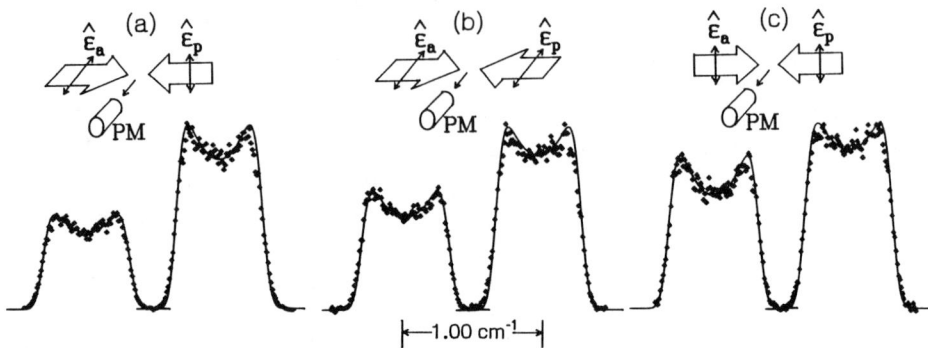

Figure 11. High-resolution Doppler profiles of the $P_2(5)$ and $^PQ_{12}(5)$ transitions of OH generated by 2^3_0 HONO photodissociation, and recorded with the three photolysis/probe/detection geometries shown on top. Superimposed on the experimental spectra (♦) are line-shapes simulated with the parameters shown in Table II.

Table II. Experimental $\Delta\bar{\nu}_D$ (cm^{-1}), β_{eff} and the bipolar moments $\beta^2_0(20)$, $\beta^0_0(22)$, $\beta^2_0(22)$ deduced from photodissociation/probe polarization dependence of the $P_2(5)$, $^PQ_{12}(5)$ Doppler profiles of OH generated by 2^3_0 photodissociation of *trans* HONO. Cases (a), (b) and (c) refer to the experimental polarization geometries shown in Figure 11(a), 11(b) and 11(c), respectively.† The bipolar moments are determined by fitting β_{eff} values for $P_2(5)$ and $^PQ_{12}(5)$ in cases (a), (b) and (c) simultaneously to equation (3). Shown at the bottom, for comparison, are bipolar moments for the idealized cases $\vec{\mu}\|\vec{J}\perp\vec{v}$ and $\vec{\mu}\perp\vec{J}\|\vec{v}$, depicted in Figures 10(b) and 10(c), respectively.

	$\Delta\bar{\nu}_D$			β_{eff}			$\beta^2_0(20)$	$\beta^0_0(22)$	$\beta^2_0(22)$
	(a)	(b)	(c)	(a)	(b)	(c)			
$P_2(5)$	0.305	0.305	0.307	−0.62	−0.53	−0.47	−0.32‡	−0.08	+0.10
							(±0.01)	(±0.01)	(±0.02)
$^PQ_{12}(5)$	0.307	0.309	0.311	−0.62	−0.70	−0.80			
$\vec{\mu}\|\vec{J}\perp\vec{v}$							−0.50	−0.50	+0.50
$\vec{\mu}\perp\vec{J}\|\vec{v}$							−0.50	+1.00	+0.50

†The simulations shown in Figure 11 are based on the assumption of a single OH recoil velocity. This is a reasonable approximation in HONO 2^3 dissociation because, as shown recently (5), OH recoil has two principal components in a ratio of 0.05:1 (these components correspond to the NO fragment being generated in v=1 and 2).

‡The $\vec{\mu},\vec{v}$ correlation parameter $\beta^2_0(20)$ is less than the limiting value of −0.50 because the parent rotates prior to fragmentation, thus blurring the fragment angular distribution (5,6).

all the available energy is channelled exclusively into the recoil of the OH and NO fragments, and NO internal excitation (5,10). The OH rotation arises predominantly from the zero-point oscillations associated with *isolated* ν_3 and ν_6 vibrations, and thus, for a given photolysis wavelength, fragments generated in all rotational states should have the same recoil velocity. By contrast, in DONO($\tilde{A}$), ν_2 is coupled to the recoil coordinate ν_4 *and* the DON bend ν_3. In other words, there is competition between $\nu_2 \to \nu_4$ (V→T) and $\nu_2 \to \nu_3$ (V→V) channels. Thus, since the ν_3 motion is the dominant contributor to the *high*-energy wing of the OD fragment rotational distribution, OD generated in *high* rotational states should have a smaller average recoil energy than the fragment with low rotational excitation. In addition, this effect should be more noticeable with increasing ν_2 content because of the ν_2-ν_3 coupling. We have observed this effect in experiments where the fragment recoil is probed by Doppler spectroscopy (6). For example, we find that the Doppler profile of the $P_2(7)$ transition (J=6.5) of OD ejected by the 2^3 state of DONO is distinctly narrower than that of the $R_2(1)$ line (J=0.5). In addition, this effect is less pronounced when fewer ν_2 quanta are excited. By contrast, we do not observe such effects in HONO photolysis. For example, the Doppler shift of the $P_2(5)$ transition (J=4.5) of OH ejected by HONO 2^3 is the same as that of the $R_2(1)$ transition. We note that the rotational energies of OD (J=6.5) and OH (J=4.5) are approximately the same, so that the narrowing of the fragment Doppler profile with increasing rotation observed in DONO 2^3_0 photodissociation is a direct consequence of ν_2-ν_3,ν_4 coupling.

All the evidence presented above leads us to conclude that the $\nu_2 \to \nu_3$ V→V transfer in DONO($\tilde{A}$) leaves a significant fingerprint on the rotational energy and motional anisotropy of the OD fragment.

Conclusions

The work summarized here represents a benchmark study of the influence of intramolecular vibrational dynamics on state-to-state photochemistry, probed *via* the motional anisotropy and energy content of a photoproduct. The prototype systems *trans* HONO and DONO are ideally suited for this purpose. The spectroscopic work of King and Moule (7) provides an excellent background. The results on the HONO and DONO photodissociation can be qualitatively understood in terms of the potential energy surfaces depicted in Figures 1 and 2. The PES shown in Figure 1 is qualitatively similar to that generated for *trans* HONO($\tilde{A}$) by Hennig et al. (9). However, a closer inspection of our results (4,5) on HONO fragmentation time-scales and OH Doppler spectroscopy, and those of Dixon and Riley (10) on NO photoejected by HONO at 355 nm suggest that the approximate (two-dimensional) PES of Hennig et al. needs improvement. In particular, we find that the exit-channel barrier is overestimated, with the consequence that the trajectory calculations (9) *considerably* overestimate the dephasing/fragmentation time-scales and the extent of by $\nu_2 \to \nu_4$ ("V→T") energy transfer in the HONO complex prior to dissociation (5). *Ab initio* calculations of reasonably sophisticated

PES for trans HONO and DONO would be very useful. Calculations on cis HONO would also be welcome because recent experiments (19,20) suggest that coupling between vibrations induced by intramolecular hydrogen bonding in this isomer leaves a discernible fingerprint on the OH photofragment's rotational energy and anisotropy. In view of the type of detailed state-to-state stereochemical experiments described here, we also feel that for a deeper understanding of photodissociation processes it would be worthwhile to go beyond obtaining the usual product energy distributions in trajectory calculations and include vector-correlations in the list of properties to be evaluated.

Acknowledgment

We are grateful to the National Science Foundation for support of this work (grant #CHE8714573).

Literature Cited

1. R. Vasudev, R.N. Zare and R.N. Dixon, *J. Chem. Phys.*, **1984**, *80*, 4863.
2. J.H. Shan and R. Vasudev, *Chem. Phys. Lett.*, **1987**, *141*, 472.
3. J.H. Shan, S.J. Wategaonkar, V. Vorsa and R. Vasudev, *J. Chem. Phys.*, **1989**, *90*, 5493.
4. J.H. Shan, S.J. Wategaonkar and R. Vasudev, *Chem. Phys. Lett.*, **1989**, *158*, 317.
5. S.W. Novicki and R. Vasudev, *J. Chem. Phys.*, in press.
6. S.W. Novicki and R. Vasudev, under preparation.
7. G.W. King and D. Moule, *Can J. Chem.*, **1962**, *40*, 2057.
8. C. Larrieu, A. Dargelos and M. Chaillet, *Chem. Phys. Lett.*, **1982**, *91*, 465.
9. S. Hennig, A. Untch, R. Schinke, M. Nonella and R. Huber, *Chem. Phys.*, **1989**, *129*, 93.
10. R.N. Dixon and H. Riley, *J. Chem. Phys.*, **1989**, *91*, 2308.
11. C.H. Green and R.N. Zare, *J. Chem. Phys.*, **1983**, *78*, 6741.
12. R.N. Dixon, *J. Chem. Phys.*, **1986**, *85*, 1866.
13. I.L. Chidsey and D.R. Crosley, *J. Quant. Spectrosc. Rad. Transfer*, **1980**, *23*, 187; D.R. Crosley and R.K. Lengel, *J. Quant. Spectrosc. Rad. Transfer*, **1977**, *17*, 59.
14. M. Alexander, P.Anderesen, R. Bacis, R. Bersohn F.J. Comes, P.J. Dagdigian, R.N. Dixon, R.W. Field, G.W. Flynn, K.H. Gericke, E.R. Grant, B.J. Howard, J.R. Huber, D.S. King, J.L. Kinsey, K. Kleinermanns, K. Kuchitsu, A.C. Luntz, A.J. McCaffery, B. Pouilly, H. Reisler, S. Rosenwaks, E.W. Rothe, M. Shapiro, J.P. Simons, R. Vasudev, J.R. Weisenfeld, C. Wittig and R.N. Zare, *J. Chem. Phys.*, **1989**, *89*, 1749.
15. S.W. Novicki and R. Vasudev, *Chem. Phys. Lett.*, **1991**, *176*, 118; S.W. Novicki and R. Vasudev, to be published.
16. D.A. Case, G.M. McClelland and D.R. Herschbach, *Mol. Phys.*, **1978**, *35*, 541; D.A. Case and D.R. Herschbach, *Mol. Phys.*, **1975**, *30*, 1537; D.S.Y. Hsu, G.M. McClelland and D.R. Herschbach, *Mol. Phys.*, **1975**, *29*, 275.
17. P.L. Houston, *J. Chem. Phys.*, **1987**, *91*, 5388.
18. R.N. Zare and D.R. Herschbach, *Proc. IEEE*, **1963**, *51*, 173.
19. J.H. Shan, S.J. Wategaonkar and R. Vasudev, *Chem. Phys. Lett.*, **1989**, *160*, 614.
20. S.W. Novicki and R. Vasudev, to be published.

RECEIVED October 3, 1991

Chapter 19

Isotopic Dependence of the Methyl-Radical Rydberg 3 s Predissociation Dynamics

S. G. Westre[1], P. B. Kelly[1], Y. P. Zhang[2], and L. D. Ziegler[2]

[1]Department of Chemistry, University of California, Davis, CA 95616
[2]Department of Chemistry, Northeastern University, Boston, MA 02115

The isotopic dependence of the predissociation rates for the first excited state of the methyl radical may be described by the quantum mechanical expression for the rate of tunneling through a one dimensional barrier. Modeling of the predissociation rates determined by resonance Raman spectroscopy for the vibrational zero point level of CH_3 and CD_3 yield estimates for the height and positon of the dissociation barrier of the Rydberg 3 s state. Vibrational excitation of the symmetric stretch and the out of plane bend are found to enhance the predissociation rate. The predissociation rates for the mixed isotopes CH_2D and CHD_2 are disscussed.

The photochemistry of a molecule is generally governed by the nature of the excited electronic states. For an electronic state that is predissociated, the isotopic dependence of the dissociation may provide information regarding the shape of the potential in the region of the barrier to dissociation. The unimolecular photochemistry of the methyl radical is of interest as it is one of the simplest free radicals and is a tractable problem for theoretical modeling. The Rydberg 3 s state of the methyl radical is a predissociated electronic state (1). The isotopic substitution can be used in the analysis of the dissociation dynamics on the excited state surface.

The photochemistry of the methyl radical has been the subject of numerous studies. The first spectroscopic study of CH_3 and CD_3 was the flash photolysis UV absorption experiment of Herzberg (2,3) which examined the lowest energy allowed electronic transition of the methyl radical - the $2p_z$ $X^2A"_2$ to Rydberg 3 s $B^2A'_1$ transition. The methyl radical absorption spectrum exhibited an isotopic dependence in that the CD_3 absorption features were narrower and better defined than those of CH_3. Herzberg attributed the apparent lifetime broadening in the excited state of the methyl radical to tunneling of the H (or D) atom through a barrier in the B state. The Rydberg 3 s state of CH_3 was also examined by resonant multiphoton ionization (4). The excitation-ionization spectrum resonant with the B-X transition is very similar to the absorption spectrum in that the CH_3 spectrum is extremely lifetime broadened while the CD_3 spectrum exhibits some rotational resolution. Danon et al. (4) used the linewidths of the excitation-ionization spectra to estimate the excited state methyl radical lifetimes at 1200 fs and 120 fs for CD_3 and CH_3, respectively.

Ab initio calculations on the methyl radical Rydberg 3 s state indicate that the barrier to dissociation is located at 1.55 Å in the H_2C-H coordinate and is approximately 4000 cm^{-1} in height (1). The *ab initio* calculations were performed with a small basis set as a benchmark for examining larger molecules. Yu et al. (1) noted that the small basis set may have resulted in an overestimation of the barrier height.

Resonance Raman spectroscopy has emerged as a powerful probe of the structure and short-time dynamics of excited electronic excited states (5-7). Selectivity is obtained by tuning the probe laser into resonance with a rovibronic absorption of the molecule of interest. The selective intensity enhancement of the Raman signal provides the capability to study the dynamics of a specific transient molecule without interference from the precursor. A great advantage of using the resonance Raman technique to examine excited state dynamics is that the resolution of the Raman spectrum is derived from the vibrational and rotational linewidths in the ground electronic state. Unlike multiphoton absorption studies, lifetime broadening due to fast dissociation processes affects only the intensity of the resonance Raman features (5-7). The analysis of the excitation wavelength dependence of the intensities in the Raman spectrum allows rotationally and vibrationally specific lifetimes to be measured in the subpicosecond time regime. The vibrational and rotational dependence of the Raman intensities provide insight into the nature of the dissociation process and the potential energy surface. Modeling of the Raman spectra of ammonia in resonance with the Rydberg 3 s state has yielded rotationally dependent subpicosecond lifetimes (5-8). The J' dependence of the ammonia lifetimes is attributed to centrifugal forces enhancing the tunneling dissociation through a potential barrier analogous to the barrier predicted for the Rydberg 3 s state of the methyl radical.

The studies presented here examine the predissociation dynamics and structure of the Rydberg 3 s state of the methyl radical. The rovibronic-specific lifetimes of CH_3 and CD_3 in resonance with the vibrationless level of the Rydberg 3 s state are modeled as an H (or D) atom tunneling through a cubic potential barrier. The inclusion of both the protonated and deuterated isotopic data in the modeling of the dissociation provides a true test of the description of the predissociation as a tunneling process through a cubic potential barrier. The dissociation dynamics obtained from the methyl radical analysis are compared to results of the previous *ab initio* calculation on the methyl radical. In order to obtain information about the dissociation processes from higher levels in the Rydberg 3 s state, the dissociation dynamics of higher vibronic levels of the methyl radical Rydberg 3 s state were also examined. Raman spectra taken in resonance with one quanta of the symmetric stretch probe the dissociation rate near the top of the potential barrier.

Resonance Raman Theory

Rotationally specific excited state lifetimes may be extracted from the analysis of resonance Raman excitation profiles (REPs) (5-8). Rotationally resolved spectra of hydrides and deuterides can be examined by resonance Raman spectroscopy due to their large rotational constants. The determination of the excited state lifetimes for the various isotopically substituted molecules relies upon measurement of the rotationally resolved intensities of the Raman spectrum. The enhancement is derived from a limited number of intermediate states, thus allowing the Kramers-Heisenberg sum-over-states approach to be employed (5-7).

The total Raman cross section can be expressed in terms of the isotropic, antisymmetric, and anisotropic scattering contributions:

$$\left(\frac{d\sigma}{d\Omega}\right)_{||+\perp} = \frac{2\pi^2\alpha^2\nu^4 P_g}{3N} \ (2\,|C_q^0|^2 + 3\,|C_q^1|^2 + 7\,|C_q^2|)^2 \quad (1)$$

where α is the fine structure constant, ν is the laser frequency, and P_g is the population of the initial level. In our experiment, the laser is tuned through resonance with individual rovibronic levels of the excited state, $|v'J'K'\rangle$ of the symmetric top. The $|C_q^k|^2$, isotropic (k=0), antisymmetric (k=1), and anisotropic (k=2) scattering contributions depend upon the total angular momentum quantum number of the initial, intermediate, and final level, the polarization direction of the resonant electronic transition moment, the detuning from rovibronic resonance and the rovibronic specific damping. The rovibronic damping constant Γ (fwhm), provides a direct measure of the dependence of the dissociation rate on the total angular momentum of a vibronically specific excited state level and allows us to characterize the isotopic dependence of the dissociation rates.

At the peak of a rotationally specific Ramanexcitation profile (REP), the observed intensities are inversely proportional to the square of the resonant excited state dephasing constant. Thus the dephasing constants are determined by both the width of the excitation profiles and the relative intensities at the peak of the REPs. The dissociation process (i.e. lifetime decay, τ) dominates over the other dephasing processes for the Rydberg 3s state of the methyl radical. The excited state lifetimes are thus given by $\tau = (2\pi c\Gamma)^{-1}$.

Experimental

The far ultraviolet resonance Raman experimental apparatus is based on a 20 Hz Nd:YAG pumped dye laser system (9,10). The second harmonic of the Nd:YAG laser (532 nm) is used to pump a dye laser. DCM or a DCM/Kiton Red mixture was used to allow tuning of the dye laser fundamental from 612 nm to 654 nm. The fundamental of the dye laser was doubled in KD*P and the resulting second harmonic was frequency summed in BBO with the remaining dye fundamental by type I phase matching to produce a tunable third output in the 204 nm to 218 nm region. The probe beam was polarized vertically by a quartz stacked plate polarizer and introduced into the sample region by a 15 cm focal length lens.

The methyl radicals were generated by 266 nm photolysis of methyl iodide. The photolysis beam was produced as the fourth harmonic of the Nd:YAG laser and was polarized ninety degrees with respect to the probe beam. Methyl iodide molecules were entrained in a methane stream which was introduced into the sample region by a modified fuel injector. The sample flow was oriented orthogonal to the probe and photolysis laser beams. Approximately 4×10^{17} molecules/cm^3 of CH_3 were produced by photolyzing 3-4% of the methyl iodide. In the CD_3 experiment, 25% of the CD_3I was photolyzed by a 0.25 mj/pulse 266 nm beam.

The probe laser arrived at the sample region 15 ns after the photolysis. The Raman scattered light was collected in backscatter geometry by a 5 cm focal length, f/0.67 spherical mirror. The Raman excitation profiles were obtained with the collection mirror stopped down to f/2.0 to minimize the finite angle correction. The collected Raman light was focussed onto the slits of a 1.0 meter monochromator. The light passed through a quartz polarization scrambler and was dispersed in third order by an 1800 groove/mm grating. The light was detected using either a solar blind photomultiplier tube or an intensified charge coupled device detector. Atomic lines from a low pressure mercury/argon lamp were used for frequency calibration.

Data points were collected for the CH_3 [0000]-[0000] experiment across the R branch absorption feature from 214.99 nm to 216.29 nm at 0.05 nm (12.5 cm^{-1}) intervals. The monochromator was operated with 300 μm slits (10 cm^{-1} FWHM) in order to resolve the S branch Raman rotational lines. The intensities of the CD_3 rotational Raman features were measured from 214.0 nm to 214.3 nm at 0.013 nm (3 cm^{-1}) intervals using 200 μm slits (7 cm^{-1}) on the monochromator. The reported line intensities represent the average of three individual measurements at each excitation wavelength.

Results and Discussion

Rotationally Resolved Dynamics of the CH_3 and CD_3 [0000] Level in the Rydberg 3 s State. Dissociation rates of CH_3 and CD_3 in the Rydberg 3s state are determined by analysis of resonance Raman intensities. The rotationally resolved S branch of the v_1 band of CH_3 and CD_3 are shown in Figures 1 and 2. The spectroscopic notation of Q, R and S indicates the changes in rotational quantum number J, by 0, 1, and 2 quanta respectively. S(J) Raman transitions (i.e. J -> J+2) are enhanced only by resonance with a rotational level in the excited state having one additional quanta of total angular momentum (i.e. the J+1 level)(5-7). The experimentally determined intensities of the individual S(J) Raman transitions are systematically measured at specific excitation wavelengths. The range of excitation wavelengths spans the [0000] absorption band of the 3s state in the methyl radical. Thus the S(J) Raman excitation profile measurements essentially map the position and breadth of the J' = J + 1 level contribution to the resonant vibronic absorption band. Representative excitation profiles for two S(J) transitions are shown in Figures 3 and 4. Modeling of the excitation profiles (5-7) determines $v_{gv,ev'}$, the band origin of the electronic transition, and $\Gamma^{J'}_{ev'}$, the dephasing constant of the transition. The best fit to the Kramers-Heisenberg intensity expression using $v_{gv,ev'}$ and $\Gamma^{J'}_{ev'}$ as adjustable parameters is denoted by a solid line in Figures 3 and 4. The rotational constants necessary for the modeling were obtained from the analysis of the CD_3 B-X absorption spectrum (2,3). The band origin and the excited state rotational constants determine the wavelength of the center for the excitation profile. The dephasing constant $\Gamma^{J'}_{ev'}$ determines the width and intensity of the excitation profile. The S branch excitation profiles were fit in this manner to within the experimental error (± 10%). Application of the Heisenberg uncertainty principle yields J' specific lifetimes, $\tau = (2\pi c\Gamma)^{-1}$, and relative dissociation rates, given by $1/\tau$, which are summarized in Table I. The lifetimes for CH_3 decrease from 82 fs for J' = 4 to 60 fs at J' = 11 of the [0000] level. The lifetimes for the deuterated methyl radical are substantially longer with lifetimes decreasing from 760 fs to 340 fs as J' increases from 2 to 15.

The difference in the lifetime of the protonated methyl radical relative to that of the perdeutero radical is indicative of a tunneling process. Thus we attribute the observed relative lifetimes to a predissociation of the B state of the methyl radical by H (or D) atom tunneling through a barrier along the dissociation coordinate. *Ab initio* calculations (1) on the methyl radical Rydberg 3 s state indicate that the predissociative barrier is due to the de-Rydbergization (11) of the carbon 3 s orbital to a molecular orbital predominantly σ* in character and ultimately resulting in a 1 s orbital on a hydrogen atom as a function of H_2C--H separation.

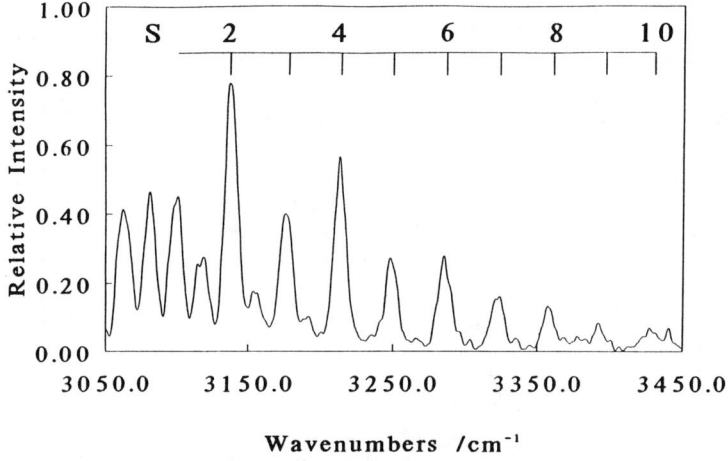

Figure 1. Resonance Raman S branch rotational structure of the CH_3 ν_1 band obtained with 216.016 nm excitation. (Adapted from ref. 8.)

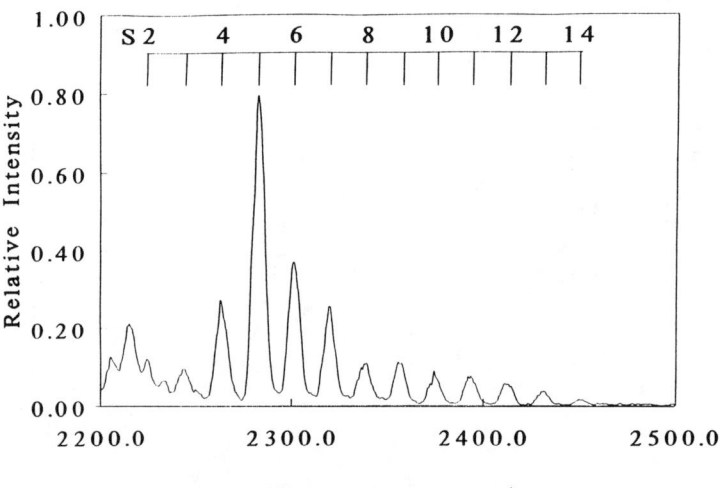

Figure 2. Resonance Raman S branch rotational structure of the CD_3 ν_1 band obtained with 214.189 nm excitation. (Adapted from ref. 8.)

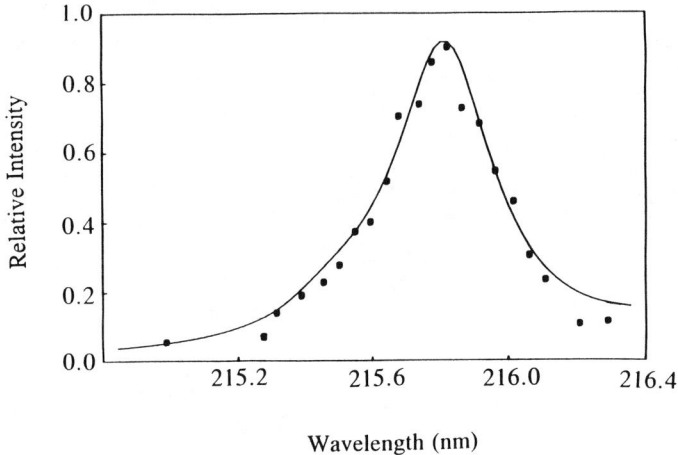

Figure 3. Raman excitation profile of the CH3 S(4) rovibrational feature. The modeled best fit to the experimental data corresponds to $\Gamma = 77$ cm^{-1}. (Adapted from ref. 8.)

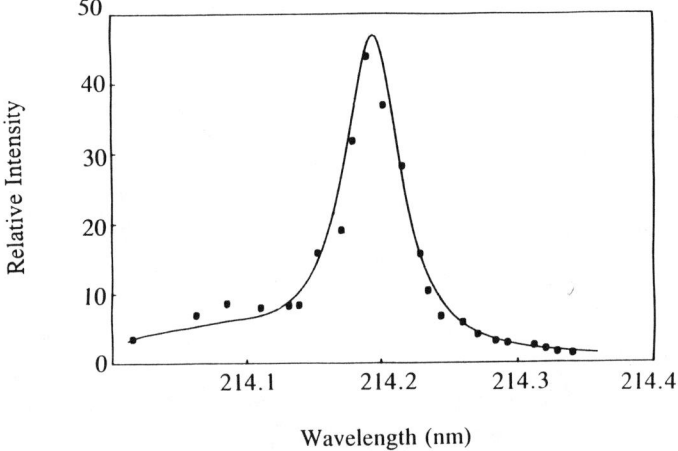

Figure 4. Raman excitation profile of the CD3 S(5) rovibrational feature. The best fit to the data yields $\Gamma = 9.8$ cm^{-1}. (Adapted from ref. 8.)

Table I. J' Specific Predissociation Rates for CH_3 and CD_3[a]

Resonant J'	CD₃		CH₃	
	Γ (cm⁻¹)	Predissociation Rate (10¹² sec⁻¹)	Γ (cm⁻¹)	Predissociation Rate (10¹² sec⁻¹)
2	7.0	1.32		
3	7.0	1.32		
4	7.0	1.32	65	12.3
5	7.2	1.36	77	14.5
6	9.8	1.85	80	15.1
7	9.8	1.85	80	15.1
8	9.8	1.85	85	16.0
9	10.0	1.88	85	16.0
10	10.5	1.98	90	17.0
11	11.0	2.07	90	17.0
12	12.0	2.26		
13	13.2	2.49		
14	14.0	2.64		
15	15.0	2.83		

[a] Absolute error is estimated to be ±10%. Estimated relative error is ±5%.
SOURCE: Adapted from ref. 8.

Isotopic Effects on the Tunneling Rates. The isotopic dependence of the tunneling rate may be described by the quantum mechanical expression for the rate of tunneling through a one-dimensional barrier. The observed dissociation rates for the protonated radical are more than five times faster than those of the corresponding J' level of the deuterated radical as shown in Table I. The rate expression for tunneling through a one-dimensional barrier (*12*):

$$\text{Rate} = \frac{3}{T_o} \exp\left\{ -\frac{4\pi}{h} \int [2m(V(x) - E)]^{1/2} dx \right\} \quad (2)$$

depends upon $1/T_o$, the vibrational frequency of the oscillator, the reduced mass of the tunneling nuclei and the area of the potential above the bound level. The energy of the tunneling state, E, is the vibrational zero point energy with the reduction in the barrier height due to rotational effects. The integration variable, $(x = r - r_b)$, is the difference between r, the length of the dissociating C-H bond, and r_b the position of the barrier maximum shown in Figure 5. The integration is taken over the range of x for which the potential barrier energy is greater than the energy of the tunneling state. The number, 3, accounts for dissociation of any of the three protons or deuterons. The functional form of the barrier to dissociation is denoted by V(x). Equation (2) shows that the observed isotope effect in the dissociation rate is due to the change in reduced mass, vibrational frequency, and zero point energy.

The cubic form approximates a potential having bound and repulsive regions. For the non-rotating molecule, V(x) is given by :

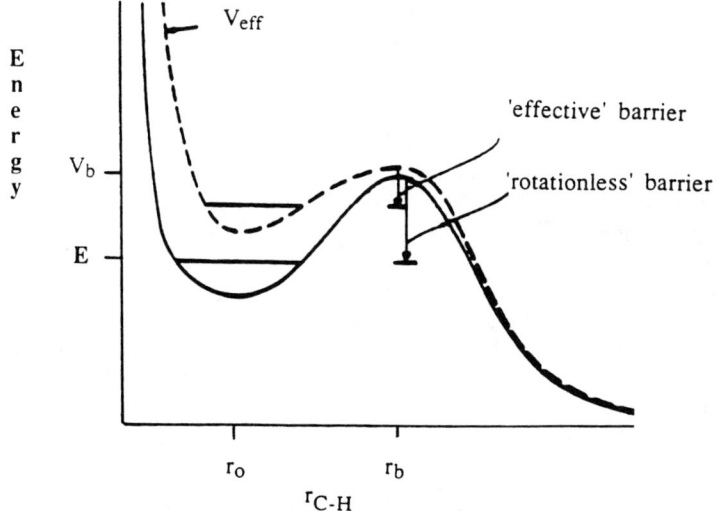

Figure 5. Model potential energy function.

$$V(x) = -ax^3 - \frac{3 ad}{2} x^2 + V_b \qquad (3)$$

where $x = (r - r_b)$, $d = (r_b - r_o)$, and V_b is the height of the barrier. The chosen values of x and d cause the potential to have a relative maximum at $r = r_b$, and a relative minimum at $r = r_o$, the C-H bond length of the Rydberg 3 s state as shown in Figure 5. There is a simple relationship between a, d, and V_b which results from the definitions of the relative minimum and relative maximum (*8*). Thus the cubic potential is a function of only two independent variables, V_b and r_b. The rate expression for the cubic potential is modified to include the reduction in the barrier height as a result of rotational excitation (i.e. centrifugal effects).

$$\text{Rate}(J) = \frac{3}{T_o} \exp\left(-\frac{4\pi}{h} \int [2m(V(x) - \Delta V_{red}(J) - E_{zpt})]^{1/2} dx\right) \qquad (4)$$

where $\Delta V_{red}(J)$ is the reduction of the barrier height due to rotational motion of the molecule. It has been shown previously that for CH_3 and CD_3 the reduction of the barrier height due to rotational excitation, $\Delta V_{red}(J)$ can be expressed in a simple functional form that depends on r_o and r_b (*8*). The $\Delta V_{red}(J)$ is averaged over the population and line strengths of the K rotational levels to yield centrifugal barrier reductions for each J, the total angular momentum value. The barrier height, V_b, and position, r_b, are determined by numerical integration of Equation (4). V_b and r_b are varied to achieve the best fit to the experimentally determined dissociation rates for CH_3 and CD_3. The cubic model of Equation (4) was applied independently to the CH_3 and CD_3 data sets. The equilibrium bond length in the B state, r_o, was taken from Herzberg's rotational analysis of the CD_3 B-X absorption spectrum (*2,3*). The zero point energy of CD_3 was estimated as 840 cm^{-1} which is approximately one half the frequency of the symmetric stretch in the Rydberg 3 s state (*13*). The isotopic relationship between the C-H and C-D oscillators yields $E_{zpt} = 1190$ cm^{-1} for CH_3. The values $V_b = 2200$ cm^{-1} and $r_b = 1.38$ Å result from fitting the cubic potential model to the observed dissociation rates for the various rotational levels of CH_3 and CD_3 shown in Table I. Our results differ from the barrier parameters predicted by the *ab initio* calculations (*1*) on the methyl radical Rydberg 3 s state which estimated a barrier height of approximately 4000 cm^{-1} at a position of 1.55 Å. Within the limitations of the one dimensional model, the difference demonstrates the difficulty of obtaining accurate *ab initio* results with small basis sets for excited states of open shell systems. Our results show that the isotopic dependence of the dissociation from the [0000] level of the methyl radical Rydberg 3 s state can be modeled by a one dimensional quantum mechanical tunneling process using a simple cubic potential model. The one dimensional model succeeds in accounting for the observed differences in the isotopic dissociation rates. Thus the fit of the model supports the hypothesis that the predissociation rate is predominantly due to a tunneling mechanism and accelerated by centrifugal effects in the origin level.

Dynamics of the [0100] Level of CD_3. The lowest energy normal mode of the methyl radical is v_2, the out of plane vibrational motion of the hydrogen atoms (457.8 cm^{-1} for CD_3). The level with one quanta of of v_2 thermally excited in the ground state [0100], has significant population in our experiments. The excitation from one quanta of v_2 in the ground state to one quanta of v_2 in the excited electronic state, [0100]-[0100], of CD_3 was observed at 211.5 nm by the absorption work of Herzberg (2,3) and later confirmed by study of Callear and Metcalfe (13). The observed intensity in the rotational features of the Raman transition originating from one quanta of v_2 and resulting in the addition of one quanta of v_1, [0100]-[1100] results from resonance with the [0100] vibrational level in the excited state. Rotationally resolved excitation profiles associated with the [0100] Rydberg 3 s level for CD_3 by measuring the intensities of the J-resolved S branch lines associated with the [0100]-[1100] Raman transition. S branch intensity measurements were performed over 20 points at even intervals between 210.989 nm and 211.638 nm, in resonance with the 211.5nm absorption band. The dephasing constants for the J' = 3 through J' = 10 levels were obtained from the best fit to the tunneling model. The J'- specific dephasing constants and the associated excited state dissociation rates are summarized in Table II. The CD_3 lifetimes decrease from 350 fs for J' = 3 to 150 fs for the J' = 10 level. The dissociation rates (given as $1/\tau$) from the [0100] level of CD_3 are more than two times faster than the corresponding dissociation rates for the [0000] level.

Table II. Predissociation Rates[a] for the [0100] Level of CD_3

J'	Γ / cm^{-1}	Predissociation Rate / 10^{12} sec^{-1}
3	15.0	2.8
4	21.0	4.0
5	19.5	3.7
6	24.0	4.5
7	26.0	4.9
8	28.0	5.3
9	27.0	5.1
10	35.0	6.6

[a] Absolute error is estimated to be ±10%. Estimated relative error is ±5%.
SOURCE: Adapted from ref. 19.

The modeling of the [0100] excitation profiles also determines the band origin for the [0100]-[0100] transition to be 47271 cm^{-1}. Using the ground state v_2 frequency of 457.8 cm^{-1} from the diode laser work of Sears et al. (14,15) and the [0000]-[0000] electronic band origin from the analysis of the [0000] Raman excitation profiles (8), the v_2 frequency in the Rydberg 3 s state is thus determined to be 1094 cm^{-1}. Our value for the Rydberg 3s state v_2 is significantly greater than that for v_2 in the ground electronic state and is consistent with the value of 1090 cm^{-1} obtained from the absorption spectrum of Callear and Metcalfe (13). It is also notable that the v_2 frequency in the Rydberg 3 s state is comparable to that in the Rydberg $3p_z$ electronic state, 1036 cm^{-1}, as determined by the (2+1) REMPI of CD_3 (16).

One of the most interesting features of the methyl radical vibrational analsis is the pseudo-Jahn-Teller interaction perturbing the out-of-plane bend vibrational manifold in the ground electronic state. The psuedo-Jahn-Teller interaction proposed by Yamada and Hirota (*17*) is a vibronic coupling of the ground and electronic states involving the out of plane bend, ν_2 normal mode. This vibronic coupling effect is evident in the observed values of the ν_2 bending frequency in the excited electronic states which are measured using resonance Raman spectroscopy. The vibronic interaction results in large anharmonicity in the ground state for the ν_2 coordinate as was observed by Yamada et al. (*17*) They found that the CH_3 ν_2 potential in the ground electronic state contains a significant quartic contribution causing the methyl radical to have an anomalously low ν_2 bending frequency. The analysis of the ν_2 ground electronic state potential performed by Yamada et al. (*17*) yields a parameterization of the vibronic coupling of the ground and excited electronic states. Our spectroscopic results for CD_3 yield a value of 1094 cm^{-1} for ν_2' which is of the order of magnitude expected from the pseudo-Jahn-Teller model (*18*). Examination of the protonated species CH_3 is required for a more thorough understanding of the interaction. The 212.5 nm absorption feature of CH_3 has been identified as the [0100]-[0100] band origin using resonance Raman spectroscopy. Experiments are in progress to determine an accurate value for ν_2' in the excited state of CH_3.

The Structure and Dynamics of the CH_3 [1000]-[0000]. Callear and Metcalfe (*13*) observed, but could not assign a vibronic feature in the methyl radical absorption spectrum having a maximum at approximately 208 nm for CH_3. Raman spectra in resonance with this CH_3 absorption feature exhibit only the presence of ν_1 and $2\nu_1$ vibrational activity. The symmetric C-H stretch (ν_1) and its first overtone ($2\nu_1$) are of approximately equal intensity. In contrast, resonance with the [0000]-[0000] electronic absorption results in the ν_1 Raman feature having 7 times greater intensity than $2\nu_1$. The change in relative Raman intensities due to Franck-Condon factors identifies the absorption feature located near 208 nm as a vibronic symmetric stretch band in the Rydberg 3 s spectrum. No Raman lines were detected in the anti-Stokes region, indicating that the absorption feature is not due to a hot band transition. Our observations lead us to identify the broad 208 nm vibronic feature as the [1000]-[0000] electronic transition of CH_3. The vibrational Raman excitation profiles of the ν_1 Q branch is shown in Figure 6. The modeling of the Raman excitation profiles yields 206.85 nm as the band origin of the [1000]-[0000] electronic absorption band. Combining the [1000]-[0000] and [0000]-[0000] electronic band origins yields a frequency for ν_1' in the Rydberg 3 s state of 2040 cm^{-1}. The value for the dephasing constant obtained from modeling the Raman excitation profile for resonance with the [1000] level is ca. 400 cm^{-1}, which corresponds to a [1000] excited state lifetime of ~ 13 fs. Excitation of one quanta of ν_1 (2040 cm^{-1}) in addition to the zero point energy places ν_1' level of the excited state above the top of the dissociation barrier as determined by the previous one dimensional model analysis. The extremely short excited state lifetime is consistent with our analysis which yielded a dissociation barrier height of 2200 cm^{-1}.

Structure and Dynamics of the CH_2D and CD_2H [0000] Level. Additionally, resonance Raman spectroscopy has been employed to study the mixed isotopes of the methyl radical. The spectra confirm that the band centers of the [0000]-[0000] electronic transition first observed by Herzberg (*2,3*) for CD_2H and CH_2D are 214.7 nm and 215.3 nm respectively. The resonance Raman spectra yield ground state vibrational frequencies for the mixed isotopes. The vibrational frequencies have been used to determine the stretching force constants for the normal

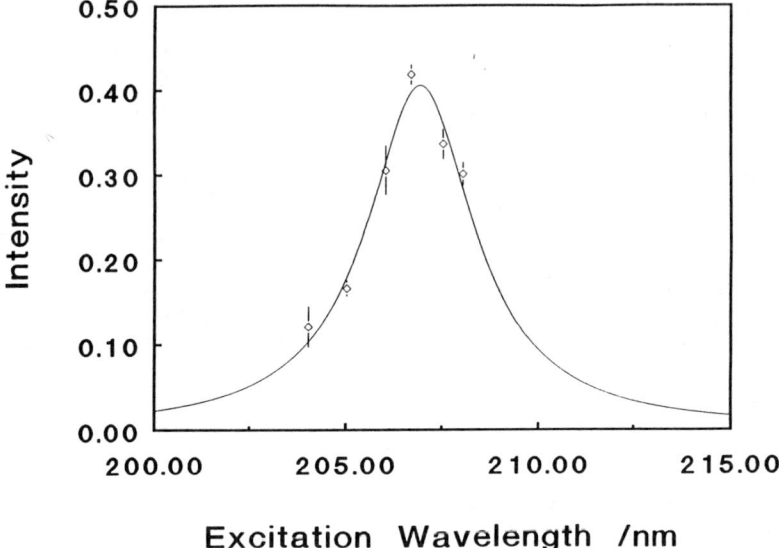

Figure 6. Q branch Raman excitation profile of ν_1 of CH_3 in resonance with the [1000] level of the Rydberg 3s state. Modeling of the Raman excitation profile yields a Γ of 400 cm^{-1} corresponding to an excited state lifetime of 13 fs.

modes of the methyl radical (19). The dissociation dynamics of the mixed isotopes are also probed by resonance Raman spectroscopy. The intensities of the resonance Raman spectra are observed to increase monotonically with increasing deuterium substitution. For excitation at the center of the REPs, the intensities of the resonance Raman transitions are inversely proportional to the square of Γ, the dephasing constant. The dissociation rate is directly proportional to the dephasing constant. Thus the correlation of increase Raman intensity with deuterium substitution is ascribed to the isotopic decrease in the tunneling rate. The resonance Raman spectra indicate that the excited state lifetime increases with increasing deuterium substitution. Further analysis of the mixed isotope intensities will yield quantitative measures of the excited state lifetimes.

Conclusion

Resonance Raman excitation profiles were utilized to provide insight into the nature of the excited electronic state of the methyl radical. The structure and dynamics of the methyl radical Rydberg 3 s state were probed by our resonance Raman technique. Isotopic dependence of the methyl radical dissociation dynamics can be described by the proposed tunneling mechanism. Detailed examination of the J dependent dissociation rates for CH_3 and CD_3 show a strong isotopic effect for the excited state lifetime. For each rotational level examined the perdeutero radical, CD_3 has an excited state lifetime that is more than five times larger than that for CH_3. The observed isotopic effect on the tunneling rate arises from the mass difference between the proton and the deuteron, the decrease in zero point energy on the excited state surface with increase dueterium substitution, and the isotopic dependent shift of the vibrational frequency of the oscillator. Raman spectra in resonance with the [0000] excited state level indicated increasing lifetime within the series of deuteration, CH_3, CH_2D, CD_2H, CD_3.

Rovibronic specific predissociation rates for the origin level of the B state were obtained by modeling the S branch Raman excitation profiles of CH_3 and CD_3. The isotopic dependence of the methyl radical dissociation from the [0000] origin level was modeled as a simple one dimensional quantum mechanical tunneling process through a cubic potential barrier. The observed dissociation rates of the [0100] level in the Rydberg 3 s state indicate that the v_2' out of plane bend level of CD_3 is below the dissociation barrier maximum and lifetimes decrease with increasing rotational excitation. The [1000] level of CH_3 exhibits an ultra-fast dissociation rate which is consistent with the expectation that the v_1' level is above the potential barrier. It is hoped that our results will stimulate a theoretical re-examination of the methyl radical Rydberg 3 s state surface.

Acknowledgments

Support for the work form the NSF (Grant #CHE-8923059), NIEHS (Grant #1P42-ES04699) and University wide Energy Research Group of the University of California (Grant #444024) is gratefully acknowledged by PBK. LDZ acknowledges the support of the NSF (Grant #CHE-8918418) and the Petroleum Research Fund, administered by the American Chemical Society.

Literature Cited

1. Yu, H.T.; Sevin, A.; Kassab, E.; Eveleth, E.M. *J. Chem. Phys.* **1984**, *80*, 2049.
2. Herzberg, G.; Shoosmith, J. *Can. J. Phys.* **1956**, *34*, 523.
3. Herzberg, G. *Proc. Roy. Soc. 262 A*, **1961**, 291.
4. Danon, J.; Zacharia, H.; Rottke, H.; Welge, K.H. *J. Chem. Phys.* **1982**, *76*, 2399.
5. Ziegler, L.D.; Chung, Y.C.; Wang, P.; Zhang, Y.P. In *Time Resolved Spectroscopy*; Clark, R.J.H.; Hester, R.E., Eds.; Wiley: New York, NY, 1989; pp 55-111, and references cited therein.
6. Ziegler, L.D.; Chung, Y.C.; Wang, P.; Zhang, Y.P. *J. Phys. Chem.* **1990**, *94*, 3394.
7. Ziegler, L.D. *J. Chem. Phys.* **1987**, *86*, 1703.
8. Westre, S. G.; Kelly, P. B.;Zhang, Y. P.; Ziegler, L. D. *J. Chem. Phys.* **1991**, *94*, 270.
9. Kelly, P.B.; Westre, S.G. *Chem. Phys. Lett.* **1988**, *151*, 253.
10. Westre, S.G.; Kelly, P.B. *J. Chem. Phys.* **1989**, *90*, 6977.
11. Mulliken, R.S. *Acc. Chem. Res.* **1976**, *2*, 7.
12. Rice, O.K. *Physic. Rev.* **1930**, *35*, 1538.
13. Callear, A.B.; Metcalfe, M.P. *Chemical Physics* **1976**, *14*, 275.
14. Sears, T.J.; Frye, J.M.; Spirko, V.; Kraemer, W. P.; *J. Chem. Phys.* **1989**, *90*, 2125.
15. Frye, J.M.; Sears, T.J. ; Leitner, D. *J. Chem. Phys.* **1988**, *88*, 5300.
16. Parker, D.H.; Wang, Z.W.; Jansen, M.H.M. ; Chandler, D.W. *J. Chem. Phys.* **1989**, *90*, 60.
17. Yamada, C.; Hirota, E.; Kawaguchi, K. *J. Chem. Phys.* **1981**, *75* , 5256.
18. Hirota, E., In *High Resolution Spectroscopy of Transient Molecules*, Series Springer-Verlag: New York, NY, 1985.
19. Westre, S. G.; Gansberg, T.E.; Kelly, P.B.; Ziegler, L.D. Submitted to *J. Phys. Chem.*

RECEIVED December 5, 1991

Chapter 20

Laser-Stimulated Selective Reactions and Synthesis of Isotopomers
New Strategies from Diatomic to Organometallic Molecules

J. E. Combariza[1,6], C. Daniel[2], B. Just[1,7], E. Kades[3], E. Kolba[1,8], J. Manz[1,7], W. Malisch[4], G. K. Paramonov[5], and B. Warmuth[1]

[1]Institut für Physikalische Chemie, Universität Würzburg, Marcusstrasse 9–11, D–8700, Würzburg, Germany [2]Laboratoire de Chimie Quantique, Universite Louis Pasteur, 4 Rue Blaise Pascal, 67000 Strasbourg, France [3]Physikalisch-chemiches Institut der Universität Zürich, CH–8057 Zürich, Switzerland [4]Institut für Anorganische Chemie, Universität Würzburg, Am Hubland, D–8700 Würzburg, Germany [5]Institute of Physics, Academy of Science of Belorussia, Minsk 220602, Belorussia

> By analogy with new strategies for laser-control of (i) state-selective transitions and (ii) vibrationally-mediated photodissociation of small molecules, we suggest (i) selective isomerization and (ii) selective separation of ligands in organometallic compounds, designed to prepare pure isotopomers. Specifically, we extend (i) our previous evaluation of optimal infrared (IR) picosecond (ps) laser pulses with analytical (e.g. Gaussian) shapes from exclusive high overtone excitation of OD or OH bonds to exclusive isomerization of $[(CO)_2(C_5H_5)FePHD]$ or $[(CO)_2(C_5H_5)FePH_2]$. Moreover, we adapt (ii) the continuous wave (cw) IR + vis/UV two-photon strategy of Crim et. al. and extend it from dissociation of OH (not OD) to fission of a metal-ligand bond in a specific organometallic isotopomer, e.g. dissociation of Ni–C_2H_4 (not Ni–C_2D_4, etc.) or $HCo(CO)_4$ (not $DCo(CO)_4$). The parameters of the relevant series of laser pulses (i) or cw lasers (ii) are optimized for selective primary photochemical processes in a single isotopomer. Further reactions of specific isomers (i) or radicals (ii) yield the desired preparation of pure isotopomers. The model simulations employ a broad variety of theoretical chemistry techniques, from quantum chemistry ab initio evaluation or modelling of potential energy surfaces and electric dipole functions, to calculations of relevant vibrational states, to simulations of laser-stimulated molecular reaction dynamics by propagation of representative wavepackets.

[6]Current address: Chemistry Department, Louisiana State University, Baton Rouge, LA 70820
[7]Current address: Freie Universität Berlin, FB Chemie, Institut für Physikalische und Theoretische Chemie, Takustrasse 3, 1000 Berlin 33, Germany
[8]Current address: Siemens-Nixdorf-Informationssysteme AG, Otto-Hahn-Ring 6, 8000 München 83, Germany

Laser assisted separation and preparation of molecular isotopomers have been considered as challenging tasks ever since the discovery of lasers. Corresponding efforts have already achieved substantial, albeit not universal success; e.g. by means of continuous wave (cw) laser induced multiphoton excitations, (for reviews see Refs *(1-3)*.) The purpose of this article is to present – in a brief survey-type manner – two novel strategies for laser assisted primary reactions in selective isotopomers of organometallic compounds, e.g. exclusively in the deuterated one (called briefly [D] below), not in the naturally abundant hydrogenated one, [H], or vice versa. Sequel reactions may yield the desired pure product isotopomers. The two methods are extensions, or adaptations of related, very promising new strategies for laser assisted selective transitions or reactions in small molecules. Briefly, (i) strong infrared (IR) ps laser pulses with analytical (typically Gaussian) shapes may induce vibrational state-selective transitions, e.g. exclusive high overtone excitation of specific diatomic isotopomers (molecules, radicals or fragments) such as [D]=OD, not [H]=OH, or vice versa *(4-6)*. This method is important, since the vibrationally excited "hot" isotopomer, OD, may serve as a preferential reagent (in comparison with "cold" OH) in a sequel deuterium (not hydrogen) transfer reaction, yielding exclusively the deuterated, not the hydrogenated products *(7)*. For example, selective consumption of vibrational energy is favoured by bimolecular reactions such as $OD + Cl \rightarrow O + DCl$, preferably if they have late barriers of their potential energy surfaces *(8-11)*. However, we center our attention on extending the fundamental methods of selective vibrational excitations to laser pulse assisted isomerization of exclusively organometallic model isotopomers, e.g. $[D]=[(CO)_2CpFePHD]$, not $[H]=[(CO)_2CpFePH_2]$, ($Cp=C_5H_5$), and vice versa.

(ii) Continuous wave (cw) IR + vis/UV two-photon excitations have been developed systematically, in particular by Crim et. al. *(13,14)* for laser control of photochemical reactions. The method has already been used for isotope separation of small molecules *(15-18)*, however we are not aware of any previous applications to organometallic compounds. The culmination of this IR + vis/UV two photon approach is presumably bond-selective two photon dissociation of small molecules, as documented in the first consistent theoretical prediction and experimental verification carried out for vibrationally mediated photodissociation of HOD in the $A(^1A_1)$ state by Imre and Zhang *(19,20)* (see also the work of Schinke et al *(21,22)*), of Crim *(23)* and of Rosenwaks and Valentini *(24)*. Shapiro and Segev *(25,26)* made similar predictions for bond selective IR + VUV photodissociation of HOD in the $B(^1B_1)$ state.

Here we suggest that essentially the same IR + vis/UV two-photon strategy *(13-26)* may be used for selective photodissociation of specific isotopomers, from small molecules e.g. OH, not OD to organometallic model compounds, e.g. [H]= Ni–C_2H_4, not [D]= Ni–C_2D_4, or [H]= $HCo(CO)_4$, not [D]=$DCo(CO)_4$. Some complementary laser-induced reactions of selective organometallic isotopomers have been presented very recently *(27)*.

The original purpose of the methods (i) *(4-6,12)*, (ii) *(13-26)* has been laser induced state mode or bond selectivity in unimolecular reactions *(28-31)*. The present extension to selective preparation of organometallic isotopomers provide novel yet straightforward applications of these techniques.

The present choice of organometallic compounds is motivated primarily by broad interest of several groups, in these model compounds, $[(CO)_2CpFePH_2]$, *(12,32)*, $[Ni(C_2H_4)]$, *(33,34)*, and $[HCo(CO)_4]$, *(35-39)*. Clearly, these specific compounds serve just as rather simple examples for the application of the new strategies (i), (ii) to other organometallic compounds with similar structural and dynamical properties. Their most obvious common property is the binding of one or more ligands to a transition metal atom, and this suggests rather simple model simulation of strategies (i), (ii) for organometallic compounds. Another strong motivation for the choice of our model systems is as follows: The heavy metal atoms should block *(40-47)* intramolecular vibrational energy redistribution (IVR) on a time scale τ_{IVR} which may be longer than the duration of the laser-induced processes (i) and (ii), i.e. longer than few ps or ca. 100 fs, respectively (see however Ref. *(48)*). As a consequence, many molecular degrees of freedom are decoupled on the time scale of laser induced processes (i), (ii), i.e. one may design reduced-dimensionality models which should account for all "active" promoting modes of the reactions, neglecting all "passive" spectator modes. In fact, we shall assume simple one dimensional (**1D**) or two dimensional (**2D**) models, describing the relevant reaction coordinate, possibly plus another important promoting mode of the systems. In particular, we shall disregard any detailed effects of rotations – this approximation is presumably more appropriate for large, slowly rotating ($\tau_{rot} > 5ps$) molecules than for small, fast rotating ones.

In practice, we shall assume that the molecular orientation is fixed, with its dominant dipole component μ_x along the axis of linearly (x-) polarized electric field of the lasers. Of course, we are aware of some inherent artifacts which may be associated with such simple **1D** and **2D** models *(49-55)*. Therefore we should emphasize from the outset that the subsequent results are to be considered as semi-quantitative, subject to tests of self-consistency in refined models with higher dimensionalities, ultimately by experimental verifications. Nevertheless, by analogy with some successful **1D** or **2D** approaches to related model systems (see the next sections), we are confident that the present predictions are quite meaningful, and stimulating.

The new strategies for laser induced isotopomer selective isomerizations (i) and bond-fissions (ii) will be presented in two subsequent sections. Each Section consists of two parts: First we give a brief account of the corresponding strategy for small molecules, second we extend the methods to organometallic compounds. The fundamental methods cover a broad range of theoretical chemistry techniques, including models or quantum chemistry ab initio calculations of potential energy surfaces and dipole functions, evaluation of molecular vibrational states, and simulation of the laser assisted molecular reaction dynamics by time dependent propagation of representative wavepackets, together with optimal design of the laser parameters. The conclusions are presented in the final section.

Isotopomer-Selective Unimolecular Processes Induced by Strong ps IR Laser Pulses with Analytical Shapes

Selective Vibrational High-overtone Excitation of OD versus OH.
Let us consider, as an example, the problem of exclusive vibrational high overtone excitation of OD, not OH. The relevant model potential energy surface V and dipole function μ_x versus bond coordinate x are adapted from Refs *(6,56-58)*, with parameters adjusted to the OH/OD bonds of HOD, as shown in figure 1a, together with the OD vibrational energies E_v and wavefunctions $\varphi_v(q) = <q \mid v>$. These are obtained by solving the time independent Schrödinger equation

$$H_{mol} \mid v> = E_v \mid v> \tag{1}$$

with molecular Hamiltonian
$$H_{mol} = T + V, \tag{2}$$

kinetic energy operator
$$T = p^2/2\mu \tag{3}$$
and reduced mass
$$\mu = m_O m_D/(m_O + m_D) \tag{4}$$
by means of finite differences *(59,60)*.

For an explicit rather simple introductory example, let us assume first that the model OD bond should be excited from an initial state $|i> = |v_i>$, typically the ground state $|v_i> = |0>$, to a moderately excited target state $|f> = |v_f>$, e.g. $|v_f> = |5>$, by means of an x-polarized IR ps laser pulse with Gaussian shape. (Such moderate excitations may also be achieved by means of cw-lasers, as described in the next sections.) In the semiclassical approximation, the pulse is represented by its electric field

$$\vec{\mathcal{E}}(t) = (\mathcal{E}_x(t), 0, 0)^T, \tag{5a}$$

where $\mathcal{E}_x(t)$ has the general form

$$\mathcal{E}_x(t) \equiv \mathcal{E}_k(t) = \mathcal{E}_k \cdot [sin\omega_k(t - t_k)] \cdot S_k(t - t_k; t_{pk}) \tag{5b}$$

with the Gaussian shape function

$$S_k(\tau, t_p) = exp[-4ln2(\tau/t_p^2)], \tag{5c}$$

and parameters $\mathcal{E}_k$ = field amplitude, ω_k = IR carrier frequency, t_k = delay time and t_{pk} = pulse duration, possibly labelled by an index k (k = 1,2,...,N for sequences of pulses.

The solution of this problem - selective excitation of a moderately excited overtone of OD - is adapted from Refs *(4-6)*, see also Ref *(61)*. Essentially, we have to evaluate the time-dependent Schrödinger equation

$$i\hbar |\dot{\Psi}_i(t)> = H|\Psi_i(t)> \tag{6}$$

subject to the initial condition

$$|\Psi_i(0)> = |v_i> \tag{7a}$$

The time-dependent Hamiltonian

$$H = H_{mol} - \mu_x \mathcal{E}_x(t) \tag{8}$$

describes the molecule plus its electric dipole interaction with the laser field. For convenience, the resulting molecular wavepacket $\Psi_i(t)$ is expanded in terms of molecular eigenstates

$$|\Psi_i(t)> = \sum_v |v> C_{vi}(t) \tag{9}$$

Insertion of eqn (9) and Dirac-bra-operation of the molecular state $<u|$ then converts eqn (6) into the corresponding algebraic version,

$$i\hbar \frac{\partial}{\partial t} \mathcal{C}_i(t) = \mathcal{HC}_i(t) \tag{10}$$

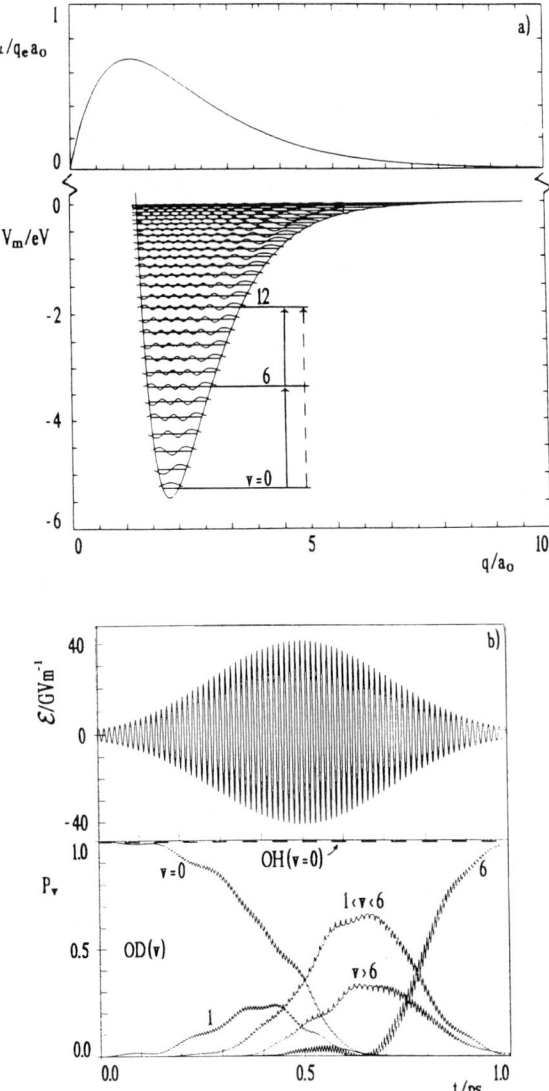

Figure 1. Selective vibrational overtone excitation of model OD by ps IR laser pulses with Gaussian shapes. **1a.** Potential energy V, vibrational energies E_v and wavefunctions ϕ_v (bottom panel) and dipole function μ (top panel). Selective vibrational transitions are indicated by vertical arrows. **1b.** Electric fields E with Gaussian envelopes S (top panels) and resulting populations P_v of vibrational states v (bottom panels) for selective $0 \to 6$ transition. The laser pulses do not affect the OH isotopomer, dashed line in bottom panel of 1b.

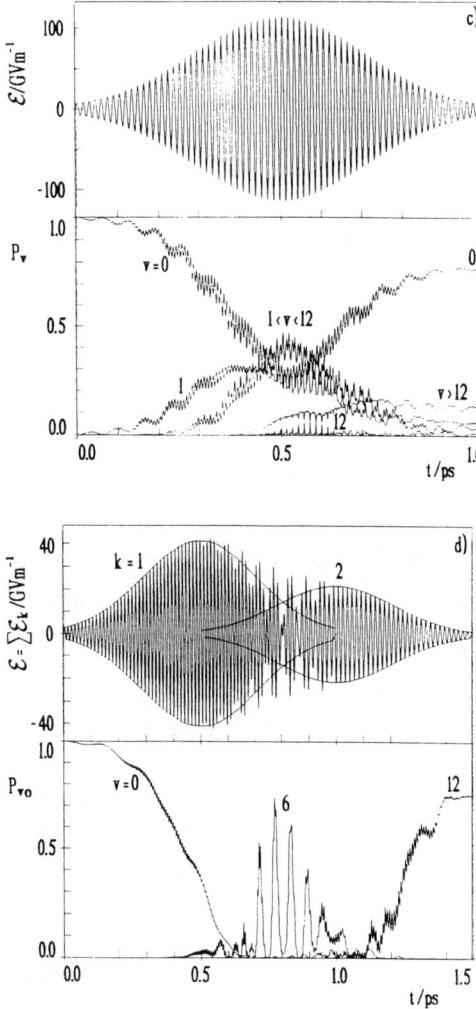

Figure 1 continued. Electric fields E with Gaussian envelopes S (top panels) and resulting populations P_v of vibrational states v (bottom panels) for (**1c**) nonselective $0 \to 12$ transition and (**1d**) selective series (k=1,2) of $0 \to 6 \to 12$ transitions.

with vector $\mathcal{C}_i$ of expansion coefficients $\mathcal{C}_{vi}$, and Hamilton matrix $\mathcal{H}$ with elements

$$H_{uv}(t) = E_v \delta_{uv} - <u\,|\,\mu_x\,|\,v> \mathcal{E}_x(t) \tag{11}$$

The initial condition (7a) is equivalent to

$$C_{vi}(0) = \delta_{vi}. \tag{7b}$$

The coupled equations (10) subject to (7b) are propagated by Runge-Kutta integration *(59,60)*. The resulting coefficients $C_{vi}(t)$ yield the time-dependent populations of molecular eigenstates

$$P_{vi}(t) = |C_{vi}(t)|^2. \tag{12}$$

In particular, the relevant population of the target state $|\,v> = |\,v_f> = |\,f>$ at the end of the pulse, i.e. at time $t = t_{pk}$, is $P_{fi}(t_{pk})$. Of course, $P_{fi}(t_{pk})$ depends (via eqns 15, 10-12) also on the other laser parameters $\mathcal{E}_k, \omega_k$, thus

$$P_{fi}(t_{pk}) = P_{fi}(t_{pk}, \omega_k, \mathcal{E}_k) \tag{13a}$$

Systematic variation of $\omega_k, \mathcal{E}_k$ yield optimal laser parameters for maximum excitation of the target state *(4-6)*; in favourable cases,

$$P_{fi}(t_{pk}, \omega_k, \mathcal{E}_k) \simeq 1, \tag{13b}$$

where $\omega_k, \mathcal{E}_k$ denote the optimal laser parameters. Details of the optimization procedures are given in Refs *(4-6)*. Suffice it here to say that the resulting frequencies are typically close to, but not identical to, zero-order frequencies $\omega_k \simeq (E_f - E_i)/(n_k \hbar)$ of $n_k = 1,2,3,..$etc photon transitions. For the purpose of isotope selectivity, it is found that these optimal populations of target states (13b) are extremely sensitive to tiny variations $\triangle \omega_k$ of the optimal frequencies ω_k, i.e. $P_{fi}(t_{pk}, \omega_k + \triangle \omega_k, \mathcal{E}_k) \to 0$ for $|\triangle \omega_k|/\omega_k > 0.001$. In contrast, the optimal $P_{fi}(t_{pk}, \omega_k, \mathcal{E}_k)$'s are rather robust even for substantial variations ($\triangle \mathcal{E}_k/\mathcal{E}_k \leq 0.1$) of the electric field strength *(61)*.

A successful model application for selective vibrational $|\,0> \to |\,6>$ excitation of OD is shown in figure 1b. The optimal laser pulse's electric field and its Gaussian envelope are shown in the top panel, the resulting time-dependent populations are in the bottom panel, yielding almost perfect excitation of the target state $|\,v_f> = |\,6>$, see eqn (13b). In contrast, the effect of the same pulse on ground state OH isotopomers is entirely negligible, see the bottom panel of figure 1b. Likewise, it is possible to design optimal laser pulses for selective excitations of OH *(6,61)*, but in turn, these pulses do not at all excite the OD isotopomer.

These model simulations clearly demonstrate the desired isotopomer-selectivity of the present laser pulses. The most important origin for this discrimination of nearly perfect OD versus entirely inefficient OH excitations, and vice versa, is the different (mass-dependent! cf. eqns (1) - (4)) eigenenergies, implying extremely narrow ranges of optimal IR transition frequencies ω_k for either OH or OD, which do not overlap (except possibly in rare accidental cases).

Unfortunately, however, selective excitation of isotopomers by a single IR ps laser pulse is not a universal tool. Depending on the systems *(4-6,61)*, its favourable domain are moderate, not high overtone excitations, or in general transfer of "few", not "many" vibrational quanta. For example, the successful application to moderate OD $|\,0>\,\to|\,6>$ excitation cannot be extrapolated to high OD $|\,0>\,\to|\,12>$ excitation, compare the positive and negative results shown in figures 1b and 1c, respectively. Nevertheless, the desired high overtone excitations, $|\,0>\,\to|\,12>$, may be achieved by an extension (see Refs *(12,61)*) of the original method *(4-6)*. One simply divides the high overtone excitation $|0>\,\to|12>$ into smaller individual steps labelled $k=1,2,...,N$, e.g. $N=2$ for $|0>\,\to|6>\,\to|12>$. In general, the sequence of N transitions from $|\,i>$ to $|\,f>$ may be written

$$|i>=|v_1>\,\to|v_2>\,\to\cdots\to|v_k>\,\to|v_{k+1}>\,\to\cdots\to|v_N>\,\to|v_{N+1}>=|f> \tag{14}$$

Then one designs optimal laser fields $\mathcal{E}_k(t)$, eqns (5), for each individual transition $|\,v_k>\,\to\,|\,v_{k+1}>$, $k=1,2,...,N$. Finally, the individual laser fields are superimposed

$$\mathcal{E}_x(t) = \sum_{k=1}^{N} \mathcal{E}_k(t) \tag{15}$$

with adequate delay times t_k, e.g. the sequel pulse k+1 should start when the preceding one k has its maximum intensity. The overall laser field (15) will then drive the molecule from its initial state $|\,i>$ via the sequence (14) of intermediate states $|\,v_k>$ towards the target state $|\,f>$. This strategy for high overtone excitation by sequences of ps IR laser pulses with analytical shapes is demonstrated in figure 1d for the sequence $|\,0>\,\to|\,6>\,\to|\,12>$, yielding substantial, (ca. 75%) high overtone excitation of the OD target state $|\,12>$. The individual pulse for the first step $|\,0>\,\to|\,6>$ has already been shown in figure 1a; a similar pulse is designed for the second step $|\,6>\,\to|\,12>$; the cumulative effect of both pulses is shown in figure 1d. Similar sequences of laser pulses (14) may be designed for any other high overtone excitation *(61)*. Furthermore, the same sequence of laser fields $\mathcal{E}_x(t)$ which yields selective high overtone excitation of OD (v = 12) have negligible effect on OH (v = 0).

Before closing this sub-section, let us point to an important technical aspect which makes the present approach to selective high overtone excitations easier than some complementary ones *(62-65)*. Specifically the present simple superposition of IR ps laser fields $\mathcal{E}_k(t)$ generate automatically, via interferences, the required, rather complex structure of the effective overall fields $\mathcal{E}_x(t)$, see figure 1d. Comparably more complex structures have been derived earlier by Rabitz et al *(62-64)* for much more demanding, i.e. non-analytical design of optimal laser pulses, see also our extension *(65)*. As an example we show in figure 2 some results for the Ni-C_2H_4, which will be described in more detail later, using an IR 100 femtosecond laser pulse with non-analytical shape, to excite the system from the ground state to a higher vibrational state $|4>$. The complexity and difficulty to evaluate the results are seen in figure 2, in contrast with the results obtained using analytical pulses as shown in figures 1b-1d.

Selective Isomerizations of [Cp(CO)$_2$FePHD] versus [Cp(CO)$_2$FePH$_2$].

In close analogy to selective vibrational high overtone excitation of OD versus OH, let us now consider the problem of exclusive isomerization of a model organometallic compound, $[Cp(CO)_2FePHD]$, not $[Cp(CO)_2FePH_2]$. The two isomers

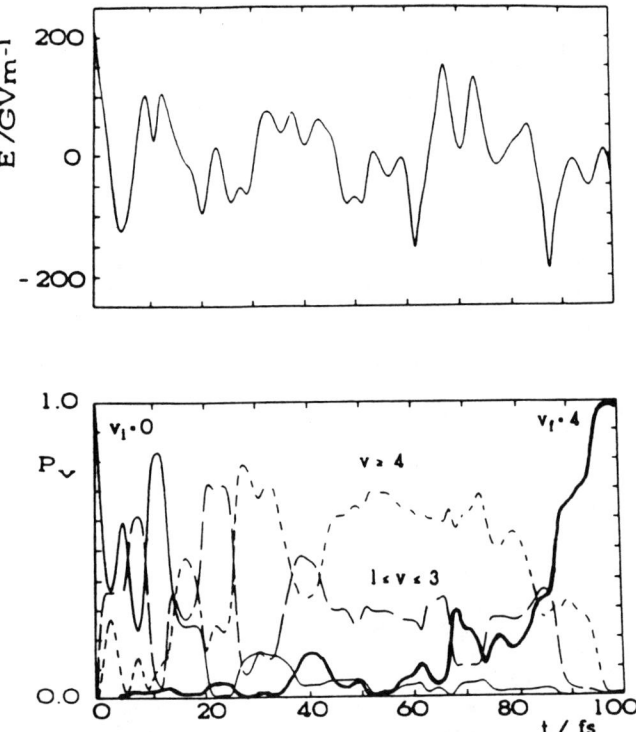

Figure 2: Example of hybrid techniques for isotopomer selective reactions of organometallic molecules. The IR excitation of $Ni-C_2H_4$ is attempted by an IR 100 fs laser pulse with non-analytical shape. Compare with simpler ps IR laser pulse excitations as in Figure 1.

have the pyramidal FePHD ferrio-phosphine group either bent towards the (CO) ligands (isomer A) or towards the cyclopentadienyl ligand (isomer B); see figure 3a. Isomerization from A to B may be considered as the inversion of the FePHD fragment. As it is well known, some important aspects of inversion processes (e.g. dynamical tunnelling) of these paradigma are well described by **1D** models, see e.g. Ref *(66)*. By analogy, and neglecting IVR among different ligands due to heavy atom blocking we employ a **1D** model for inversion of $[(CO)_2CpFePHD]$. The corresponding potential energy surface V and dipole function μ_x versus inversion coordinate σ are shown in figure 3a, together with the vibrational energies E_v and wavefunctions $\varphi_v(\sigma) = <\sigma|v>$. The evaluation of V and μ_x by quantum chemistry ab initio methods has been described in Ref *(12)*. The energy barrier between the two isomers is ca. 20 Kcal/mol in fortuitous agreement with experimental results for similar type of ferrio-complexes *(32)*. The difference in energy between the two configurations is ca. 4 Kcal/mol.

For convenience, the vibrational states $|v>$ may be re-labelled as $|Av_A>$ for isomers A, $|Bv_B>$ for isomers B and $|Cv_C>$ for the group of highly excited, delocalized states which spread over the common ("C") domain of isomers A and B, see figure 3a. For example, the ground states of isomers A and B are labelled $|v> = |0>$, $|3>$ or $|A0>$, $|B0>$, respectively. Likewise, the lowest delocalized state $|C0>$ is equivalent to $|v> = |28>$, with energy E_{C0} just above the barrier of the potential energy surface V.

The detailed strategy for selective laser-pulse-assisted isomerizations has been derived and described in detail in Ref *(12)*, in close analogy to the strategy for selective high overtone excitations in OD. Essentially, we determine the initial state, typically the ground state $|i> = |A0>$ of isomer A, and the target state $|f>$, a highly excited state such as $|B9>$ of isomer B. Then we divide the demanding overall transition, $|A0> \rightarrow |B9>$, into a series of small individual ones, for example

$$|i>=|A0>\rightarrow|A3>\rightarrow|A6>\rightarrow|A9>\rightarrow|A12>\rightarrow|C1>\rightarrow|B9>=|f>, \qquad (16)$$

see figure 3a, similar to the previous scheme (14). For each individual transition $|v_k> \rightarrow |v_{k+1}>$ we design an optimal laser field $\mathcal{E}_k(t)$, as in eqns (5). Finally, the individual $\mathcal{E}_k(t)'s$ are superimposed, yielding the overall laser field $\mathcal{E}_x(t)$. For clarity of presentation and interpretation, the results are shown separately for transitions k = 1-4 and 5,6 in figures 3b and 3c, respectively. Both forms of presentation and interpretation are analogous to the results shown in figure 1d for selective high overtone excitation of OD.

The sequence of transitions (16), which is achieved by the corresponding series of laser pulses, is seen in the resulting probabilities shown in the bottom panels of figures 3b,3c. The transitions k = 1-4 yield high overtone excitation of isomer A, whereas transitions k = 5,6 yield the isomerization from states $|A12>$ to $|B9>$ via the delocalized state $|C1>$. This "detour" via the highly excited state $|C1>$ is essential since it provides indirect coupling of states $|A12>$, and $|B9>$ via significant dipole matrix elements $<A12|\mu_x|C1>$ and $<C1|\mu_x|B9>$. In contrast, any direct transitions from $|A12>$ to $|B9>$ (or from any other states of isomers A to B) are prohibited by entirely negligible matrix elements $<Av_A|\mu_x|Bv_B>$, due to negligible overlap of the domain of wavefunctions $|Av_A>$ and $|Bv_B>$. Still, the "detour" transition $|A12> \rightarrow |C1>$ turns out to be a "bottleneck", reducing the overall probability of isomerization to

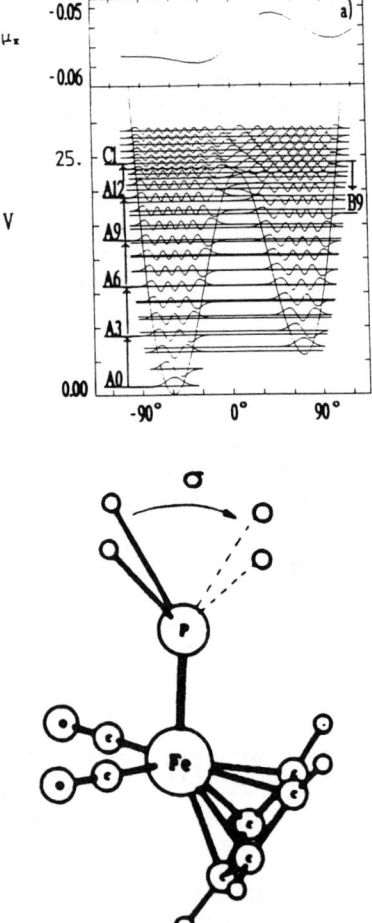

Figure 3. Selective isomerization of $Cp(CO)_2$ FePHD from isomer A to B by series of ps IR laser pulses with Gaussian shapes. **3a.** Model with series of selective transitions $A0 \to ... \to B9$. V in Kcal/mol and μ_x in atomic units.

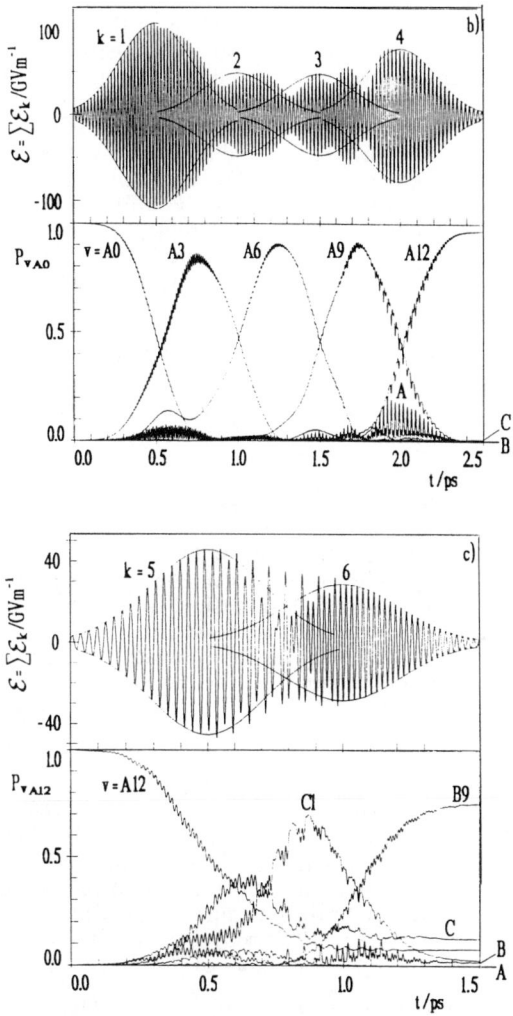

Figure 3 continued. 3b. Selective vibrational excitation of isomer A (transitions k=1-4). **3c.** Selective isomerization from state A12 of isomer A via delocalized state C1 to state B9 of isomer B (transitions k=5,6).

~0.75, evaluated by adding the final probabilities $P_{Bv_B A0}$ for transitions from state $| A0 >$ of isomer A to all states $| Bv_B >$ of isomer B. However, this obstacle may be overcome by few (~3) applications of the series of laser pulses k = 1-6, yielding ultimately pure isotopomers B.

In close analogy to the results for selective high overtone excitation of OD, and for equivalent reasons, the series of laser pulses shown in figures 3b,3c yield exclusive isomerization of $[D] = [Cp(CO)_2FePHD]$, not $[H] = [Cp(CO)_2FePH_2]$. As a consequence, we may use the present strategy to switch all [D]'s into isomers B, whereas at low temperatures, the vast majority of [H]'s exists as thermodynamically preferred isomer A. In fact, the small fraction of [H]'s that may exist as isomers B may also be converted into isomers A by an appropriate series of laser pulses. The laser-pulse induced primary reaction turns different isotopomers into different isomers, [D] = B and [H] = A. In favourable cases, sequel reactions which are isomer-selective will yield specific product isotopomers. This possibility is indicated in figure 4: a bulky, substituted methyliodide ICR_3 (e.g. R = benzene) approaches preferably isomer B, whereas nucleophilic attack of invertomer A is hindered sterically by the large Cp-ligand (or even larger ones, C_5Me_5, C_5Ph_5 etc, Me=CH_3, Ph=C_6H_6). As a consequence, the sequel S_{N2} reactions yield exclusively the deuterated product salt, $Cp(CO)_2Fe(PHD)(CR_3)^+I^-$, not $Cp(CO)_2Fe(PH_2)(CR_3)^+I^-$. This hypothetical example should indicate the type of synthesis of pure organo-metallic isotopomers that may be achieved by selective primary isomerization processes induced by series of ps IR laser pulses with Gaussian shapes.

Isotopomer-Selective Photodissociations Induced by Weak Continuous Wave IR + vis/UV Two-Photon Laser Excitation

Vibrationally Mediated Dissociation of OH versus OD. Let us now consider, as an example, the problem of exclusive photodissociation of OH, not OD. For convenience and consistency of presentation, we use the same model system, with potential energy surface V_X of the electronic ground state X, vibrational energies E_v and wavefunctions $\varphi_v(q)$, as in section 2.1, see figures 1a and 5. We assume a simple model potential curve V_A for the repulsive electronically excited state A,

$$V_A(q) = V_A^0 \exp(-q/q_0) \qquad (16)$$

with parameters $V_A^0 = 3.64 eV$ and $q_0 = 0.590 a_o$, fitted accordingly to the potential energy surface of the electronically excited A state of H_2O, as used by Imre and Zhang *(19, 20)*, adapted from Refs. *(21, 22)*. This simple OH/OD model is appropriate for schematic explanations and semiquantitative evaluations of the fundamental solution of our problem, which is based on the approach of Refs. *(13 - 18)*; implicitly it also points to some analogy of selective photodissociation of isotopomeres such as OH versus OD and selective fissions of OH versus OD bonds of HOD, as pioneered by the authors of Refs. *(19 - 26)*.

Isotopomer-selective photodissociation of OH, not OD, may be achieved by means of vibrationally mediated IR + vis/UV two photon excitations, using two cw lasers as in Refs. *(13 - 26)*, as follows (see figure 5): The first IR photon yields moderate vibrational overtone excitation from the initial (ground) state $| i >$ to the "mediating" vibrational state $| m >$, e.g. $| 0 > \longrightarrow | m > = | 4 >$ for OH. The second (vis or) UV photon lifts $| m >$ from the electronic ground state by a vertical Franck-Condon (FC) transition to the repulsive potential

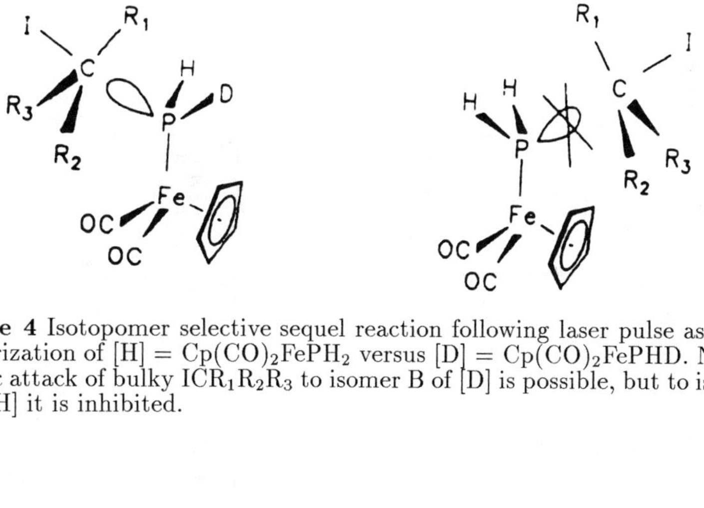

Figure 4 Isotopomer selective sequel reaction following laser pulse assisted isomerization of [H] = $Cp(CO)_2FePH_2$ versus [D] = $Cp(CO)_2FePHD$. Nucleophilic attack of bulky $ICR_1R_2R_3$ to isomer B of [D] is possible, but to isomer A of [H] it is inhibited.

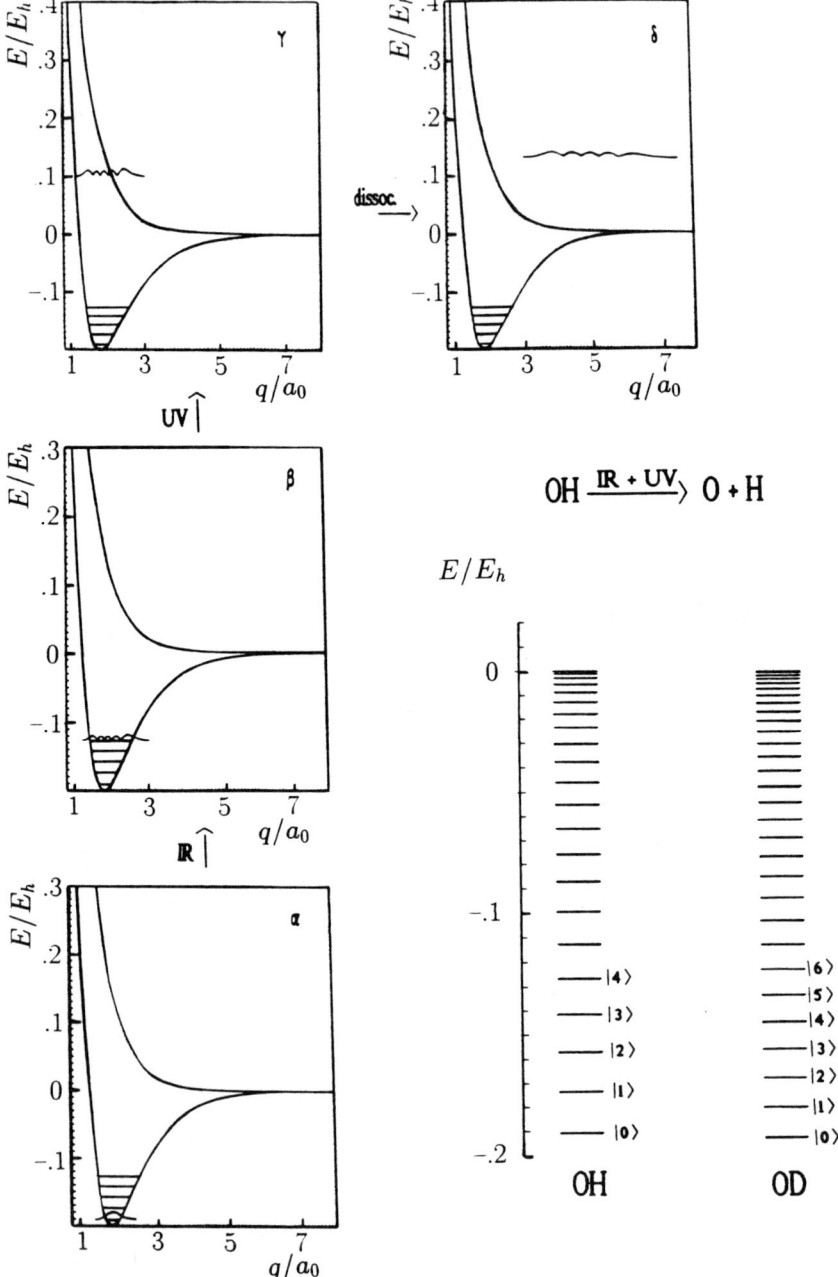

Figure 5: Selective cw IR + UV photodissociation of OH, not OD. Sequence of vibrational (IR) plus electronic (UV) photodissociations (t= 0 - 20 fs) indicated by density of corresponding wavepackets superimposed on model potential energy curves in panels $\alpha - \delta$. Selective vibrationally mediating level $|4>$ of OH versus other levels of OH and OD are also indicated.

energy surface of the electronically excited state, where it dissociates as time-dependent wavepacket $| \psi(t) >$. The three steps of this strategy - IR excitation $| i > \longrightarrow | m >$, vis or UV excitation $| m > \longrightarrow | \psi(0) >$ and dissociation $| \psi(0) > \longrightarrow | \psi(t) >$ - are illustrated for selective IR + UV photodissociation of OH in figure 5. The time evolution of the dissociative wavepacket $| \psi(t) >$ is simulated by Fast Fourier Transform propagation (67-70), as in Refs. (6,19,20,28,34,71). The alternative method of solving the time dependent Schrödinger equation by expansion in terms of molecular eigenstates (without continuum states!), see eqns. (6) - (10), is restricted to non-dissociative isomerization or intramolecular processes.

The superb selectivity of vibrationally mediated IR + UV photodissociations, in comparison with traditional single-photon excitations, is based on three conditions (see figure 5): (i) The IR + UV excitation prepares the molecule in a very repulsive domain of the potential energy surface of the electronically excited state, inducing dissociation on ultra-short time scales ($\leq 100 fs$), faster than any competing process such as IVR. (ii) The energy gap ΔV between the potentials of the electronically excited and the ground state (in the relevant FC-region(s)) is divided into the two IR + UV photons' energies,

$$\Delta V = \hbar\omega_{IR} + \hbar\omega_{UV}. \quad (17)$$

Neither of the individual photon energies $\hbar\omega_{IR}$, $\hbar\omega_{UV}$ is large enough to excite the target electronic state. (iii) Both individual IR and UV transitions are efficient, i.e. the corresponding electric dipole transition elements, e.g. $< m | \mu | i >$ for direct overtone excitation, as well as the overlaps $< d | \psi(0) > = < d | m >$ of the mediating vibrational and dissociative wavepacket $| m >$ and $| d >$ should be large.

Isotopomer selectivity may be achieved if conditions (i) - (iii) are satisfied exclusively by one specific isotopomer. For example, the selective IR + UV photodissociation of OH, illustrated in figure 5 is obtained by IR photons which excite exclusively the vibrationally mediating state $| m > = | 4 >$ of OH, whereas all vibrational levels of OD are off-resonance, see also figure 1a. Moreover, neither $\hbar\omega_{IR}$ nor $\hbar\omega_{UV}$ suffice for direct single photon dissociation of OD. As in the previous systems, by tuning the IR frequency to resonant overtone excitation of OD, one may achieve similar selective IR + UV two-photon dissociation of OD, not OH.

Vibrationally Mediated Dissociations of Organometallic Isotopomers.

A) $Ni - C_2H_4$ versus $Ni - C_2D_4$. In close analogy to selective IR + UV two-photon dissociation of OH, not OD, let us now consider selective vibrationally mediated fissions of isotopically labelled metal ligand bonds. For our first model system, $Ni - C_2H_4$ versus $Ni - C_2D_4$, we employ a simple **2D** model with two coordinates q_a, q_b describing essentially the $Ni-C_2H_4(C_2D_4)$ bond distance and the symmetric out-of-plane vibration ν_7 of $C_2H_4(C_2D_4)$, see figure 6. We assume that the C_{2v}-symmetry of the equilibrium configuration of $Ni-C_2H_4$ is conserved along the reaction path, such that the photodissociation of $Ni - C_2H_4(C_2D_4)$ may be described by a simple collinear model $ABC^* \longrightarrow A + BC$, where A = Ni, and B, C represent the centers of masses of the carbon (C_2) and hydrogen (H_4) or deuterium (D_4) atoms, respectively. The distances between A-B and B-C are defined as coordinates q_a and q_b of the present **2D** model. Essentially, q_a describes

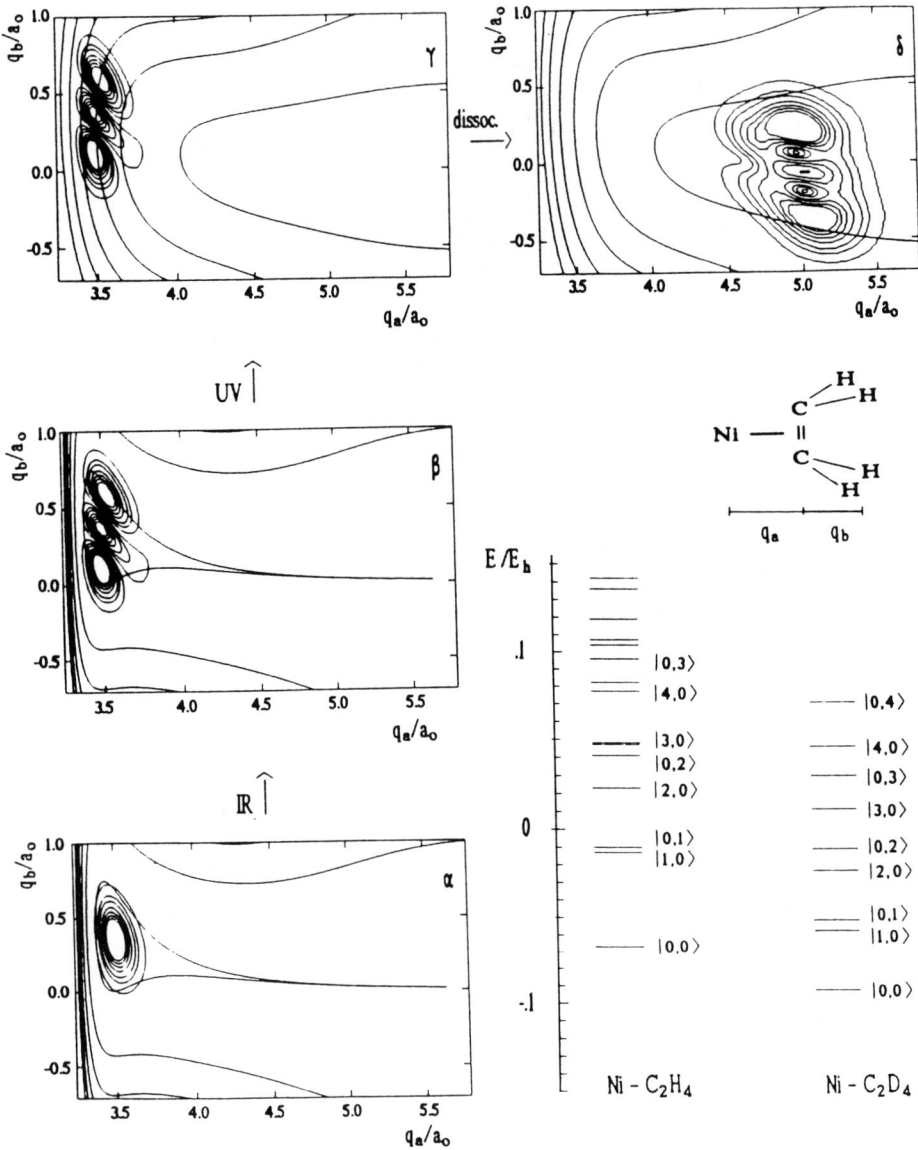

Figure 6: Selective IR + UV photodissociation of Ni-C_2H_4, not Ni-C_2D_4. Potential energy surfaces of electronic ground (α, β) and excited states (γ, δ) together with densities of corresponding wavepackets are shown as contour plots. States labelled $|0, v>$ denote $0 + v$ vibrational quantas in the Ni-ligand bond and in the C_2H_4 (C_2D_4) bond ν_7, respectively.

the dissociative metal-ligand distance, whereas q_b accounts for the C_2H_4/C_2D_4 bending mode ν_7. All other degrees of freedom are neglected in this zero-order approximation.

Our **2D** model for $Ni - C_2H_4/Ni - C_2D_4$ is motivated by a variety of structurally related systems. In particular, Amirav et al *(72)* explained their photodissociation spectra of $Cd(CH_3)_2$ and $Cd(CD_3)_2$ by exclusive consideration of the $Cd-C$ dissociative bonds and the symmetric out-of-plane ("umbrella") vibrations of $CH_3(CD_3)$. Moreover, Zewail et al *(73)* have shown that photodissociation of He-trans-stilbene van der Waals molecules are accelerated by low-energy out-of-plane bending modes, in contrast with inefficient high-energy in-plane vibrations. The inefficiency of in-plane bends has also been documented for photodissociation of Ar-benzene complexes by Parmenter et al *(74)*, although these modes should of course be incorporated in rigorous, quantitative models, as suggested by Hutson, Clary, and Beswick *(75)* for photodissociation of Ar,Ne-ethylenes. Last but not least, Shapiro et al *(76,77)* and very recently Schatz et al *(78)* described photodissociation of methyliodide by a collinear ABC model (A = I, B = C, C = center of mass of H_3). Likewise, it is reasonable that the ultrafast ($\sim$ 20 fs, see below) photodissociation of $Ni - C_2H_4(C_2D_4)$ will transfer sudden energy release preferably into $Ni + C_2H_4(C_2D_4)$ translational energy plus vibrational energy of the ν_7 mode, i.e. by pushing the C_2 atoms towards the plane of the H_4 atoms, opposite of the Ni atom. Subsequently, the carbon atoms may transfer momentum partially towards their hydrogen neighbours, thus exciting other vibrational modes such as the symmetric stretch ν_1 of the nascent ethylene. For simplicity we assume that the type of IVR from ν_7 towards other modes ν_1, ν_2, etc. proceeds on a much longer time scale, i.e. during dissociation, the translational mode $Ni + C_2H_4(C_2D_4)$ is coupled exclusively to ν_7. Let us point out that this mechanism yields quasi-coherent (or "transient" *(79)*) excitations of the ν_7 mode, not vibrational eigenstates of the nascent ethylene molecules, which could be monitored by femto-chemical pump-and-probe experiments *(80,81)*, yielding isotopically selective quantum beats (35.1 fs and 46.3 fs for C_2H_4 and C_2D_4, respectively).

The relevant potential energy surfaces of the electronic ground state and excited state, $(^1A_1)$ and $(^3A_1)$, are shown in figure 6. These are modelled by analytical functions fitted to quantum chemistry ab initio results of Siegbahn et al *(33)*; further details are given in Ref. *(34)*. Of course, we anticipate that the corresponding quantum yields of electronic $^1A_1 \longrightarrow {}^3A_1$ transitions (induced by weak spin-orbit couplings of the metal atom) will be rather low, but this does not affect the issue of isotopomer selectivity. Similar model simulations could be carried out for IR + UV two-photon transitions to even more excited electronic states, as soon as (repulsive!) potential energy surfaces are available.

The vibrational wavefunctions $\varphi_v(q_a, q_b) = <q_a q_b | v >$ and energies E_v are evaluated by means of imaginary time Chebyshev propagations, using the method of Ref. *(82,83)*. The resulting vibrational ground $| 0 >$ and mediating $| 4 >$ wavefunctions, are illustrated by equidensity contours versus coordinates in figure 6, panels α and β, respectively. From the nodal structure, one readily sees that $| 4 >$ accumulates two vibrational quanta in the C_2H_4 bend along q_b, but zero quanta in the Ni-ligand stretch along q_a. Correspondingly, state $| 4 >$ may be relabeled as $| v_a = 0, v_b = 2 >$, with corresponding notations for similar excitations of local bending modes $| 0, v_b >$ or local Ni-ligand stretches $| v_a, 0 >$. The corresponding energies $E_v = E_{0v_b}$ or $E_{v_a 0}$ of those states and several other non-local states are shown in figure 6 for both $Ni - C_2H_4$ and $Ni - C_2D_4$ complexes.

Selective IR + UV two-photon dissociation of $Ni - C_2H_4$ (not $Ni - C_2D_4$) is illustrated in figure 6. These model simulations indicate the possibility of exclusive fission of the $Ni - C_2H_4$ bond. In practical applications, this elimination of $Ni - C_2H_4$ complexes might support indirect enrichment of deuterated Ni-ethylene isotopomers. Also, the same strategy, but with IR photons tuned to vibrational overtones of $Ni - C_2D_4$, would produce exclusively perdeuterated ethylenes.

B) $HCo(CO)_4$ versus $DCo(CO)_4$. As our second example, we simulate isotopomer selective IR + UV two-photon dissociation of $HCo(CO)_4$, not $DCo(CO)_4$. For this purpose, we employ a simple **2D** model, similar to that of $Ni-C_2H_4/Ni-C_2D_4$ or of the related system, ICH_3 *(76-78)*. The model is based on the ground state geometry of $HCo(CO)_4$ which is a distorted trigonal bipyramid, and we assume that its C_{3v} symmetry is conserved during photodissociation, similar to $ICH_3^* \longrightarrow I + CH_3$ *(76-78)*. Correspondingly, vibrationally mediated bond fissions

$$HCo(CO)_4^* \longrightarrow H + Co(CO)_4 \qquad (18a)$$

are modelled - again - as a collinear process

$$ABC^* \longrightarrow A + BC \qquad (18b)$$

where A = H, B = Co, and C = center of mass of the axial CO ligand. The equatorial CO ligands were not included in the dynamics. The distances between A-B and B-C correspond to the HCo bond length and to the distance between Co and the center of mass of the axial CO ligand; these are described by coordinates q_a and q_b of the present **2D** model, respectively. All other degrees of freedom are neglected.

In principle, this model also allows one to describe the competing dissociation

$$HCo(CO)_4^* \longrightarrow HCo(CO)_3 + CO \qquad (18c)$$

i.e.

$$ABC^* \longrightarrow AB + C \qquad (18d).$$

Both channels of photodissociations (18a), (18c) have been observed by Sweany *(39)*, however our subsequent simulation centers attention on selective homolysis of the HCo bond. Vibrationally mediated control of HCo versus axial $Co - CO$ bond fission, similar to HO versus OD bond fission of HOD *(19-26)* is being currently studied in our group.

The potential energy surfaces of the electronic ground state and the excited state, $(^1A_1)$ and $(^3A_1)$ are modelled - again - by fitting analytical functions to quantum chemistry ab initio results, adapted from Refs. *(35-38)*. Specifically, the ground state $^1A_1$ potential is approximated by double Morse oscillators, similar to Refs. *(28,68,69,84,85)*. In contrast, the excited state $^3A_1$ is represented by rotated Morse oscillators *(86)* fitted to the ab initio potential energy curves versus the HCo and axial $Co - CO$ bonds of Refs. *(35,36)*. These fitted potential energy surfaces are illustrated by **2D** contour plots versus q_a, q_b in figure 7. The competing product channels (18a) and (18c) correspond to two exit valleys of

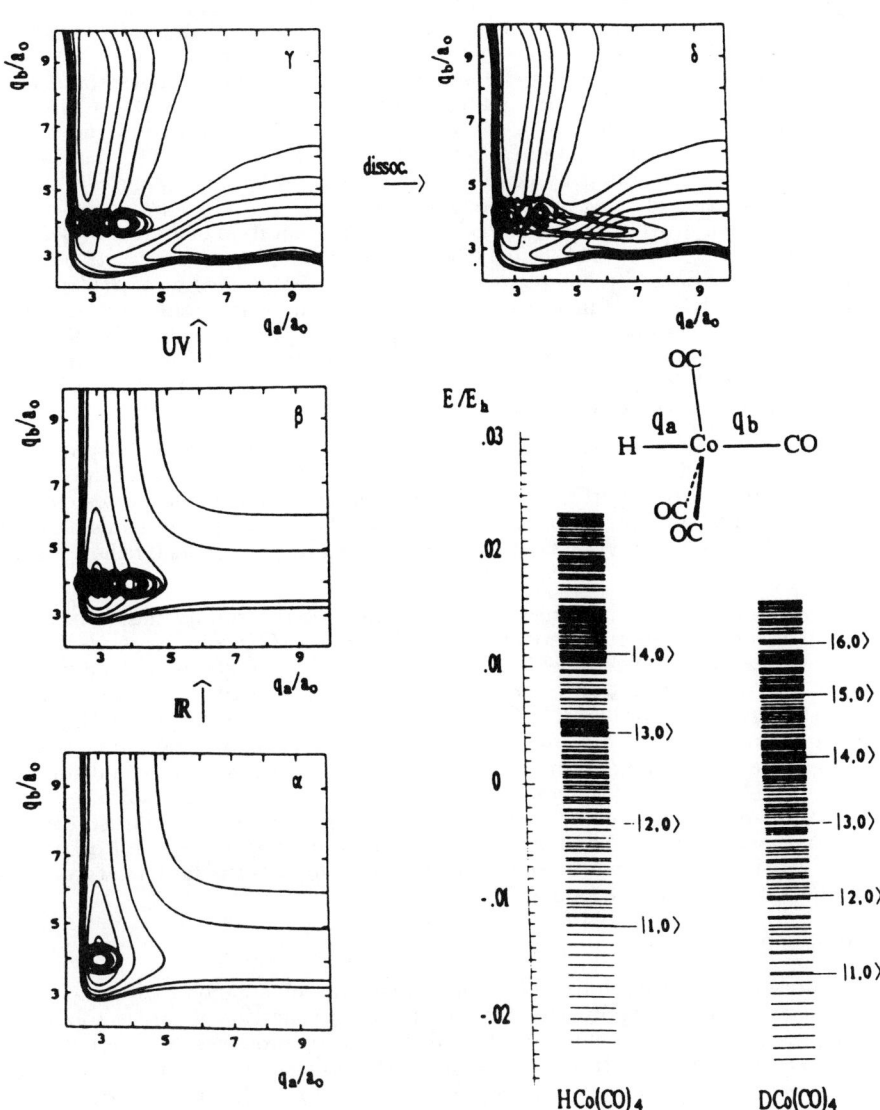

Figure 7: Selective IR + UV photodissociation of $HCo(CO)_4$, not $DCo(CO)_4$. Notations are analogous to Figures 5 and 6. States labelled $|v, 0 >$ denote v + 0 vibrational quantas in th H-Co (D-Co) and axial Co-CO bonds, respectively.

the potential energy surface, in figure 7. In contrast, figure 6 shows a single exit valley for photodissociation of Ni-ethylene.

The vibrational states $|v>$ and energies E_v are evaluated by expansions in terms of Morse oscillator basis functions for the individual $H-Co$ and axial $Co-CO$ distances, using the techniques of Ref. (28). Preferential fission of the HCo bond requires that the dissociative frontier lobes of the mediating state $|m>$ extend into the repulsive domain of the exit valley of the potential energy surface of the excited electronic state, leading towards products $H + Co(CO)_4$. This condition is satisfied by $|m> = |114>$, which describes the fourth excited vibrational state of the HCo local mode in $HCo(CO)_4$. The wavepacket $|114>$ is also illustrated by equidensity contours in figure 7, panels β and γ. The accumulation of $v_a = 4$ vibrational quanta in the $H-Co$ bond versus $v_b = 0$ quanta in the $Co-(CO)$ mode is visible in this presentation of state $|v = 114>$; correspondingly, $|v = 114>$ may be relabelled as $|v_a = 4, v_b = 0>$, with analogous notations for complementary local modes $|v_a, v_b>$ of $HCo(CO)_4$, compare with the wavefunctions of Ni-ethylene. The corresponding vibrational energies E_v of the model $HCo(CO)_4$ and $DCo(CO)_4$ are also shown in figure 7, with marks $|v_a, 0>$ of the vibrationally excited HCo or DCo local stretches. By analogy with local mode overtone excitations (14), we assume that it will be possible to excite these stretches selectively.

As in the case of Ni-ethylene, we also assume that UV excitation deposits the vibrationally mediating state $|4, 0>$ on the potential energy surface of the excited state without further perturbations. This approximation should be tested, and possibly refined, in more rigorous evaluations of the excitation mechanism, which may involve transition via interfering electronic states, in particular the ^1E state (35-37).

The results shown in figure 7 are analogous to the results shown in figure 5, and 6 for our previous model systems, HO, not OD and $Ni-C_2H_4$, not $Ni-C_2D_4$, respectively, yielding also equivalent explanations and interpretations. Accordingly, it will be possible to photolyze selectively $HCo(CO)_4$, not $DCo(CO)_4$ (or similarly, $DCo(CO)_4$, not $HCo(CO)_4$), by means of IR + UV two photon excitation. In practical applications, this may yield enrichment of the deuterated isotopomer, in particular by sequel reactions which could stabilize the $Co(CO)_4$ radical, e.g. by bonding to Cr, Mo, or W centered radicals or even by isotope exchange, e.g. in deuterium containing matrices. Similar isotopomer selective primary reactions may also be used to analyze and control other secondary reactions, last not least the cobalt based hydroformylation catalysis via the Heck-Breslow mechanism (87,88).

Conclusions

The present strategies for laser-induced reactions of organometallic molecules should stimulate new approaches for the selective preparation of pure isotopomers. Our present examples include selective isomerization processes by series of ps IR laser pulses, as well as IR + vis/UV two-photon dissociations, with applications to simple model compounds, $[Cp(CO)_2FePH_2]$, Ni - ethylene and $HCo(CO)_4$ and their deuterated isotopomers. Clearly, these techniques are more demanding than previous approaches to laser control of the reactions of organometallic molecules, i.e. mainly control of branching ratios by single photon excitations (see e.g. Refs (39,89-91)) or in principle also by IR multiple photon absorption (1,2,92) induced by monochromatic cw lasers. The required investment should be payed back by larger flexibility, efficiency and, most importantly, selectivity of the new strategies, in comparison with the traditional ones *(1,2,39,89-92)*. Our confidence is based on similar progresses which have been achieved recently for small molecules, both theoretically *(17,19-23,25,26)* and experimentally *(13-16,18,24)*. In fact, the present strategies for organometallic molecules have been developed in close analogy to corresponding novel techniques for small molecules *(13-26)*.

The present extensions of strategies for laser control of isotopomer selectivity, from diatomic to organometallic molecules, is of course not trivial. Let us re-emphasize that the present promising successes for simple models have to be tested for self-consistency in more realistic, much more demanding multi-dimensional simulations. Ultimately they should guide the experimentalist to verifications of the predicted effects, and hopefully to efficient applications. In addition, the extensions from small to organometallic molecules involve also some non-trivial modification of the basic concepts. In particular, the "ladder climbing" strategy for high overtone excitation of diatomic molecules is replaced by a "climb-and-descend" strategy for selective isomerizations, and this is achieved by subtle yet important differences in the parameters of - otherwise similar - series of ps IR laser pulses. Likewise, the parameters of IR + UV cw lasers for selective photodissociations have to be retuned from appropriate vibrations and dissociation channels of the diatomic to those of organometallic molecules.

The present strategies should be considered as examples for a variety of alternative strategies for laser control of selective organometallic isotopomers, beyond the simplistic traditional single-photon approaches *(39,89-91)*. For example, very recently Garvey *(27)* has pioneered novel pathways to selectively produce isotopomers by multiple photon excitation of organometallic molecules embedded in clusters. Likewise, one may extend the photo-chemical pump-and-dump strategy, e.g. for transitions from electronic ground to excited states and back to the ground state, from selective dissociation *(94,95)* or isomerization *(96)* of small or large organic molecules, to organometallic ones. There are several other, related techniques which lend themselves to similar extensions *(28-31,97-104)*. It is possible that each of these strategies will have its own preferential domain of applications, extending the present examples to larger, novel classes of organometallic molecules, e.g. specific preparation of enantiomers (or diastereomers) with isotopomer-selective sequel reactions, or even to related selective reactions in clusters or at surfaces. We consider the exploration of this field as stimulating and promising task.

Acknowledgement

We should like to express our thanks to Dipl. Chem. S. Görtler and F. Seyl for their enthusiastic preparations of video-movies illustrating part of the present laser-assisted molecular dynamics. Generous financial support by DFG (project Ma 515/9-1), SFB 347 (projects B2 and C3) and the Fonds der Chemischen Industrie is also gratefully acknowledged. E.K. thanks the Studienstiftung des deutschen Volkes for a scholarship. The computations have been carried out on the Cray Y-MP 4/432 of the Leibniz-Rechenzentrum der Bayerischen Akademie der Wissenschaften, the Vax 6000 of the Rechenzentrum der Universität Würzburg and our Sun Sparcstations 2.

References

[1] Bagratashvili, V. N.; Letokhov, V. S.; Makarov, A. A.; Ryaboy, E. A. *"Multiple Photon Infrared Laser Photophysics and Photochemistry"*; Harwood, Chur, **1984**; Chapter 8.
[2] Ambartzumian, R. V. in *"Multiple-Photon Excitation and Dissociation of Polyatomic Molecules"*; Cantrell, C. Ed.; Springer, Berlin, **1986**; Chapter 4.
[3] Greenland, P. T. *Contemp. Phys.* **1990**, 31, 405.
[4] Paramonov, G. K.; Savva, V. A. *Phys. Lett. A* **1983**, 97, 340.
[5] Paramonov, G. K. *Chem. Phys. Lett.* **1990**, 169, 573, and references therein.
[6] Jakubetz, W.; Just, B.; Manz, J.; Schreier, H.-J. *J. Phys. Chem.* **1990**, 94, 2294.
[7] Blackwell, B. A.; Polanyi, J. C., Sloan, J. J. *Chem. Phys.* **1977**, 24, 25.
[8] Polanyi, J. C. *Acc. Chem. Res.* **1972**, 5, 161.
[9] Levine, R. D.; Manz, J. *J. Chem. Phys.* **1975**, 63, 4280.
[10] Manz, J. in *"Molecules in Physics, Chemistry, and Biology"*, Maruani, J. Ed.; Vol. 3; Kluwer, Dordrecht, 1989 p. 365.
[11] Polanyi, J. C. *Science* **1987**, 236, 680.
[12] Combariza, J. E.; Just, B.; Manz, J.; Paramonov, G. K. *J. Phys. Chem.* **1991**, 95, 10351.
[13] Likar, M. D.; Baggott, J. E.; Sinha, A.; Ticich, T. M.; Vander Wal, R. L.; Crim, F. F. *J. Chem. Soc. Faraday Trans. 2* **1988**, 84, 1483.
[14] Crim, F. F. *Science* **1990**, 249, 1387.
[15] Zittel, P. F.; Little, D. D. *J. Chem. Phys.* **1979**, 71, 713.
[16] Zittel, P. F.; Lang, V. I. *J. Photochem. Photobiol. A* **1991**, 56, 149.
[17] Shapiro, M. Private Communication, **1991**.
[18] Valentini, J. J. Private Communication, **1991**
[19] Zhang J.; Imre, D. G. *Chem. Phys. Lett.* **1988**, 149, 233.
[20] Imre, D. G.; Zhang, J. *Chem. Phys.* **1989**, 139, 89.
[21] Engel, V.; Schinke, R. *J. Chem. Phys.* **1988**, 88, 6831.
[22] Engel, V.; Schinke, R.; Staemmler, V. *J. Chem. Phys.* **1988**, 88, 129.
[23] Vander Wal, R. L.; Scott, J. L.; Crim, F. F. *J. Chem. Phys.* **1990**, 92, 803.
[24] Bar, I.; Cohen, Y.; David, D.; Rosenwaks, S.; Valentini, J. J. *J. Chem. Phys.* **1990**, 93, 2146.
[25] Segev E.; Shapiro, M. *J. Chem. Phys.* **1982**, 77, 5604.
[26] Shapiro, M. This volume.
[27] Garvey, J. F. This volume.
[28] Hartke, B.; Manz, J.; Mathis, J. *Chem. Phys.* **1989**, 139, 123.
[29] *"Selectivity in Chemical Reactions"* Whitehead, J. C. Ed.; Reidel, Dordrecht, **1988**.
[30] Letokhov, V. S. *Appl. Phys. B* **1988**, 46, 237
[31] Schäfer, F. P. *Appl. Phys. B* **1988**, 46, 199.
[32] Angerer, W.; Sheldrick, W. S.; Malisch W. *Chem. Ber.* **1985**, 118, 1261.
[33] Widmark, P. O.; Roos, B. O.; Siegbahn, P. E. M. *J. Phys. Chem.* **1985**, 89, 2180.
[34] Warmuth, B. Diplomarbeit, Universität Würzburg, **1990**.
[35] Veillard, A.; Strich, A. *J. Am. Chem. Soc.* **1988**, 110, 3793.

[36] Daniel, C.; Hyla-Kryspin, I.; Demuynck, J.; Veillard, A. *Nouv. J. Chim.* **1985**, 9, 581.
[37] Veillard, A.; Daniel, C.; Rohmer, M. M. R. *J. Phys. Chem.* **1990**, 94, 556.
[38] Veillard, A.; Daniel, C.; Strich, A. *Pure Appl. Chem.* **1988**, 60, 215.
[39] Sweany, R. L. *Inorg. Chem.* **1982**, 21, 752.
[40] Rogers P.; Montague, D. C.; Frank, J. P.; Tyler, S. C.; Rowland, F. S. *Chem. Phys. Lett.* **1982**, 89, 9.
[41] Rowland, F. S. *Faraday Discussions Chem. Soc.* **1983**, 75, 158.
[42] Jonas, J.; Peng, X. *Ber. Bunsenges. Phys. Chem.* **1991**, 95, 243.
[43] Lopez, V.; Marcus, R. A. *Chem. Phys. Lett.* **1982**, 93, 232.
[44] Swamy, K. N.; Hase, W. L. *J. Chem. Phys.* **1985**, 82, 123.
[45] Marshall, K. T.; Hutchinson, J. S. *J. Phys. Chem.* **1987**, 91, 3219.
[46] Lederman, S. M.; Lopez, V.; Fairen, V.; Voth, G. A.; Marcus, R. A. *Chem. Phys.* **1989**, 139, 171.
[47] Uzer, T.; Hynes, J. T. *Chem. Phys.* **1989**, 139, 163.
[48] Wrigley, S. P.; Oswald, D. A.; Rabinovitch, B. S. *Chem. Phys. Lett.* **1984**, 104, 521.
[49] Makri, N.; Miller, W. H. *J. Chem. Phys.* **1987**, 86, 1451.
[50] Okuyama, S.; Oxtoby, D. W. *J. Chem. Phys.* **1988**, 88, 2405.
[51] Shida, N.; Barbara, P. F.; Almlöf *J. Chem. Phys.* **1989**, 91, 4061.
[52] Sekiya, H.; Nagashima, Y.; Nishimura, Y. *J. Chem. Phys.* **1990**, 92, 5761.
[53] Redington, R. L. *J. Chem. Phys.* **1990**, 92, 6447.
[54] Bosch, E.; Moreno, M; Lluch, J. M.; Bertran, J. *J. Chem. Phys.* **1990**, 93, 5685.
[55] Meyer, R.; Ernst, R. R. *J. Chem. Phys.* **1990**, 93, 5518.
[56] Wright, J. S.; Donaldson, D. J. *Chem. Phys.* **1985**, 94, 15.
[57] Lawton, R. T.; Child, M. S. *Molec. Phys.* **1980**, 40, 773.
[58] Jakubetz, W.; Manz, J.; Mohan, V. *J. Chem. Phys.* **1989** 90, 3686.
[59] Press, W. H.; Flannery, B. P.; Teukolsky, S. A.; Vetterling, W. T. *"Numerical Recipes"*, 1st ed., Cambridge University Press: Cambridge, **1986**.
[60] IMSL Library. *"Fortran subroutines for mathematics and statistics"* Version 9.0, **1985**.
[61] Just, B. Diplomarbeit, Universität Würzburg **1991**.
[62] Shi, S.; Woody, A.; Rabitz, H. *J. Chem. Phys.* **1988**, 88, 6870.
[63] Shi, S.; Rabitz, H. *J. Chem. Phys.* **1990**, 92, 2927.
[64] Dahleh, M.; Peirce, A. P.; Rabitz, H. *Phys. Rev. A* **1990**, 42, 1065.
[65] Jakubetz, W.; Manz, J.; Schreier, H.-J. *Chem. Phys. Lett.* **1990**, 165, 100.
[66] Flygare, W. H. *"Moleculare Structure and Dynamics"*, Prentice-Hall: Englewood Cliffs, **1978**.
[67] Kosloff, D.; Kosloff, R. *J. Comput. Phys.* **1983**, 52, 35.
[68] Bisseling, R. H.; Kosloff, R.; Manz, J. *J. Chem. Phys.* **1985**, 83, 993.
[69] Joseph, T.; Manz, J. *Molcc. Phys.* **1986**, 57, 1149.
[70] Kosloff, R. *J. Phys. Chem* **1988**, 92, 2087.
[71] Hartke, B.; Kolba, E.; Manz, J.; Schor, H. H. R. *Ber. Bunsenges. Phys. Chem.* **1990**, 94, 1312.
[72] Amirav, A.; Penner, A.; Bersohn, R. *J. Chem. Phys.* **1989**, 90, 5232.
[73] Semmes, D. H.; Baskin, J. S.; Zewail, A. H. *J. Chem. Phys.* **1990**, 92, 3359.
[74] O, H.-K.; Parmenter, C. S.; Su, M. C. *Ber. Bunsenges. Phys. Chem.* **1988**, 92, 253.

[75] Hutson, J. M.; Clary, D. C.; Beswick, J. A. *J. Chem. Phys.* **1984**, 81, 4474.
[76] Shapiro, M. *Chem. Phys. Lett.* **1977**, 46, 442.
[77] Shapiro, M.; Bersohn, R. *Annu. Rev. Phys. Chem.* **1982**, 33, 403.
[78] Schatz, G. C.; Hammerich, A. D. *J. Chem. Phys.*, in the press.
[79] Metiu, H. *Faraday Discuss. Chem. Soc.* **1991**, 91, in the press.
[80] Rosker, M. J.; Dantus, M.; Zewail, A. H. *Science* **1988**, 241, 1200.
[81] Zewail, A. H.; *Faraday Discuss. Chem. Soc.* **1991**, 91, in the press.
[82] Tal-Ezer, H.; Kosloff, R. *J. Chem. Phys.* **1984**, 81, 3967.
[83] Kosloff, R.; Tal-Ezer, H. *Chem. Phys. Lett.* **1986**, 127, 223.
[84] Rosen, N. *J. Chem. Phys.* **1933**, 1, 319.
[85] Thiele, E.; Wilson, D. J. *J. Chem. Phys.* **1961**, 35, 1256.
[86] Connor, J. N. L.; Jakubetz, W.; Manz, J. *Molec. Phys.* **1975**, 29, 347.
[87] Heck, R. F.; Breslow, D. S. *J. Am. Chem. Soc.* **1961**, 83, 4023.
[88] Forster, D.; Schaefer, G. F. *J. Molec. Catalysis* **1991**, 64, 283.
[89] Parnis, J. M.; Ozin, G. A. *J. Phys. Chem.* **1989**, 93, 4023.
[90] Cartland, H. E.; Pimentel, G. C. *J. Phys. Chem.* **1990**, 94, 536.
[91] Stuke, M. private communication **1990**.
[92] Perutz, R. N.; Belt, S. T.; McCamley, A.; Whittlesey, M. K. *Pure Appl. Chem.* **1990**, 62, 1539.
[93] Quack, M. *Adv. Chem. Phys.* **1982**, 50, 395.
[94] Tannor, D. J.; Rice, S. A. *J. Chem. Phys.* **1985**, 83, 5013.
[95] Kosloff, R.; Rice, S. A.; Gaspard, P.; Tersigni, S.; Tannor, D. J. *Chem. Phys.* **1989**, 139, 201.
[96] Repinec, S. T.; Sension, R. J.; Hochstrasser, R. M. *Ber. Bunsenges. Phys. Chem* **1991**, 95, 248.
[97] Ben-Shaul, A.; Haas, Y.; Kompa, K. L.; Levine, R. D. *"Lasers and chemical change"*, Springer, Berlin **1981**.
[98] Bondybey, V. E. *Ann. Rev. Phys. Chem* **1084**, 35, 591.
[99] Crim, F. F. *Ann. Rev. Phys. Chem.* **1984**, 35, 647.
[100] Frei, H.; Pimentel, G. C. *J. Chem. Phys. Annu. Rev. Phys. Chem.* **1985**, 36,491.
[101] Asaro, C.; Brumer, P.; Shapiro, M. *Phys. Rev. Letters* **1988**, 60, 1634.
[102] Bandrauk, A. D.; ed."Atomic and molecular processes with short intense laser pulses",Plenum Press, New York **1988**.
[103] Gaubatz, U.; Rudecki, P.; Schiemann, S.; Bergmann, K. *J. Chem. Phys.* **1990**, 92, 5363.
[104] Breuer, H. P.; Dietz, K.; Holthaus, M. *J. Phys. B* **1990**, 24, 1343.

RECEIVED September 4, 1991

Chapter 21

Multiphoton Ionization Dynamics Within $(CH_3OH)_n Cr(CO)_6$ van der Waals Clusters
Isotope Effects in Intracluster Energy Transfer

William R. Peifer and James F. Garvey[1]

Department of Chemistry, Acheson Hall, State University of New York at Buffalo, Buffalo, NY 14214

> We have recently examined the multiphoton dissociation and ionization dynamics of mixed van der Waals heteroclusters of $Cr(CO)_6$ solvated by methanol, and have inferred from photoion fragmentation branching ratios that CD_3OD is more efficient than CH_3OH in relaxing excess internal energy of the nascent photoion via intracluster energy transfer. Multiphoton ionization is suggested to proceed via single-photon photodissociation of the solvated $Cr(CO)_6$, followed by two-photon ionization of the coordinatively unsaturated photoproduct. Excess energy in the internal modes of the nascent photoion appears to be disposed of via non-statistical, mode-specific transfer to the internal degrees of freedom within the cluster.

Perhaps one of the most fundamental of challenges at the interface of physics and chemistry is the unraveling of the detailed sequence of events which ensues when two molecules collide (*1*). For a macroscopic system composed of many molecules, one may observe the phenomenological consequences of a very large number of collisions averaged over time, energy, and collision geometry. The chemical dynamicist, who wishes to study individual collisions, may start with a prior knowledge of the chemical identities of reactants and products, and attempt to probe the influence of reactant quantum state and geometry of approach on reaction probability, as well as the redistribution of available energy amongst the various degrees of freedom throughout the encounter. Indeed, reasonably complete ab initio treatment of some of the simpler atom transfer reactions is currently within the realm of computational tractability (*2*), and the experimental study of more complicated systems will no doubt provide a stringent proving ground for the refinement of quantum theoretical approaches.

[1]Corresponding author

Given a sufficiently detailed understanding of the collision dynamics for a particular set of reactants, one could in principle deduce the behavior and properties of a macroscopic ensemble of reactant molecules in the condensed phase (*3*). Unfortunately, our understanding of the dynamics for even the simplest of systems is generally not of sufficient detail to permit such an *a priori* deduction. If we wish to bring the power of a chemical dynamics approach to bear upon the study of chemistry within condensed phases, thereby shedding light on the evolution of physical and chemical properties from those of isolated pairs of molecules undergoing collision to those of macroscopic collections of such molecules, we must develop appropriate models for experimental inquiry.

An emergent sub-discipline of physical chemistry directed toward the development and study of such models is the field of van der Waals cluster research (*4*). The study of bimolecular chemistry within these van der Waals clusters provides a conceptual link between bimolecular reaction dynamics in the gas phase, and processes of equal or higher molecularity in condensed phases. Clusters of a few to several hundred molecules may be easily generated in the free-jet expansions of molecular beams, and subsequently interrogated by a variety of powerful laser spectroscopic and mass spectrometric techniques. These clusters are sufficiently small to be amenable to theoretical treatment, yet sufficiently large to serve as sophisticated models for the study of such condensed-phase phenomena as aerosol formation, crystallization, and structural and mechanistic aspects of solvation. Intracluster bimolecular chemistry may be induced through the creation of reactive species within clusters via electron impact, optical excitation, or photodissociation of appropriate precursor molecules. The experimentalist has control over not only the internal energies of these cluster-bound reactants, but in many cases the collision geometry as well, since the arrangement of molecules within van der Waals clusters is often highly ordered (*5,6*).

A matter of considerable interest to the cluster research community involves energy disposal processes within van der Waals clusters (*7*). How does a cluster in which an exothermic bimolecular reaction has just taken place dispose of the excess energy? How is the internal energy which is localized on a single molecule, perhaps even in a single vibrational or rotational mode, transferred to the remaining molecules within the cluster, and on what timescale does such intracluster energy redistribution take place? How well can the cluster accommodate this redistributed energy, and by what mechanisms does the entire cluster ultimately dissipate that portion of redistributed energy which cannot be accommodated? It has been well-established from the study of the fragmentation of cluster ions, following electron impact excitation of size-selected neutral clusters, that an important and general energy disposal mechanism is the sequential ejection, or evaporation, of individual molecules from the cluster. This process is accurately modeled by Boltzmann statistics and can be likened to the macroscopic process of evaporative cooling in liquids and solids. Alternatively, one may view this process within the microscopic context of chemical dynamics, wherein the relaxation of excess energy in the electronic, vibrational, and/or rotational degrees of freedom of a given molecule within the cluster, correlates ultimately with the appearance of excess energy in the translational degrees of freedom of the remaining molecules: such processes can be referred to as E-T, V-T, or R-T energy transfer.

Does the relaxation of any given excited molecule within a van der Waals cluster always proceed by means of a purely statistical transfer of energy to translations of the surrounding molecules? Recent experiments within our own research group suggest that this is not always the case; that, in fact, excitation in the internal modes of a given molecule can in some cases be transferred preferentially to a restricted number of modes in the surrounding molecules (for example, to a specific bending or stretching motion of the acceptor molecule). We have recently examined the multiphoton dissociation and ionization dynamics of $Cr(CO)_6$ bound within van der Waals clusters of either CH_3OH or CD_3OD (8,9). In addition to uncovering some rather unusual photophysics, we have gained some insight regarding the mode-specific character of intracluster energy transfer processes occurring between excited photoproducts of $Cr(CO)_6$ and the surrounding "solvent" molecules. In this chapter, we review our study of the multiphoton ionization dynamics for two systems of heterogeneous van der Waals clusters: $(CH_3OH)_nCr(CO)_6$, and the perdeuterated isotopomers, $(CD_3OD)_nCr(CO)_6$. We discuss differences in the observed photofragment cluster ion yields, and their implications for intracluster energy transfer and relaxation processes.

Rationale for Selection of the Model Heterocluster Systems

Of the myriad intracluster energy transfer processes which one might choose to examine, we have undertaken a study of energy transfer from excited $Cr(CO)_6$ photoproducts to surrounding solvent molecules of either CH_3OH or CD_3OD. Our motivation for the selection of these particular systems stems from fundamental interests in three principal areas: namely, the single- and multiple-photon photochemistry of coordinatively saturated transition metal carbonyls, the chemical significance of coordinatively unsaturated metal carbonyls in stoichiometric and catalytic reactions, and the development of techniques for spectroscopic characterization of these reactive, unsaturated species.

Photophysics of Transition Metal Carbonyls. Coordinatively unsaturated transition metal carbonyls play a central role in mechanistic organometallic chemistry (10). Many of these species are thought to be important intermediates in industrially significant catalytic schemes (11,12). These highly reactive molecules can be conveniently synthesized in the laboratory by pulsed UV laser photolysis of coordinatively saturated precursors (13). In condensed phases, absorption of a single UV photon by the precursor molecule results in the loss of a single ligand, independent of photon wavelength (*14*), and sequential absorption of several photons is likewise accompanied by sequential loss of additional ligands. In the gas phase, the extent of ligand loss is highly wavelength-dependent, and single-photon absorption may lead to multiple-ligand loss (*15-17*). Absorption of multiple photons by isolated metal carbonyl molecules in the gas phase results primarily in complete ligand stripping, or multiphoton dissociation (MPD), rather than molecular ionization. (Such photophysical behavior is typical of organometallic compounds in general.) Consequently, atomic metal ions are often the only photoproduct ions observed following multiphoton ionization (MPI) of organometallic compounds (*18*). To understand this transition in photophysical behavior from that typical of isolated molecules in the gas phase to that of solvated molecules in condensed phases, we need to develop an understanding of energy transfer processes between internally excited metal carbonyls and adjacent molecules in the surrounding environment.

Spectroscopic Probes of Relaxation Dynamics in Metal Carbonyls. A variety of spectroscopic techniques have been utilized by several groups to examine the relaxation dynamics of the excited species produced following UV photolysis of transition metal carbonyls in both the gas phase and condensed phases. Weitz and co-workers (*19-22*), Rosenfeld and co-workers (*23-25*), and Rayner and co-workers (*26*) have employed a transient infrared absorption technique to probe the photolysis of Group VIB hexacarbonyls in the gas phase. From these studies have come a wealth of information on the vibrational spectra, geometry, metal-ligand bond strengths, reactivity, and vibrational relaxation rates for various unsaturated metal carbonyls in their electronic ground states. The dynamics of UV photolysis of $Cr(CO)_6$ in the liquid phase, as well as the solvation dynamics for the nascent $Cr(CO)_5$ photoproduct, have been studied on the picosecond and sub-picosecond timescales in the transient visible absorption experiments of Nelson (*27*), Lee and Harris (*28*), and Simon and Xie (*29-32*); the transient IR absorption experiments of Spears and co-workers (*33,34*) and Hochstrasser and co-workers (*35*); and the transient Raman scattering experiments of Hopkins and co-workers (*36*). From these studies, we have developed an understanding of the microscopic details of the solvation process and the timescale for vibrational relaxation of $Cr(CO)_5$ in its electronic ground state following solvation.

The spectroscopic pump-probe techniques described above are all suitable for studying vibrational relaxation of metal carbonyls in the ground electronic state. Relaxation of excited electronic states in solution occurs on a timescale too rapid to probe, even with the <100-femtosecond timescale (*27*) of the fastest transient visible absorption technique. Upper electronic states of naked (isolated, gas-phase) metal carbonyls are difficult to study directly, using conventional UV laser absorption and emission techniques, since excited states of these species generally undergo rapid and efficient internal conversion to repulsive surfaces. UV absorption bands are consequently broad and quite diffuse. If we hope to advance our understanding of the upper electronic structure of metal carbonyls, relaxation dynamics of molecules in these excited states, and the significance of electronic excitation in mechanistic organometallic chemistry, we need to develop gas-phase spectroscopic probes of organometallic electronic structure.

MPI of van der Waals Clusters Containing $M(CO)_6$ (M = Cr, Mo, W). Very early in the course of our own studies of the photophysics and photochemistry of organometallic species within van der Waals clusters, we discovered that MPI, combined with mass spectrometry, might in fact be a suitable probe of the intracluster bimolecular reactivity (if not the electronic spectroscopy) of coordinatively unsaturated Group VIB carbonyls. We observed that the photoproduct ions following 248-nm MPI of homogeneous van der Waals clusters of either $Mo(CO)_6$ or $W(CO)_6$ are not exclusively atomic metal ions, as one might expect by analogy with the photophysics of the naked hexacarbonyls, but include a significant yield of totally unexpected metal oxide ions as well (37,38). These photoions could not be attributed to reactions of $M(CO)_6$ photoproducts with oxygen containing impurities (e.g., H_2O, O_2, etc.) present in the molecular beam. Since we did not observe these metal oxide ions in the electron impact cluster mass spectrum, we reasoned that the metal oxide ions were not the products of an intracluster ion-molecule reaction.

We suggested that these metal oxide ions were instead daughter ions which arise following photoionization of the binuclear product of some reaction between a neutral coordinatively unsaturated metal carbonyl and an adjacent hexacarbonyl "solvent" molecule within the cluster. This type of neutral bimolecular chemistry is analogous to that which takes place in the gas phase, generally at or near gas-kinetic rates for spin-allowed reactions, between unclustered partners (39). We considered the mechanistic implications of our unusual observations, borrowing from insights in molecular orbital theory, synthetic organometallic chemistry, and surface science, and predicted that the inferred intracluster chemistry would take place for coordination compounds of metals with large d-orbitals (such as Mo and W), but not for those of metals with more contracted d-orbitals (such as Cr). This prediction was, in fact, borne out as we failed to observe evidence of the analogous chemistry following MPI of $Cr(CO)_6$ van der Waals clusters.

To further understand the unusual photophysics of these van der Waals clusters of Group VIB hexacarbonyls, and to unravel the intracluster chemistry, we decided that we needed to examine the MPD and MPI of these molecules within clusters of more inert "solvent" molecules. We chose $Cr(CO)_6$, since the photophysics for naked $Cr(CO)_6$ has been studied in greater depth and is better-characterized than those of the other two hexacarbonyls. We chose methanol as a solvent molecule for several reasons. First, it possesses a vapor pressure and ionization potential which are both conveniently high (making it possible to generate fairly large heteroclusters whose solvent molecules are not easily ionized). Second, it is a moderate σ-donor and a poor π-acceptor, so we would not expect significant perturbations of the ligand field states due to the proximity of methanol molecules within the clusters. Finally, vibronic relaxation of $Cr(CO)_5$ by liquid-phase methanol has been extensively studied (27-29,36). We describe in the remainder of this chapter our most recent research involving the MPI of $Cr(CO)_6$-containing heteroclusters.

Time-of-flight Mass Spectrometry of $Cr(CO)_6$-containing Heteroclusters Following Multiphoton Ionization

Experimental Apparatus. A detailed description of the experimental apparatus and protocol has appeared elsewhere (9), so we shall give only a brief description here. Heterogeneous van der Waals clusters containing $Cr(CO)_6$ and methanol were produced in the free-jet expansion of a pulsed beam of seeded helium. The resulting cluster beam was skimmed and admitted into the differentially pumped ion source of a commercially-available time-of-flight mass spectrometer (R.M. Jordan Company). The cluster beam was irradiated within the ion source by the output from a pulsed UV laser, which was triggered to fire so that molecules within the center of the molecular beam pulse were photoionized. These photoions were then extracted in a direction perpendicular to both the molecular beam and the laser beam and accelerated to a nominal kinetic energy of about 4 keV. Ions which completed successful trajectories through the 1.5-m flight tube were detected by a dual microchannel plate detector. Mass spectra were collected by irradiating the neutral beam at a constant laser wavelength while acquiring and averaging detector signals following each shot with a digital storage oscilloscope (LeCroy 9400). Optical (MPI) spectra were collected by slowly scanning the wavelength of the pulsed UV laser while using a gated integrator (EG&G/PAR 4420) to acquire and average signal from ions of a given m/z.

Immediately after passing out of the ion source and into the flight tube, the photoions experience a small electric field, normal to the ion beam axis, which is of sufficient magnitude to compensate for any transverse component of photoion kinetic energy. Since neutrals in the molecular beam have the same *forward* component of mean velocity, photoions in the ion beam will have the same *transverse* component of mean velocity (assuming photofragment recoil, if any, is negligible). Consequently, larger cluster photoions will have higher transverse kinetic energy and will reach the detector only if a transverse deflecting field of greater magnitude is employed. The mass spectrometer therefore acts as a crude bandpass filter, efficiently transmitting a narrow range of ions centered about an optimum m/z value which is directly proportional to the magnitude of the deflecting field. A noteworthy point is that ions of smaller-than-optimal m/z may be efficiently transmitted if they have a larger-than-nominal transverse velocity (for example, to photofragment recoil in the direction of the neutral beam). Likewise, larger-than-optimal ions with smaller-than-nominal transverse velocity (due, for example, to photofragment recoil in the direction opposite the neutral beam) may also reach the detector.

Photoionization Lasers. The MPI of $Cr(CO)_6$-containing heteroclusters was examined in two different regions of the UV spectrum. Two distinct laser systems were used as photoionization sources. The first was an excimer laser (Lambda Physik EMG-150) we operated on the KrF transition at 248 nm. This wavelength is very close to the absorption maximum of $Cr(CO)_6$ assigned as the $^1A_{1g} \rightarrow {}^1T_{1u}$ MLCT band (*40*). $Cr(CO)_4$ is the primary photoproduct of the 248-nm single-photon photodissociation of naked $Cr(CO)_6$ in the gas phase (*41*). Copious photoion yields can be realized following irradiation at this wavelength even with mildly focused laser light (corresponding to power densities on the order of 10^7 W/cm^2).

The second laser we utilized as a photoionization source was an excimer-pumped dye laser (Lambda Physik FL3002) operated with 2-methyl-5-t-butyl-quaterphenyl dye in the range of 346-377 nm. This is near the less-intense $^1A_{1g} \rightarrow {}^1T_{1g}$ LF band of $Cr(CO)_6$ (*40*). $Cr(CO)_5$ is the primary photoproduct of single-photon photodissociation of naked, gas-phase $Cr(CO)_6$ at wavelengths within this range (*42*). An excimer laser operating on the XeF transition at 351 nm accesses the same portion of the UV spectrum. However, we chose to use the dye laser as the second system for two reasons. First, it is necessary to use strongly focused laser light (power densities on the order of 10^{12} to 10^{13} W/cm^2) to produce reasonable photoion yields in this wavelength region, and because of the differing beam divergences of the two lasers, it is possible to get nearly as large of power densities from the tightly focused dye laser as from the tightly focused XeF excimer laser. Second, the tunability of the dye laser allows us to collect resonance enhanced MPI (REMPI) spectra, which give us optical signatures of the neutral species undergoing ionization.

Photoion Yields

Mass Spectra Following MPI at 248 nm. Portions of the mass spectra collected following 248-nm irradiation of $Cr(CO)_6/CH_3OH$ and $Cr(CO)_6/CD_3OD$ cluster beams at moderate power density (10^7 W/cm^2) are shown in Figures 1 and 2, respectively. One immediately notices that the mass spectra are not dominated by signals due to solvated metal ions, as one might expect by analogy with the multiphoton dissociation and ionization dynamics of naked, unclustered $Cr(CO)_6$. Instead, repetitive sequences of solvated molecular ions appear, corresponding to several different chromium carbonyl species in varying states of coordinative unsaturation. Somewhat more careful examination reveals that the sequences of photoions which appear following multiphoton ionization of $Cr(CO)_6/CH_3OH$ heteroclusters are not identical to

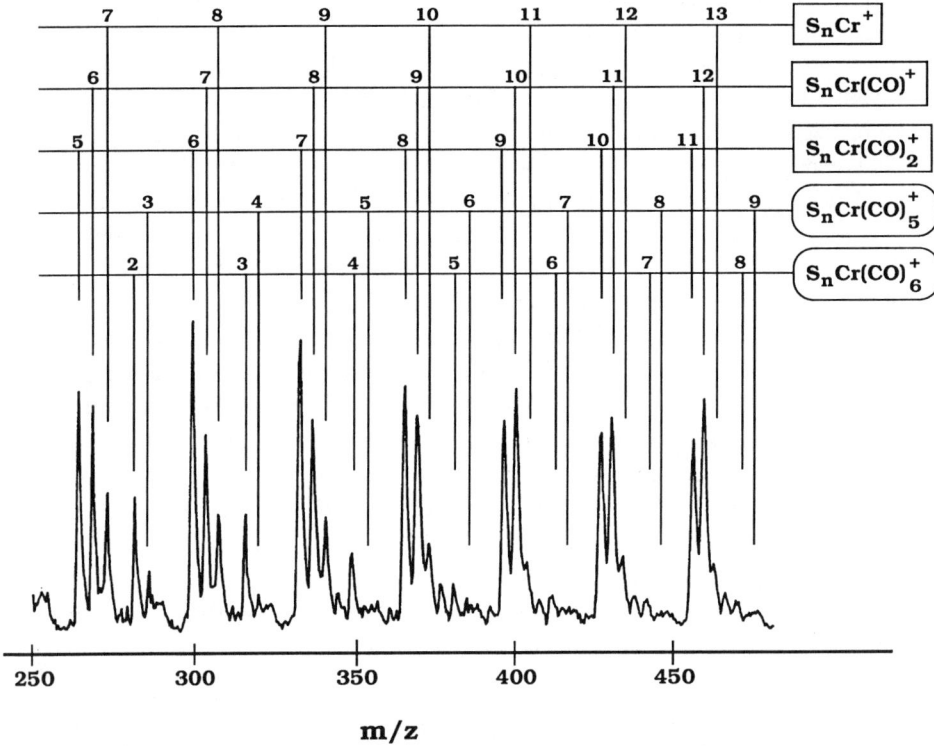

Figure 1. Mildly focused 248-nm MPI mass spectrum of $Cr(CO)_6/CH_3OH$ heteroclusters where S=CH_3OH and the numbers above the bar represent n. (Reproduced with permission from ref. 9. Copyright 1991 American Institute of Physics.)

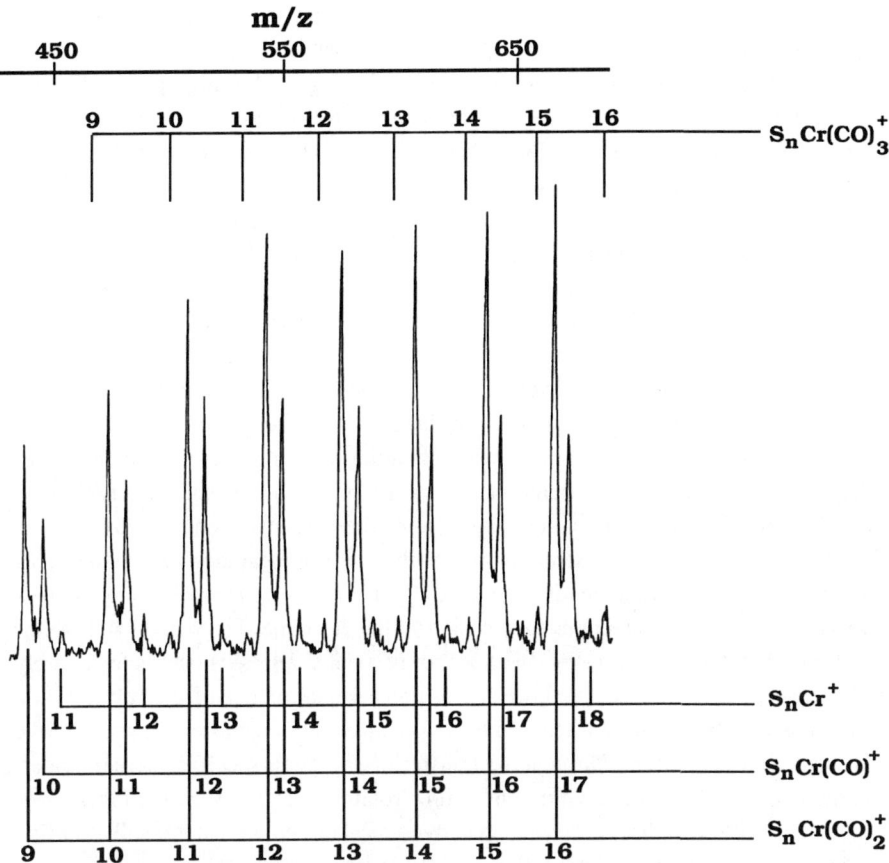

Figure 2. Mildly focused 248-nm MPI mass spectrum of $Cr(CO)_6/CD_3OD$ heteroclusters where $S=CD_3OD$ and the numbers above the bar represent n. Peaks corresponding to $S_nCr(CO)_6^+$ appear as trailing shoulders on the more intense peaks assigned to $S_nCr(CO)_2^+$. (Reproduced with permission from ref. 9. Copyright 1991 American Institute of Physics.)

those which appear following MPI of the perdeuterated isotopomers. While photoionization of the CH_3OH-solvated heteroclusters leads to sequences of empirical formula $(CH_3OH)_nCr(CO)_x^+$, x=0,1,2,5, and 6 (but not 3), photoionization of the perdeuterated heteroclusters leads to sequences of empirical formula $(CD_3OD)_nCr(CO)_x^+$, where x takes on the values 0,1,2,3, and 6 (but not 5).

Following 248-nm MPI at extremely high power density (10^{13} W/cm^2), the extent of fragmentation increases, as one might expect. As seen in the mass spectra in Figures 3a and 3b, photoions corresponding to solvated $Cr(CO)_x^+$, where x now takes on the values 0,1, and 2, are observed following MPI at high power density. Under these conditions, the same sequences appear following photoionization of either the CH_3OH- or CD_3OD-solvated heteroclusters. However, the branching ratios are now significantly different, with photoionization of the CD_3OD-solvated heteroclusters favoring production of larger proportions of the more extensively ligated species. A prominent sequence of peaks corresponding to $(CH_3OH)_nCr(H_2O)^+$ also appears in the spectrum in Figure 3a. These ions arise not from the presence of trace impurities of water in the gas mixture, but apparently via an intracluster bimolecular reaction, mediated perhaps by the presence of a coordinatively unsaturated chromium carbonyl species within the cluster. The corresponding sequence in the mass spectrum of the perdeuterated heteroclusters is obscured by isobaric interference.

Assuming the identities of the primary photoions do not depend on the identity of the solvent, these observed differences in the mass spectra of the isotopomeric heteroclusters suggest that the rates for relaxation of the nascent parent ions by the surrounding solvent bath, and therefore probabilities for fragmentation to daughter ions, are dependent on solvent identity. Since CH_3OH and CD_3OD are expected to have virtually identical cross sections for collision with a given metal carbonyl ion, it seems unlikely that the intracluster relaxation of the nascent parent ions occurs by a purely statistical process of collisional transfer to translational modes of the solvent bath. Based on our observations, it seems more reasonable that intracluster relaxation of the nascent photoions occurs through the transfer of energy to specific internal modes, perhaps a restricted set of vibrations and/or rotations, of the surrounding solvent molecules. We will discuss this apparent dynamical effect in greater detail after first considering the photoion yields following MPI at other wavelengths.

Mass Spectra Following MPI at 350 nm. The mass spectrum collected following 350-nm MPI of $Cr(CO)_6/CH_3OH$ heteroclusters at extremely high power density (10^{12} W/cm^2) is shown in Figure 4a, and an expanded portion of this spectrum appears in Figure 4b. Features in the mass spectrum are attributed to three sequences of cluster photoions: $(CH_3OH)_nCr(CO)_4H^+$, $(CH_3OH)_nCr(CO)_5H^+$, and $(CH_3OH)_n(H_2O)Cr(CO)_5H^+$. (The $Cr(CO)_5$ moiety is isobaric with the methanol pentamer; however, our assignment was confirmed by comparison with the mass spectrum of the perdeuterated heteroclusters.) Ion trajectory calculations suggest that

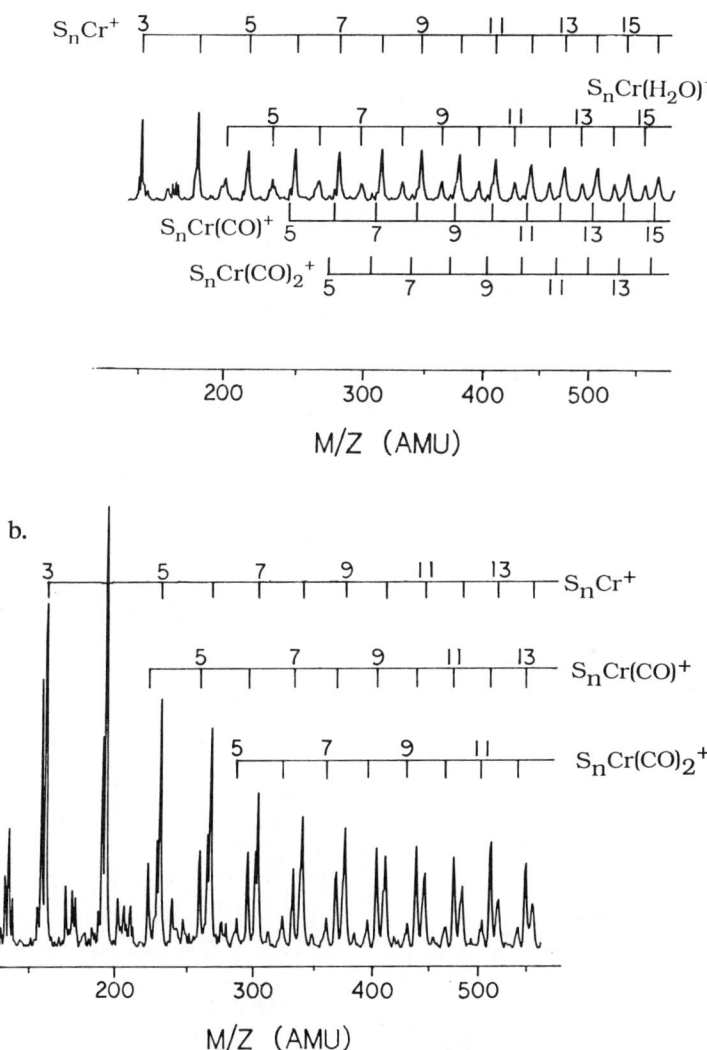

Figure 3a. Tightly focused 248-nm MPI mass spectrum of $Cr(CO)_6/CH_3OH$ heteroclusters where $S=CH_3OH$ and the numbers above the bar represent n.
b. Tightly focused 248-nm MPI mass spectrum of $Cr(CO)_6/CD_3OD$ heteroclusters where $S=CD_3OD$ and the numbers above the bar represent n.

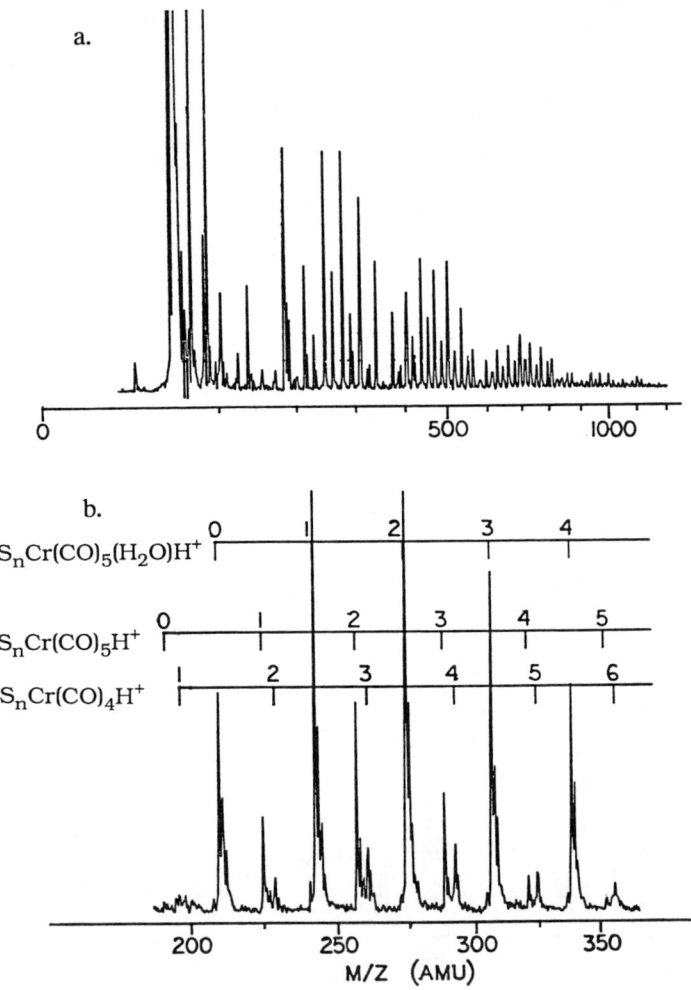

Figure 4a. Tightly focused 350-nm MPI mass spectrum of $Cr(CO)_6/CH_3OH$ heteroclusters where $S=CH_3OH$ and the numbers above the bar represent n.
b. Expanded portion of 4a.

the ensemble of cluster ions in the spectrum of Figure 4a which is centered around 460 amu corresponds to cluster ions with zero photofragment recoil kinetic energy, while those centered at 260 amu and at 690 amu correspond to ions with recoil vectors of about 1.7 eV in magnitude, oriented either parallel or antiparallel to the molecular beam vector.

The mass spectrum collected following 350-nm MPI of $Cr(CO)_6/CD_3OD$ heteroclusters at extremely high power density (10^{12} W/cm^2) is shown in Figure 5a, and an expansion of this spectrum is shown in Figure 5b. At 350 nm, the contribution to the photoion yield from homogeneous solvent cluster ions is negligible. In this spectrum, sequences of ions assigned as $(CD_3OD)_nCr(CO)_5D^+$ and $(CD_3OD)_n(D_2O)Cr(CO)_5D^+$ are present, but the sequence corresponding to $(CD_3OD)_nCr(CO)_4D^+$ is missing. Again, the fragmentation of the nascent parent ions appears to occur to a lesser extent when they are surrounded by CD_3OD, and to a greater extent when they are surrounded by CH_3OH. There is no evidence of photoions in this mass spectrum with non-negligible photofragment recoil, as was the case with the non-deuterated heteroclusters. As shown in Figure 5c, irradiation at 360 nm (the wavelength of maximum lasing efficiency for the dye) leads primarily to ionization of the homogeneous solvent clusters, while heterocluster photoion yields have dropped by nearly an order of magnitude (relative to yields following irradiation at 350 nm).

Mass-resolved REMPI Spectra: 346-377 nm. Optical spectra collected by plotting yields for $Cr(CO)_5D^+$ (a representative heterocluster photoion), Cr^+, and $(CH_3OH)_7H^+$ (a representative homogeneous solvent cluster photoion) against MPI laser wavelength are shown in Figures 6a, 6b, and 6c, respectively. These REMPI spectra are not corrected for the wavelength dependence of the laser pulse energy, which is shown in Figure 6d. Mass-resolved REMPI spectra provide optical signatures of the neutral species which, upon undergoing MPI, give rise to the monitored photoions. The fact that the heterocluster photoion spectrum does not display the same features as either the Cr^+ spectrum or the solvent cluster ion spectrum can be taken as definitive evidence that the heterocluster ion does not arise from photoionization of either methanol or atomic chromium, but rather some other neutral molecule, presumably a cluster-bound chromium carbonyl species.

Mass spectral evidence strongly suggests that the heterocluster photoions observed following irradiation at 248 nm also arise from MPI of some chromium carbonyl species. Do the heterocluster photoions which appear following 248-nm MPD/MPI, and heterocluster photoions which appear following 350-nm MPD/MPI, arise from ionization of the same neutral precursor? Suppose for the moment that this is, in fact, the case. How would we expect the heterocluster mass spectra following MPI at two different wavelengths to compare?

$Cr(CO)_6$ has a much larger absorption cross section at 248 nm than at 350 nm,

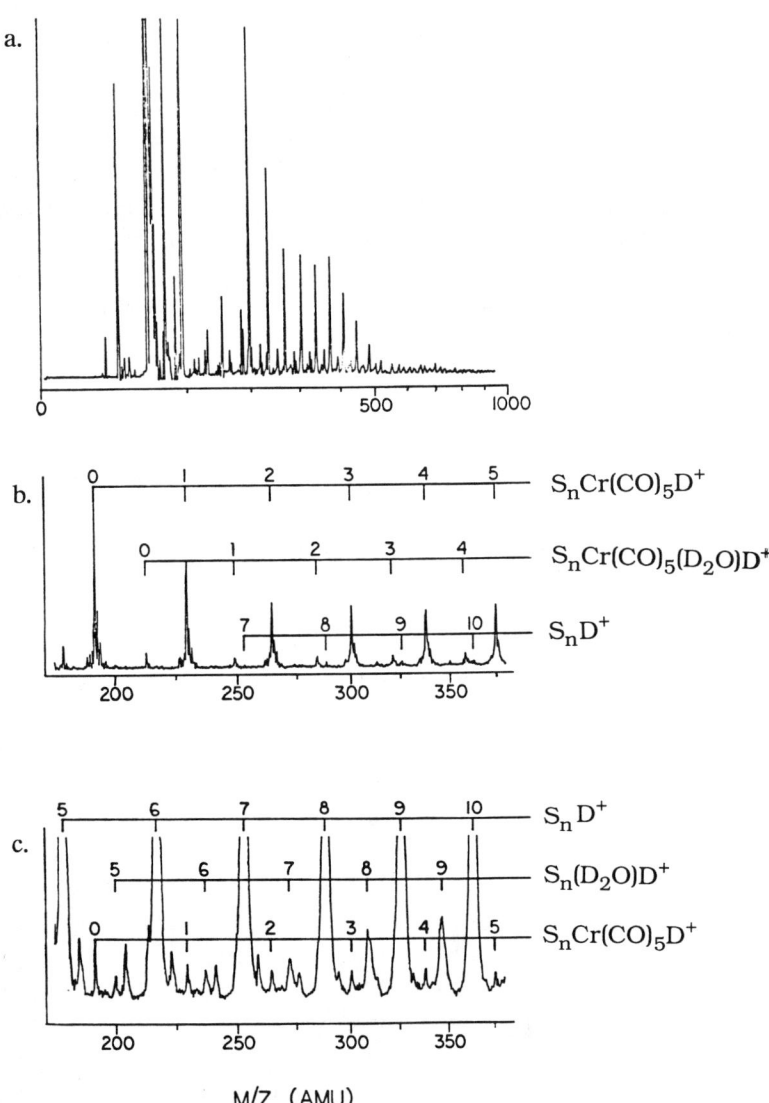

Figure 5a. Tightly focused 350-nm MPI mass spectrum of $Cr(CO)_6/CD_3OD$ heteroclusters where $S=CD_3OD$ and the numbers above the bar represent n.
b. Expanded portion of 5a. c. Expanded portion of tightly focused 360-nm MPI mass spectrum of $Cr(CO)_6/CD_3OD$ heteroclusters where $S=CD_3OD$ and the numbers above the bar represent n.

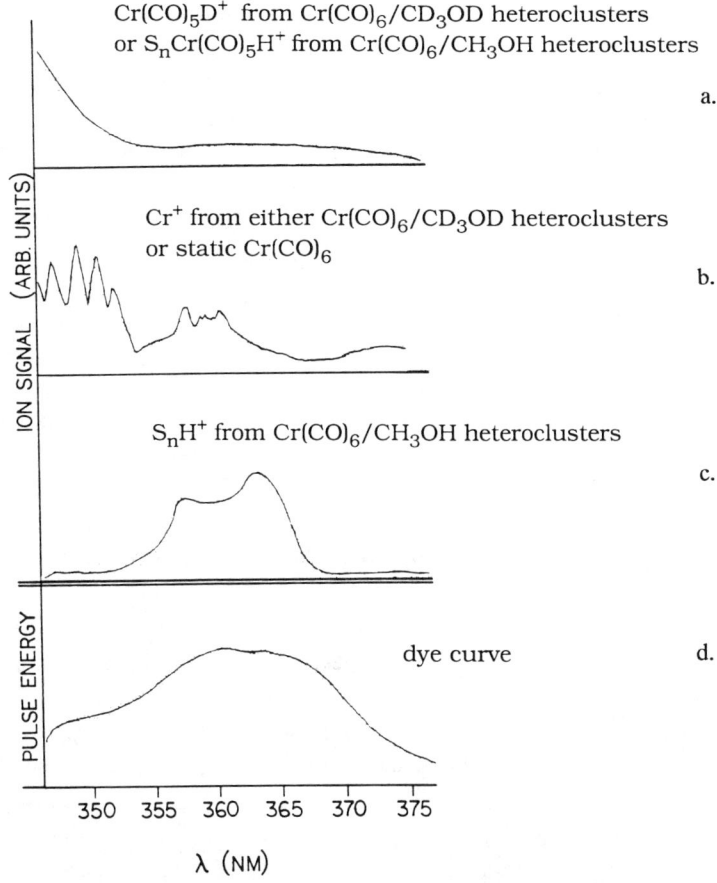

Figure 6. Resonance enhanced MPI spectra of $Cr(CO)_6/CH_3OH$ heteroclusters. a. $Cr(CO)_5D^+$ signal vs laser wavelength produced following MPI of the $Cr(CO)_6/CD_3OD$ cluster beam. b. Cr^+ signal vs laser wavelength, produced following MPI of the $Cr(CO)_6/CH_3OH$ cluster beam. c. $(CH_3OH)_7H^+$ signal vs laser wavelength produced following MPI of the $Cr(CO)_6/CH_3OH$ cluster beam. d. Experimentally measured dye laser intensity vs wavelength.

and one would expect that under conditions where transitions are not saturated, photoion yields following MPI at 248 nm would be larger, and the extent of ion photofragmentation would be greater. Under such circumstances, the 248-nm mass spectrum would not be expected to resemble the 350-nm mass spectrum. However, under conditions of sufficiently high power densities (such as the 10^{12} to 10^{13} W/cm^2 used in our high-intensity MPI experiments) essentially all one-photon processes, and many two-photon processes, are expected to be saturated. The extent of ionization and fragmentation should therefore depend not on the energy of the individual photons employed, but on the total number of photons absorbed by a given molecule. Extensive fragmentation would be expected at any laser wavelength employed, and photofragment branching ratios would be fairly insensitive to laser wavelength. This is typical for MPI of many polyatomic organic molecules, such as benzene, for example (43). We have suggested that our 248-nm, high-fluence MPI mass spectrum does not resemble the 350-nm, high-fluence spectrum because the primary photoproduct ion at 248 nm is not the same as the primary photoproduct ion at 350 nm. This would most likely be the case if the neutral chromium carbonyl species undergoing MPI at 248 nm were different from the one undergoing MPI at 350 nm. Furthermore, we can explain not only the photofragment branching ratios, but also the apparent solvent-dependent relaxation phenomenon, if we assume that photodissociation along the neutral ladder, prior to MPI, is analogous to the single-photon photophysics of $Cr(CO)_6$.

Photodissociation and Ionization Dynamics of Cluster-bound $Cr(CO)_6$

Dynamical Scheme. The results of our high-fluence MPI mass spectrometric experiments suggest that the solvated neutral chromium carbonyl species which we are probing via MPI at 248 nm is not identical to the neutral species which we are probing via MPI at 346-377 nm. It seems quite likely, then, that we are spectroscopically probing, via MPI, the neutral products of one-photon photodissociation of cluster-bound $Cr(CO)_6$, and that the coordinatively unsaturated species we are preparing depends on the wavelength we utilize. A not unreasonable extrapolation from the known photophysics of naked $Cr(CO)_6$ in the gas phase suggests a dynamical scheme for the photophysics of cluster dissociation and ionization which is also consistent with the apparent intracluster relaxation processes evidenced in the mass spectra.

The 248-nm photodissociation of naked $Cr(CO)_6$ is known to give rise to an excited photoproduct, $Cr(CO)_4^*$, whose internal energy is estimated to be as much as 38 kcal/mol; while the 351-nm photodissociation is known to give rise to a $Cr(CO)_5^*$ photoproduct, whose internal energy is estimated between 30 and 45 kcal/mol (42). The ionization potentials for the ground states of these primary photoproducts can be estimated from known bond dissociation energies and ion appearance potentials, and we have estimated the ionization potentials for naked $Cr(CO)_4$ and $Cr(CO)_5$ to be approximately 167 and 173 kcal/mol, respectively (9). Let us take the photophysical description of $Cr(CO)_6$ and the thermochemical description of the coordinatively unsaturated photoproducts to be appropriate for the description of the cluster-bound analogs. We can then propose the schemes shown below to account for photodissociation, MPI, and subsequent fragmentation.

248 nm

$$S_nCr(CO)_6 + h\upsilon \rightarrow S_nCr(CO)_4^* + 2\ CO$$
$$S_nCr(CO)_4^* + 2h\upsilon \rightarrow [S_nCr(CO)_4^*]^+ \rightarrow [S_{n-1}\cdots S^*\cdots Cr(CO)_4]^+$$
$$[S_nCr(CO)_4^*]^+ \rightarrow [S_nCr(CO)_x]^+ + 4\text{-}x\ CO$$

350 nm

$$S_nCr(CO)_6 + h\upsilon \rightarrow S_nCr(CO)_5^* + CO$$
$$S_nCr(CO)_5^* + 2h\upsilon \rightarrow [S_nCr(CO)_5^*]^+ \rightarrow [S_{n-1}\cdots S^*\cdots Cr(CO)_5]^+$$
$$[S_nCr(CO)_5^*]^+ \rightarrow [S_nCr(CO)_4]^+ + CO$$

In the initial step of each scheme, single-photon absorption by the cluster-bound $Cr(CO)_6$ leads to photodissociation and loss of one or more carbonyl ligands. Subsequent two-photon absorption leads to ionization of the single-photon photoproduct and production of a nascent photoion with a certain degree of excess energy. Assuming the estimated I.P.'s given above for the unsaturated chromium carbonyls are accurate, internal energy for the nascent cluster-bound $[Cr(CO)_4^*]^+$, created via 248-nm MPI, is expected to be as large as 103 kcal/mol. An ion with this much internal energy would be expected to undergo facile loss of additional ligands, giving rise to ions such as the ones we observe following 248-nm MPI. Tyndall and Jackson have observed that low-energy electron impact of the nascent neutral $Cr(CO)_4$ photoproduct resulting from 248-nm photodissociation of $Cr(CO)_6$ results in the production of ions of the formula, $Cr(CO)_x^+$ (x=0,1,2), which have undergone extensive fragmentation (44). Two-photon ionization of the $Cr(CO)_5^*$ photoproduct at 350 nm, however, should only give rise to a photoion with an internal energy of between 21 and 36 kcal/mol, which would most likely be insufficient to permit prompt loss of any more than one additional CO ligand.

We must also allow for some mechanism for production of cluster-bound $Cr(CO)_6^+$ and $Cr(CO)_5^+$ ions in the 248-nm MPI scheme, although we have omitted these additional steps for the sake of clarity. Direct two-photon ionization of the solvated $Cr(CO)_6$ should yield a nascent ion having as much as 60 kcal/mol. This ion probably has sufficient internal energy to undergo prompt loss of one additional ligand, giving rise to a cluster-bound $Cr(CO)_5^+$ daughter ion in a manner analogous to that shown in the other two schemes.

Intracluster Energy Transfer. If the internal energies of the nascent photoproduct ions can be efficiently relaxed by transfer to an adjacent solvent molecule within the cluster, denoted as S in the schemes above, the extent to which the ions undergo loss of additional ligands via unimolecular decay will be somewhat reduced. It appears that CD_3OD is generally a more efficient acceptor of internal energy transferred from the nascent photoions. 248-nm MPI of CD_3OD-containing heteroclusters appears to be accompanied by a lesser degree of fragmentation of nascent, cluster-bound $Cr(CO)_6^+$ (and consequently lower yields of the $Cr(CO)_5^+$ daughter fragment), as well as less extensive fragmentation of the nascent $Cr(CO)_4^+$ (and consequently higher yields of the less unsaturated daughter, $Cr(CO)_3^+$). 350-nm MPI of CD_3OD-containing clusters proceeds with a lesser degree of fragmentation of the nascent cluster-bound $Cr(CO)_5^+$ (and consequently no detectable yield of the $Cr(CO)_4^+$ daughter). This observed isotope (or isotopomer) effect may be a consequence of non-statistical, mode-specific intracluster energy transfer between excited chromium carbonyl ions and adjacent solvent molecules within the van der Waals cluster. CD_3OD may be a more efficient acceptor molecule than CH_3OH because of fortuitous overlaps with the vibrational frequencies and symmetries of the various metal carbonyl species.

Although we do not at present understand the details of the intermolecular vibrational coupling mechanisms inferred from our data, it is possible to offer some speculation concerning the coupling between methanol and $Cr(CO)_6^+$. Complete vibrational analyses for open-shell polyatomics such as $Cr(CO)_6^+$ are generally not available, but we expect the corresponding vibrational frequencies of the ion to be very nearly equal to those of the neutral. Normal coordinate analyses for methanol (45) and $Cr(CO)_6$ (46) have been performed. The 12 normal vibrations of methanol belong to the species 8A' + 4A". The 33 normal vibrations of $Cr(CO)_6$ belong to the species $2A_{1g} + 2E_g + 4T_{1u} + 2T_{2g} + T_{1g} + 2T_{2u}$ and can be classified as C-O stretching modes (three fundamentals), M-C stretching modes (three fundamentals), M-C-O bending modes (four fundamentals), and C-M-C bending modes (three fundamentals), for a total of 13 fundamental frequencies. None of these 13 fundamentals are resonant with any of the 12 fundamental frequencies of any methanol isotopomer. However, three combination bands of moderate intensity: $v_6 + v_{11}$ (C-O stretch plus C-M-C bend, 2089 cm^{-1}), $v_5 + v_7$ (M-C-O bend plus C-M-C bend, 1032.2 cm^{-1}), and $v_5 + v_{12}$ (M-C-O bend plus C-M-C bend, 875.0 cm^{-1}), overlap with respect to both frequency and symmetry with three fundamental vibrations of CD_3OD: v_3 (C-D stretch, 2080 cm^{-1}),v_6 (C-O-D bend, 1029 cm^{-1}),and v_{11} (O-C-D bend, 888 cm^{-1}). Only one of the $Cr(CO)_6$ combination bands is resonant with any of the fundamentals of CH_3OH (v_8, C-O stretch, 1034 cm^{-1}), but overlap is expected to be poor on the basis of symmetry considerations.

Conclusions

We have examined the multiphoton dissociation and ionization dynamics of van der Waals heteroclusters of $Cr(CO)_6$ and methanol generated in a pulsed free-jet expansion of seeded helium. We find that the multiphoton photophysics of $Cr(CO)_6$ solvated within van der Waals clusters is strikingly different from that of the naked molecule in the gas phase, in that molecular photoionization prevails over complete ligand stripping. We observe two principal series of cluster ions following 248-nm irradiation at moderate laser fluence (10^7 W/cm^2): a minor series corresponding to $S_nCr(CO)_x^+$ (x=5,6); and a major series corresponding to $S_nCr(CO)_x^+$ (x=0,1,2). We note that fragmentation is much more extensive under conditions of extremely high fluence (10^{12} to 10^{13} W/cm^2), and that fragmentation branching ratios are highly wavelength dependent. At 248 nm, irradiation at extremely high fluence leads to the appearance of the series, $S_nCr(CO)_x^+$ (x=0,1,2); while at 350 nm, the series corresponding to $S_nCr(CO)_xH^+$ (x=4,5) appears. At high fluence, evidence of extensive intracluster ion-molecule chemistry (i.e., proton transfer reactions, and solvent-solvent reactions leading to production of cluster-bound water) is observed. For all fluences, solvation by CD_3OD correlates with a less extensive degree of cluster ion photofragmentation than solvation by CH_3OH. The observation of a strong wavelength dependence in the fluence regime where virtually all one-photon processes are expected to be saturated suggests either that a certain degree of wavelength-dependent photodissociative ligand loss precedes cluster ionization, or that cluster ion photofragmentation is not correctly described by a statistical model. We suggest that wavelength-dependent, one-photon photodissociation of the solvated $Cr(CO)_6$ takes place initially in the neutral manifold, and that the cluster-bound primary one-photon photoproduct subsequently undergoes MPI. Furthermore, we propose that the dependence of the extent of photoion fragmentation on the isotopomeric identity of the solvent can be accounted for in terms of a dynamical scheme involving mode-specific, intracluster energy transfer. CD_3OD appears to be more efficient than CH_3OH in its ability to accommodate the energy transferred from internal modes of the nascent metal carbonyl photoion.

Acknowledgment

We gratefully acknowledge the financial support of this work provided by the Office of Naval Research. JFG also acknowledges the Alfred P. Soan Foundation for a Research Fellowship (1991-1993).

Literature Cited

1. Levine, R. D.; Bernstein, R. B. *Molecular Reaction Dynamics and Chemical Reactivity*; Oxford University Press: New York, NY, 1987.
2. *Resonances in Electron Molecule Scattering, van der Waals Complexes, and Reactive Chemical Dynamics*; Truhlar, D. G., Ed.; ACS Symp. Series No. 263; American Chemical Society: Washington, D. C., 1984.
3. Oxtoby, D. W. *J. Phys. Chem.* **1990**, *87*, 3028.
4. Castleman, A. W., Jr.; Keesee, R. G. *Annu. Rev. Phys. Chem.* **1986**, *37*, 525.
5. Echt, O.; Morgan, S.; Dao, P. D.; Stanley, R. J.; Castleman, A. W., Jr. *Ber. Bunsenges. Phys. Chem.* **1984**, *88*, 217.
6. Wei, S.; Shi, Z.; Castleman, A. W., Jr. *J. Chem. Phys.* **1991**, *94*, 3268.
7. Even, U.; Amirav, A.; Leutwyler, S.; Ondrechen, M. J.; Berkovitch-Yellin, Z.; Jortner, J. *Faraday Discuss. Chem. Soc.* **1982**, *73*, 153.
8. Peifer, W. R.; Garvey, J. F. *J. Phys. Chem.* **1991**, *95*, 1177.
9. Peifer, W. R.; Garvey, J. F. *J. Chem. Phys.* **1991**, *94*, 4821.
10. Hoffmann, R. *Angew. Chem., Intl. Ed. Engl.* **1982**, *21*, 711.
11. Meyer, T. J.; Caspar, J. V. *Chem. Rev.* **1985**, *85*, 187.
12. Poliakoff, M.; Weitz, E. *Adv. Organomet. Chem.* **1986**, *25*, 277.
13. Geoffroy, G. L.; Wrighton, M. S. *Organometallic Photochemistry*; Academic: New York, NY, 1979.
14. Perutz, R. N.; Turner, J. J. *J. Am. Chem. Soc.* **1975**, *97*, 4800.
15. Nathanson, G.; Gitlin, B.; Rosan, A. M.; Yardley, J. T. *J. Chem. Phys.* **1981**, *74*, 361.
16. Yardley, J. T.; Gitlin, B.; Nathanson, G.; Rosan, A. M. *J. Chem. Phys.* **1981**, *74*, 370.
17. Tumas, W.; Gitlin, B.; Rosan, A. M.; Yardley, J. T. *J. Am. Chem. Soc.* **1982**, *104*, 55.
18. Hollingsworth, W. E.; Vaida, V. *J. Phys. Chem.* **1986**, *90*, 1235.
19. Ouderkirk, A.; Weitz, E. *J. Chem. Phys.* **1983**, *79*, 1089.
20. Ouderkirk, A.; Werner, P.; Schultz, N. L.; Weitz, E. *J. Am. Chem. Soc.* **1983**, *105*, 3354.
21. Seder, T. A.; Church, S. P.; Ouderkirk, A. J.; Weitz, E. *J. Am. Chem. Soc.* **1985**, *107*, 1432.
22. Seder, T. A.; Church, S. P.; Weitz, E. *J. Am. Chem. Soc.* **1986**, *108*, 4721.
23. Fletcher, T. R.; Rosenfeld, R. N. *J. Am. Chem. Soc.* **1983**, *105*, 6358.
24. Holland, J. P.; Rosenfeld, R. N. *J. Chem. Phys.* **1988**, *89*, 7217.
25. Ganske, J. A.; Rosenfeld, R. N. *J. Phys. Chem.* **1989**, *93*, 1959.
26. Ishikawa, Y.; Brown, C. E.; Hackett, P. A.; Rayner, D. M. *J. Phys. Chem.* **1990**, *94*, 2404, and references therein.
27. Joly, A. G.; Nelson, K. A. *J. Phys. Chem.* **1989**, *93*, 2876.

28. Lee, M.; Harris, C. B. *J. Am. Chem. Soc.* **1989**, *111*, 8963.
29. Simon, J. D.; Xie, X. *J. Phys. Chem.* **1986**, *90*, 6751.
30. Simon, J. D.; Xie, X. *J. Phys. Chem.* **1989**, *93*, 291.
31. Simon, J. D.; Xie, X. *J. Phys. Chem.* **1989**, *93*, 4401.
32. Xie, X.; Simon, J. D. *J. Am. Chem. Soc.* **1990**, *112*, 1130.
33. Wang, L.; Zhu, X.; Spears, K. G. *J. Am. Chem. Soc.* **1988**, *110*, 8695.
34. Wang, L.; Zhu, X.; Spears, K. G. *J. Phys. Chem.* **1989**, *93*, 2.
35. Moore, J. N.; Hansen, P. A.; Hochstrasser, R. M. *J. Am. Chem. Soc.* **1989**, *111*, 4563, and references therein.
36. Yu, S.-C.; Xu, X.; Lingle, R., Jr.; Hopkins, J. B. *J. Am. Chem. Soc.* **1990**, *112*, 3668.
37. Peifer, W. R.; Garvey, J. F. *J. Phys. Chem.* **1989**, *93*, 5906.
38. Peifer, W. R.; Garvey, J. F. *Int. J. Mass Spectrom. Ion Proc.* **1990**, *102*, 1.
39. Dearden, D. V.; Hayashibara, K.; Beauchamp, J. L.; Kirchner, N. J.; van Koppen, P. A. M.; Bowers, M. T. *J. Am. Chem. Soc.* **1989**, *111*, 2401.
40. Hay, P. J. *J. Am. Chem. Soc.* **1978**, *100*, 2411.
41. Fletcher, T. R.; Rosenfeld, R. N. *J. Am. Chem. Soc.* **1985**, *107*, 2203.
42. Fletcher, T. R.; Rosenfeld, R. N. *J. Am. Chem. Soc.* **1988**, *110*, 2097.
43. Bernstein, R. B. *J. Phys. Chem.* **1982**, *86*, 1178.
44. Tyndall, G. W.; Jackson, R. L. *J. Chem. Phys.* **1989**, *91*, 2881.
45. Gebhardt, O.; Cyvin, S. J.; Brunvoll, J. *Acta Chem. Scand.* **1971**, *25*, 3373.
46. Jones, L. H.; McDowell, R. S.; Goldblatt, M. *Inorg. Chem.* **1969**, *8*, 2349.

RECEIVED October 3, 1991

APPLICATIONS OF ISOTOPE EFFECTS

Chapter 22

Deuterium Fractionation in Interstellar Space

E. Herbst[1]

Department of Physics, Duke University, Durham, NC 27706

Strong deuterium fractionation effects occur in the chemistry of interstellar clouds, which are giant accumulations of gas and dust located in between stars. Upwards of eighty different molecules up to thirteen atoms in size have been observed in the gas phase of interstellar clouds; for at least fifteen of these molecules, deuterated isotopomers have also been detected. Although the abundance ratio of deuterium to hydrogen is equal to a few parts in 10^5, singly deuterated isotopomers of polyatomic molecules can have concentrations approaching 10% of the normal hydrogen isotopomers. Fractionation is driven by both gas phase reactions and reactions on the surfaces of tiny dust particles. Current chemical models of interstellar clouds can account for most of the observed fractionation. Fractionation in regions of active star formation is more difficult to understand.

Matter in the universe is located primarily in galaxies, which are large accumulations of matter in the form of stars and interstellar material. The interstellar material is typically cool and diffuse, and is concentrated into regions labeled clouds. Interstellar clouds can range in size up to 100 light years or more in length and consist of matter in both gaseous form and in the form of dust particles, typically 0.1 μ in size. Clouds are labeled "diffuse" or "dense" according to the gas density; in a typical "dense" interstellar cloud, the gas density is in the range 10^3 - 10^4 molecules cm^{-3}. Note that this density corresponds to a nearly perfect vaccum in the terrestrial laboratory!

Although perhaps 99% of the matter is in the form of gas, the dust-like ("grain") component in interstellar clouds is responsible for scattering and absorbing external radiation, so that in dense clouds, visible and ultra-violet radiation cannot penetrate and these sources are dark to the human eye. Not surprisingly therefore, low temperatures prevail, a temperature of 10 K - 50 K being typical. Radio, micro-

[1]Current address: Department of Physics, Ohio State University, 174 West 18th Avenue, Columbus, OH 43210

wave, and infrared radiation can penetrate dense clouds since their wavelengths exceed the size of the dust particles, and clouds can be studied by astronomers at these wavelengths if sources of radiation are available. The coldness of clouds, however, constrains their thermal emissions to lie longward of the infrared for the most part. The gas has been well characterized by radioastronomical investigations of numerous high resolution spectral features seen in emission and caused by rotational transitions of molecules. At present, between eighty and ninety molecules have been discovered, the largest being a thirteen-atom unsaturated nitrile $HC_{10}CN$ (*1*), and most being organic in nature. In addition to a wide variety of standard organic molecules, there are many radicals as well as molecular positive ions. All molecules with the exception of H_2 are trace constituents of the clouds; carbon monoxide, the second most abundant molecule, has a fractional concentration or "fractional abundance" (with respect to H_2) of 10^{-4}, whereas an organic molecule such as methanol has a fractional abundance of $\leq 10^{-6}$, dependent somewhat on the specific source. The dominance of molecular hydrogen stems from the dominance of the element hydrogen. Like most of the universe, interstellar clouds follow the standard stellar elemental abundance pattern, in which most matter is in the form of hydrogen and helium, and carbon, nitrogen, and oxygen are four orders of magnitude lower in abundance than hydrogen.

The study of high resolution molecular spectral features allows astronomers to probe the physical conditions of the gas such as its density and temperature and to determine heterogeneities in these physical conditions (*2*). The most important heterogeneities are due to the slow collapse and heating up of material to form stars; the collapsing and warming material is referred to as a protostar or star formation region, and can be seen emitting in the infrared once the temperature has risen sufficiently. As far as is known, dense interstellar clouds are the only birthplaces of stars and so are important objects for study by astronomers.

Although the gas of interstellar clouds can be studied in some detail, the same cannot be said of interstellar dust particles. From the study of their scattering properties and some broad resonances, it has been established that in dense clouds, the dust particles are best characterized as having core-mantle structures (*3*), with the core consisting of perhaps graphite and silicates or some other refractory material blown out of previous generations of stellar objects, and the mantle consisting of ices and other adsorbates derived from the gas phase of the clouds. From some broad infra-red features observed from regions around which actual stars have already formed, it has been suggested that a third phase exists in interstellar clouds, with perhaps 0.01 - 0.1% of the mass. This third phase is said to consist of polycyclic aromatic hydrocarbons, or PAH's for short, which can be thought of as very large gas phase molecules (20 - 50 carbon atoms) or very tiny dust particles (*4*).

Chemistry of Dense Interstellar Clouds. The gaseous molecules detected in dense interstellar clouds are produced locally. Although the cores of dust particles (and possibly the PAH's as well) are synthesized in the atmospheres of a variety of stellar objects and either blown out gently or explosively into space, gas phase molecules are too fragile under the harsh radiation conditions of interstellar space to survive (*5*). Rather, the molecules are synthesized from precursor atomic material of stellar origin via gas phase processes and processes occurring on the surfaces of the dust particles.

The gas phase processes are strongly constrained by the low density and temperature to binary collisions that occur without activation energy. From an initial gas consisting of atoms, diatomic molecules can only be formed by the process of two atoms sticking together. Although such an event can occur via the emission of excess energy in the form of a photon ("radiative association"), this process is exceedingly inefficient for the collision of two atoms (*6*). It certainly cannot account for the almost total conversion of H into H_2 seen in dense clouds. So, it would appear that the formation of molecular hydrogen occurs on dust particle surfaces (7).

Although the surface chemistry of interstellar dust particles has been studied theoretically for many years (3,6-8), it is still poorly understood. It is clearly important, however, since sticking probabilities onto cold surfaces are close to 100 % and at typical interstellar conditions, sticking occurs within 10^6 yr. One process all investigators agree on is the formation of molecular hydrogen and of simple ices (H_2O, NH_3, CH_4) via H atom addition reactions with atoms and radicals occurring preferentially due to the rapid migration of atomic hydrogen from site to site via tunneling (3,6,9). Of these species, only H_2 thermally evaporates within a reasonable time at the low temperatures of the ambient interstellar clouds. Thus, the formation of other species on grain surfaces is probably not detected via gas phase rotational spectra except, as we shall see, in star formation regions where the temperature warms up. Once H_2 desorbs backs into the gas phase, it serves as the initiator of a rapid and complex gas phase chemistry which seemingly can explain most if not all of the observed gas phase molecules outside of star formation regions.

The low temperature gas phase chemistry consists of exothermic reactions without activation energy, since a glance at the simple Arrhenius form for the bimolecular rate coefficient k

$$k = A(T)\exp(-E_a/k_B T)$$

where k_B is the Boltzmann constant and E_a the activation energy should convince the reader that at a temperature of 10 K, even very small activation energies ($E_a \approx 0.1$ eV = 2.3 kcal mol^{-1}) lead to extraodinarily small rate coefficients compared with the collision frequency A(T). Note that astronomers prefer to measure energy in K (1 kcal mol^{-1} = 503 K); for example, an activation energy of 0.1 eV would correspond to E_a/k_B = 1160 K. Although atom-radical and radical-radical reactions probably occur without activation energy, the most important class of reactions occurring in interstellar clouds are ion-molecule reactions, in which the ions are positively charged. These reactions are known to occur rapidly and without activation energy in the vast majority of instances (10) and many have been studied in the laboratory at temperatures under 100 K (11-12). Ion-non-polar neutral reactions tend to occur at the temperature-independent Langevin rate (10) whereas ion-polar neutral reactions tend to increase in rate as the temperature is lowered (12). Ionization comes from cosmic ray bombardment, chiefly of H_2. Thousands of ion-molecule reactions, many of which have been studied in the laboratory, have been included in models of the gas phase chemistry of clouds to synthesize molecules as complex as 10 atoms (13). Below, a rather simple synthesis of interstellar water from atomic oxygen and H_2 via ion-molecule reactions is shown as a paradigm of such syntheses:

$$H_2 + \text{Cosmic Ray} \longrightarrow H_2^+ + e^- + \text{Cosmic Ray}$$
$$H_2^+ + H_2 \longrightarrow H_3^+ + H$$
$$O + H_3^+ \longrightarrow OH^+ + H_2$$
$$OH^+ + H_2 \longrightarrow H_2O^+ + H$$
$$H_2O^+ + H_2 \longrightarrow H_3O^+ + H$$
$$H_3O^+ + e^- \longrightarrow H_2O + H.$$

The last step in the synthesis is referred to as dissociative recombination; rate coefficients for these processes are known to be very rapid and to lead to dissociated products (14). Only recently have experiments been performed to actually determine partially the assorted neutral product branching ratios (15,16). In the absence of experimental investigations or detailed theory, simple statistical theories have been attempted (*17*) with mixed success.

Similar syntheses to the one for water, some involving radiative association processes when important ion-molecule reactions are endothermic, have been devised for many neutral molecules (*13*) and used in models of the chemistry, in which gas phase reactions both synthesize and destroy molecules. The simplest types of models are called "pseudo-time-dependent"; in these the chemistry of homogeneous regions evolve under fixed physical conditions. In more complex models, physical heterogeneity can be assumed (*18*) and/or hydrodynamic collapse can be incorporated (*19*). The simplest models, which have been used with the largest chemical networks, show best agreement with observed concentrations at cloud ages of 10^5 - 10^6 yr, before most of the gas phase molecules have been adsorbed onto the dust particles (*13,20*). While molecules are being synthesized and destroyed in the gas, both adsorption onto the grain and grain chemistry are occurring simultaneously. Unless the heavy molecules formed on the dust particle surfaces desorb back into the gas via some non-thermal mechanism (or in the vicinity of star formation), the surface chemical syntheses will not contribute to the observed gaseous abundances. Non-thermal desorption mechanisms (e.g., photo-desorption, cosmic ray effects) have been discussed (*6,8,21*) but are still controversial.

Deuterium Fractionation

Singly deuterated (and in one case doubly deuterated) isotopomers of a significant number of interstellar molecules have been observed, often with surprisingly large abundances given the small overall ratio of deuterium to hydrogen. Some well-known interstellar deuterium-containing molecules are: HD, HDO, NH_2D, CCD, DCN, DNC, HDCO, CH_3OD (and possibly CH_2DOH), DC_3N, DC_5N, DCO^+, and DN_2^+. The doubly deuterated species is D_2CO, which has been seen in only one source. Just as the dominant repository for hydrogen is the molecule H_2, the dominant respository for deuterium is HD, with an abundance ratio $HD/H_2 \approx 3 \times 10^{-5}$, which is essentially twice the elemental D to H ratio. Unlike the case of H_2, the formation of HD occurs most efficiently in the gas via cosmic ray ionization of deuterium atoms followed by reaction of D^+ with H_2 (*6,22*).

Gas Phase Fractionation. The primary fractionation mechanism in interstellar clouds consists of exothermic gas phase reactions between assorted ions and HD to produce deuterated ions and H_2. The most important such reaction at temperatures under 50 K is

$$H_3^+ + HD \longrightarrow H_2D^+ + H_2$$

which is calculated to be exothermic by 227 K (*23*) due to zero-point energy differences between the reactants and products and the forbiddenness of the ground rotational state of H_3^+ in perfect analogy to the absence of the lower inversion state in NH_3. This forbiddenness arises from the fact that no combination of the ground (J = 0, K = 0) rotational state wave function multiplied by a suitable electronic spin wave function is antisymmetric to exchange of two protons. At temperatures significantly below 200 K, the backward reaction becomes much slower than the forward reaction due to its endothermicity, and the equilibrium shifts dramatically to the right. Although both forward and backward reactions have been measured in the laboratory (*24*), the rate of the forward reaction is pretty much independent of temperature (with a slight inverse dependence on temperature characteristic of slightly exothermic systems) and the backward reaction rate at very low temperatures is best obtained via calculation of the equilibrium constant (*23,25*).

The calculated equilibrium constant at 10 K, obtained by considering both enthalpy and entropy effects via the laws of statistical mechanics, is 4.7×10^8, which, given a normal collision rate coefficient of 1.7×10^{-9} cm^3 s^{-1} for the forward reaction, leads to a much smaller rate constant of 3.6×10^{-18} cm^3 s^{-1} for the backward reaction. Assuming that equilibrium is reached, the relation

$$[H_2D^+][H_2] / \{[H_3^+][HD]\} = 4.7 \times 10^8$$

holds at 10 K which, with the abundance ratio $[HD]/[H_2]$ at its standard (unchanging) value of 3×10^{-5}, leads to a predicted abundance ratio $[H_2D^+]/[H_3^+]$ >> 1 in dense clouds. This unphysical result contradicts observation, and it is obvious that equilibrium is not reached because the backward reaction is so slow that H_2D^+ is depleted more rapidly by other reactions (e.g., with electrons or CO). The abundance ratio at 10 K has been calculated in some detail to be 0.1 (26), which is still almost 4 orders of magnitude in excess of the $[HD]/[H_2]$ value. At higher temperatures, equilibrium is a better approximation, and the calculated $[H_2D^+]/[H_3^+]$ abundance ratio declines toward the $[HD]/[H_2]$ value. Even at 70 K, however, the former is calculated to be 2.1×10^{-4} if equilibrium prevails and 2.0×10^{-4} considering other linked reactions (26).

Interestingly, many exothermic ion-molecule reactions of the general type

$$XH^+ + HD \longrightarrow XD^+ + H_2$$

are surprisingly slow due to activation energy barriers, as discussed by Henchman et al. (27). Although long-range attractive forces correlate with long-range complexes in both entrance and exit channels, the interchange of H and D atoms between the entrance and exit channel complexes leads to a transition state which, if it occurs at sufficiently high energy, can choke off reaction. The potential energy surface for this situation is shown in Figure 1 below.

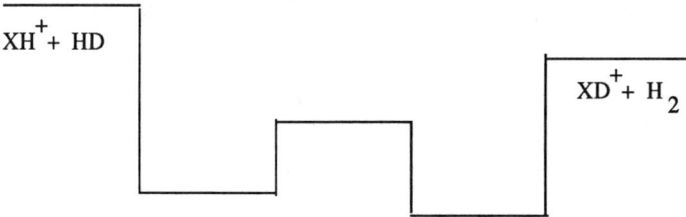

Figure 1. The potential energy surface for reactions of the type $XH^+ + HD \longrightarrow XD^+ + H_2$ with a transition state between entrance channel and exit channel complexes. The energy of the transition state shown here is not high enough to choke off reaction. See Henchman et al. (27).

Thus, only certain ions which are known to be abundant in interstellar clouds, are also known to exchange rapidly and exothermically with HD. Besides H_3^+, these include the simple hydrocarbon ions CH_3^+ and $C_2H_2^+$:

$$CH_3^+ + HD \longrightarrow CH_2D^+ + H_2$$
$$C_2H_2^+ + HD \longrightarrow C_2HD^+ + H_2;$$

these reactions are calculated and/or measured to be exothermic by 370 K and 550 K respectively (28). Given the relatively large exothermicities of the reactions, it can be seen that the backwards reactions will be slow up to much higher temperatures than the corresponding $H_2D^+ + H_2$ reaction, which is endothermic by only 227 K. Although the above hydrocarbon ion-HD reactions are important in deuteration up to a higher temperature than the H_3^+ - HD system, this latter system is much more efficient at low temperatures due to the large abundance of H_3^+. One exchange reaction that does *not* proceed is that between the abundant interstellar ion HCO^+ and HD.

The large abundance ratio between H_2D^+ and H_3^+ in cold dense clouds communicates itself through ion-molecule reactions to a wide variety of species. Perhaps the most salient deuterated isotopomer, due to its easy observability in space, is DCO^+, which is produced by the ion-molecule reaction between H_2D^+ and CO (among other reactions):

$$H_2D^+ + CO \longrightarrow DCO^+ + H_2$$

on presumably 1/3 of reactive collisions. This statistical factor of less than unity lowers the DCO^+/HCO^+ abundance ratio below that of H_2D^+/H_3^+ by a factor of 3 if DCO^+ is produced primarily by the above reaction (as is the case for clouds at 10 K). Analogous reactions lead to H_2DO^+, HDO, NH_3D^+, NH_2D and a wide variety of other deuterated isotopomers from H_2D^+ (26).

Both the ions H_2D^+ and DCO^+ react dissociatively with electrons to produce D atoms and other products. These reactions lead to a rather large atomic D/H abundance ratio, which has been calculated to be 0.005 at 10 K (26). As we shall see below, this large ratio has important ramifications for deuterium surface chemistry once the D atoms adsorb onto the dust particle surfaces. In the gas, D atoms also play an important role in fractionation through ion-molecule exchange reactions in analogy to those of HD, some of which have been studied in the laboratory (29). However, the D atom exchange reactions are typically more exothermic than their HD counterparts and one can generally ignore the backward endothermic reactions in interstellar clouds. For example, the reaction

$$D + HCO^+ \longrightarrow DCO^+ + H$$

is exothermic by 800 K. In addition to ion-molecule exchange reactions, D atoms can probably undergo neutral-neutral exchange reactions with radicals at low temperature. One example of importance is

$$D + OH \longrightarrow OD + H$$

which leads to a large OD/OH abundance ratio at low temperatures if it occurs without activation energy, as seems likely since the reaction can occur on the ground potential energy surface of water (26, 30). The reactions undergone by the two other ions which are important in fractionation - CH_2D^+ and C_2HD^+ - and the resulting fractionation are described in some detail by Millar *et al.* (26). In general, these ions fractionate a wide variety of organic molecules noticeably at temperatures up to 70 K.

Detailed Models of the Fractionation. The most complete gas phase model of deuterium fractionation has been undertaken by Millar *et al.* (26) following earlier work of Brown and Rice (31,32). The model of Millar *et al.* is of the pseudo-time-dependent variety and contains a large number (1700) of gas phase reactions linking 225 normal and deuterated isotopomers. Deuterium fractionation occurs principally via the exchange reactions discussed above and similar, although less important reactions. The deuterated species then react with other molecules to widen the effects of fractionation considerably. An important simplifying assumption, that of statistical branching ratios, is made for these latter reactions, which consist mainly of ion-molecule and dissociative recombination processes. For example, in the ion-molecule reaction between C_3^+ and HD, it is assumed that the two possible sets of products - $C_3H^+ + D$ and $C_3D^+ + H$ - are produced at equal rates, whereas in the dissociative recombination reaction between NH_3D^+ and electrons, it is assumed that the $NH_2D + H$ exit channel is three times as important as the $NH_3 + D$ exit channel. Using these assumptions can be criticized, although as a practical matter, it is hard to do otherwise. For ion-molecule reactions, phase space theory (33) would favor product channels that are more exothermic and have higher densities of states; however, it is impractical to carry out phase space calculations for all reactions of importance, and phase space theory is itself only approximate. For dissociative recombination reactions, there is little evidence against the statistical approximation although experiments on similar systems show opposing results (34). Improvements in models cannot feasibly be made until these assumptions are scrutinized.

Some results of the model of Millar et al. (26) are shown in Table I. In this table, calculated abundance ratios R for singly deuterated isotopomers to normal isotopomers are shown and compared with observed results for the well-studied interstellar cloud TMC-1 ("Taurus Molecular Cloud 1") in the constellation Taurus. TMC-1 has a temperature of 10 K and a gas density of $\approx 10^4$ molecules cm^{-3}, and these numbers are used in the pseudo-time-dependent gas phase model calculation. The calculated results are those for the time at which the absolute calculated abundances of the normal isotopomers are in best agreement with observation. (The calculated results shown are also for a particular assumption regarding the products of dissociative recombination reactions, which we deem to be most likely at this

Table I. Abundance Ratios R For TMC-1

Species	Observed R	Calculated R
HCO^+	0.015(2)	0.031
HCN	0.023(1)	0.011
HNC	0.015	0.018
CCH	0.01	0.014
HC_3N	0.015(5)	0.019
HC_5N	0.013(4)	0.020

Note: Numbers in parentheses uncertainties in last digit.

time.) It can be seen that the agreement is generally quite good, implying that the basic assumptions underlying the model are correct in this source. A similar conclusion is drawn from the comparison of observed and calculated abundance ratios in the extended cloud in Orion (26), another well-studied interstellar source, which is at a warmer temperature than TMC-1 (70 K). Here the fractionation that occurs is more a function of the very exothermic exchange reactions rather than the $H_3^+ + HD$ system, which dominates for sources at 10 K. Orion has some well-known regions in which star formation occurs, and in these regions the simple gas phase fractionation model is inadequate in several ways. The topic of fractionation in star-forming regions is discussed next.

Star Formation and Fractionation

In many dense interstellar clouds, there is strong evidence that star formation is well advanced. Indeed, newly formed bright stars are also found in interstellar clouds, as any amateur observer can detect by looking at the Trapezium stars in the Orion Nebula. The actual process of star formation and what initiates it are far from well understood, and high resolution observations directed towards star forming regions indicate great complexity and heterogeneity (2,35). For example, while most of the Orion Molecular Cloud is cool ($\approx$ 70 K) and at constant density (10^3 - 10^4 molecules cm^{-3}), star forming regions abound. Perhaps the best known such region is called the Kleinmann-Low Nebula; inside this region there are at least three spatially distinct sources, in which the abundances of polyatomic molecules are also quite distinct from one another. One source - the so-called "Plateau" source - shows indications of being perturbed by a strong shock wave. The other sources - the "Hot Core" and the "Compact Ridge" are both more quiescent, having temperatures in the range 100 - 200 K and gas densities in the range 10^6 - 10^7 cm^{-3}.

What does one expect the deuterium fractionation to be in quiescent star formation regions? At temperatures of 100 - 200 K, the gas phase ion-molecule models predict much less fractionation than at the lower temperatures of ambient regions. This result is intuitive because as the endothermicities of the backward (right-to-left) reactions of the exchange systems discussed above become smaller in relation to the temperature, their rates increase and the equilibria no longer lie on the extreme right. In other words, the equilibrium coefficients K, given by the well-known formula

$$K(T) = \exp(-\Delta G^0/k_B T),$$

rapidly decrease with increasing temperature.

Most surprisingly, the abundance ratios of deuterated to normal isotopomers for selected species in the two quiescent star forming regions discussed above are not in accord with this view. Although a variety of values have been deduced from observations for these abundance ratios, it does appear that the fractionation is far greater than can be expected. For example, very recent observations by Turner (36) of the Compact Ridge source indicate that $[HDCO]/[H_2CO] \approx 0.14$ and $[NH_2D]/[NH_3] \approx 0.062$. At a temperature of 70 K, the model results of Millar et al. (26) for these two abundance ratios are 0.019 and 5.4 x 10^{-4}, respectively. The calculated ammonia abundance ratio at 70 K, the highest temperature at which results are presented, is already two orders of magnitude low, whereas the formaldehyde result is not as low because it is fractionated by reactions based on the very exothermic organic ion - HD exchange reactions. Even at 10 K, however, the model results are too low compared with Turner's observations.

A possible explanation for the discrepancy derives from a chain of evidence which has been obtained in recent years favoring the idea that surface chemistry is important in star formation regions such as the Compact Ridge.

Surface Chemistry and Fractionation. In the Hot Core and Compact Ridge sources, the abundances of certain molecules, which tend to be relatively saturated or hydrogen-rich, are orders of magnitude larger than can be accounted for by standard gas phase models. Examples of molecules for which this is true in one source and/or the other are: NH_3, C_2H_5CN, CH_3OH, CH_3OCH_3, and $HCOOCH_3$. Even a non-standard gas phase model, in which the special physical conditions of the Compact Ridge and its time dependence are taken into account, appears to fail

(37). A school of thought holds that the molecules with unusually large abundances are produced on grain surfaces during a previous, cool phase and then desorbed into the gas phase when the temperature rises in the act of star formation, during which the high gas density may act as a brake on gas phase ion-molecule chemistry so that the large abundances need not decrease rapidly *(9).* This scenario is easiest to visualize for the case of NH_3, since H atom addition reactions to N atoms (H + N -> NH) followed by addition to NH (H + NH -> NH_2) and NH_2 (H + NH_2 -> NH_3) radicals are among the least controversial of grain surface reactions, given the high surface mobility of H atoms. However, a detailed model calculation for the Hot Core using this mechanism overpredicts the resultant gas phase abundance of ammonia *(9).* The cases of the oxygen-containing organic molecules, found principally in the Compact Ridge, are harder to fathom and only some suggestions have been made *(2,37,38).* One attractive possibility is for methanol to be formed on the grains, desorb into the gas when the temperature rises, and then act as a precursor for more complex molecule synthesis.

If it is reasonable to suggest that grain surface reactions during the pre - star formation stage offer an explanation for some of the chemistry observed in the gaseous portion of star formation sources, then such an explanation may also help to explain the deuterium fractionation observed. The initial NH_2D observations on the Orion star forming regions were first explained by simple gas phase chemistry pertaining to an earlier cool stage followed by adsorption onto the dust particles, and then desorption into the gas when the temperature rose *(39).* The problem with this explanation is that it does not explain the large *absolute* abundances of NH_3 and NH_2D. The grain production hypothesis for the large NH_3 abundance *(9)* soon led to such a hypothesis for the large NH_2D abundance *(40).* Building on earlier work of Tielens *(41),* Brown and Millar *(40)* invoked grain surface reactions involving D atoms as a fractionation mechanism capable of producing copious amounts of NH_2D on the surfaces of dust particles. The rising temperatures caused by star formation would then release at least some of the NH_2D from the surface. The model of Brown and Millar can indeed reproduce the 0.003 abundance ratio detected for NH_2D/NH_3 in the Hot Core *(39)* although not the value of 0.062 quoted for the Compact Ridge *(36).*

Brown and Millar *(42)* later extended their model to estimate the abundance of doubly deuterated ammonia - NHD_2 - in the Hot Core region of Orion, and predicted it to be large enough for observation. Although Turner *(36)* subsequently found some spectroscopic evidence for this species in the Compact Ridge (the observed two spectral lines were rejected as being too strong), he also detected D_2CO and analyzed his result in terms of the D atom surface chemistry approaches of Tielens *(41)* and Brown and Millar *(40,42).* Turner utilized an atomic D/H abundance ratio in the gas of 0.029, which appears to be much higher than achievable by current gas phase models, even at the lowest utilized temperatures. In this connection, it must be re-emphasized that the high relative D atom abundances on grain surfaces can only result from previous low temperature production in the gas.

It would seem that although the surface chemistry explanation of the unusually large deuterium fractionation effects seen in the Hot Core - Compact Ridge complex of star forming regions in Orion is promising, much more work remain to be done before it will be compelling. On the theoretical side, the actual rates of D and H atom migration on surfaces should be looked at more closely; if a tunneling mechanism is dominant, should not D atoms be much slower at going from site to site? On the observational side, the question of the NH_2D abundance needs to be reinvestigated and a wider variety of deuterated and doubly deuterated isotopomers detected.

Acknowledgments. I would like to acknowledge the support of the National Science Foundation for my research program in astro-chemistry.

Literature Cited

(1) Herbst, E. *Ang. Chem. Inter. Ed.* **1990**, *29*, 595-608.
(2) Blake, G. A.; Sutton, E. C.; Masson, C. R.; Phillips, T. G. *Astrophys. J.* **1987**, *315*, 621-645.
(3) Tielens, A. G. G.M.; Allamandola, L. J. In *Interstellar Processes;* Hollenbach, D. J.; Thronsen, H. A., Eds.; Reidel: Dordrecht, 1987; pp 397-469.
(4) Allamandola, L. J.; Tielens, A. G. G. M.; Barker, J. R. *Astrophys. J. Suppl.* **1989**, *71*, 733-775.
(5) Stief, L. J.; Donn, B.; Glicker, S.; Gentieu, E. P.; Mentall, J. E. *Astrophys. J.* **1972**, *171*, 21-30.
(6) Watson, W. D. *Rev. Mod. Phys.* **1976**, *48*, 513-552.
(7) Hollenbach, D. J.; Salpeter, E. E. *Astrophys. J.* **1970**, *163*, 166-164.
(8) d'Hendecourt, L. B.; Allamandola, L. J.; Baas, F.; Greenberg, J. M. *Astron. Astrophys.* **1982**, *109*, L12-L14.
(9) Brown, P. D.; Charnley, S. B.; Millar, T. J. *Mon. Not. Roy. Astron. Soc.* **1988**, *231*, 409-417.
(10) Anicich, V G.; Huntress, W. T., Jr. *Astrophys. J. Suppl.* **1986**, *62*, 553-672.
(11) Hawley, M.; Smith, M. A. *J. Amer. Chen. Soc.* **1989**, *111*, 8293-8294.
(12) Rowe, B. R. In *Rate Coefficients in Astrochemistry;* Millar; T. J.; Williams, D. A., Eds.; Kluwer: Dordrecht, 1988; pp 135-152.
(13) Herbst, E.; Leung, C. M. *Astrophys. J. Suppl.* **1989**, *69*, 271-300.
(14) Bates, D. R. *Phys. Rev.* **1950**, *78*, 492-493.
(15) Herd, C. R.; Adams, N. G.; Smith, D. *Astrophys. J.* **1990**, *349*, 388-392.
(16) Adams, N. G.; Herd, C. R., Geoghegan, M.; Smith, D.; Canosa, A.; Gomet, J. C.; Rowe, B. R.; Queffelec, J. L.; Morlai, M. *J. Chem. Phys.* **1991**, *94*, 4852-4857.
(17) Galloway, E. T.; Herbst, E. *Astrophys. J.* **1991**, August 1.
(18) Chièze, J. P., Pineau des Forêts, G. *Astron. Astrophys.* **1989**, *221*, 89-94.
(19) Tarafdar, S. P.; Prasad, S. S.; Huntress, W. T., Villere, K. R.; Black, D. C. *Astrophys. J.* **1985**, *289*, 220-237.
(20) Winnewisser, G.; Herbst, E. *Topics in Current Chem.* **1987**, *139*, 119-172.
(21) Léger, A.; Jura, M.; Omont, A. *Astron. Astrophys.* **1985**, *144*, 147-160.
(22) Black, J. H.; Dalgarno, A. *Astrophys. J.* **1973**, *184*, L101-L104.
(23) Herbst, E. *Astron. Astrophys.* **1982**, *111*, 76-80.
(24) Adams, N. G.; Smith, D. *Astrophys. J.* **1981**, *248*, 373-379.
(25) Smith, D.; Adams, N. G.; Alge, E. *Astrophys. J.* **1982**, *263*, 123-129.
(26) Millar, T. J.; Bennett, A.; Herbst, E. *Astrophys. J.* **1989**, *340*, 906-920.
(27) Henchman, M. J.; Paulson, J. F.; Smith, D.; Adams, N. G.; Lindinger, W. In *Rate Coefficients in Astrochemistry;* Millar, T. J.; Williams, D. A., Eds.; Kluwer: Dordrecht, 1988; pp 201-208.
(28) Herbst, E.; Adams, N. G.; Smith, D.; DeFrees, D. J. *Astrophys. J.* **1987**, *312*, 351-357.
(29) Adams, N. G.; Smith, D. *Astrophys. J.* **1985**, *294*, L63-L65.
(30) Crosswell, K.; Dalgarno, A. *Astrophys. J.* **1985**, *289*, 618-620.
(31) Brown, R. D.; Rice, E. H. N. *Phil. Trans. Roy. Soc. London A* **1981**, *303*, 523-533.

(32) Brown, R. D.; Rice, E. H. N. *Mon. Not. Roy. Astron. Soc.* **1986**, *223*, 429-442.

(33) Light, J. C. *Disc. Faraday Soc.* **1967**, *44*, 14-29.

(34) Gellene, G. L.; Porter, R. F. *J. Phys. Chem.* **1984**, *88*, 6680-6684.

(35) Genzel, R.; Harris, A. I.; Jaffe, D. T.; Stutzki, J. *Astrophys. J.* **1988**, *332*, 1049-1057.

(36) Turner, B. E. *Astrophys. J.* **1990**, *362*, L29-L33.

(37) Millar, T. J.; Herbst, E; Charnley, S. B. *Astrophys. J.* **1991**, *369*, 147-156.

(38) Brown, P. D. *Mon. Not. Roy. Astron. Soc.* **1990**, *243*, 65-71.

(39) Walmsley, C. M.; Hermsen, W.; Henkel, C., Mauersberger, R.; Wilson, T. L. *Astron. Astrophys.* **1987**, *172*, 311-315.

(40) Brown, P. D.; Millar, T. J. *Mon. Not. Roy. Astron. Soc.* **1989**, *237*, 661-671.

(41) Tielens, A. G. G. M. *Astron. Astrophys.* **1983**, *119*, 177-184.

(42) Brown, P. D.; Millar, T. J. *Mon. Not. Roy. Astron. Soc.* **1989**, *240*, 25P-29P.

RECEIVED September 4, 1991

Chapter 23

Deuterium in the Solar System

Yuk L. Yung and Richard W. Dissly

Division of Geological and Planetary Sciences, California Institute of Technology, Pasadena, CA 91125

A survey of the abundances of deuterium in planetary atmospheres and small bodies has been carried out. The observed pattern of D/H ratios in the solar system may be interpreted in terms of a few simple concepts: origin, fractionation, and dilution. There appear to be two distinct reservoirs of hydrogen in the solar nebula: the bulk of hydrogen as H_2, and a smaller amount in ices and organics. The latter reservoir is characterized by a higher D/H ratio than the former, and may be the principal source of hydrogen to the terrestrial planets and small bodies. The evolution of planetary atmospheres over the age of the solar system has resulted in substantial changes in the D/H ratio in the atmospheres of the terrestrial planets. In the giant planets the abundance of D is dominated by the primordial HD, and there has been negligible chemical evolution since formation. Quantitative modeling of the D/H ratio in the solar system remains hampered by the lack of appropriate chemical kinetics data.

Deuterium in the universe is believed to have been synthesized in the first few minutes of the expansion of the universe when the ambient temperature exceeded 10^9 K (1). Subsequent nuclear reactions may destroy deuterium by conversion to heavier elements, but would not be a significant source of fresh deuterium (2). This fact, in addition to its chemical reactivity and its larger mass with respect to hydrogen, makes deuterium one of the most useful chemical tracers for studying the origin and evolution of planetary atmospheres (3).

To restrict the scope of this review, we shall focus on three essential aspects of deuterium in the solar system. First, we will trace the origin of deuterium in the solar system to its precursor in the solar nebula and its placental molecular cloud. Second, we will examine the physical and chemical mechanisms that lead to fractionation of D relative to H in various reservoirs from the molecular cloud down to atmospheres of planets. Finally, perhaps the most exciting question we will examine concerns the size of hydrogen reservoirs in the planets and the evolutionary history of these reservoirs. An understanding of the first two aspects of deuterium, together with present day observations, may provide new insights into this important question.

The extensive observations of deuterium in the solar system and molecular clouds are summarized in the next section. This is followed by a survey of the important physical and chemical fractionation mechanisms. The next section attempts to explain the observations using the known fractionation processes. Finally, we list a number of outstanding problems to challenge future laboratory and observational programs, as well as theoretical models.

Observations

It is convenient for this review to divide the relevant astronomical objects into four categories: (i) terrestrial planets, (ii) giant planets, (iii) small solar system bodies, and (iv) interstellar molecular clouds. The motivation behind this choice is as follows. The interstellar molecular clouds yield clues to the initial composition of the protosolar nebula. The small bodies may have preserved a pristine record of the chemical state of condensables in the solar nebula. The giant planets reflect the bulk composition of the solar nebula and have undergone little change since formation, while the terrestrial planet atmospheres have evolved extensively over the age of the solar system. The overall picture of D/H ratios in the solar system is summarized in Figure 1.

(i) Terrestrial Planets. The level of deuterium fractionation in terrestrial volatile envelopes is subject to extensive change over the age of the solar system. An understanding of the degree of fractionation can provide clues to the past chemical evolution of these atmospheres. The bulk of terrestrial hydrogen resides in ocean water. Its D/H ratio is known as SMOW (Standard Mean Ocean Water) and has the value of $1.5576 \pm 0.0005 \times 10^{-4}$ (4, 5). It is convenient to measure the D/H ratio in a sample relative to this standard. The deviation, δD, is defined as

$$\delta D = \{ [(D/H)_{sample} / (D/H)_o] - 1\} \times 1000$$

where $(D/H)_o$ refers to SMOW.

Pioneer Venus mass spectrometers provided the bulk of the data for Venus atmospheric D/H. McElroy et al. (6) used the ion mass spectrometer to give a homopause (130 km) D/H of about 1×10^{-2}, further refined by Kumar and Taylor (7) to 2.5×10^{-2} and 1.4×10^{-2} for two different orbital data sets. Donahue et al. (8) report a value of $1.6 \pm 0.2 \times 10^{-2}$ for D/H below the cloud base of the atmosphere (< 63 km) using the probe neutral mass spectrometer to compare HDO to H_2O abundances. This value was confirmed by the groundbased observations of HDO/H_2O by deBergh et al. (9), who report D/H = $1.9 \pm 0.6 \times 10^{-2}$, or 120 ± 40 times $(D/H)_o$.

The D/H ratio in the Martian atmosphere was measured by Owen et al. (10) to be $9 \pm 4 \times 10^{-4}$, or 6 ± 3 times $(D/H)_o$, using groundbased HDO/H_2O integrated column observations. The accuracy of this measurement was improved by Bjoraker

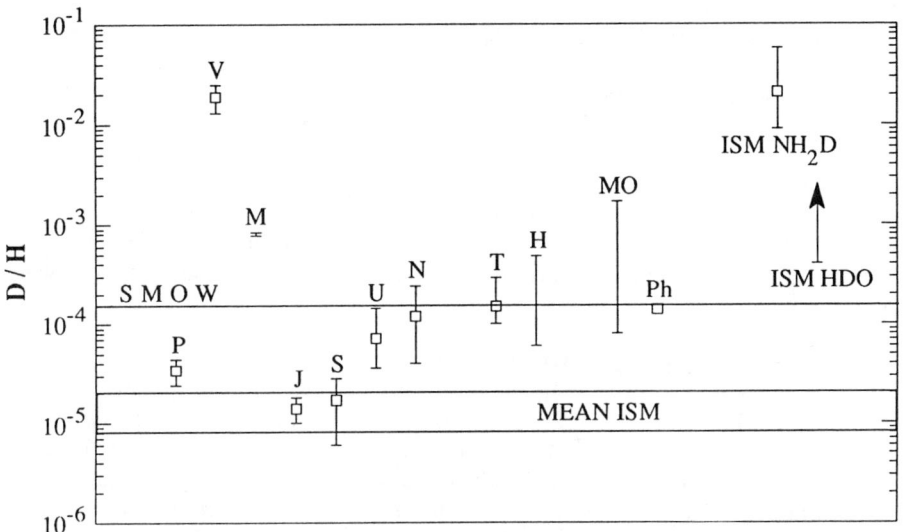

Figure 1. D/H ratios in the solar system and interstellar molecular clouds. The uncertainties in the measurements are enclosed by vertical bars. References: SMOW (5), Mean ISM (35), (P)rotosolar (37), (V)enus (9), (M)ars (11), (J)upiter (17), (S)aturn (22), (U)ranus (24), (N)eptune (15), (T)itan (27), (H)alley (28), (M)eteoritic (O)rganics, (Ph)yllosilicates (30), ISM NH_2D (42), ISM HDO (39).

et al. (11) to D/H = 8.1 ±0.3 x 10^{-4}, or 5.2 ±0.2 times $(D/H)_o$, again comparing HDO to H_2O column abundances, but measured using an airborne Fourier Transform Spectrometer.

(ii) Giant Planets. The D/H ratio in the atmospheres of the giant planets has been the subject of extensive observations. Because the mass of these planets is dominated by atmospheres that have undergone relatively little chemical processing, they should reflect the bulk composition of the gaseous solar nebula. Thus elemental and isotopic abundance ratios in these bodies provide a fundamental framework for understanding origin processes. By far the largest reservoir of hydrogen is in H_2, with an atmospheric mole fraction ($[H_2]$/total) of 0.897 ±0.030 for Jupiter (12). The cores of these planets, thought to be partly composed of reduced ices that can condense at relevant temperatures and pressures in the outer solar system (primarily H_2O, NH_3, and CH_4), provide a secondary reservoir for hydrogen.

The observed molecules that are used as diagnostics for D/H in the giant planets are HD and CH_3D. The HD is a remnant of the gaseous part of the solar nebula, while the CH_3D was more likely derived by outgassing of methane from ices in the core of the planet. If we adopt the common assumption that vertical mixing occurs in the giant planet atmospheres, so all parts of the atmosphere pass through high enough temperatures to kinetically allow rapid deuterium exchange between molecules, then fractionation equilibrium should hold, represented by the following equality:

$$[D]/[H] = [HD]/2[H_2] = [CH_3D]/4f[CH_4]$$

The numerical factors 2 and 4 arise because there are two equivalent positions for deuterium in HD and four in CH_3D. The fractionation factor f arises because the deuterium atom is more tightly bound than H, and becomes systematically more concentrated in CH_3D than in HD, so that $f > 1$ (see following section). This fractionation factor is strongly temperature dependent, so calculating a global average of f for a planet is sensitive to the choice of model for the temperature profile in the planet's atmosphere. The most detailed thermophysical model to date for Jupiter and Saturn is that of Fegley and Prinn (13), who give values of $f = 1.205$ for Jupiter and $f = 1.229$ for Saturn. We adopt these values to normalize past observations of these bodies, and conservatively assume that $f = 1.25$ ±0.05 for Uranus and Neptune.

The first observation of a deuterated molecule in an extraterrestrial body was the detection of CH_3D on Jupiter by Beer and Taylor (14). Many observations have been made since then, using both HD and CH_3D, for all four giant planets. (For reviews of past observational work see, e.g., deBergh et al., 1990 (15), Gautier and Owen, 1989 (16), or Bjoraker et al., 1986 (17)). In this paper, we present only the observations thought to give the best current estimate of D/H for each giant planet. Only observations using CH_3D are considered, as the observed HD lines are probably severely blended with weak lines from higher hydrocarbons present in the upper atmospheres of these planets (16,18). Observations using CH_3D suffer from uncertainties as well, as abundance determinations are somewhat model dependent. For example, the observer has to assume what depth in the atmosphere from which the observations are being made. Such an inference is only as accurate as the model for the atmosphere it is compared to, and although the Voyager data for the giant planets have greatly improved such models, uncertainties are still present.

The most recent determination of CH_3D abundance in Jupiter was made by Bjoraker et al. (17), at 5 μm and high spectral resolution (0.5 cm^{-1}) from airborne observations, giving a mixing ratio of $CH_3D = 2.0 \pm 0.4 \times 10^{-7}$. Combined with their observed mole fraction for CH_4 of $3.0 \pm 1.0 \times 10^{-3}$, they report D/H = $1.4 \pm 0.5 \times 10^{-5}$. Previous determinations of Jovian D/H using Voyager IRIS data were made by Drossart et al. (19) at 4.5 μm and Kunde et al.(20), combining data at 4.7 μm and 8.5 μm. Using their reported mole fractions of CH_3D, and the methane mixing ratio of $2.0 \pm 0.2 \times 10^{-3}$ from Voyager IRIS (21), these authors give D/H = $1.9^{+1.5}_{-1.0} \times 10^{-5}$ and $3.6^{+1.1}_{-1.4} \times 10^{-5}$, respectively.

For Saturn, recent groundbased observations by Noll and Larson (22) of CH_3D at 5 μm assign it a mole fraction of $3.3 \pm 1.5 \times 10^{-7}$. When combined with the methane mixing ratio of $4 \pm 2 \times 10^{-3}$ (16), they report D/H = $1.7 \pm 1.1 \times 10^{-5}$. Voyager IRIS results for Saturn by Courtin et al. (23) at 8.6 μm yield a similar value of D/H = $1.6^{+1.3}_{-1.2} \times 10^{-5}$.

The observations of Uranus and Neptune are less numerous and less certain. Groundbased observations at 1.6 μm by deBergh et al. yield values of D/H = $7.2^{+7.2}_{-3.6} \times 10^{-5}$ for Uranus (24) and D/H = $1.2^{+1.2}_{-0.8} \times 10^{-4}$ for Neptune (15), adopting the value of $f = 1.25 \pm 0.05$ for these two bodies. More accurate determinations for all four bodies will probably have to wait until space-based observations are possible in the submillimeter, where the R(0) and R(1) lines of HD in giant planet atmospheres should be readily detected and relatively free from unknown continuum blends and modeling uncertainties (25).

(iii) Small Solar System Bodies. In addition to D/H observations of terrestrial and giant planet atmospheres, measurements of the atmosphere of Titan, Saturn's largest satellite, meteorites, and Halley's comet supplement our knowledge of D/H in the solar system. Titan is the only satellite in the solar system with a substantial atmosphere, with a surface pressure of 1.5 bar and a methane mole fraction of 1-3%. Thus, the primary deuterated species on Titan is CH_3D, so that [D]/[H] = $[CH_3D]/4[CH_4]$ (there is no factor f here because CH_4 is the major hydrogen species). Recent observations include the ground-based detection of CH_3D by deBergh et al. (26) at 1.6 μm, comparing CH_3D and CH_4 abundances to give D/H = $1.65^{+1.65}_{-0.8} \times 10^{-4}$. Analysis of Voyager data from at 8.6 μm by Coustenis et al. (27) is in excellent agreement with this value, yielding D/H = $1.5^{+1.4}_{-0.5} \times 10^{-4}$.

The best observation to date made for a comet was the in situ neutral mass spectrometric measurement of HDO by the Giotto spacecraft on comet Halley, finding $0.6 \times 10^{-4} \leq$ D/H $\leq 4.8 \times 10^{-4}$ (28). Although the uncertainty in this measurement is quite large, it gives a higher value of D/H than the best estimates for Jupiter and Saturn, and is consistent with the value for Titan's atmosphere, implications to be discussed later in this paper.

Deuterium abundances in meteorites are difficult to characterize. Wide ranges of deuterium enrichments are seen, even in the analysis of a single meteorite. A compilation of recent experimental D/H values given by Zinner (29) yields a maximum range for ordinary chondrites of $8 \times 10^{-5} \leq$ D/H $\leq 1.05 \times 10^{-3}$. Yang and Epstein (30) give a similar range of $8 \times 10^{-5} \leq$ D/H $\leq 1.7 \times 10^{-3}$ for hydrogen in meteoritic organic matter, and they report consistent values for the water of hydration in phyllosilicates (clay minerals) at δD = -110‰ (D/H = 1.39×10^{-4}). It has been suggested, however, that the upper limit of 1.7×10^{-3} for the acid soluble H lost during treatment is an exaggerated value, as the processing involved can expose the meteoritic sample to terrestrial contamination (31). Kerridge suggests a more conservative upper limit for D/H in carbonaceous chondrites of 6×10^{-4} (Kerridge, J. F., University of California at Los Angeles, personal communication, 1991). Interplanetary dust particles (IDP's) that are collected intact from the stratosphere can yield high deuterium enrichments up to D/H = 1.6×10^{-3} (32). With such heterogeneity, it is difficult to derive a systematic scheme for meteoritic histories.

(iv) Interstellar Molecular Clouds. The average D/H in interstellar gas (representative of bulk hydrogen) is a difficult quantity to define, because of large scale inhomogeneities in the chemical state of the interstellar medium. Vidal-Madjar et al. (33) suggest D/H can vary by a factor of two on scales of 15 pc or less. Recent compilations along many observational lines of sight give a range of values. Murthy et al. (34) suggest that D/H = 2.0×10^{-5} is consistent with most of the observations from the IUE satellite, although lower values toward many hot stars remain likely. Borsgaard and Steigman (35) summarize several recent observations and propose a range for D/H of 0.8-2.0×10^{-5} for interstellar gas. The D/H ratios are greatly enhanced for organic molecules in molecular clouds. Recent observations suggest the following ratios (38): DCN/HCN = 0.002-0.02, DCO^+/HCO^+ = 0.004-0.02, C_3HD/C_3H_2 = 0.03-0.15.

Protostellar D/H can be inferred from solar evolution models and a knowledge of the He^3/He^4 ratio in the present sun. Measurements of this value from the solar wind are reported by Geiss and Boschler (36), and further refined by Anders and Grevesse (37) to give D/H = $3.4 \pm 1.0 \times 10^{-5}$ for the protosun. This value is systematically higher than the mean interstellar medium values, as would be expected from deuterium processing in the local ISM during the age of the solar system.

Water, methane and ammonia ices are thought to be the most abundant hydrogen containing solids available for making planets. Unfortunately, we have little information on the D/H ratio for these ices in molecular clouds. Recent measurements by Knacke et al. (39) suggest $[HDO]/[H_2O] \geq 0.8$-3.0×10^{-3} for the gas phase, and $[H_2O (gas)]/[H_2O (ice)] \leq 0.06$. The latter number is uncertain by a factor of 5. Combining the two results we may conclude that the D/H in interstellar ice is at least 0.4-1.5×10^{-3}. For gas-phase ammonia, we have $NH_2D/NH_3 \sim 3 \times 10^{-3}$ - 0.14 (40,41), and a more recent value of $6.2^{+11.1}_{-3.6} \times 10^{-2}$ by Turner (42). There is no information on NH_3 ice. No deuterium fractionation information is available on methane.

Mechanisms of Fractionation

We provide a brief survey of the important physical and chemical mechanisms that can lead to local deuterium/hydrogen fractionation. This is by no means a complete list. There may be hitherto unknown reactions that are or were important in the solar system. Of all fractionation mechanisms to be discussed below, it is generally agreed that mechanism (viii), ion chemistry, is most important for molecular clouds, and that mechanism (ii), Jeans escape, is most important for planetary atmospheres. (For a more detailed description of fractionation mechanisms and their effect on other elements in planetary atmospheres see, e.g., Kaye (*43*)).

(i) Phase Change. When material undergoes a phase change there is usually a slight difference in the vapor pressures of the normal and the deuterated species, leading to isotopic fractionation. For example, when water vapor condenses into ice the fractionation factor

$$\alpha = \frac{(D/H)_{ice}}{(D/H)_{vapor}}$$

is 1.13 at 273 K and can be as large as 1.23 at 233 K (*44*).

(ii) Jeans Escape. For molecules of molecular mass m in a planetary atmosphere to escape into space, the flux (molecules cm^{-2} s^{-1}) due to thermal evaporation is given by the Jeans formula (*45*),

$$\phi = n_c <v>$$

$$<v> = \frac{u}{2\sqrt{\pi}}(1+\lambda)e^{-\lambda}$$

where n_c is the number density of the escaping molecules at the critical level (at radial distance r_c from the center of the planet), $<v>$ is the effusion velocity, and $u = (2kT_c/m)^{1/2}$ is the most probable velocity of a Maxwellian distribution of the molecules at T_c, the temperature at the critical level. The dimensionless parameter λ is given by

$$\lambda = \frac{GMm}{kT_c r_c}$$

where GM/r_c is the planetary gravitational potential. Note that λ measures the gravitational binding energy of a molecule relative to its thermal kinetic energy. Since λ is proportional to m, the difference in $<v>$ between H and D is usually very large whenever λ is large. This can lead to profound changes in D/H in planetary evolution. We should point out, however, that this is not the only escape mechanism that can result in a fractionation for D/H. Escape via charge exchange may also discriminate between D and H (*46*).

(iii) Diffusive Separation. The bulk of the atmosphere is usually well mixed due to dynamical wave activity. However, above a certain level named the homopause, diffusive separation between light and heavy species becomes important. The pressure of the homopause, P_h, in a planetary atmosphere is determined from observations of chemical speciation, and P_h is of the order of magnitude of 10^{-6} bars. The functional behavior of the number density of a nonreacting species i above the homopause is approximately described by

$$n(z) = n_h \exp\left[-\int_h^z \frac{dz}{H_i(z)}\right]$$

$$H_i(z) = \frac{kT(z)}{m_i g(z)}$$

where n_h is the number density at the homopause altitude h, and H_i is the "scale height" for a species with molecular mass m_i. T and g refer to ambient temperature and gravity, respectively. Note that m_i exerts its influence via an exponential function, and can greatly enhance the abundance of a lighter isotope relative to the heavier species (47).

(iv) Rayleigh Distillation. This is not a new mechanism for isotopic fractionation. Nevertheless, it is an extremely important process by which cumulative fractionation can be achieved. Consider the example of water condensation described in (i). As water vapor is depleted from the atmosphere by ice formation the remaining water vapor is systematically depleted in deuterium. Subsequent condensation will result in even more depletion of deuterium. The isotopic ratio (R) of the vapor is given by

$$R = R_o f^{\alpha-1}$$

where R_o is the initial D/H value in the vapor, f is the fraction of vapor remaining, and α is the fractionation factor defined in part (i) of this section. A similar formula may be derived for computing the cumulative enrichment in D/H caused by Jeans escape.

(v) Thermochemical Equilibrium. The exchange of deuterium between different reservoirs is governed by equilibrium reactions such as

$$CH_4 + HD \rightleftarrows CH_3D + H_2$$
$$H_2O + HD \rightleftarrows HDO + H_2$$
$$NH_3 + HD \rightleftarrows NH_2D + H_2$$

At low temperatures (T < 500 K) the equilibrium favors the right hand side of the above reactions and deuterium tends to become more concentrated in CH_3D, HDO, and NH_2D with respect to HD (48). However, the time constants for these reactions to reach equilibrium may be excessively long (compared with the age of the solar system), especially at lower temperatures where we expect the largest fractionation. One possible way to speed up these reactions is by surface catalysis on dust grains, the subject of the next paragraph.

(vi) Surface Catalysis. Chemical reactions which proceed very slowly in the gas phase have been observed to proceed with greater rates on catalytic surfaces. Perhaps the best known of such reactions is the industrial Fischer-Tropsch synthesis, which produces gasoline from CO and H_2. The best surfaces are associated with metallic nickel and magnetite. Isotopic exchange is expected to take place on such surfaces. For example, when HDO and H_2 are both adsorbed on a metallic surface, we can have the following simple reaction scheme (49) for the adsorbed species X (ad),

$$H_2O(ad) \rightleftarrows H_2(ad) + O(ad)$$
$$HD(ad) + O(ad) \rightleftarrows HDO(ad)$$

The net result is equivalent to
$$H_2 + HD \rightleftarrows HDO + H_2$$

(vii) Photochemistry. Exchange between D and H can be greatly facilitated if a stable molecule is first dissociated into radicals and then reconstituted. For example, absorption of ultraviolet radiation at wavelengths < 1975 Å can lead to photolysis of H_2O

$$H_2O + h\nu \rightarrow OH + H$$

The radical OH can now attack HD

$$OH + HD \rightarrow H_2O + D$$

followed by

$$D + OH \rightarrow OD + H$$

and

$$OD + H_2 \rightarrow HDO + H$$

The net result is equivalent to

$$H_2O + HD \rightarrow HDO + H_2$$

These reactions once initiated by photolysis can proceed rapidly even at 150 K, leading to large fractionation of HDO relative to HD (50).

(viii) Ion Chemistry. Ion-molecule reactions generally have little or no activation energy and can proceed rapidly at low temperatures. Reactions such as

$$H_3^+ + HD \rightarrow H_2D^+ + H_2$$
$$H_2D^+ + CO \rightarrow DCO^+ + H_2$$

can readily produce species enriched in deuterium. In fact, most reactions of the form

$$XH^+ + HD \rightleftarrows XD^+ + H_2 + \Delta E$$

where X is a neutral molecule, favor the right hand side (51). The reason is that the deuterated molecule XD^+ is usually more tightly bound due to a reduced zero point energy, and the reaction proceeding from the left to the right is slightly exothermic. At low temperatures (T < < 100 K), $\Delta E/kT$ becomes large, and deuterated species are produced in greater abundance.

Origin, Fractionation and Dilution

The pattern of D/H ratios in the solar system and molecular clouds as shown in Figure 1 reveals four general features: (a) the bulk reservoirs of hydrogen in the interstellar medium, the protosun, Jupiter, and Saturn have similar D/H ratio $\sim 10^{-5}$, (b) the "minor constituents" in the molecular clouds have greatly enhanced D/H ratios $\sim 10^{-3}$ or higher, (c) there is a gradual increase in the D/H ratio from the large giant planets (Jupiter and Saturn) to minor giant planets (Uranus and Neptune) and finally to small bodies (Titan, Halley's comet) in the outer solar system, and (d) the terrestrial planets (Mars and Venus) have D/H ratios significantly higher than that of the Earth. We will attempt to interpret these patterns in terms of the simple concepts of origin, fractionation, and dilution.

A schematic diagram showing the principal pathways by which our solar system is formed is shown in Figure 2. The subject matter associated with this figure is quite extensive. The interested reader is referred to an excellent conference proceedings (52). Here we will restrict our attention to deuterium. Deuterium preserves the memory of the origin of the universe in the Big Bang. Subsequent to formation, deuterium went through stars and interstellar molecular clouds prior to the formation of the solar system. The D/H enhancements observed in the outer solar system may be traced back to the molecular clouds. On the other hand, the high D/H ratios in the terrestrial planets are more likely to be the result of chemical evolution. Intermediate values of the D/H ratio may be explained by dilution between different reservoirs of similar materials.

The primary reservoir of deuterium in the molecular cloud and the protosolar nebula is HD, with a D/H ratio $\sim 10^{-5}$. However, there is a smaller reservoir of deuterium in the form of ices such as HDO, CH_3D, NH_2D, and organics in molecular clouds with a much higher D/H ratio (at least 10^{-3}). The origin of the deuterium enhancement due to ion molecular reactions is now generally accepted (51, 53, 54). The precise size of the reservoir and the D/H ratio of the icy material in the protosolar nebula are unknown. However, from the D/H ratios of the small bodies which may not have undergone appreciable chemical evolution (such as comets, meteorites, and Titan) we infer an ice D/H ratio in the solar nebula of $\sim 10^{-4}$, similar to that of SMOW and the mean of meteorites. This value is significantly lower than that of the molecular clouds. There are at least four explanations for this difference.

First, the ice in the solar nebula might have equilibrated its deuterium by exchange with the larger hydrogen reservoir. Hubbard and MacFarlane (48) computed the D/H fractionation for H_2O, CH_4, and NH_3 in the solar nebula. As shown in Figure 3, the fractionation factors are largest at the lower temperatures (T < 300 K). As pointed out by Grinspoon and Lewis (49), the kinetic rate of exchange for

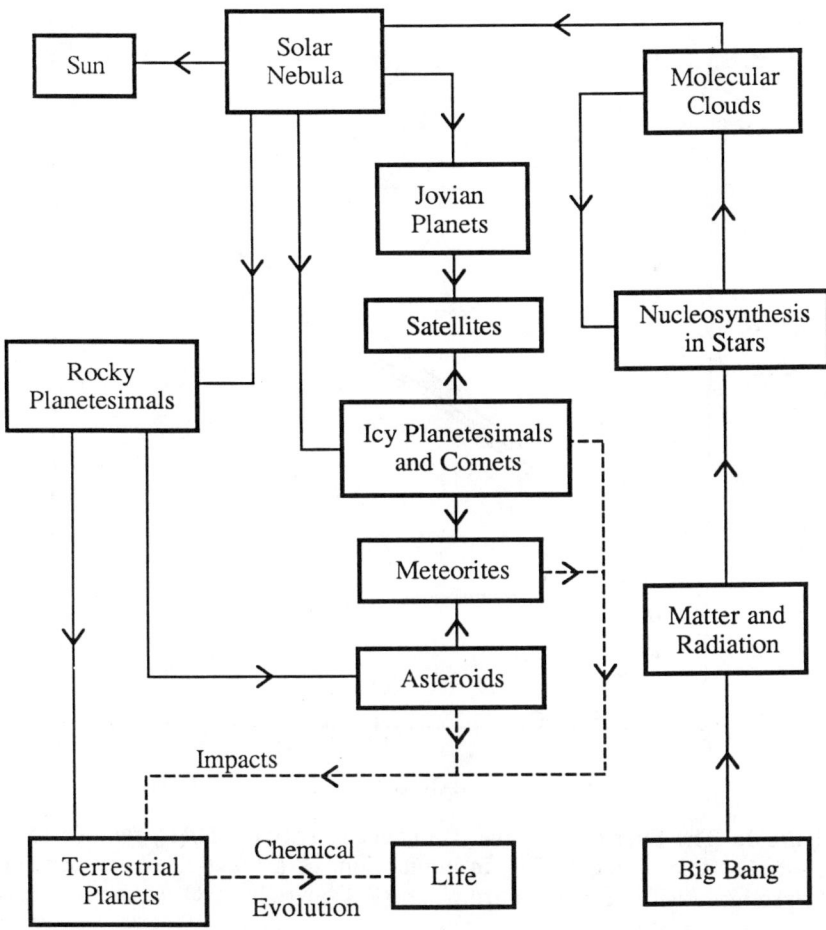

Figure 2. Schematic diagram showing the principal pathways of the origin and evolution of the solar system.

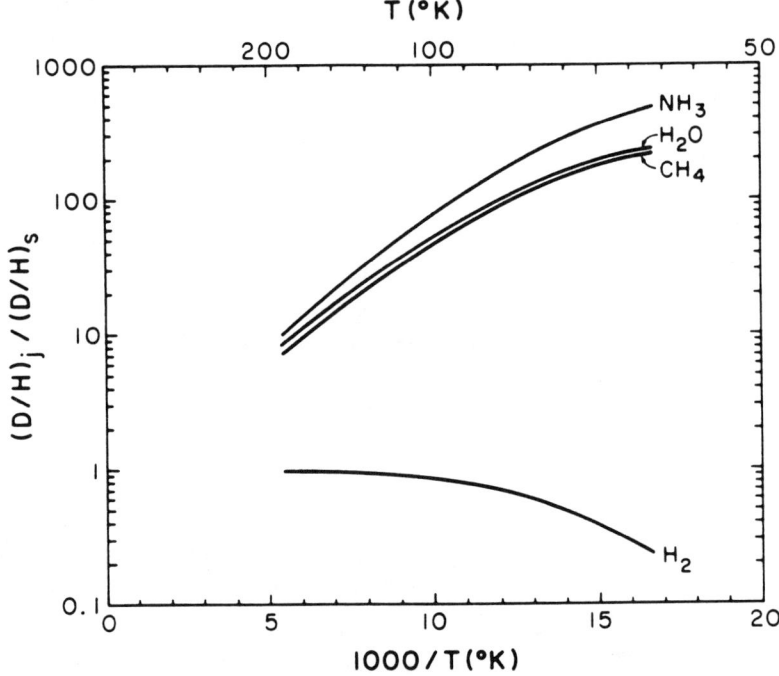

Figure 3. The D/H ratio in the jth molecular species $(D/H)_j$ relative to the primordial solar value $(D/H)_s$ in equilibrium as a function of temperature. (Reproduced with permission from ref. 48. Copyright 1980 Academic Press.)

deuterium between stable molecules in the solar nebula is too long, even in the presence of catalytic grain surfaces. The efficiency of exchange might be increased by photochemical processes (50), but the ultraviolet flux in the nebula is uncertain (55). In retrospect, the conclusions of Grinspoon and Lewis (49) and Prinn and Fegley (55) may have been unduly pessimistic. The recent discovery of interstellar doubly deuterated formaldehyde (D_2CO) by Turner (42) brilliantly confirmed Tielens' model (56) of surface catalysis on grains at low temperature. Also disequilibrium chemical processes in the solar nebula may have been initiated by shocks (57), especially by the generation of H atoms. These atoms could also be transported from the inner solar nebula (close to the sun) to other parts of the nebula (Yelle, R., University of Arizona, personal communication, 1991). H atoms, being reactive radicals, might be able to drive kinetic exchange reactions proposed in Yung et al.'s photochemical model (50) of the solar nebula.

Second, it is possible is that large amounts of ices were chemically synthesized in situ in the solar nebula by Fischer-Tropsch processes (58). A likely source of H_2O would be reduction of CO by H_2 on grain surfaces. Recent meteoritic evidence downplays the importance of Fischer-Tropsch type synthesis for the production of organic material (59,60) however, favoring instead interstellar origin (61), placing serious doubt on the importance of such catalysis to manufacture ices.

Third, it is possible to transport H_2O from the inner solar nebula to the outer solar system (62). In this model, the outer solar system at about the orbit of Jupiter would act as a "cold finger" for the large amounts of H_2O in the inner solar system. Extraction and condensation of H_2O may have increased the local density of ice by ~ 75 times. In the latter two cases, the ice would have a D/H ratio typical of HD ($\sim 10^{-5}$). Therefore, the mixing of the locally produced ice and pre-nebula ice may provide the explanation for the D/H ratio in the solar system ice observed today in the primitive bodies.

Finally, it is quite possible that the high fractionation levels observed in the interstellar medium are not indicative of the initial chemical state of the solar nebula. If ion-molecule reactions continue during the infall of the nebula, temperatures may get high enough to drive deuterium back into HD, pushing the fractionation of "minor" species such as H_2O into equilibrium with the bulk reservoir of hydrogen. An even simpler explanation is that the D/H in the local interstellar medium at the time of solar system formation was less than that of observed clouds with high degrees of fractionation, a consequence of the inhomogeneities in the interstellar medium mentioned previously.

Given that these two primordial reservoirs of deuterium (unfractionated HD and highly fractionated ices) exist (63), one can explain the observed D/H ratios of the outer solar system merely by mixing these two reservoirs in the right initial proportions for each body. Titan and comet Halley appear to represent the ice endmember, as all of their volatile envelopes are derived from vaporization and outgassing of this enriched ice. Jupiter and Saturn are representative of the gaseous endmember. Although the cores of these bodies are thought to consist in part of the enriched ice component, the atmospheres of these two planets are so large that the deuterium enhancement of the ice is overwhelmed, even if the ice in the core and the predominantly H_2 atmosphere are well mixed. Uranus and Neptune represent an intermediate case. In these planets, the core contributes at least two-thirds of the total mass of the body, as compared to the upper limit of about 25% for Jupiter and Saturn (64). Hubbard and MacFarlane (48) predicted that such a high core/atmosphere ratio would leave an enriched D/H signature in the atmospheres of Uranus and Neptune, due to mixing of the outgassed icy core component and the primordial hydrogen envelope. It would appear that the D/H ratios of the giant planets can be explained by this simple dilution effect, almost demanding two separate primordial deuterated reservoirs to explain the observations.

The terrestrial planets have probably retained none of the HD from the solar nebula. The primary source of deuterium is probably derived from impact of solar system ices (see Fig. 2). Hence, the primordial D/H ratio in these atmospheres may be close to the terrestrial SMOW value. The question arises as to the exact value of the "primordial" D/H on Earth. The δD value in deep-sea tholeiite glass containing primordial He and Ne was measured to be −77‰ (65), close to the value −110‰ inferred for phyllosilicates in meteorites (30).

Perhaps the most interesting question one can ask from the D/H ratios in the terrestrial planets concerns the size of the hydrogen reservoirs in these planets and the history of chemical evolution. Table I summarizes the known hydrogen contents of the terrestrial planets. For Mars the only observed reservoir is atmospheric H_2O, about 20 precipitable microns (highly variable). The quantity of subsurface water is unknown, but must significantly exceed the atmospheric values. For convenience we also express the H_2O reservoir as the equivalent thickness of water if it were uniformly deposited on the surface.

Table I. Hydrogen Reservoirs and Escape Rates in the Terrestrial Planets

	Earth	Venus	Mars
Total H_2O (g)	1.39×10^{24a}	$0.8\text{-}4.0 \times 10^{19}$	$>>2.9 \times 10^{15}$
Equivalent thickness of water layer (m)	2720^b	0.02-0.09	$>>2 \times 10^{-5}$
Escape rate of hydrogen (10^8 atoms cm^{-2} s^{-1})	3.0	0.27	1.7
Efficiency factor for D escape (E)	0.74	0.08	0.32
Total H_2O lost (m)	6.4^c	0.57	3.6
$(D/H)/(D/H)_o$	1.006	6.4-25	—

[a] The Martian H_2O value refers to that observed in the atmosphere. The subsurface reservoir must be much greater. For Venus, the lower value is based on 40 ppmv H_2O and the higher value is based on 200 ppmv H_2O.
[b] Globally averaged value.
[c] Assuming constant escape rate over the age of the solar system, 4.5×10^9 yrs.

The escape rates of hydrogen in the current epoch are also shown in Table I, along with the efficiency (E) of escape for deuterium relative to hydrogen (*6,46,66-73*). The factor E depends on the detailed mechanism of escape. In Figure 4 we illustrate how Jeans escape discriminates between D and H and would strongly influence the value of E. Note that at present in Mars the effusion velocity of H is about 100 times that of D. Other factors such as atmospheric diffusion (*73*), may contribute to determine the magnitude of E. The values of E quoted in Table I represent what is tentatively known for the current terrestrial atmospheres. If we assume that the escape rates were uniform over the age of the solar system, we can compute the total amount of H_2O that has been lost from the planet assuming H_2O is the major reservoir of D (it is easy to dispose of the oxygen by oxidation of crustal material). The results are equivalent to 6.4, 0.57, and 3.6 meters of H_2O for Earth, Venus, and Mars, respectively. Based on arguments given by McElroy and Yung (*47*), the fractionation of the heavy isotope relative to the lighter isotope is determined by the initial reservoirs of the respective species and the escape efficiency. If we further assume a constant (over time) relative efficiency factor (E) for deuterium escape relative to hydrogen escape, we arrive at the fractionation factors due to chemical evolution: $(D/H)/(D/H)_o$ = 1.006, and 6.4-25, respectively for Earth and Venus. The observed values of $(D/H)/(D/H)_o$ = 1.08 (*65*) and 120 (*9*) for Earth and Venus, respectively, are much higher than this, assuming that the primordial level of fractionation for both bodies is given by $\delta D = -77‰$. It is not possible to compute a fractionation factor for Mars with any confidence, as the available H_2O reservoir is unknown. However, if we use the observed degree of fractionation on Mars, and assume a primordial $\delta D = -77‰$ as on Earth, then the present escape rate of hydrogen implies that there is only 0.3 meters of total exchangeable water in the crust of Mars today (*72*). This conclusion is challenged by the geological community, who prefer a much higher crustal H_2O reservoir (*74,75*).

In order to account for the δD measurements of primordial water on Earth, as much as 300-1200 m of H_2O may have been lost (*46*), implying that the escape rate of hydrogen was much greater in the past, about 50-200 times the present rate. For Venus the average hydrogen escape rate must have been at least an order of magnitude higher than the present (*76,77*). In addition, an easy resolution to the conflict over the size of the H_2O reservoir on Mars is that the escape rate of hydrogen was much greater in the past.

What could be responsible for faster escape rates of hydrogen from the terrestrial planets in the past? We speculate on a number of possibilities. First, there is circumstantial astrophysical evidence that the sun's ultraviolet radiation was stronger in the past (*78*). Since photochemical processes in planetary atmospheres are primarily driven by solar ultraviolet photons, this would imply a greater rate of chemical evolution than that computed on the basis of a uniform sun (as in Table I). The other reasons for greater chemical evolution are specific to the individual terrestrial planets. On Earth the bulk of solar ultraviolet light is absorbed by O_2 and O_3, which are ultimately biogenic. In the absence of an O_2 and O_3 shield in the primitive Earth's atmosphere, short wavelength photons can penetrate to the troposphere, where they will be absorbed mainly by CO_2 and H_2O. The

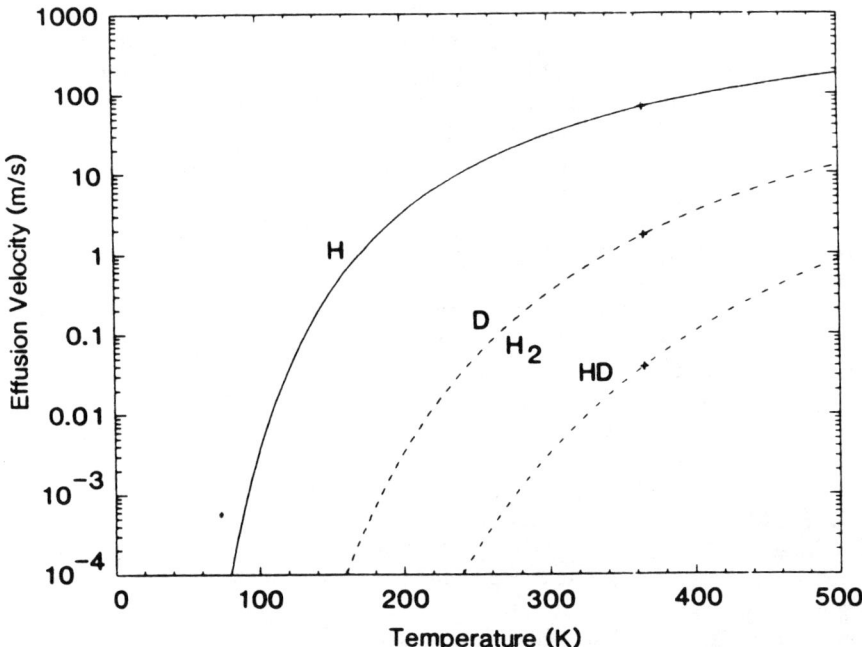

Figure 4. The Jeans effusion velocities (in m s^{-1}) for Mars. The values for the current atmosphere are marked by crosses. (Reproduced with permission from ref. 72. Copyright 1988 Academic Press.)

photochemistry of this hypothetical atmosphere has been investigated by Kasting et al. (79) and Wen et al. (80). The tentative conclusion of both investigations is that the escape rate of hydrogen in this case could be 100 times that of the present. On Venus Yatteau (81) has proposed an ingenious argument for greater hydrogen escape rates in the past. In the present atmosphere the mixing ratio of H_2O is about 40-200 ppmv below the cloud tops, but drops to about 1 ppmv above the clouds due to drying by the H_2SO_4 cloud. The abundance of sulfur on Venus is independent of H_2O, and the drying mechanism works as long as $[H_2SO_4] > [H_2O]$. While the above condition holds today, Venus was wetter in the past and once we have $[H_2O] > [H_2SO_4]$, the dehydration by H_2SO_4 clouds will fail. According to Yatteau, if Venus were just a little (factor of ~3) wetter than today, the hydrogen escape flux would have been higher by a factor of ~100. On Mars today most of its water is frozen as ice at the poles or in the regolith. However, if the climate were warmer in the past, there would have been more atmospheric H_2O. Consequently there would be greater dissociation of H_2O, followed by greater escape of hydrogen. We must emphasize that all the above possibilities are highly speculative and no definitive work has been carried out to study the comparative chemical evolution of the three terrestrial planets.

Conclusions

A major achievement of the last two decades of extensive study of the solar system is the set of D/H measurements summarized in Figure 1. For the first time we have a glimpse of the global pattern of D/H ratios from the sun to the outer solar system, as well as the interstellar molecular clouds from which the solar system was originally derived. A summary of our current understanding of the origin and fractionation of deuterium is presented in Figure 5. The precursor of the solar nebula, the molecular cloud, is the source of both unfractionated and highly fractionated deuterated material. The primary reservoir was gaseous, with HD as the largest deuterated component, and the smaller, secondary reservoir was in the form of icy solids, with monodeuterated water, methane, and ammonia as the deuterated species, and with an overall D/H at least an order of magnitude higher than the gaseous component. The observed pattern in the D/H ratio from the small bodies to the giant planets reflects varying degrees of dilution of deuterium in ices by HD from the primary reservoir.

The terrestrial planets probably did not retain any significant amounts of the primordial HD in the solar nebula. The bulk of deuterium in these planets was derived from smaller solar system bodies. The observed D/H fractionation in the terrestrial planets with respect to the solar system small bodies and ices is the result of atmospheric evolution. The escape rates of hydrogen from these planets may have been significantly higher than that today, implying that except for the Earth the bulk of hydrogen in the terrestrial planets may have been lost to space.

We conclude this review by asking a few fundamental questions related to deuterium:

(a) Why are the D/H values of meteoritic water, comets, Titan and SMOW so similar?

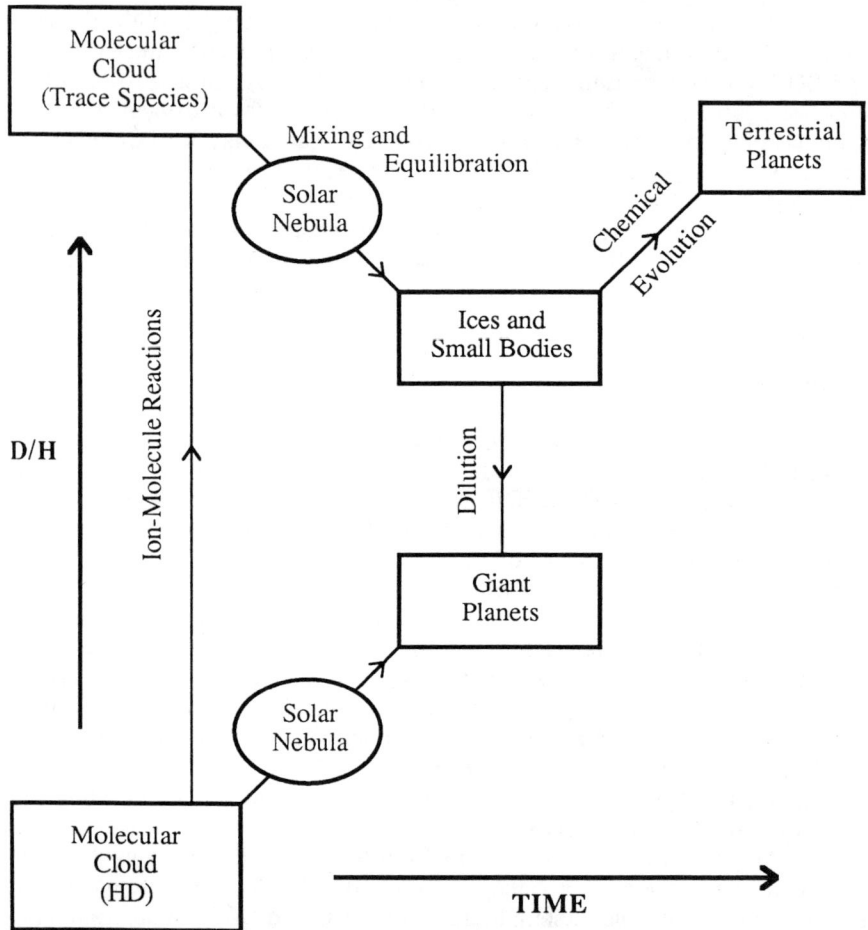

Figure 5. Simplified schematic showing the origin of deuterium and important fractionation processes. The values of D/H in the rectangular boxes near the top are higher than those below. Time evolution is from the left to the right.

(b) How did ices derived from molecular clouds equilibrate with the bulk hydrogen reservoir in the solar nebula?

(c) Was there mixing in the solar nebula? If so, to what extent was it responsible for resetting D/H values of primordial objects?

(d) What caused the greater rates of hydrogen escape from the terrestrial planets? Was the solar UV flux higher in the past?

(e) What were the initial reservoirs of hydrogen in the terrestrial planets?

Acknowledgments

We thank Mark Allen, Dave Stevenson and Geoff Blake for critically reading the manuscript, and Bill Langer for drawing our attention to recent developments in deuterium chemistry in the interstellar medium. The research is supported in part by NASA grants NAGW 1509 and NAGW 1538 to the California Institute of Technology. One of us (RWD) acknowledges support of a NASA graduate fellowship.

Literature Cited

1. Wagoner, R.V., Fowler, W.A., and Hoyle, F. *Astrophys. J.*, **1967**, 148, 3-49.
2. Schramm, D.N., and Wagoner, R.V. *Physics Today*, **1974**, 41-47.
3. Geiss, J., and Reeves, H. *Astron. Astrophys.*, **1981**, 93:189-199.
4. IUPAC Commission on Atomic Weights and Isotopic Abundances. *Pure Appl. Chem.*, **1983**, 55:1119-1136.
5. Hagemann, R., Nief, G., and Roth, E. *Tellus*, **1970**, 22:712-715.
6. McElroy, M.B., Prather, M.J., and Rodriguez, J. *Science*, **1982**, 215:1614-1615.
7. Kumar, S. and Taylor, H. A. *Icarus*, **1985**, 62:494-504.
8. Donahue, T.M., Hoffman, J.H., Hodges, R.R., Jr., and Watson, A.J. *Science*, **1982**, 216:630-633.
9. deBergh, C., et al. *Science*, **1991**, 251:547-549.
10. Owen, T., Maillard, J.P., deBergh, C., and Lutz, B.L. *Science*, **1988**, 240:1767-1770.
11. Bjoraker, G.L., Mumma, M.J., and Larson, H.P. *BAAS*, **1989**, 21(3):991.
12. Gautier, D., and Owen, T. *Nature*, **1983**, 302:215-218.
13. Fegley, B., Jr., and Prinn, R.G. *Astrophys. J.*, **1988**, 326:490-508.
14. Beer, R., and Taylor, F.W. *Astrophys. J.*, **1973**, 179:309-327.
15. deBergh, C., Lutz, B.L., Owen, T., and Maillard, J.P. *Astrophys. J.*, **1990**, 355:661-666.
16. Gautier, D., and Owen, T. In *Origin and Evolution of Planetary and Satellite Atmospheres*; Editors: S.K. Atreya, J.B. Pollack, and M.S. Matthews; University of Arizona Press: Tucson, Arizona, 1989, pp 487-512.
17. Bjoraker, G.L., Larson, H.P., and Kunde, V.G. *Icarus*, **1986**, 66:579-609.

18. Smith, W.H., Schempp, W.V., Simon, J., and Baines, K.H. *Astrophys. J.*, **1989**, 336:962-966.
19. Drossart, P., et al. *Icarus*, **1982**, 49:416-426.
20. Kunde, V. et al., *Astrophys. J.*, **1982**, 263:443-467.
21. Gautier, D., and Owen, T. *Nature*, **1983**, 304:691-694.
22. Noll, K.S. and Larson, H.P. *Icarus*, **1990**, 89:168-189.
23. Courtin, R., et al. *Astrophys. J.*, **1984**, 287:899-916.
24. deBergh, C., et al. *Astrophys. J.*, **1986**, 311:501-510.
25. Bezard, B., Gautier, D., and Marten, A. *Astron. Astrophys.* **1986**, 161:387-402.
26. deBergh, C., Lutz, B.L., Owen, T., and Chauville, J. *Astrophys J.* **1988**, 329:951-955.
27. Coustenis, A., Bezard, B., and Gautier, D. *Icarus*, **1989**, 82:67-80.
28. Eberhardt, P., et al. *Astron. Astrophys.*, **1987**, 187:435-437.
29. Zinner, E. In *Meterorites and the Early Solar System*, Editors, J.F. Kerridge and M.S. Matthews; University of Arizona Press: Tucson, Arizona, 1988; pp. 956-983.
30. Yang, J., and Epstein, S. *Geochim. Cosmochim. Acta*, **1983**, 47:2199-2216.
31. Kerridge, J. F. *Lunar Planet. Sci.*, **1985**, XVI:432-433.
32. McKeegan, K.D. et al. *Lunar Planet. Sci.*, **1987**, XVIII:627-628.
33. Vidal-Madjar, A., Ferlet, R., Gry, C., and Lallement, R. *Astron. Astrophys.*, **1986**, 155:407-412.
34. Murthy, J., et al. *Astrophys. J.*, **1987**, 315:675-686.
35. Borsgaard, A.M., and Steigman, G. *Ann. Rev. Astron. Astrophys.*, **1985**, 23:319-378.
36. Geiss, J., and Bochsler, P. In *Solar Wind IV*, Editor, H. Rosenbauer; Max-Plank-Institut für Aeronomie: Lindau, 1981; pp 403-413.
37. Anders, E., and Grevesse, N. *Geochim. Cosmochim. Acta*, **1989**, 53:197-214.
38. Irvine, W.M., and Knacke, R.F. In *Origin and Evolution of Planetary and Satellite Atmospheres*; Editors, S.K. Atreya, J.B. Pollack, and M.S. Matthews; University of Arizona Press: Tucson, Arizona, 1989; pp 3-34.
39. Knacke, R.F. Larson, H.P. and Noll, K.S. *Astrophys. J.*, **1988**, 335: L27-L30.
40. Olberg, M., et al. *Astron. Astrophys.*, **1985**, 142:L1-L4.
41. Walmsley, C.M., et al. *Astron. Astrophys.*, **1987**, 172:311-315.
42. Turner, B.E. *Astrophys. J.*, **1990**, 362:L29-L33.
43. Kaye, J. A. *Rev. Geophys.*, **1987**, 25(8):1609-1658.
44. Merlivat, L., and Nief, G. *Tellus*, **1967**, 19:122-127.
45. Chamberlain, J.W., and Hunten, D.M. *Theory of Planetary Atmospheres*, **1987**, Academic Press, Orlando; p. 332.
46. Yung, Y. L. et al. *J. Geophys. Res.*, **1989**, 94:14971-14989.
47. McElroy, M.B., and Yung, Y.L. *Planet. Space Sci.*, **1976**, 24:1107-1113.
48. Hubbard, W.B., and MacFarlane, J.J. *Icarus*, **1980**, 44:676-682.
49. Grinspoon, D.H., and Lewis, J.S. *Icarus*, **1987**, 72:430-436.
50. Yung, Y.L., et al. *Icarus*, **1988**, 74:121-132.
51. Watson, W.D., *Rev. Mod. Phys*, **1976**, 48:513-522.

52. Black, D.C., and Matthews, M.S. *Protostars and Planets II*, University of Arizona Press: Tucson, Arizona, 1985; pp 1293.
53. Dalgarno, A., and Lepp, S. *Astrophys. J.*, **1984**, 287:L47-L50.
54. Millar, T.J., Bennett, A., and Herbst, E. *Astrophys. J.*, **1989**, 340:906-920.
55. Prinn, R.G., and Fegley, B., Jr. In *Origin and Evolution of Planetary and Satellite Atmospheres*, Editors, S.K. Atreya, J.B. Pollack and M.S. Matthews; University of Arizona Press: Tucson, Arizona, **1989**; pp 78-136.
56. Tielens, A.G.G.M. *Astron. Astrophys.*, **1983**, 119:177-184.
57. Hollenbach, D., and McKee, C.F. *Astrophys. J.*, **1989**, 342:306-336.
58. Anders, E., Hayatsu, R., and Studier, M.H. *Astrophys. J.*, **1974**, 192: L101-L105.
59. Yuen, G. U., et al. *Lunar Planet. Sci.*, **1990**, XXI:1367-1368.
60. Cronin, J. R., and Pizzarello, S. *Geochim. Cosmochim. Acta*, **1990**, 54:2859-2868.
61. Kerridge, J. F. *Origins Life Evol. Biosph.*, **1991**, 21:19-29.
62. Stevenson, D.J., and Lunine, J.I. *Icarus*, **1988**, 75:146-155.
63. Owen, T., Lutz, B.L., and deBergh, C. *Nature*, **1986**, 320:244-246.
64. Pollack, J.B., and Bodenheimer, P. In *Origin and Evolution of Planetary and Satellite Atmospheres*; Editors, S.K. Atreya, J.B. Pollack, and M.S. Matthews; University of Arizona Press: Tucson, Arizona, 1989; pp 564-602.
65. Craig, H. and Lupton, J. E. *Earth Planet. Sci. Lett.*, **1976**, 31:369-385.
66. McElroy, M.B. *Science*, **1972**, 175:443-445.
67. Hunten, D.M., and Strobel, D.F. *J. Atmos. Sci*, **1974**, 31:305-317.
68. Liu, S.C., and Donahue, T.M. *J. Atmos. Sci*, **1974**, 31:1118-1136.
69. Liu, S.C., and Donahue, T.M. *J. Atmos Sci.*, **1974**, 31:1466-1470.
70. Liu, S.C., and Donahue, T.M. *J. Atmos Sci.*, **1974**, 31:2238-2242.
71. Kumar, S. Hunten, D. M. and Pollack, J. B. *Icarus,* **1983**, 55:369-389.
72. Yung, Y.L., et al. *Icarus*, **1988**, 76:146-159.
73. Hunten, D. M. and Donahue, T. M. *Ann. Rev. Earth Planet. Sci.*, **1976**, 4:265-292.
74. Carr, M. H. *Icarus*, **1990**, 87:210-227.
75. Jakosky, B.M. *J. Geophys. Res.*, **1990**, 95:1475-1480.
76. Kasting, J. F., and Pollack, J. B. *Icarus*, **1983**, 53:479-508.
77. Zahnle, K. J. and Kasting, J. F. *Icarus*, **1986**, 68:462-480.
78. Canuto, V.M., Levine, J.S., Augustsson, T.R., and Imhoff, C.L. *Nature*, **1982**, 296:816-820.
79. Kasting, J.F., Pollack, J.B., and Crisp, D. *J. Atmos. Chem.*, **1984**, 1:403-428.
80. Wen, J.S., Pinto, J.P., and Yung, Y.L. *J. Geophys. Res.*, **1989**, 94:14957-14970.
81. Yatteau, J.H. Some issues related to evolution of planetary atmospheres, Ph.D. Thesis, Division of Applied Sciences, Harvard University **1983**.

RECEIVED January 6, 1992

Chapter 24

Kinetic Isotope Effects and Their Use in Studying Atmospheric Trace Species

Case Study, $CH_4 + OH$

Stanley C. Tyler

National Center for Atmospheric Research, 1850 Table Mesa Drive, Boulder, CO 80303

The principal sink for methane in the atmosphere is from its reaction with OH radicals. We have measured the carbon kinetic isotope effect for this reaction and have reported a value of 1.0054±0.0009 (2σ) for the ratio of the rate coefficients k_{12}/k_{13}. This value improves on earlier determinations made in the laboratory by ourselves and others. It is important because the $\delta^{13}CH_4$ measurements of background methane and its sources and sinks are used along with companion data of fluxes to help determine relative source strengths of methane. A mass-weighted average composition of all sources should equal the mean $\delta^{13}C$ of atmospheric methane corrected for any isotopic fractionation effects in methane sink reactions. A discussion of experimental techniques and difficulties is presented here. A mass-weighted hydrogen isotope balance similar to that described for carbon could also be very useful in constraining the methane budget. The utility of carbon and hydrogen kinetic isotope effect data for methane studies is illustrated.

Methane is the most abundant hydrocarbon in the atmosphere. Because of its importance in tropospheric and stratospheric chemistry (*1-5*) and its role as a greenhouse gas (*6-7*), measurements of its concentration and studies of its budget (i.e. production and loss processes which account for its concentration in the atmosphere) have been made for a number of years (*8-9*). The principal sink for methane is reaction with OH radicals as in reaction 1.

$$CH_4 + OH \Rightarrow CH_3 + H_2O \qquad (1)$$

NOTE: The National Center for Atmospheric Research is sponsored by the National Science Foundation.

Table I shows a summary of budget data for atmospheric methane. One feature that stands out in the table is the fact that methane in the atmosphere is not at steady state. Methane concentration in the atmosphere has increased at a rate of about 1% per year since at least 1978 (*10-12*), although the rate of increase appears to be slowing down (Blake, D., Univ. of Calif. at Irvine, personal communication, 1991). Significant increase began much earlier as ice core data show that it has more than doubled overall in the last few hundred years after being relatively stable around a value of about 0.65 ppmv during the time since the last glaciation (*13-14*). Explanations for the increase may include increases in some methane sources, decreases in methane sink processes such as reaction with OH radicals, or combinations of the two types of possibilities (*9, 15*).

Table I. Summary of Data for Atmospheric Methane: Concentrations, Residence Time, and Budgets

Average Tropospheric Concentration	1.72 ppmv
Total Atmospheric Burden	3.05×10^{14} mole or 4.87×10^{15} grams
Residence Time in Years	10-15 years
Steady State Source or Sink Assuming 12.5 Year Residence Time	390×10^{12} grams
Rate of Increase of Net Sources Over Sinks Assuming 1%/Year Increase in Concentration	48×10^{12} grams
Known Sinks Reaction with OH in Troposphere (about 85%) Reaction with Soil Microorganisms (~0 to 10 %) Escape to Stratosphere with Further Chemical Reaction with OH, O(^{1}D), and Cl (about 15%)	~332×10^{12} grams 2 to 58×10^{12} grams ~59×10^{12} grams

In 1982, Stevens and Rust (*16*) proposed that by measuring the stable isotopes of carbon in atmospheric methane and its sources and sinks, the global methane budget could be determined better. They reasoned that a mass-weighted carbon isotope balance of all the sources of methane must equate to the globally averaged isotopic value of atmospheric methane. If this calculation could be made based on available measurement data then the relative source strengths of methane sources contributing to the atmosphere could be better known. Rust and Stevens correctly pointed out that the isotopic balance must be corrected for isotope fractionation in any sink processes. They had made a measurement of the carbon kinetic isotope effect in reaction 1 in an earlier study and found it to be relatively unimportant (*17*). In their study, they compared the relative rates of reactions of OH with $^{12}CH_4$ and $^{13}CH_4$ as in reactions 2 and 3.

$$^{12}CH_4 + OH \Rightarrow {}^{12}CH_3 + H_2O \qquad (2)$$

$$^{13}CH_4 + OH \Rightarrow \ ^{13}CH_3 + H_2O \qquad (3)$$

Table II shows how the above concept works using a simplified example. In it a reasonable $\delta^{13}CH_4$ value for each of several broad categories of methane sources (18) has been assigned with companion data on the relative source strengths for each category. δ is the delta notation, given in per mil (‰), which is defined from equation 1 (19). The globally averaged atmospheric methane $\delta^{13}CH_4$ value is then the weighted sum of the sources corrected by a shift in $\delta^{13}CH_4$ from the sink processes. The principal sink, reaction 1 above, enriches atmospheric methane in ^{13}C by its kinetic isotope effect and so shifts the atmospheric value somewhat toward less negative values of $\delta^{13}CH_4$ than its weighted sources would indicate.

$$\delta^{13}C = [^{13}C/^{12}C|_{sample} / \ ^{13}C/^{12}C|_{standard} - 1] \times 1000 \qquad (E1)$$

Table II. Simplified Example of Weighted Averaging of Methane Sources and Sinks Using $^{13}C/^{12}C$ Ratios

Source or Sink	Fraction of Total CH_4	$\delta^{13}CH_4$	Contribution in ‰
Sources			
Ruminants	.20	-57	-11
Paddy Fields	.15	-60	-9
Natural Wetlands	.25	-68	-17
Fossil Fuels	.20	-38	-8
Biomass Burning	.10	-25	-3
Solid Waste	.10	-52	-5
$\delta^{13}CH_4$ for weighted sum of source fluxes			-53
Sinks			
Reaction with OH in the Atmosphere			+5
Reaction with Oxidizing Bacteria in Soil			+1
$\delta^{13}CH_4$ for weighted sum of the sinks			+6
Globally Averaged Atmospheric Methane $\delta^{13}CH_4$			
Inferred from Sources and Sinks			-53 + 6 = -47
From Measurements			-47

Presently, methane budget calculations which use isotopic data are much more complex. A precise value for the kinetic isotope effect in reaction 1 is critical in any evaluation of the atmospheric methane budget that uses isotopic data. For this reason, the value has been re-determined since the Rust and Stevens study to increase the precision and accuracy of the measurement. The most recently determined value was measured by Cantrell and co-workers (20) who reported a value of 1.0054±0.0009 (2σ) for the ratio k_{12}/k_{13} where k_j is the ratio of the rate coefficients from reaction 1. This value is an improvement on the original Rust and Stevens study as well as an earlier determination made by Davidson and co-workers (21). A discussion of experimental techniques and difficulties in making this measurement is presented in the following pages. The hydrogen/deuterium kinetic isotope effect in the CH_4 + OH reaction is outlined as are methods for its use in methane budget studies. Additional discussion shows how isotopic data are used to study methane and what kind of results can be had from interpreting this data.

Chemical System

Kinetic isotope effects in the reaction of CH_4 with OH are of two kinds. The primary kinetic isotope effect is illustrated by contrasting reactions 4 and 5.

$$CH_4 + OH \Rightarrow CH_3 + H_2O \qquad (4)$$

$$CH_3D + OH \Rightarrow CH_3 + HDO \qquad (5)$$

Here the success of OH abstraction of methane hydrogen is directly affected by the difference between hydrogen-carbon and deuterium-carbon bonds in the methane. This leads to a relatively large isotopic effect because the transition state vibrational mode is directly affected by which types of atoms are transferred in the bond breaking and bond formation.

The system which has undergone the most study and the one that will be discussed in detail is the secondary isotope effect illustrated by contrasting reactions 2 and 3. In this case, the difference between the two types of carbon atom affects the transition state of the reaction only indirectly (i.e., the isotopic substitution is at a bond not being broken in the reaction). A smaller isotope effect is expected from this kind of reaction kinetics.

In considering the secondary effect the time rate of change for each methane species can be written as equations 2 and 3.

$$d[^{12}CH_4]/dt = -k_{12}[OH][^{12}CH_4] \qquad (E2)$$

$$d[^{13}CH_4]/dt = -k_{13}[OH][^{13}CH_4] \qquad (E3)$$

By manipulating these expressions and using the definition of δ, a formula is derived which can be tested experimentally (21). The formula reproduced in equation 4 where A is the fraction of CH_4 remaining after reaction.

$$k_{12}/k_{13} =$$

$$\ln(A) / [\ln(A) + \ln\{(\delta_t + 1000)/(\delta_o + 1000)\}] \qquad (E4)$$

The utility of the formula for experimental purposes becomes apparent when one realizes that only precise measurements of $\delta^{13}CH_4$ at times $t = 0$ and $t = t'$ as well as a measurement of the fraction of CH_4 reacted are needed to get an accurate value for k_{12}/k_{13}. The ratio k_{12}/k_{13} is calculated directly without actually measuring the absolute rate of the CH_4 + OH reaction. By calculating k_{12}/k_{13} from easily measurable parameters, difficulties which arise in trying to measure OH loss directly are avoided (22).

Since the secondary isotope effect is relatively small, it is important to design an experiment where reaction 1 can be studied without interfering reactions. The measurement of $k_{12}/k_{13} = 1.0054 \pm 0.0009$ determined by Cantrell and co-workers as well as the two previous determinations were measured in reaction cells in a laboratory environment. Studies of this type require a suitable OH source and allow for the opportunity to isolate the reactions of interest from potential interfering reactions. In the next section a comparison of the reaction systems studied by Rust and Stevens (17), Davidson and co-workers (21), and Cantrell and co-workers (20) is made. From this one can see that the evolution of the experimental design leads to a more precise determination of k_{12}/k_{13}.

Reaction System Design

Rust and Stevens chose for their OH source the photolysis of H_2O_2. Using a Hg arc lamp and Hg filter they expected to photolyze H_2O_2 with the 2537 A line from the lamp while filtering out another strong Hg line at 1842 A with a Hg vapor filter. The reaction generally took on the order of about 40 hours to convert about 10% of the methane to CH_3 and subsequent products. Using these experimental conditions, the value of k_{12}/k_{13} was determined to be 1.0028. However, a further analysis of this system by the Davidson group indicated that undesirable chemical reactions were taking place (21).

Typical concentrations in the cell used by Rust and Stevens were $[H_2O_2] = 4.9 \times 10^{16}$ molecules/cm^3 (vapor pressure above 90% H_2O_2), $[CH_4] = 1.6 \times 10^{16}$ molecules/cm^3, $[O_2] = 9.7 \times 10^{17}$ molecules/cm^3, and $[H_2O] = 1 \times 10^{17}$ molecules/cm^3 (v.p. above 90% H_2O_2). Hydroxyl radicals are not used very efficiently in this system because reaction with H_2O_2 is fast compared with CH_4. Therefore, Rust and Stevens were constrained to work at very low methane conversion fractions because of the inefficient use of OH radicals. Figure 1 shows that at very low fractional conversions, there is greater error in determining k_{12}/k_{13} from the isotopic measurements. In addition, CH_3 radicals may react with H_2O_2 to reform CH_4. This would mix the isotopic signal of the remaining methane from reaction 1 with an unseparable one from the methane made by $CH_3 + H_2O_2$.

Davidson and co-workers' suggest that another potential problem in the system was that by using a Hg lamp, in spite of using a Hg vapor filter, it is likely that O_2 was also photolyzed, resulting in the formation of ozone and ultimately $O(^1D)$. This comes about because pressure and temperature-broadened Hg radiation in the Schumann-Runge band range would still photolyze O_2. The low kinetic isotope effect they obtained might then be a combination of a low kinetic isotope effect in reaction 6 coupled with a larger effect in reaction 1, the reaction of interest. However, under closer scrutiny, it appears that only a small fraction of the methane loss observed by Rust and Stevens could have been due to reaction 6 between CH_4 and $O(^1D)$ because reaction 7 will dominate reaction 6 if the concentration of H_2O is high compared to CH_4 concentration (Tyndall, G., NCAR, Boulder, Colorado, personal communication, 1991).

$$CH_4 + O(^1D) \Rightarrow CH_3 + OH \qquad (6)$$

$$O(^1D) + H_2O \Rightarrow 2\,OH \qquad (7)$$

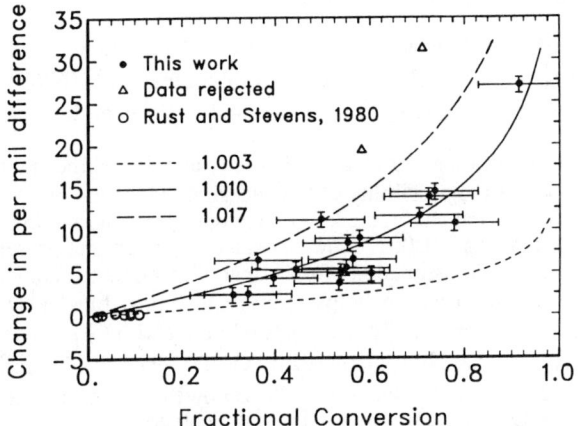

Figure 1. Change in the magnitude of final minus initial methane $\delta^{13}CH_4$ value versus fractional conversion of methane in the $CH_4 + OH$ reaction. "This work" refers to Davidson and co-workers' research. (Reproduced with permission from ref. 21. Copyright 1987 American Geophysical Union.)

Davidson and co-workers (21) analyzed the photochemistry and kinetics of the chemical system more fully than did Rust and Stevens. The OH source was from photolysis of ozone in the Hartley band to produce oxygen atoms (see reaction 8) which react with water vapor to produce OH (see reaction 7).

$$O_3 + h\nu \Rightarrow O_2 + O(^1D) \qquad (8)$$

Four key conditions for a successful experiment were applied. They are summarized as the following:

1. the ratio H_2O/CH_4 is kept relatively high (~1000) so that $O(^1D)$ reacts with H_2O and not with CH_4
2. N_2O (inert toward OH) has been added at the same $O(^1D)$ reactivity as CH_4 to look for N_2O loss by $O(^1D)$
3. O_3 is added continuously to maintain a steady level
4. sufficient O_2 is added to react with CH_3 radicals and eliminate them from further reaction in the system.

With the four conditions above maintained, typical cell concentrations are $[H_2O] = 5 \times 10^{17}$ molecules/cm^3, $[CH_4] = 5 \times 10^{14}$ molecules/cm^3, $[N_2O] = 2.5 \times 10^{14}$ molecules/cm^3, $[O_2] = 3 \times 10^{17}$, and $[O_3]_{ss} = 1 \times 10^{13}$ molecules/cm^3 where ss means steady state.

Because the rate coefficient for reaction 7 is only about 1.4 times that for reaction 6, the first condition keeps the possibility of the undesirable reaction of $CH_4 + O(^1D)$ to a minimum. In the Davidson study, the kinetic isotope effect in reaction 6 was found to be on the order of $k_{12}/k_{13} = 1.001$. Therefore, as suspected, the presence of reaction 6 would indeed cause underestimation of the effect of reaction 1 if not suppressed.

The second condition above helps monitor any possible $O(^1D)$ reactions by observing N_2O. N_2O reacts readily with $O(^1D)$ but not with OH. Loss of N_2O would indicate that CH_4 was also being depleted by $O(^1D)$. The third condition is needed because of the large rate of production of $O(^1D)$ by the Xenon arc lamp used in the experiment. To use O_3 efficiently it is fed into the cell continuously to maintain a low steady state concentration. Otherwise the OH made reacts with O_3 rather than CH_4. Also a constant supply of O_3 means a constant supply of OH throughout the reaction period. This decreases the total reaction time needed for reaction 1 to occur and allows a greater range of methane fractional conversions to be investigated in a reasonable period of time. The fourth condition prevents recombination of methyl radicals or reaction with RH impurities. If they recombine the C_2H_6 produced is difficult to separate from CH_4 and can bias the k_{12}/k_{13} determination. If they react with RH, methane is formed, and isotopic integrity of the methane is compromised.

In addition to running experiments under these conditions, Davidson's group also did some experiments in which 1) an amount of C_3H_8 equal to the CH_4 was added to the mixture, 2) $O(^3P)$ was purposely formed, and 3) kinetics were allowed to proceed without O_3 added to the mixture. Experiments using the first condition were the ones that allowed the Davidson group to determine that the kinetic isotope effect of $O(^1D)$ attack on methane was only $k_{12}/k_{13} = 1.001$. This is because propane reacts about 140 times faster than methane with OH but both compounds are attacked equally fast by $O(^1D)$. Therefore, the first condition makes the predominant loss of methane reaction with $O(^1D)$. Experiments using the second condition were accomplished by omitting H_2O from the cell and adding extra N_2 to quench $O(^1D)$ to $O(^3P)$. No methane was reacted in this test, leading to the conclusion that under normal test conditions, any $O(^3P)$ formed was not afffecting the methane. In the third test condition, photoysis of the mixture without O_3 was made to check for photosensitized or other unidentified photolytic CH_4 loss processes. No other loss processes were found.

Referring to Figure 1, one can see that the biggest problem in getting a precise value for k_{12}/k_{13} was in the errors of the methane concentration measurements. Although fractional conversions were determined by both gas chromatography and infrared absorption, the errors in measurement of starting and ending methane concentrations in the cell were relatively large. The Davidson study (21) determined that the value of $k_{12}/k_{13} = 1.010 \pm 0.007$ (2σ).

The Cantrell (20) study was a re-determination of the carbon kinetic isotope effect by the same group that published the Davidson et al. results. Several pitfalls have been avoided in the Cantrell group's experiments. Improvements in the experiments which make the data more precise in comparison to earlier experiments include changes in the methane concentration measurements and sample gas preparations and measurements for isotope ratio analysis. A Hewlett-Packard Model 5880 gas chromatograph fitted with a flame ionization detector (FID) was used for all concentration measurements. Two calibrated standard gas mixtures were used, 3.88 ppmv (NBS SRM 1660a) and 71.83 ppmv (Specialty Gas Products). These were measured against each other and indicated a good linearity of the FID for samples analyzed throughout the high and low ends of concentration range in the experiment.

Experiments were performed using a temperature controllable cell multiple reflection optics (23). The cell included a stainless steel bellows pump to circulate the mixture, a manifold for sample removal, and a vacuum system for gas handling and cell evacuation. The design of the cell also allowed the effect of temperature on the ratio k_{12}/k_{13} to be studied. No temperature dependence was found for the kinetic isotope effect in reaction 1 between 273 and 353°K. Figure 2a shows the results of the Cantrell study in which the value 1.0054±0.0009 (2σ) was obtained. Figure 2b shows the lack of a temperature dependence in the temperature range above.

Combustion of the sample gas mixture to convert CH_4 to CO_2 for stable isotopic analysis was done using a new high-volume, fast flow combustion train with a low blank (24). In principle the procedure follows that described by Tyler (25), a procedure adhered to in the Davidson study (21), but several key differences are noted. The sample inlet includes a flow integrator for better accuracy in determining recovery of converted samples. The flow is set for 1.0 liter/min. Four multi-loop traps in series strip out CO_2 from the sample gas stream cryogenically at liquid nitrogen temperature.

The oven catalyst has been changed to Platinized-Alumina (1% loading) at 800°C. At the time of development of the vacuum line, the catalyst was conditioned for several days by flushing it with very dry zero air at 3.0 liter/min. Although Pt-Al is porous and subject to high carbon blanks, the conditioning noted above with routine re-conditioning using a slower overnight flow of zero air on a regular basis keeps the carbon blank at < 1 μliter of CO_2 or below for 300 liters of clean zero air processed. With the low blank and flow integrator, a precision of expected conversion of CH_4 to CO_2 of 100±1% was obtained. This was true even for the very small sample quantities of CH_4 recovered as CO_2 in this experiment (sample size 20-50 μliters).

Another improvement in the isotope determinations over the previous studies resulted from the use of a Finnigan-Mat Delta-E isotope ratio mass spectrometer for all measurements of $\delta^{13}C$. Overall precision for individual values was < 0.10‰ using a cold finger inlet system to freeze the relatively small samples of CO_2 into the analyzer.

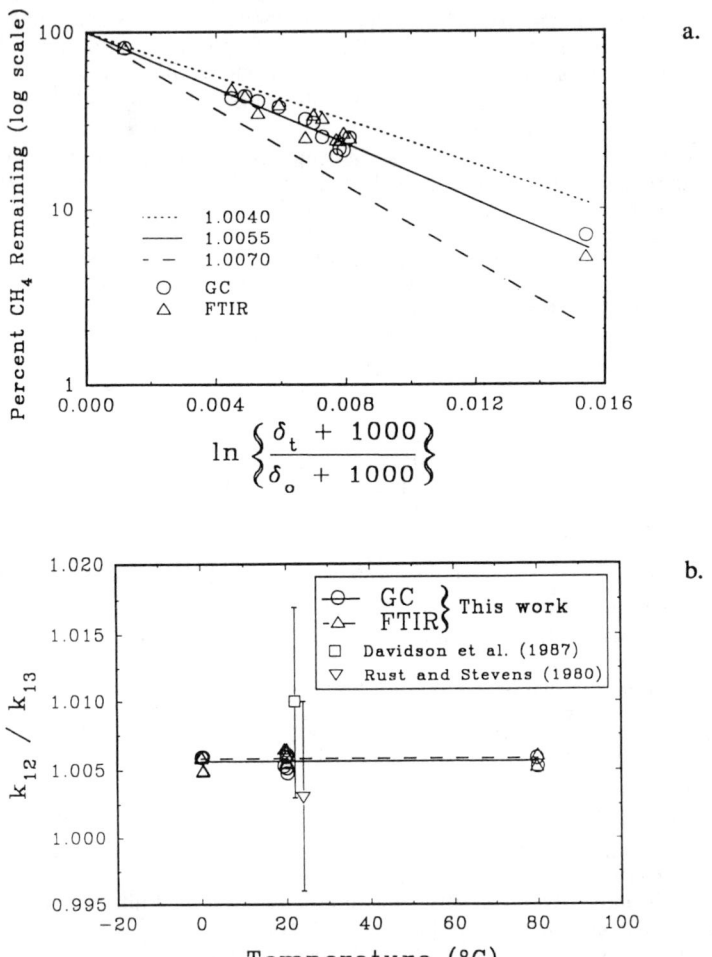

Figure 2. a) Alternative form of equation 3 which shows how secondary kinetic isotope effect value was determined in Cantrell and co-workers' research. Derivation of alternative form is described in their work. Best fit line for data points corresponds to a k_{12}/k_{13} of $1.0054 \pm 0.0009(2\sigma)$. Lines corresponding to k_{12}/k_{13} values of 1.0040 and 1.0070 are shown for comparison. b) Secondary kinetic isotope effect value (k_{12}/k_{13}) versus temperature. "This work" refers to Cantrell and co-workers' research. (Reproduced with permission from ref. 20. Copyright 1990 American Geophysical Union)

The working standard gas was Oztech-002 (Oztech Gas Co.) which is -30.011‰ with respect to PDB carbonate.

Other Determinations of k_{12}/k_{13} in the Reaction of CH_4 with OH

There are two other determinations of the carbon kinetic isotope effect in reaction 1. One is a calculated value for k_{12}/k_{13} from ab initio perturbation molecular orbital (PMO) theory by Lasaga and Gibbs (26). The other is a field measurement of k_{12}/k_{13} by Wahlen and co-workers (27).

As suggested by the name, ab initio calculations (ab initio means from the beginning) is not based on experimentally determined parameters. An approximate representation of the molecular orbitals in the transition state is made using wavefunctions which describe the energy of the electron orbitals used in the reaction.

Successive improvements in the wavefunctions by Lasaga and Gibbs (26) took into account electron correlation and tunneling. They found that there is a systematic trend in the reduction of the secondary kinetic isotope effect of reaction 1 as the calculations become more sophisticated. Their value for k_{12}/k_{13} at 275°K is 1.0072 with little temperature dependence in the range 225 to 350°K. The calculations indicate that an expected experimental kinetic isotope effect at all temperatures encountered in the atmosphere will be relatively small.

Another approach to measuring the carbon kinetic isotope effect in reaction 1 is that taken by Wahlen and co-workers (27). They reported a value for k_{12}/k_{13} from near-tropopause measurements of methane in the atmosphere. The reaction of OH on methane should result in an isotopic enrichment of ^{13}C in methane with altitude and decreasing CH_4 concentration. In Wahlen's study upper troposphere and lower stratosphere air samples were collected to obtain a vertical profile of methane concentration and $\delta^{13}C$ data. Their measurements were used to calculate a value that $k_{12}/k_{13} = 1.0143 \pm 0.0008$ at atmospheric temperatures of about 220 to 230°K. Wahlen and co-workers' then assumed a temperature dependence based on the rate constant for reaction 1 which leads to a $k_{12}/k_{13} = 1.0109$ at 297°K.

Possible problems in a field study of this kind include errors in estimating the vertical profile of methane, measuring CH_4 over a short time span when seasonality may be important, and in estimating the isotopic enrichment impact of interfering reactions such as Cl and $O(^1D)$ reacting with CH_4. Nevertheless, the field measurement study is an important cross check to the laboratory studies. It may point the way toward processes overlooked or misinterpreted in a laboratory systems study.

Discussion of Carbon Kinetic Isotope Effect and the Methane Budget

Stevens and Rust (16) made $\delta^{13}CH_4$ measurements of background methane in air and some important methane sources and compiled data from studies by others. They also designed an experiment to measure the carbon kinetic isotope effect in the principal sink process for atmopheric methane, that being reaction of CH_4 with OH in the atmosphere (17). Because the majority of measured methane sources were of biological origin, the weighted sum of the sources was very much depleted in ^{13}C with respect to atmospheric methane. When combined with the information that only a relatively small correction (about 3‰ from their k_{12}/k_{13} value) need be applied to the source data, the calculations indicated that one or more overlooked or greatly underestimated sources of relatively ^{13}C-enriched methane were needed to balance to budget.

Since then, many researchers have made $\delta^{13}CH_4$ measurements of CH_4 sources and methane in remote background air. Collectively, these include first-time measurements of sources such as burning biomass, termites, rice paddies, tropical swamps and marshes, and boreal wetlands (25, 28-33). Table III shows the ranges of $\delta^{13}CH_4$ for the methane sources where individual sources are grouped into several broad categories. It is still true that most sources of atmospheric methane are of biogenic origin and are relatively depleted in ^{13}C methane with respect to atmospheric methane (9, 18). While there does not seem to be much difference between k_{12}/k_{13} values between 1.003 and 1.010 when compared to the uncertainty in the weighted $\delta^{13}CH_4$ of all methane sources, there is a consensus among researchers that a kinetic isotope effect in the principal sink for atmospheric methane which is larger than 1.0028 makes sense based on the isotopic balance between methane sources and the atmospheric methane.

Table III. Methane Sources and Their Carbon Signatures

Source	$\delta^{13}C$ (IN ‰)	^{14}C (IN pMC)
Biogenic Recent	-50 to -90	~110 to 130
Biogenic Natural Gas	-50 to -110	~0
Thermogenic Natural Gas and Other Fossil Fuels	-25 to -50	~0
Biomass Burning	-22 to -34	~110 to 130

The unit pmC is per cent modern carbon in which comparisons are made against comtemporary carbon.

δ is the delta value, given in per mil (‰), which is defined from equation 1.

That a larger carbon kinetic isotope effect fits the available data better can be seen from relatively simple models. Stevens (34), Lowe and co-workers (35), Wahlen and co-workers (33), and Quay and co-workers (36) have all used isotopic data along with companion data of fluxes from methane sources and sinks to determine relative source strengths and to provide inputs into models of atmospheric methane mixing. These indicate that a k_{12}/k_{13} value larger than the original determination of 1.0028 fits the available source and background atmospheric methane data. In the studies above, the value $k_{12}/k_{13} = 1.010 \pm 0.007$ was used for calculations in all but Quay and co-workers' model.

Stevens's 2-box model (34) studied $\delta^{13}CH_4$ source flux trends in each hemisphere by fitting methane source distributions to northern and southern hemispheric average $\delta^{13}CH_4$ over the period 1978 to 1987. One conclusion he reached was that ^{13}C-enriched fluxes are increasing after 1983. This seems reasonable in light of later studies by Wahlen's group (33) and Lowe's group (35). Ehhalt had determined that the fossil fuel contribution to atmospheric methane was at most 20% (8). The two later studies by Wahlen's and Lowe's groups each used $^{14}CH_4$ measurements along with $^{13}CH_4$ measurements to demonstrate that 20 to 30% of the methane released annually may be from fossil fuels. They used a two-source function in which methane is either radiocarbon dead fossil fuel (i.e., no radiocarbon ^{14}C methane, enriched ^{13}C methane, see Tyler (18) for discussion of radiocarbon) or biologically produced (i.e., modern level of $^{14}CH_4$, depleted $^{13}CH_4$) to match the source flux with background atmospheric measurements.

Rather than use a measured k_{12}/k_{13} value to arrive at likely methane source strengths and distributions, Quay and co-workers (36) made calculations which tend to corroborate the use of the k_{12}/k_{13} value of 1.0054±0.0009 in reaction 1. They used a time series of atmospheric $\delta^{13}CH_4$ and CH_4 concentration measurements to arrive at an overall kinetic isotopic fractionation of CH_4 during oxidative loss. Recent studies by Steudler and co-workers (37) and Born and co-workers (38) indicate that soil oxidation of methane may have been underestimated in the past and may be as much as 10% of the total sink. To account for this, Quay's group calculated a value of k_{12}/k_{13} = 1.007±.002 assuming that 7% of the CH_4 oxidation is from microbial oxidation in soils. Although the fractionation effect of soil is based on few measurements thus far (39-40), it appears that their choice of the value k_{12}/k_{13} =1.020 for soils is representative. Under these conditions the contribution to the overall kinetic isotope effect from OH oxidation would be 1.006 from their study.

Impact of the Value of the Kinetic Isotope Effect on Modeling the Methane Budget

As a way to illustrate the impact of the carbon kinetic isotope effect on the isotopic methane budget, a two-box model of the atmosphere is described. It can be used to show the sensitivity of carbon isotopic data on atmospheric methane and its source fluxes to the kinetic isotope effect in the CH_4 plus OH sink reaction.

In a simple two-box model of the troposphere, continuity equations for the abundance of a trace species in each hemisphere can be written as equations 5 and 6.

$$dM_n / dt = S_n - M_n / \lambda_n - (M_n - M_s) / \lambda_T \quad (E5)$$

$$dM_s / dt = S_s - M_s / \lambda_s + (M_n - M_s) / \lambda_T \quad (E6)$$

M_n and M_s are total hemispheric abundances (concentrations) in the Northern and Southern Hemispheres, S_n and S_s are hemispheric release rates (source fluxes), λ_n and λ_s are trace species' removal times (lifetimes based on sink processes), and λ_T is the average interhemispheric mixing time (1.25 years in the following calculations).

One can treat $^{13}CH_4$ and $^{12}CH_4$ methane as different species and write equations for each species like equations 5 and 6 above. The individual source fluxes and concentrations of $^{13}CH_4$ and $^{12}CH_4$ are related to the total source fluxes and methane concentrations by the equation which defines the delta per mil value for carbon isotopes (see equation 1). The total source strength and carbon isotope value chosen as inputs for each hemisphere are thus broken down into ^{13}C and ^{12}C molecule abundances, each with unique lifetimes based on the ratio k_{12}/k_{13} applied to the currently accepted average lifetime of 12.5 years (22). To solve these equations numerically, the time step can be iterated to arrive at new $^{13}CH_4$ and $^{12}CH_4$ values in each hemisphere. Alternatively, these equations can be used to solve for a particular set of solutions at steady state by solving them algebraically.

Once a framework for a box is described as above, the 2-box model can be used in a manner in which the sources are derived from the concentrations. An example is described below, in which a steady state algebraic solution is found for the four equations described by writing the basic equations 5 and 6 for both $^{13}CH_4$ and $^{12}CH_4$. Integral expressions which solve the basic equations are evaluated for a particular set of source functions. In these calculations, CH_4 sources are assumed to be constant for present day values of methane concentration and $\delta^{13}CH_4$ value in each hemisphere. The starting hemispheric distribution of sinks (reaction with OH radicals only) is assumed to be constant and the kinetic isotope effect and sink strength is assumed to be constant and identical in both hemispheres. These conditions are not strictly correct, but for illustrative purposes, the situation is simplified by disregarding the calculations needed to take this into account. For example, the present day growth of approximately 1% per year is not factored in. However, in this example, the difference between constant sources and those that cause a growth of 1% per year in methane concentration is negligible for the analysis being presented here.

Table IV shows the sensitivity of the calculated north/south difference of $\delta^{13}CH_4$ in the source flux methane to the k_{12}/k_{13} value. The kinetic isotope effect value is varied for two sets of north/south distributions of concentration and $\delta^{13}CH_4$. In each case the values used for M_n and M_s concentrations of $^{13}CH_4$ and $^{12}CH_4$ are calculated from isotopic measurements by Quay and co-workers (36) and assumed starting concentrations for total methane in each hemisphere of 1.72 ppmv (northern) and 1.67 ppmv (southern). Quay and co-workers' measurements indicate a northern hemispheric average value of $\delta^{13}CH_4$ of -47.4‰ and a southern hemispheric value of -47.0‰. Three kinetic isotope effect values are considered.

The calculations in the simplified example show that quite different average isotopic source fluxes are needed in each hemisphere to account for the average hemispheric abundances and $\delta^{13}CH_4$ values depending on the k_{12}/k_{13} chosen. As the kinetic isotope effect becomes larger, source fluxes in each hemisphere must become more ^{13}C-depleted to arrive at the background hemispheric values. For the most plausible scenario, with a k_{12}/k_{13} of 1.0054, the weighted average source flux would would have a $\delta^{13}CH_4$ of -52.0‰ if 75% of the sources are in the northern hemisphere. One might conclude that although the northern hemisphere has more ^{13}C-enriched fossil fuel sources, it also must have many more ^{13}C-depleted sources such as livestock, boreal wetlands, and rice paddies in comparison to the southern hemisphere. Correspondingly, the southern hemisphere might have relatively more ^{13}C-enriched sources such as biomass burning from land clearing as a proportion of its total methane sources.

Primary Kinetic Isotope Effect in CH_4 + OH Reaction

A mass-weighted deuterium/hydrogen isotope balance similar to that described for carbon could also be very useful in constraining the methane budget. This is because $\delta(D/H)$ values can help characterize methane sources in two ways, 1) by differentiating methane sources isotopically where $\delta^{13}CH_4$ values of sources overlap, and 2) by differentiating between methane formation pathways of $CO_2 + H_2$ reduction and acetate fermentation (41). Although few measurements of $\delta(D/H)$ values exist for background atmospheric methane and its sources and sinks (41-45), the data are adequate to illustrate the usefulness of $\delta(D/H)$ data. The range of $\delta(D/H)$ for broad categories of the sources are listed in Table V.

Table IV. 2-Box Model Calculation of Changing Northern and Southern Hemispheric $\delta^{13}CH_4$ of Source Fluxes Versus Size of Carbon Kinetic Isotope Effect

	$\delta^{13}CH_4$ of Background Mixed Air	$\delta^{13}CH_4$ of Source Flux if R = 1.0028	$\delta^{13}CH_4$ of Source Flux if R = 1.0054	$\delta^{13}CH_4$ of Source Flux if R = 1.010
N. Hemis.	-47.4‰	-52.0‰	-53.3‰	-56.6‰
S. Hemis.	-47.0‰	-44.4‰	-48.1‰	-57.2‰

R = k_{12}/k_{13} in reaction $jCH_4 + OH \Rightarrow jCH_3 + H_2O$ where j is a superscript indicating either 12C or 13C methane.
CH_4 concentration is assumed to be 1.72 ppmv and 1.67 ppmv in northern and southern hemispheres respectively.
Lifetime for methane based on CH_4 + OH reaction is 12.5 years with OH distributed evenly between the hemispheres.
Mixing time between hemispheres is 1.25 years.
Methane source is assumed to be constant.

Table V. Methane Sources and $\delta^{13}C$ and $\delta(D/H)$ Signatures

Source	$\delta^{13}C$ (in ‰)	$\delta(D/H)$ (in ‰)
Biogenic Recent		
Acetate Fermentation	-50 to -70	-300 to -400
$CO_2 + H_2$ Reduction	-60 to -100	-150 to -250
Biogenic Natural Gas	-50 to -110	-180 to -280
Thermogenic		
Natural Gas and		
Other Fossil Fuels	-25 to -50	-130 to -250
Biomass Burning	-22 to -34	-10 to -50

δ is the delta value, given in per mil (‰), which is defined from the formulas

$$\delta^{13}C = [^{13}C/^{12}C|_{sample} / ^{13}C/^{12}C|_{standard} - 1] \times 1000 \text{ and}$$

$$\delta(D/H) = [D/H|_{sample} / D/H|_{standard} - 1] \times 1000 .$$

Table VI shows a simplified example for the $\delta(D/H)$ isotope budget which assumes ranges for the background methane and source methane isotopic values to arrive at an inferred deuterium/hydrogen kinetic isotope effect value. The range for the weighted sum of the sources comes from data summarized in Tyler (18) and a calculation made by Senum and Gaffney (46). The kinetic isotope effect value inferred in Table 6 is then about K_{CH_4}/K_{CH_3D} = 168 to 252‰ for the range of background atmospheric values measured by Ehhalt (42) and Wahlen and co-workers (45).

Table VI. Rough Estimate of $\delta(D/H)$ Isotopic Balance for Atmospheric Methane and Its Sources and Sinks

	Range	Median
$\delta(D/H)$ Weighted Sum of Source Fluxes	-262 to -322‰	-292‰
$\delta(D/H)$ of Background Atmospheric Methane	-72 to -94‰	-82‰
Inferred Isotopic Shift from Principal Sink Reaction of CH_4/CH_3D + OH	+168 to +252‰	+210‰

A single set of laboratory measurements contrasting reactions 4 and 5 was made by Gordon and Mulac (47) at a temperature of 416°K. They got a primary kinetic isotope effect of 1.500 at 416°K. Gordon and Mulac studied the two reactions in separate experiments using pulse radiolysis of H_2O to generate OH and monitoring the decay of OH by means of absorption spectrophotometry. Aside from using a relatively high temperature in the study which makes direct comparison to tropospheric temperatures difficult, there are potential problems with this result. Reaction cell impurities and interfering wall reactions may be different in the separate experiments involving first CH_4 and then CH_3D. These problems can be eliminated in a study in which the ratio k_{CH_4}/k_{CH_3D} is measured directly in a system where both CH_4 and CH_3D are present. In addition, it seems clear that reactions 2 and 3, either separately or together, need to be studied at tropospheric temperatures.

It is also possible to make a computer simulation of the transition state for this reaction which allows the k_{CH_4}/k_{CH_3D} to be estimated by calculation. The BEBOVIB-IV computer program described by Burton and co-workers (48) provides simulations of the transition state which can be used in the calculation of primary kinetic isotope effects in simple systems. Inputs of data to the calculation include vibrational frequencies for bonds in the transition state complex. An example of its use is the study of H_2 and HD reacting with OH by Ehhalt and co-workers (49). In that study, the results were identical when comparing experimental results to the computer calculation leading to a $k_{H_2}/k_{HD} = 1.65 \pm .05$.

In the case of the primary kinetic effect in reaction 1, it is possible to estimate the vibrational frequencies with some confidence. More importantly, the difference between the vibration with D substituted for H in the activated complex is large enough that errors in the vibrational frequency chosen lead to relatively small errors in the overall computation. Using the BEBOVIB-IV computer program, the D/H kinetic isotope effect in reaction 1 should be about 180‰ (Cantrell, C., NCAR, Boulder, Colorado, unpublished data, 1989). This is in reasonable agreement with what is expected from a rough analysis such as appears in Table VI.

The BEBOVIB-IV program is less reliable for a secondary kinetic isotope effect. In reaction 1 for instance, calculations for k_{12}/k_{13} ranged from about 1.000 to about 1.040 when studied by Cantrell and co-workers (20). This is because it is more difficult to estimate differences between vibrational frequencies in the activated complex when differences between atoms not directly transferred are involved. These uncertainties overwhelm the secondary kinetic isotope effect which is small compared to the primary kinetic effect.

Conclusions

A determination of k_{12}/k_{13} in the reaction of CH_4 with OH has proved useful for weighting relative source strengths of methane with carbon isotope ratios. However, the calculations involving relative source strengths are somewhat empirical. Several sources of methane overlap in their $\delta^{13}CH_4$ values, making an accurate determination of source strength for those sources solely dependent on flux measurements and ecological data.

Conversely, although 1.0054 ± 0.0009 at 297°K is the most precisely determined laboratory value for k_{12}/k_{13}, available methane data can't be used to prove that it is absolutely correct. Other recent determinations which indicate a higher effect of around 1.010 at 297°K can not be ruled out based on available data. As additional studies add information on both methane fluxes and $\delta^{13}CH_4$ isotopes, distributions and strengths for methane sources will be better known. When coupled with accurate models of methane mixing, these measurements may tend to corroborate a particular determination of k_{12}/k_{13} over the others.

Nevertheless, the kinetic isotope effect measurements should be used to help validate potential methane source distributions and not the other way around. Efforts should continue to measure k_{12}/k_{13} in either a laboratory or field experiment in a manner which gains a consensus of approval from the community of scientists using isotope data in methane studies. These could include more laboratory studies of CH_4 + OH which are made with a different OH source and a wider range of temperature dependence and/or field studies in which the transport of species and potential interfering reactions are given a more rigorous treatment in the calculations that lead to the k_{12}/k_{13} value.

Key areas for continued methane study include an accurate determination of the k_{CH_4}/k_{CH_3D} of the primary kinetic isotope effect in the CH_4 + OH reaction as well as a thorough consideration of methane oxidizing bacteria in soils and additional measurements of $\delta(D/H)$ values in methane. Measurements of $\delta(D/H)$ in methane should include atmospheric background measurements as well as sources and sinks on the same scope as are already being made for $\delta^{13}CH_4$. The recent measurement by Vaghjiani and Ravishankara (22) indicates a tropospheric lifetime of at least 12 years for methane based on reactions with OH. This necessitates an adjustment downward in total annual source strength of methane when compared to earlier source budgets. However, the adjustment downward may be lessened if soil bacteria are being underestimated as a methane sink. Studies of fractionation effects on both carbon and hydrogen in methane by bacterial oxidation could become very important in using isotopic balances to interpret the methane budget.

Last of all, it is noted that trace gases other than methane are also studied using isotopic measurements. It will be important to better understand kinetic isotope effects in these gases just as it is for methane. For instance, CO (50-52) and CH_2O (53) are studied using isotopes and both of these react with OH. Stevens and co-workers made determinations of the carbon and oxygen kinetic isotope effects in the reaction of CO and OH (54). Recent improvements in experimental methods could be used to corroborate or improve upon their work. And as is the case with methane, measurements of radiocarbon CO are also made (51-52). However, until the precision of measurements of pmC values of $^{14}CH_4$ and ^{14}CO can be improved past the current best of about ±0.5 pmC (per cent modern carbon), a value for k_{12}/k_{14} of the same magnitude as that for k_{12}/k_{13} can't be measured accurately. Nevertheless, the utility of measuring kinetic isotope effects in atmospheric trace gases and applying the results to additional trace gas cycling studies is clear.

Acknowledgements

The author acknowledges Chris Cantrell, John Orlando, Rick Shetter, and Geoff Tyndall, all of the National Center for Atmospheric Research (NCAR), Boulder, Colorado, for discussions regarding kinetics and photochemistry and Martin Manning of the Department of Scientific and Industrial Research in New Zealand for insight into modelling atmospheric methane. I would also particularly like to thank Ralph Keeling and Geoff Tyndall, both of NCAR, for reading and discussing all versions of the manuscript. The National Center for Atmospheric Research is sponsored by the National Science Foundation. This work has also been supported by a grant from the National Aeronautics and Space Administration under order W-16,184, mod. 5.

Literature Cited

1. Levy II, H. *Science*. **1971**, *173*, 141-143.
2. Logan, J. A.; Prather, M. J.; Wofsy, S. C.; McElroy, M. B. *J. Geophys. Res.* **1981**, *86(C8)*, 7210-7254.

3. Levine, J. S.; Rinsland, C. P.; Tenille, G. M. *Nature.* **1985**, *318*, 254-257.
4. Thompson, A. M.; Cicerone, R. J. *J. Geophys. Res.* **1986**, *91(D10)*, 10853-10864.
5. Crutzen, P. J. In *The Geophysiology of Amazonia.*; Dickinson, R. E., Ed.; John Wiley: New York, New York, 1987, 107-130.
6. Ramanathan, V.; Cicerone, R. J.; Singh, H. B.; Kiehl, J. T. *J. Geophys. Res.* **1985**, *90(D3)*, 5547-5566.
7. Dickinson, R. E.; Cicerone, R. J. *Nature.* **1986**, *319*, 109-115.
8. Ehhalt, D. H. *Tellus.* **1974**, *26*, 58-70.
9. Cicerone, R. J.; Oremland, R. S. *Global Biogeochem. Cycles.* **1988**, *2*, 299-327.
10. Rasmussen, R. A.; Khalil, M. A. K. *J. Geophys. Res.* **1981**, *86(C10)*, 9826-9832.
11. Steele, L. P.; Fraser, P. J.; Rasmussen, R. A.; Khalil, M. A. K.; Conway, T. J.; Crawford, A. J.; Gammon, R. H.; Masarie, K. A.; Thoning, K. W. *J. Atmos. Chem..* **1987**, *5*, 125-171.
12. Blake, D. R., Rowland; F.S. *Science.* **1988**, *239*, 1129-1131.
13. Craig, H.; Chou, C. C. *Geophys. Res. Lett..* **1982**, *9*, 1221-1224.
14. Rasmussen, R. A.; Khalil, M. A. K. *J. Geophys. Res.* **1984**, *89(D7)*, 11599-11605.
15. Khalil, M. A. K.; Rasmussen, R. A. *Atmos. Environ.* **1985**, *19*, 397-407.
16. Stevens, C. M.; Rust, F. E. *J. Geophys. Res.* **1982**, *87(C7)*, 4879-4882.
17. Rust, F.; Stevens, C. M. *Int. J. Chem. Kinet.* **1980**, *12*, 371-377.
18. Tyler, S. C. In *Microbial Production and Consumption of Greenhouse Gases: Methane, Nitrogen Oxides, and Halomethanes*, Rogers, J. E.; Whitman, W. B., Eds.; Amer. Soc. for Microbiology: Washington D. C., 1991, 7-31.
19. Craig, H. *Geochim. Cosmochim. Acta.* **1957**, *12*, 133-149.
20. Cantrell, C. A.; Shetter, R. E.; McDaniel, A. H.; Calvert, J. G.; Davidson, J. A.; Lowe, D. C.; Tyler, S. C.; Cicerone, R. J.; Greenberg, J. P. *J. Geophys. Res.* **1990**, *95(D13)*, 22455-22462.
21. Davidson, J. A.; Cantrell, C. A.; Tyler S. C.; Shetter, R. E.; Cicerone, R. J.; Calvert, J. G. *J. Geophys. Res.* **1987**, *92(D2)*, 2195-2199.
22. Vaghjiani, G. L.; Ravishankara, A. R. *Nature.* **1991**, *350*, 406-409.
23. Shetter, R. E.; Davidson, J. A.; Cantrell, C. A.; Calvert, J. G. *Rev. Sci. Instr.* **1987**, *58*, 1427-1428.
24. Lowe, D. C.; Brenninkmeijer, C.; Tyler, S. C.; Dlugokencky, E. *J. Geophys. Res.* **1991**, *96(D8)*, 15455-15467.
25. Tyler, S. C. *J. Geophys. Res.* **1986**, *91(D12)*, 13232-13238.
26. Lasaga, A. C.; Gibbs, G. V. *Geophys. Res. Lett.* **1991**, *18*, 1217-1220.
27. Wahlen, M.; Deck, B.; Henry, R.; Tanaka, N.; Shemesh, A.; Fairbanks, R.; Broecker, W.; Weyer, H.; Marine, B.; Logan, J. *EOS Trans. Amer. Geophys. Union.* **1989**, *70*, 1017.
28. Tyler, S. C.; Blake, D. R.; Rowland, F. S. *J. Geophys. Res.* **1987**, *92(D1)*, 1044-1048.
29. Devol, A. H.; Richey, J. E.; Clark, W. A.; King, S. L.; Martinelli, L. A. 1988. *J. Geophys. Res.* **1988**, *93(D2)*, 1583-1592.
30. Stevens, C. M.; Engelkemeir, A. *J. Geophys. Res.* **1988**, *93(D1)*, 725-733.
31. Tyler, S. C.; Zimmerman, P. R.; Cumberbatch, C.; Greenberg, J. P.; Westberg, C.; Darlington, J. P. E. C. *Global Biogeochem. Cycles.* **1988**, *2*, 349-355.
32. Quay, P. D.; King, S. L.; Lansdown, J. M.; Wilbur, D. O. *Global Biogeochem. Cycles.* **1988**, *2*, 385-397.

33. Wahlen, M.; Tanaka, N.; Henry, R.; Deck, B.; Zeglen, J.; Vogel, J. S.; Southon, J.; Shemesh, A.; Fairbanks, R.; Broecker, W. *Science.* **1989**, *245*, 286-290.
34. Stevens, C. M. *Chem. Geol..* **1988**, *71*, 11-21.
35. Lowe, D. C.; Brenninkmeijer, C. A. M.; Manning, M. R.; Sparks, R.; Wallace, G. *Nature.* **1988**, *332*, 522-525.
36. Quay, P. D.; King, S. L.; Stutsman, J.; Wilbur, D. O.; Steele, L. P; Fung, I.; Gammons, R. H.; Brown, T. A.; Farwell, G. W.; Grootes, P. M.; Schmidt, F. H. *Global Biogeochem. Cycles.* **1991**, *5*, 25-47.
37. Steudler, P. A.; Bowden, R. D.; Melillo, J. M.; Aber, J. D. *Nature.* **1989**, *341*, 413-416.
38. Born, M.; Dorr, H.; Levin, I. *Tellus.* **1989**, *42B*, 2-8.
39. King, S. L.; Quay, P. D.; Lansdown, J. M. *J. Geophys. Res.* **1989**, *94(D15)*, 18273-18277.
40. Tyler, S. C.; Keller, M.; Brailsford, G.; Crill, P.; Stallard, R.; Dlugokencky, E. *EOS Trans. Amer. Geophys. Union.* **1990**, *71*, 1260.
41. Whiticar, M. J.; Faber, E.; Scholl, M. *Geochim. Cosmochim. Acta.* **1986**, *50*, 693-709.
42. Ehhalt, D. H. In *Carbon and the Biosphere*; Woodwell, G. M.; Pecan, E. V., Eds.; U. S. Atomic Energy Commission: Oak Ridge, Tennessee, 1973, 144-158.
43. Schoell, M. *Geochim. Cosmochim. Acta.* **1980**, 44, 649-661.
44. Burke, Jr., R. A.; Sackett, W.M. In *Org. Mar. Chem.*; Sohn, M. L., Ed.; Amer. Chem. Soc.: Washington D. C., 1986; 297-313.
45. Wahlen, M.; Tanaka, N.; Henry, R.; Yoshinari, T.; Fairbanks, R. G.; Shemesh, A.; Broecker, W. S. *EOS Trans. Amer. Geophys. Union.* **1987**, *68*, 1220
46. Senum, G. I.; Gaffey, J. S. In *The Carbon Cycle and Atmospheric CO_2: Natural Variations Archean to Present*; Sundquist, E. T.; Broecker, W. S., Eds.; Geophysical Monograph Series, American Geophysical Union: Washington D. C., 1985, 32, 61-69.
47. Gordon, S.; Mulac, W. A. *Int. J. Chem. Kinet. Symp.* **1975**, *7(1)*, 289-299.
48. Burton, G. W.; Sims, L. B.; Wilson; Fry, A. *J. Amer. Chem. Soc.* **1977**, *99*, 3371-3379. Quantum Chemistry Program Exchange, Indiana University, Chemistry Department, Bloomington, Indiana, Program No. 337.
49. Ehhalt, D. H.; Davidson, J. A.; Cantrell, C. A.; Friedman, I.; Tyler, S. *J. Geophys. Res.* **1989**, *94(D7)*, 9831-9836.
50. Stevens, C. M.; Krout, L.; Walling, D.; Venters, A.; Engelkemeir, A.; Ross, L. E. *Earth and Planet. Sci. Lett.* **1972**, *16*, 147-165.
51. Brenninkmeijer, C. A. M.; Manning, M. R.; Lowe, D. C.; Mak, J. E. *EOS Trans. Amer. Geophys. Union.* **1991**, *72*, 103.
52. Mak, J. E.; Brenninkmeijer, C. A. M.; Manning, M. R.; Brasseur, G. P. *EOS Trans. Amer. Geophys. Union.* **1991**, *72*, 103.
53. Johnson, B. J.; Dawson, G. A. *Environ. Sci. Technol.* **1990**, *24*, 898-902.
54. Stevens, C. M.; Kaplan, L.; Gorse, R.; Durkee, S.; Compton, M.; Cohen, S.; Bielling, K. *Int. J. Chem. Kinet.* **1980**, *12*, 935-948.

RECEIVED March 2, 1992

Chapter 25

Key Sulfur-Containing Compounds in the Atmosphere and Ocean

Determination by Gas Chromatography–Mass Spectrometry and Isotopically Labeled Internal Standards

Alan R. Bandy, Donald C. Thornton, Robert G. Ridgeway, Jr.[1], and Byron W. Blomquist

Chemistry Department, Drexel University, Philadelphia, PA 19104

Gas chromatographic/mass spectrometric (GC/MS) methods using isotopically labeled internal standards (GC/MS/ILS) are described for determining atmospheric sulfur dioxide (SO_2), dimethyl sulfide (DMS), carbon disulfide (CS_2), dimethyl sulfoxide (DMSO), dimethyl sulfone ($DMSO_2$) and carbonyl sulfide (OCS) and aqueous phase dimethyl sulfide and dimethyl sulfoxide. GC/MS/ILS has great immunity to variations in sampling efficiency and changes in detector sensitivity. Using cryogenic preconcentration and integration times of three minutes, lower limits of detection are below one part per trillion for these gas phase species. Lower limits of detection for aqueous phase measurements are better than one picomole. Measurement precision is limited by either the lower limit of detection or the repeatability of the addition of the standard. Accuracy is determined primarily by the accuracy of the standards. GC/MS/ILS appears to have the sensitivity and precision to make real time isotopic ratio measurements.

Isotopic dilution methods are widely employed for improving the reliability of difficult determinations (1). Despite their advantages, however, they are rarely used in atmospheric and oceanic science.

We describe in this paper isotopic dilution methods for several key sulfur species present in the atmosphere and ocean. We use a variation of isotope dilution in which isotopically labeled analyte is added to the sample as an internal standard. Only stable isotopes are used. Analyses are carried out by GC/MS.

[1]Current address: Air Products and Chemicals, Inc., Allentown, PA 18195

Advantages of GC/MS/ILS

Compared to other methods the isotopic dilution GC/MS method, GC/MS/ILS, has several important advantages:

- Insensitivity to sampling losses
- Insensitivity to changes in GC/MS sensitivity
- Isotopically labeled standard trapped with each sample
- Large linear dynamic range
- High Sensitivity - low detectable limits
- High manifold analyte concentration due to labeled standard

Insensitivity to sampling losses and changes in GC/MS sensitivity are demonstrated by considering two isotopomers simultaneously monitored. Instrument responses for the two isotopomers are given by the expressions:

$$H_1 = \alpha_1 k_1 C_1 \tag{1}$$

$$H_2 = \alpha_2 k_2 C_2 \tag{2}$$

Here H, α, k and C are the instrument response, sampling efficiency, instrument sensitivity and analyte concentration respectively. If isotopomer 1 is in ambient air only and isotopomer 2 in the standard only, the ambient air concentration is given by the expression

$$C_1 = \frac{H_1}{H_2} C_2 = R C_2 \tag{3}$$

We have assumed that the sampling efficiencies and the instrument responses are the same for the two isotopomers. For ^{34}S labeled isotopomers we found no differences in sampling efficiencies for standard and ambient isotopomers. For deuterated standards we observed no differences in sampling efficiencies or instrument sensitivities for the ambient and standard isotopomers, however, the retention time of DMS-d_6 was less than the retention time of DMS by a few seconds. The retention time was shorter for DMS-d_6 than for DMS apparently because DMS forms stronger hydrogen bonds to the stationary phase than DMS-d_6. The retention time difference was observable because of the large number of equilibrium steps (theoretical plates) in the separation which is very sensitive to small differences in interaction with the stationary phase.

Our studies confirm that the calculated ambient concentration is independent of the instrument sensitivity and the sampling efficiency as the above argument suggests. A slightly more complex equation presented in a subsequent section applies when isotopomers 1 and 2 are in both ambient air and the standard. As in the simple case the sampling efficiency and instrument sensitivity are absent from the final expression in this more complex case.

An internal calibration is included in every sample. Since the internal standard is an isotopomer of the analyte it has the same chemistry as the analyte, at least within the measurement precision of the common GC/MS systems.

Inclusion of the isotopically labeled internal standard eliminates the effect of the sample matrix on precision and accuracy. Estimates of sampling efficiencies and instrument calibrations using test atmospheres prepared in "zero grade" air ignore the matrix effect on calibration. This approach often fails in atmospheric sampling except for the most inert atmospheric constituents.

Standard addition calibrations are a partial solution. In standard addition calibrations using nonisotopically labeled standards, instrument response is assumed to be proportional to analyte concentration over the entire dynamic range of the analyte. Furthermore, the proportionality coefficient is assumed to be constant. For reactive atmospheric species these conditions may not be met, especially at low analyte concentrations.

Standard addition calibration using isotopically labeled standards circumvents all of these problems. Accurate calculations of the lower limit of detection and sensitivity for every sample and no need of a special calibration sequence that would decrease the sampling rate are also important advantages.

The mass spectrometer, MS, is an extremely sensitive and specific detector. The lower limit of detection is about 4 femtomoles per second, which is at least 10 times better than the flame photometric detector and very close to the lower limit of detection of the electron capture detector.

High sensitivity and the collection of large samples make measurements possible even in high loss conditions. Analyte losses are typically less than a few percent for gas phase sulfur dioxide, SO_2, carbonyl sulfide, OCS, and carbon disulfide, CS_2. Because of oxidation in the trap, losses are higher and more variable for gas phase dimethyl sulfide, DMS, dimethyl sulfoxide, DMSO, and dimethyl sulfone, $DMSO_2$. Oxidant scrubbers installed in the manifold just after the air driers and before the cryogenic trap reduce these losses to manageable levels. Variable trapping and conversion efficiencies exist in the determinations of aqueous DMS, DMSO and $DMSO_2$. The isotopically labeled internal standard, however, makes the GC/MS/ILS immune to these types of the potential errors.

Because the standard concentration in the manifold is maintained at about 500 pptv, the manifold always contains a high concentration of analyte. To illustrate the advantage, consider a manifold that will remove 10 pptv. If no internal standard is present and the ambient concentration is 20 pptv, the instrument will yield a result of 10 pptv. If 500 pptv of isotopically labeled standard analyte is present in the manifold the total manifold concentration will be 520 pptv. As in the previous example 10 pptv will be removed by the manifold and the instrument will yield 510 pptv. However, the standard and ambient air analyte will be lost in the same proportion, therefore, the instrument will yield (510/520)*20 = 19.6 pptv for ambient air representing only a 2 percent error. This result should be compared to the 100 percent error that results if the manifold contains no standard analyte.

High concentrations of analyte in the manifold also decrease the manifold equilibration time. Adsorptive sites can cause long equilibration times, especially at low concentrations where an appreciable portion of the analyte is adsorbed in passing through the measurement system. For Teflon manifolds connected with machined stainless steel fittings, there are relatively small numbers of these sites. In the GC/MS/ILS manifold such sites are occupied primarily by isotopically labeled analytes, which are typically in large excess in the manifold. Therefore, fluctuations in the ambient air concentration are small fractions of the total concentration in the manifold and are transmitted more quickly and efficiently and with less distortion and delay through the manifold.

Data Reduction Algorithm

The ion intensities for m/z of the molecular ion from the naturally abundant compound and for m/z of the molecular ion from the isotopically enriched standard of the same compound are monitored. These ion intensities are used to determine the concentration of the compound in ambient air. For the typical operating case where the mass filter is tuned to pass only the ion of interest, the general expression for the signal intensity of this ion, I_i, is *(2-3)*

$$I_i \propto C_a \sum_j K_{j,a} F_{j,i} + C_s \sum_k K_{k,s} F_{k,i} \qquad (4)$$

The terms C_a and C_s are the ambient and internal standard analytical concentrations, respectively. The K terms are the fractional abundances of the various isotopomers of the analyte. They are calculated from a knowledge of the isotopic abundances of the analyte in ambient air and the standard. The F_{ji} terms are the fragmentation factors for the parent molecules present in ambient air which can contribute to the intensity at m/z_i. The F_{ki} terms are the fragmentation factors for the parent molecules in the isotopically labeled standard gas added to the ambient sample. These terms are determined from mass spectrometric studies of the fragmentation of the analyte under the ionization conditions used in the analysis. Algorithms for SO_2 *(4)*, DMS *(3)*, OCS *(5)* and CS_2 *(2)* have been developed. Procedures for modifying the algorithm for small changes in ionization conditions are described in these references.

For DMS *(3)* and CS_2 *(2)* the standard isotopomers monitored, DMS-d_6 and $C^{34}S_2$, are not present in the ambient air and the ambient isotopomers monitored, DMS and $C^{32}S_2$, not present in the standard at measurable levels. Consequently, the atmospheric and the standard concentrations are proportional, making calibrations and calculations simple. Because the ^{32}S content of our older OCS and SO_2 standards was a few percent, typically 4.5%, the calibration curve had an intercept because the ambient and standard isotopomers are present in both the standard and ambient air. Recently OCS and SO_2 standards were prepared with sulfur containing less than 0.01% ^{32}S. These standards make the ratio of ambient and standard concentrations for OCS and SO_2 strictly proportional to R at R-values below 0.2. This greatly improves the precision of the measurement of these species at low concentration.

Atmospheric Measurements

The concentrations of sulfur gases in a relatively unpolluted atmosphere are below 1 part per billion by volume (ppbv). Carbonyl sulfide, OCS, is the most abundant species, having an average concentration of about 500 pptv *(6)*. Because the OCS atmospheric lifetime is more than one year, its fluctuations are small. The analytical challenge, therefore, is to determine the small variations in its concentration with high precision. Short term fluctuations are less than 5% except in polluted air masses. The precision afforded by GC/MS/ILS makes such measurements possible.

Sulfur dioxide and dimethyl sulfide are present in the unpolluted atmosphere at concentrations below 50 to 200 pptv. In many areas both species are present below 20 pptv. The lower limit of detection must be about 1 pptv for determining these species.

Because the atmospheric lifetimes of DMS and SO_2 are short, hours to a few days, significant and informative fluctuations occur on time scales of a few minutes. To capture this information sampling times must be comparably short.

Carbon disulfide is present in the background atmosphere at 0.2 to 10 pptv. Its low concentration and short lifetime require a lower limit of detection below 0.2 pptv and sampling times and frequencies of a few minutes or less. Because CS_2 is ubiquitous in nonpristine atmospheres, contamination is a common problem. Otherwise its unreactivity makes sampling simple.

Sampling System. Because of the generally high reactivity of some of the atmospheric sulfur gases and their low concentration, special attention is needed to reduce sampling losses and contamination. The manifold developed for collecting and analyzing samples is shown in Figure 1. We use this manifold for collecting and analyzing air samples in real time and for collecting grab samples. These grab samples are refrigerated with liquid nitrogen and returned to the laboratory for analysis.

The main manifold, constructed from perfluorinated ethylene-propylene Teflon tubing, transports ambient air into the aircraft or laboratory and then to a pump located near the instrument. The pump is constructed from stainless steel and Teflon and is a bellows type (Metal Bellows Corp.). Operating the manifold after the pump at about 10 psi reduces contamination from leaks of ambient air into the manifold. The manifold air flow rate, maintained at about 15 L min^{-1}, is monitored by a mass flow meter. The manifold air is exhausted through a needle valve, which is used to control the manifold pressure.

The main manifold flow is sampled by a secondary Teflon manifold at a flow rate of 200 to 1200 mL min^{-1}. This air is dried by a Nafion dryer and passed though an unpacked Teflon trap cooled by liquid argon. The trap removes the sulfur gases from the air stream. A mass flow meter is used to monitor the trap flow rate from which the total volume of air sampled is computed.

Air passing through the mass flow meter and trap pump is returned to the main manifold through tubing coaxial with the Nafion dryer. Because much of the water has been removed by the cryogenic trap and the pressure is lower because of the pressure drop across the sampling valves, the water mixing ratio in this air is much lower that the incoming ambient air. This difference in mixing ratio provides the driving force for drying the incoming air stream in the Nafion drier. Pretraps to remove ozone and other oxidants are required for the determination of DMS, DMSO and $DMSO_2$.

Normally the trap contents are volatilized by hot water and analyzed by a GC/MS attached directly to the sampling system. In the grab sampling mode the trap is removed, stored under liquid nitrogen and returned to the laboratory where the contents also are volatilized using hot water and analyzed by GC/MS. During storage the grab samples are maintained at liquid nitrogen temperatures.

Calibrations are carried out by standard addition of the isotopically labeled analyte to ambient air at a point very near the inlet of the manifold using another Teflon manifold. Thus, losses in the main manifold or instrument affect the standard and ambient analyte proportionately, thereby having no affect on the accuracy of the method unless they are so large that the detection limit is approached. Since the MS can separately and simultaneously monitor the labeled (standard) and unlabeled (atmospheric) analyte, a standard addition calibration is included in every sample, thereby eliminating the need for a separate calibration sequence. Using this approach, very precise determinations of the analyte can be made.

When sampling speed is not critical the pump can be placed after the point where the manifold is sampled to reduce the risk that the analyte would be removed in the pump and the risk that nonambient analyte may enter the manifold through leaks in the pump. Well maintained pumps do not have these problems, but pump maintenance in the field can be a serious and sometimes unnecessary burden. Usually we pressurize the manifold when aircraft platforms are used to maintain manifold and trap flow rates while at high altitude.

Mass chromatograms for SO_2 are shown in Figure 2 for a sample obtained south of Barbados in September, 1989. The peak height ratio was 0.064 and the standard concentration was 1576 pptv. Using an algorithm based on the distribution of isotopomers in the sample and standard

$$C(ambient\ SO_2) = \frac{(0.918R - 0.042)}{(0.9457 - 0.0435R)} C(standard\ SO_2) = 28 pptv \quad (5)$$

Measurements of DMSO and $DMSO_2$. These species are of interest in the atmosphere because they are formed in one channel of the oxidation of DMS that does not lead directly to SO_2 (7). Sampling of these species is complicated by the conversion of DMS to DMSO and DMSO to $DMSO_2$ in the trapping phase. We

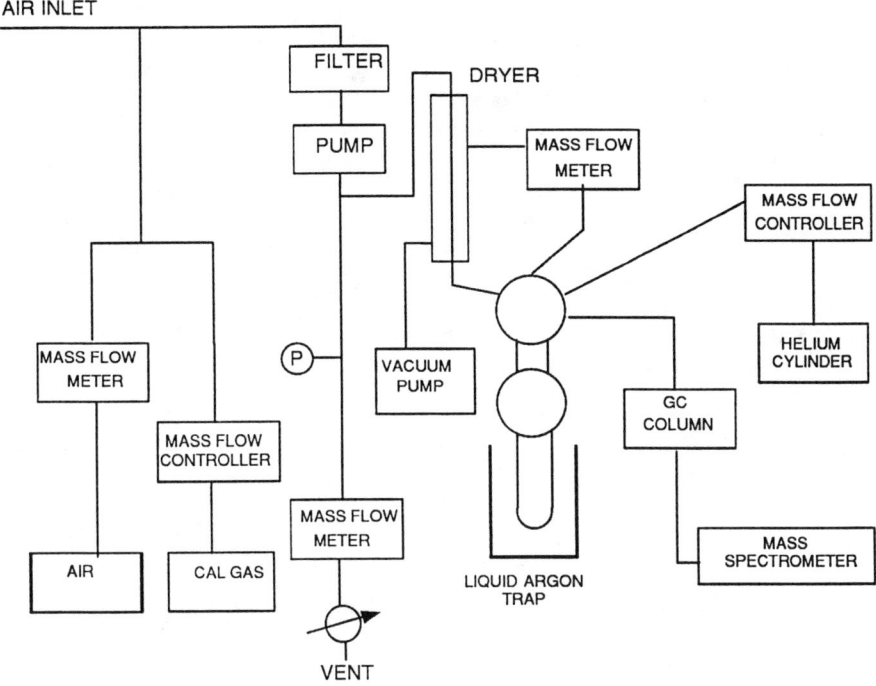

Figure 1. Pressurized manifold system for sampling ambient sulfur gases.

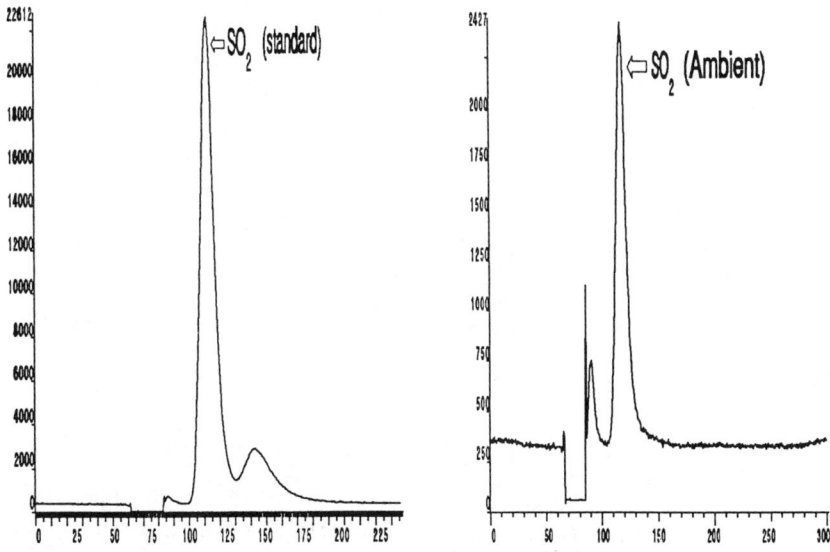

Figure 2. Chromatogram of ambient SO_2 with $^{34}SO_2$ added.

are developing the following GC/MS/ILS scheme for accounting for these conversions as well as losses of these species during sampling and analysis.

We use the manifold system shown in Figure 1. A standard containing known amounts of d_6-DMSO, d_3-DMSO$_2$ and $(CD_3)_2{}^{34}S$ is added to the manifold at its entrance. DMS, DMSO and DMSO$_2$ are trapped and brought to the gas chromatograph/mass spectrometer for analysis. The DMS is cryogenically trapped, whereas the DMSO and DMSO$_2$ are trapped on solid adsorbents.

The molecular ions monitored are m/z 62, and 70 for DMS, m/z 78, and 84 for DMSO and m/z 94, and 97 for DMSO$_2$. Conversion of DMS to DMSO and DMSO$_2$ can be accounted for using the peak areas for each of the isotopomers. Mass balance among the isotopomers is shown schematically in Figure 3.

The experimentally determined parameters are the peak areas S_{62}, S_{70}, S_{78}, S_{84}, S_{94}, and S_{97}. The standard concentrations, [DM^{34}S-d_6], [DMSO-d_6] and [DMSO$_2$-d_3], are known. Mass balance considerations lead to the following relationships for unambiguously determining the concentrations of ambient DMS, DMSO and DMSO$_2$:

$$[DMS]_{air} = \left(\frac{S_{62}}{S_{70}}\right)\left(\frac{\alpha^s_{70}}{\alpha^{air}_{62}}\right)[DMS]_s \tag{6}$$

$$[DMSO]_{air} = \left[\left(\frac{S_{78}}{S_{84}}\right) - \left(\frac{S_{86}}{S_{84}}\right)\left(\frac{S_{62}}{S_{70}}\right)\right]\left(\frac{\alpha^s_{84}}{\alpha^{air}_{78}}\right)[DMSO]_s \tag{7}$$

$$[DMSO_2]_{air} = \left[\left(\frac{S_{94}}{S_{97}}\right) - \left(\frac{S_{100}}{S_{97}}\right)\left(\frac{S_{78}}{S_{84}}\right)\right]\left(\frac{\alpha^s_{97}}{\alpha^{air}_{94}}\right)[DMSO_2]_s \tag{8}$$

where

$$S_{62} = \gamma_{DMS}(1-\delta_{DMS})\beta_{DMS}\alpha^{air}_{62}[DMS]_{air} \tag{9}$$

$$S_{70} = \gamma_{DMS}(1-\delta_{DMS})\beta_{DMS}\alpha^s_{70}[DMS]_s \tag{10}$$

$$S_{78} = \gamma_{DMSO}(1-\delta_{DMSO})(\beta_{DMSO}\alpha^{air}_{78}[DMSO]_{air} + \beta_{DMS}\delta_{DMS}\alpha^{air}_{62}[DMS]_{air}) \tag{11}$$

$$S_{84} = \gamma_{DMSO}(1-\delta_{DMSO})\beta_{DMSO}\alpha^s_{84}[DMSO]_s \tag{12}$$

$$S_{86} = \gamma_{DMSO}\delta_{DMS}\beta_{DMS}\alpha^s_{70}[DMS]_s(1-\delta_{DMSO}) \tag{13}$$

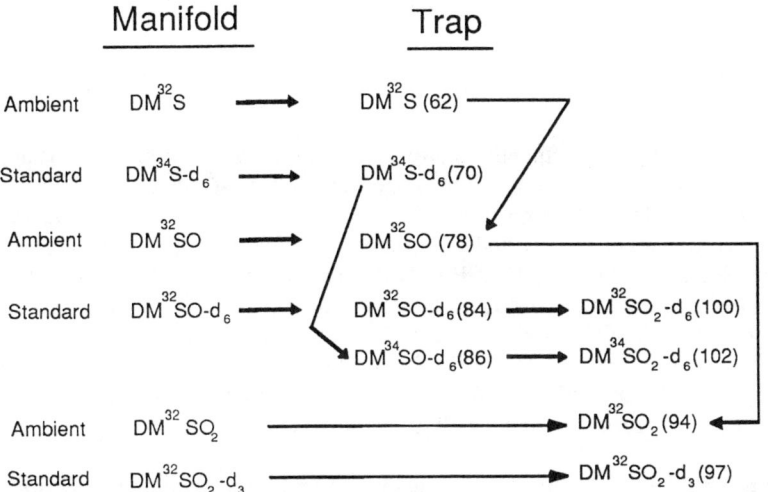

Figure 3. Calibration using isotopically labeled variants of DMS, DMSO, and $DMSO_2$.

$$S_{94} = \gamma_{DMSO_2}\beta_{DMSO_2}\alpha_{94}^{air}[DMSO_2]_{air} + \gamma_{DMSO_2}\delta_{DMSO}(\beta_{DMSO}\alpha_{78}^{air}[DMSO]_{air} +$$
$$\delta_{DMS}\beta_{DMS}\alpha_{62}^{air}[DMS]_{air}) \quad (14)$$

$$S_{97} = \gamma_{DMSO_2} \beta_{DMSO_2} \alpha_{97}^{air} [DMSO_2]_s \quad (15)$$

$$S_{100} = \gamma_{DMSO_2} \delta_{DMSO} \beta_{DMSO} \alpha_{84}^s [DMSO]_s \quad (16)$$

Here δ_i is the conversion efficiency of the ith species from a reduced species prior to the ith species in the oxidation sequence, β_i is the trapping efficiency of the ith species, γ_i is the overall analytical efficiency excluding trapping efficiency, α^s_m is the fraction of the analyte molecules in the standard having mass m, and α^{air}_m is the fraction of the analyte molecules in the air having mass m.

Ocean Measurements

Dimethyl sulfide, dimethyl sulfoxide, and dimethyl sulfoniopropionate, DMSP, are the sulfur containing ocean species of greatest interest in atmospheric chemistry *(8)*. DMS has been measured by many investigators using a purge and trap gas chromatographic method with flame photometric detection *(8-12)*. DMSP is determined by converting DMSP to DMS by adding base *(11)*. DMSO is determined by converting DMSP to DMS and then purging the DMS from the solution. The DMSO then is reduced to DMS by sodium borohydride *(13)*. The DMS produced in the DMSP and DMSO analysis is analyzed by the purge and trap method. We use isotopic dilution methods to improve the analysis for these species and to investigate potential interferences.

Dimethyl Sulfide. The precision of this purge and trap technique for DMS without isotopically labeled internal standard is estimated to be 3-10%, which is adequate for most purposes. The isotopic dilution method we developed *(14)* improves the precision to better than 1%. This improved precision includes corrections for problems in the purge and trap method not previously recognized.

Standards of DMS prepared in ethylene glycol were found by GC/MS/ILS to be in error by as much as 20%. Plots of the ratio of the signals for solutions containing DMS and DMS-d_6 as a function of the ratio of the concentrations should have a slope of 1.00 *(14)*. The slope of calibration curves using gravimetric standards was 0.87 representing a 13% error. Employing a titrimetric technique with potentiometric end point detection for determining DMS, the gravimetrically prepared standards were shown to be too low. Using the concentrations obtained titrimetrically the slope of the GC/MS/ILS calibration curve was 0.99 in statistical agreement with theory. Further investigation revealed that errors in the DMS standards prepared gravimetrically was caused by volatilization from the ethylene glycol.

Using isotopic dilution techniques we investigated how much DMS is generated from DMSP during sampling. The DMS-d_6 standard was added to the sample when the sample was collected. In repetitive purge and trap analyses, the ratio of DMS and DMS-d_6 should remain constant unless the concentration of one of these species changes.

The constancy of the ratio during extraction is demonstrated in Figure 4 for the extraction of DMS and DMS-d_6 from distilled water. The ratio of the signals for DMS and DMS-d_6 is plotted as a function of purge volume. Notably the ratio does not vary with purge volume, indicating that as the concentrations of two species change during extraction, they remain in the same proportion.

A plot of the calculated concentration ratio for an ocean sample is plotted as a function of extraction volume in Figure 5. For this sample the ratio increases markedly with extraction volume. Formation of DMS during extraction is the likely cause. The correct ratio can be obtained by extrapolation to zero extraction volume. The ratio at zero extraction volume is the ratio for the unperturbed sample.

The increases in ratio for this analysis began at low extraction volumes. Large overestimates of aqueous DMS content are likely if the DMS is exhaustively extracted, as it is for purge and trap methods not using isotopic dilution.

Dimethyl Sufoniopropionate. Measurements of DMSP are greatly enhanced by isotopic dilution. DMSP-d_6 is added to the sample as an internal standard. DMS in the sample is purged and base added to convert the DMSP to DMS. The DMSP-d_6 converts with the same efficiency as DMSP so isotopic dilution methods do not require that all the DMSP be converted. In most cases small efficiencies are consistent with milder extraction conditions, primarily as the result of lower flow rates and total volume of the extraction fluid used in the extraction.

Dimethyl Sulfoxide. Isotopic dilution methods have provided the few reliable data available for this species in rain, cloud and ocean (15). Presently there are no reliable direct methods. Attempts to reduce DMSO to DMS were not encouraging because of variable conversion efficiency.

Isotopic dilution methods easily circumvent this problem. Using DMSO-d_6 as an internal standard, variable conversion efficiencies have little influence on precision or accuracy. The first step in this analysis is to add DMSO-d_6 standard to the sample. Following the addition of base to convert the DMSP to DMS, the DMS is exhaustively purged. The DMS can be completely removed because no DMSP is present to be converted to DMS by the purging process. After removal of the DMSP, the solution is acidified and the DMSO (including the isotopically labeled standard) is reduced to DMS by sodium borohydride. The DMS and DMS-d_6 are determined by purge and trap GC/MS.

The lower limit of detection of this method is about 1 picomole. For one liter samples the lower limit of detection is 1 picomole per liter. This is adequate for sea water samples where the concentration is a few nanomoles per liter. Rain water samples are generally limited to about 20 ml thus the lower limit of detection is about 50 picomoles per liter. However, DMSO is typically a few nanomoles per liter so this lower limit of detection is also adequate for this measurement.

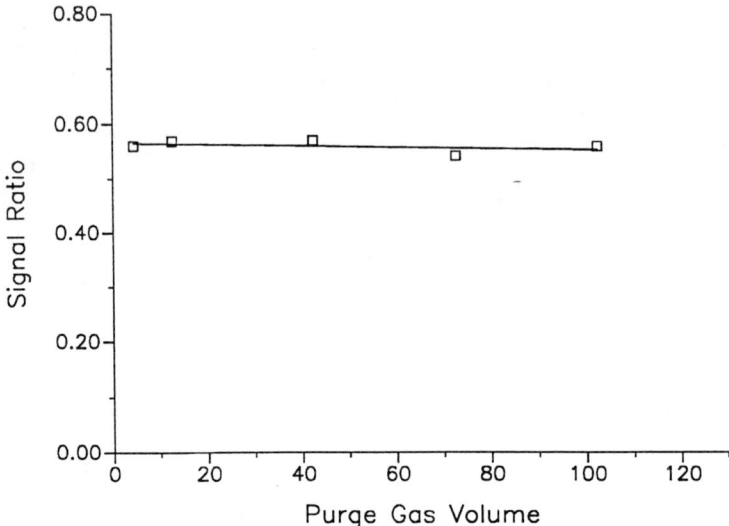

Figure 4. A plot of the ratio of the ambient and standard DMS concentration ratio as a function of purge volume for distilled water.

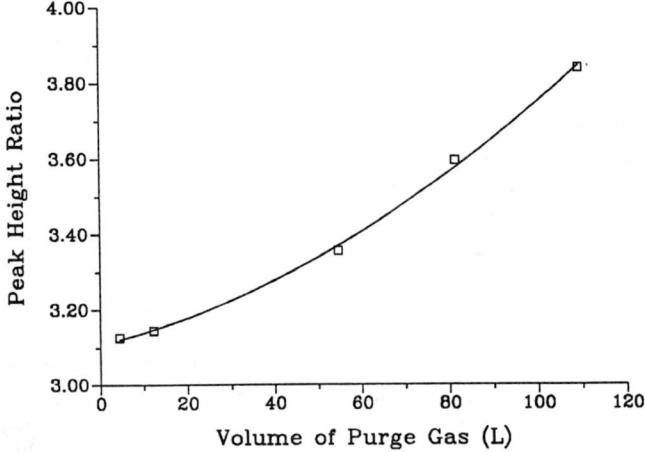

Figure 5. A plot of the ratio of the ambient and standard DMS concentration ratio as a function of purge volume for an ocean water sample.

Real Time Isotopic Distribution Measurements

GC/MS/ILS has the sensitivity and precision to make isotope ratio measurements on individual sulfur species on a time scale of a few minutes. These isotope ratio measurements can be useful in studying the relative importance of sources. For example, such measurements might provide information on how much of the SO_2 in the background atmosphere comes from transport from anthropogenic sources compared to that produced by the oxidation of DMS. This may be possible because the isotopic composition of anthropogenic SO_2 is different than SO_2 produced from the oxidation of oceanically derived DMS *(16)*.

Ratio measurements are presently carried out by trapping the sulfur gas by some means and then converting it to either SO_2, sulfur hexafluoride, SF_6, or some other relatively inert sulfur species. The sulfur isotope ratio work is carried out on the inert species. Because speciation is normally very crude the isotope ratio's reported to date relate to one compound only when that compound is the predominate species present during sampling. GC/MS would provide isotope ratio data on specific molecules with high temporal resolution. Typically we would not use the internal standard for these studies. Because of the high separation power of the gas chromatograph, the degree of speciation is very high. The mass spectrometer adds to the degree of speciation and provides the very high sensitivity required.

Although our studies indicate that isotope ratio measurements are possible in real time we have not made actual measurements in the field. Note that GC/MS provides isotopomer ratio measurements not the conventional isotope ratio for an element such as sulfur. Therefore, fluctuations in the distributions of carbon, hydrogen and oxygen may be significant problems.

To illustrate the application of isotopomer distrubution data we use the standard definition of the del-value for ^{34}S:

$$\delta^{34}S = \frac{\left(\frac{^{34}S}{^{32}S}\right)_a - \left(\frac{^{34}S}{^{32}S}\right)_s}{\left(\frac{^{34}S}{^{32}S}\right)_s} \times 1000$$

This equation applies only for elemental sulfur and not for the isotopomer distribution but the idea is similar. The $\delta^{34}S$ for ocean sulfate is 20‰ and varies little. $\delta^{34}S$ for coal is 24 to -30 ‰ whereas in petroleum it is in the range -8 to 32 ‰. Although $\delta^{34}S$ varies a lot for coal and petroleum on the average, it varies over a much smaller range for a specific source for the coal or petroleum. Hence information concerning the specific source of the coal or petroleum may exist in the $\delta^{34}S$ data. It is this type of problem in atmospheric studies that GC/MS/ILS may be applied to in the future.

Summary

The use of isotopic dilution methods has improved the precision of measurements of atmospheric DMS, OCS and CS_2 and aqueous DMS and DMSP. We expect the same level of improvement for the determination of atmospheric, DMSO and $DMSO_2$. It has made possible the measurement of atmospheric SO_2 at low pptv levels, and aqueous phase DMSO. Many other applications of this approach remain in atmospheric and oceanic science. Real time isotope ratio ratio measurements for individual sulfur containing molecules may be a very important future application of GC/MS/ILS.

Acknowledgments

This research was supported by NSF grant ATM-8515000-01.

Literature Cited

(1) Webster, R.K, In *Mass Spectrometric Isotope Dilution Analysis*; Smales A.A. and L.R. Wager, Eds.; John Wiley and Sons, New York, NY, 1960; 202-246.
(2) Bandy, A.R.; Tucker B.J.; Maroulis P.J.; *Anal. Chem.* **1985**, *547*, 1310-1314.
(3) Thornton D.C.; Bandy, A.R.; Ridgeway, R.G.; Driedger III; A.R.; Lalevic, M. *J. Atmos. Chem.* **1990**, *11*, 299-308.
(4) Driedger III, A.R.; Thornton, D.C.; Lalevic M.; Bandy A.R. *Anal. Chem.*, **1987**, *59,*1196-1200.
(5) Lewin, E.E.; Taggart R.L.; Lalevic M.; Bandy A.R. *Anal. Chem.*, **1987**, *59* 1296-1301.
(6) Maroulis, P.J.; Torres A.L.; Bandy A.R. *Geophys. Res. Lett.* **1977**, *4,* 510-512.
(7) Yin F.; Grosjean D.; Seinfeld J.H. *J. Atmos. Chem.* **1990**, *11*, 309-364.
(8) Andreae, M.O. *Limnol. Oceanogr.* **1985**, *30*, 1208-1218.
(9) Rasmussen, R.A. *Amer. Lab.* **1972**, *4*, 55-61.
(10) Andreae, M.O.; Barnard W.R. *Anal. Chem.* **1983**, *55*, 608-612.
(11) Barnard, W.R.; Andreae M.O.; Iverson, R.L. *Continental Shelf Res.*, **1984**, *3*, 103-113.
(12) Wakeham,. S.G., B.L. Howes and J.H. Dacey *Nature*, **1984**, *310*, 770-772.
(13, Andreae, M.O. *Anal. Chem.* **1980**, 150-153.
(14) Ridgeway,Jr., R.G.; Bandy, A.R.; Thornton D.C. *Marine Chem.*, **1991**, *11,* 321-334.
(15) Ridgeway, Jr., R.G., Ph.D. dissertation, Drexel University, 1991.
(16) Faure, G. *Principles of Isotope Geology*; Wiley: New York, NY, **1977**, 403-423.

RECEIVED February 4, 1992

INDEXES

Author Index

Anderson, S. M., 155
Armentrout, Peter B., 194
Arseneau, Donald, 111
Baer, Susan, 111
Bandy, Alan R., 409
Bersohn, R., 66
Blomquist, Byron W., 409
Bowman, Joel M., 37
Brickhouse, Mark D., 246
Carman, Howard S., Jr., 181
Combariza, J. E., 310
Compton, Robert N., 181
Daniel, C., 310
Dateo, Christopher E., 225
Dissly, Richard W., 369
Fleming, Donald, 111
Garvey, James F., 335
Gellene, Gregory I., 210
Gonzalez, Alicia, 111
Gonzalez-Lafont, Angels, 16
Graul, Susan T., 246
Griffith, K. S., 210
Harding, Lawrence B., 48
Herbst, E., 358
Hobbs, R. H., 225
Hynes, A. J., 94
Just, B., 310
Kades, E., 310
Katz, B., 66
Kaye, Jack A., 1
Kelly, P. B., 297
Klots, Cornelius E., 181
Koizumi, Hiroyasu, 37
Kolba, E., 310
Larsen, N. Wessel, 167
Liu, Yi-Ping, 16
Lu, Da-hong, 16
Lynch, Gillian C., 16
Malisch, W., 310

Manz, J., 310
Mauersberger, K., 155
Maurice, David, 16
Michael, J. V., 80
Michels, H. H., 225
Morris, Robert A., 225
Morton, J., 155
Nicovich, J. M., 94
Novicki, S. W., 279
Paramonov, G. K., 310
Paulson, John F., 225
Pedersen, T., 167
Peifer, William R., 335
Ridgeway, Robert G., Jr., 409
Schatz, George C., 37
Schueler, B., 155
Sehested, J., 167
Senba, Masayoshi, 111
Shan, J. H., 279
Shapiro, Moshe, 264
Squires, Robert R., 246
Thiemens, M. H., 138
Thornton, Donald C., 409
Truhlar, Donald G., 16
Truong, Thanh N., 16
Tucker, Susan C., 16
Tyler, Stanley C., 390
van Doren, Jane M., 225
Vasudev, R., 279
Viggiano, A. A., 225
Wagner, Albert F., 48
Warmuth, B., 310
Wategaonkar, S. J., 279
Westre, S. G., 297
Wine, P. H., 94
Yung, Yuk L., 369
Zhang, Y. P., 297
Zhao, Xin Gui, 16
Ziegler, L. D., 297

Affiliation Index

Academy of Science of Belorussia, 310
Argonne National Laboratory, 48,80
Augsburg College, 155
Ben Gurion University
 of the Negev (Israel), 66
California Institute of Technology, 369
Columbia University, 66
Drexel University, 409
Duke University, 358
Emory University, 37
Georgia Institute of Technology, 94
National Aeronautics and Space
 Administration, 1
National Center for Atmospheric
 Research, 390
Northeastern University, 297
Northwestern University, 37
Oak Ridge National Laboratory, 181
Phillips Laboratory, 225
Purdue University, 246
Rutgers University, 279
State University of New York at Buffalo, 335
United Technologies Research Center, 225
Universitat Gottingen, 138
Universität Würzburg, 310
Universität Zürich, 310
Universite Louis Pasteur, 310
University of British Columbia, 111
University of California—Davis, 297
University of California—Santa Barbara, 225
University of Copenhagen, 167
University of Minnesota, 16,155
University of Notre Dame, 210
University of Utah, 194

Subject Index

A

Acetonitrile, reactions with OH and OD, 101–104f
Acetonitrile–water cluster ions, isotope exchange reactions, 259f
Addition reactions in combustion
 isotope effects, 48–62
 representative reaction, 48
Alkyl radical reactions with HBr and DBr, mechanism studies using deuterium substitution, 103,105–107
Ammonia clusters, isotope fractionation, 249–255
Ammonia–water cluster ions, isotope exchange reactions, 256–258f
Applications, mass-independent isotope fractionations, 151–153
Atmospheric measurements, sulfur-containing compounds, 413–418
Atomic ions, isotope effects in reactions with H_2, D_2, and HD, 194–208

B

Bimolecular reactions, nonequilibrium isotope effects, 7–9
Branching among competing channels for reactions, isotope effects, 3

C

Carbon disulfide
 electron affinity, 183
 electron attachment rate constants, 183
 Rydberg electron transfer, 183–184
CD_3SD, reactions with Cl, 99–101
Centrifugal dominant small-curvature semiclassical adiabatic ground-state approximation, development, 21
$C_2H + C_2H_2 \rightarrow C_4H_2 + H$, isotope effects on rate constants, 83,85t
$CH + H_2$
 isotope effects on rate constants, 58–62
 reaction path, 52,53f

INDEX

$CH + H_2 \leftrightarrow CH_3^* \leftrightarrow CH_2 + H$
  $CH + H_2$ reaction path, 52,53f
  $CH_2 + H$ reaction path, 52,54–56
  dynamics calculation procedure, 56–57
  electronic structure calculation procedure, 49–50
  isotope effects on $CH + H_2$ rate constants, 58–62
  kinetics, 58–62
  out-of-plane motion, characterization, 51
  properties of H_2, CH, CH_2, and CH_3, calculated and observed, 50,51t
  reaction enthalpies, calculated and observed, 50,51t
  theoretical studies, 49
$CH_2 + H$, reaction path, 52,54–56
$CH_4 + OH$ reaction, kinetic isotope effect, 402,404t,405
CH_2CD_2, hydrogen isotope effects in photodissociation, 77–78
$C_6H_5CH_2D$, hydrogen isotope effects in photodissociation, 78
CH_nD_{4-n}, hydrogen isotope effects in photodissociation, 76–77
t-CHDCHD, hydrogen isotope effects in photodissociation, 77–78
Chemical kinetics research, applications, 94
Chemical mass-independent fractionations, state of knowledge, 150–151
Chemistry, dense interstellar clouds, 359–361
$(CH_3OH)_nCr(CO)_6$ van der Waals cluster selection for multiphoton ionization dynamics studies
  multiphoton ionization dynamics, 335–352
  multiphoton ionization of van der Waals clusters containing $M(CO)_6$, 339–340
  photophysics of transition metal carbonyls, 337–338
  relaxation dynamics in metal carbonyls, spectroscopic probes, 338–339
CH_3SH, reactions with Cl, 99–101
$^*Cl^- + CD_3Cl \rightarrow CD_3^*Cl + Cl^-$
  kinetic isotope effects and factors, 27,29t
  vibrational contributions to kinetic isotope effects vs. frequency ranges, 29,30t
Cl reactions with atmospheric reduced sulfur compounds, mechanism studies using deuterium substitution, 99–101

$^*Cl(D_2O)^- + CH_3Cl \rightarrow CH_3^*Cl + Cl(D_2O)^-$
  kinetic isotope effects and factors, 30,31t
  structures of reactants and transition state, 30,32f,33
Cluster ions, isotope exchange reactions, 246–259
Collision dynamics for set of reactants, behavior and properties of macroscopic ensemble, 336
Collision of molecules, sequence of events, 335
Combustion, isotope effects in addition reactions, 48–62
Complex-forming bimolecular and termolecular reactions, nonequilibrium isotope effects, 7–9
Conventional transition-state theory for kinetic isotope effects, 17–18
$Cp(CO)_2FePH_2$ and $Cp(CO)_2FePHD$, selective isomerizations, 317,319–323
Cyclic ozone, 167,168f

D

D_2, isotope effects in reactions with atomic ions, 194–208
$D + D_2O \rightarrow OD + D_2$, isotope effects on rate constants, 85t,87
$D + H_2 \rightarrow HD + H$, isotope effects on rate constants, 85t,89–91
D–H branching in photodissociation, influencing factors, 11
δ notation, definition, 139
$D + O_2 \rightarrow OD + O$, isotope effects on rate constants, 85t,86
Data reduction algorithm, GC–MS–ILS, 412–413
DCO^+, production, 363
$DCo(CO)_4$, vibrationally mediated dissociations, 328–330
De-NO_x process, 37
Dense interstellar clouds
  chemistry, 359–361
  gas-phase processes, 359–360
  ion–molecule reactions, 359
  models of chemistry, 361
  water synthesis, 360

Deuterium fractionation in interstellar
 space
 deuterium-containing molecules, 361
 gas-phase fractionation, 361–363
 models, 363–365
 star-formation regions, 365–366
Deuterium in solar system
 chemical evolution, changes in rate,
  383,385
 D/H ratio, 370,371f
  explanation, 381
  ice in solar nebula, 378,380f,381
 escape, efficiency, 382t–384f
 fractionation mechanisms, 375–378
 fractionation processes, 385,386f
 fractionation vs. temperature,
  378,380f,381
 hydrogen contents of terrestrial
  planets, 382t
 in giant planets, 372–373
 in interstellar molecular clouds, 374
 in small solar system bodies, 373–374
 in terrestrial planets, 370,372
 origin, 378,385,386f
 questions, 385,387
 time of synthesis, 369
 use as chemical tracer for planetary
  atmospheric studies, 369
Deuterium substitution, use for mechanism
 studies of gas-phase free radical
 reactions, 94–107
DI, photoabsorption, 265,266f
Diatomic molecules, isotope effect, 264–269
Diffusive separation, D–H fractionation
 mechanism in solar system, 375–376
Dimethyl sulfide
 function, 95
 ocean measurements using GC–MS–ILS,
  418–420f
 reactions with NO_3, 97–99
 reactions with OH, 95–99
Dimethyl sulfone, atmospheric measurements
 using GC–MS–ILS, 414,416–418
Dimethyl sulfoniopropionate, ocean
 measurements using GC–MS–ILS, 419
Dimethyl sulfoxide, atmospheric and ocean
 measurements using GC–MS–ILS,
 414,416–419
Direct behavior, reactions of atomic ions
 with H_2, D_2, and HD, 203–205

Dissociative recombination, reaction, 360
DN_2
 ground-state elastic transition probability,
  D + N_2 vs. total energy, 40,41f
 lifetime, 38
 lifetime vs. total energy, semilog plot, 43,45f
 potential energy surface vs. Jacobi
  coordinates, contours, 38–40f
 resonance energies, widths, and
  lifetimes, 43,44t
 scattering calculations, 39–40
 stabilization calculations, 40
 stabilization plot of energy vs.
  nonlinear parameter α, 43,46f
 state-to-state transition probabilities,
  D + N_2 vs. total energy, 40,42f,43
 stationary point properties, 38,39t
DONO, state-selected dissociation, 279–295
D_2S, reactions with Cl, 99–101
Dynamic calculations of addition reactions,
 procedure, 49–50

E

Edge effect, description, 264
Electronic states of ozone, energy
 determination, 163–165t
Electronic structure calculations of
 addition reactions, procedure, 49–50
Equilibrium isotope effects
 applications, 4–5
 occurrence, 2
 representation, 4
Exchange reactions, hydrogen isotope
 effects, 66–74
Excited electronic states, structure and
 dynamics studies using resonance Raman
 spectroscopy, 298

F

F + HD, isotopic competition, 72–74
Factorization, kinetic isotope effect, 22
Flash photolysis shock tube technique
 activation energy vs. barrier height on
  potential energy surface, 80
 development and applications, 80
 high-temperature isotope effect studies
  apparatus, schematic diagram, 81,82f

INDEX

Flash photolysis shock tube technique—
Continued
$C_2D + C_2D_2 \to C_4D_2 + D$, 83,85$t$
$C_2H + C_2H_2 \to C_4H_2 + H$, 82,85$t$
$D + D_2O \to OD + D_2$, 85t–87
$D + H_2 \to HD + H$, 85t,89–91
$D + O_2 \to OD + O$, 85t,86
experimental procedure, 81,83
$H + D_2 \to HD + D$, 85t,89–91
$H + H_2O \to OH + H_2$, 85t–87
$H + O_2 \to OH + O$, 85t,86
$O + C_2D_2 \to$ products, 83,85t,86
$O + C_2H_2 \to$ products, 83,85t,86
$O + D_2 \to OD + D$, 85t,87–90f
$O + H_2 \to OH + H$, 85t,87–90f
rate constant expressions for reactions studied, 83,85t
raw data and derived first-order plot, 83,84f
isotope effect, calculation, 81
Free electron model, description, 182–183
Free radical reactions, gas phase, mechanism studies using deuterium substitution, 94–107

G

Gas chromatography–mass spectrometry using isotopically labeled internal standards (GC–MS–ILS) for determination of sulfur-containing compounds
advantages, 410–412
data reduction algorithm, 412–413
dimethyl sulfide ocean measurement, 418–420f
dimethyl sulfone atmospheric measurements, 414,416–418
dimethyl sulfoniopropionate ocean measurement, 419
dimethyl sulfoxide atmospheric measurements, 414,416–418
dimethyl sulfoxide ocean measurement, 419
high manifold analyte concentration due to labeled standard, 411–412
high sensitivity and low detectable limits, 411
insensitivity to changes in GC–MS sensitivity, 410
insensitivity to sampling losses, 410

Gas chromatography–mass spectrometry using isotopically labeled internal standards (GC–MS–ILS) for determination of sulfur-containing compounds—
Continued
isotopically labeled standard trapped with each sample, 410–411
linear dynamic range, 411
real-time isotopic distribution measurements, 421
sampling system in atmospheric measurements, 413–414,415f
species present in atmosphere, 413
species present in ocean, 418
Gas-phase chemical reactions, isotope effects, 1–10
Gas-phase deuterium fractionation in interstellar space
DCO$^+$ production, 363
D/H abundance ratio, 363
equilibrium constant, 362
ions exchanging rapidly and exothermically with HD, 362–363
models, 363–365
occurrence in star-formation regions, 365–366
potential energy surface for exothermic ion–molecule reactions, 362f
reaction, 361
Gas-phase free radical reactions, mechanism studies using deuterium substitution, 94–107
Gas-phase protonated cluster ions, isotope exchange reactions, 246–259
Giant planets, observations of deuterium, 372–373

H

H_2, isotope effects in reactions with atomic ions, 194–208
$H + CD_3H \to H_2 + CD_3$
experimental vs. theoretical kinetic isotope effects, 27,28t
kinetic isotope effects and factors, 22,23t
model testing, 24,27
potential energy vs. zero point energy, 23,25f

H + CD$_3$H → H$_2$ + CD$_3$—*Continued*
  reaction path parameters, 23,24t
  reliability of kinetic isotope effect results, 27
  vibrational modes, contributions to
    predicted kinetic effects, 23,24,26f
H + D$_2$, hydrogen isotope effects, 67,68f
H + D$_2$ → HD + D, isotope effects on rate
  constants, 85t,89–91
H + DCCD and C$_2$D$_4$, hydrogen isotope
  effects, 67
H + H$_2$O → OH + H$_2$, isotope effects on rate
  constants, 85t,87
H + MD$_4$, hydrogen isotope effects, 67,69–72
H + MHD$_3$, hydrogen isotope effects,
  67,69–72
H + O$_2$ → OH + O, isotope effects on rate
  constants, 85t,86
HBr and DBr, reactions with alkyl
  radicals, 103,105–107
HCCD, hydrogen isotope effects in
  photodissociation, 76
HCo(CO)$_4$, vibrationally mediated
  dissociations, 328–330
HD, isotope effects in reactions with
  atomic ions, 194–208
HDCO, hydrogen isotope effects in
  photodissociation, 76
HDO, hydrogen isotope effects in
  photodissociation, 75–76
Heavy ozone, isotope enrichment, 156–165
HI
  branching ratio and total absorption vs.
    vibrational state, 265,267–269
  photoabsorption, 265,266f
High temperatures, isotope effect studies
  by flash photolysis shock tube, 80–91
HN$_2$
  ground-state elastic transitions,
    probability for H + N$_2$ vs. total
    energy, 40,41f
  lifetime studies, 38
  lifetime vs. total energy, semilog plot, 43,45f
  potential surface, 38,39t,41f
  resonance energies, widths, and
    lifetimes, 43,44t
  role in thermal De–NO$_x$ process, 37
  scattering calculations, 39–40
  stabilization calculations, 40
  stabilization plot of energy vs.
    nonlinear parameter α, 43,45f,46

HN$_2$—*Continued*
  state-to-state transition probabilities
    for H + N$_2$ vs. total energy,
    40,42f,43
  stationary point properties, 38,39t
  vibrational-state studies, 38
HONO, state-selected dissociation, 279–295
H$_2$S, reactions with Cl, 99–101
Hydrogen, isotopes, 112
Hydrogen isotope effects
  exchange reactions and attack of neutral
    atoms on HD
    experimental procedure, 66–67
    H + D$_2$, 67,68f
    H + DCCD and C$_2$D$_4$, 67
    H + MD$_4$ and H + MHD$_3$, 67–72
    isotopic competition in F + HD, 72–74
  information obtained from studies, 111
  photodissociations
    CH$_2$CD$_2$, 77
    C$_6$2H$_5$CH$_2$D, 78
    CH$_n$D$_{4-n}$, 76–77
    t-CHDCHD, 77–78
    HCCD, 76
    HD$_2$CO, 76
    HDO, 75–76
    H/D ratios, 74,75t
    influencing factors, 74
    potential surface vs. isotope effects, 78

I

Impulsive behavior, reactions of atomic
  ions with H$_2$, D$_2$, and HD, 205–207
In situ mass spectrometer measurements,
  procedure, 157,158f
Intensity of rotational line, definition, 285
Intermolecular isotope effects, reactions
  of atomic ions with H$_2$, D$_2$, and HD,
  195–196
Interstellar clouds
  composition, 359
  density and size, 358
  observations of deuterium, 374
  stellar elemental abundance, 359
Interstellar molecules, deuterium
  fractionation, 361–365
Interstellar space, deuterium
  fractionation, 358–366

INDEX

Intracluster energy transfer,
(CH$_3$OH)$_n$Cr(CO)$_6$ van der Waals clusters, 352
Intramolecular isotope effects, reaction of atomic ions with H$_2$, D$_2$, and HD, 196–198
Ion chemistry, D–H fractionation mechanism in solar system, 377–378
Isotope(s), temperature, kinetic energy, and rotational energy effects, 226–243
Isotope dependence, methyl radical Rydberg 3s predissociation dynamics, 297–308
Isotope-dependent rate constants for CS$_2^-$ formation using Rydberg atoms
 bending vibration role in electron capture process, 189
 electron affinity of CS$_2$, 183
 experimental procedure, 184–185
 isotope ratio vs. effective principal quantum number, 188f,189
 negative ion intensity ratio vs. delay times, 186–188
 negative ion intensity ratio vs. effective principal quantum number, 186f
 negative ion mass spectra, 185f,186
 Rydberg electron transfer, 182t–184
 symmetry constraints on isotope effects, 189–191
Isotope dilution methods, applications, 409
Isotope effects
 addition reactions in combustion, 48–62
 branching effects, 3
 dependence on mass, 138–139
 equilibrium, See Equilibrium isotope effects
 intracluster energy transfer, multiphoton ionization dynamics within van der Waals clusters, 335–352
 isotopic composition of natural systems, 1
 nature of chemical reactions, 1
 nonequilibrium, See Nonequilibrium isotope effects
 non-mass-dependent effects in O$_4^+$ formation, 210–222
 photodissociation of small molecules, 3
 potential surface effect, 78
 types, 2
Isotope effects in reactions of atomic ions with H$_2$, D$_2$, and HD
 endothermic reactions
  direct behavior, 203,204f

Isotope effects in reactions of atomic ions with H$_2$, D$_2$, and HD—*Continued*
 endothermic reactions—*Continued*
  impulse behavior, 205–207
  mixed behavior, 207,208f
  orbital angular momentum conservation, 203,205
 exothermic reactions
  intermolecular isotope effects, 195–196
  intramolecular isotope effects, 196–198
  orbital angular momentum conservation, 196,198
 experimental procedure, 195
 rotational energy effect, 198
 statistical behavior
  high energies, 202–203
  low energies, 199,201f,202t
 translational energy effect, 196,197f
 zero-point energy effects for endothermic reactions, 198–200f
Isotope enhancement, definition, 157
Isotope enrichment
 heavy ozone history, 156–157
 laboratory ozone isotope measurements, 159–162f
 pathways to symmetry-selective isotope effects, 160,162f
 relevant electronic states, 163–165t
 stratospheric ozone isotope measurements, 157–159t
 mode specificity, 264
 photochemical reactions
  calculated absorption spectrum and isotopic ratio for HOD, 272,274–276f
  diagonal channesl potentials and vibrational wave functions, 275–277
  diatomic molecules, 264–269
  photofragmentation maps of polyatomic molecules, 269–272
  vibrational excitation effect on HOD systems, 272
Isotope-exchange reactions within gas-phase protonated cluster ions
 acetonitrile–water cluster ion reactions, 259f
 ammonia cluster formation, 249–250
 collision-induced dissociation
  partially deuterated ammonia clusters, 252–255
  partially deuterated water clusters, 249–251f

Isotope-exchange reactions within gas-phase protonated cluster ions—*Continued*
  collision-induced dissociation—*Continued*
    singly deuterated methanol clusters, 256
    displacement reaction for methanol clusters, 255
    experimental procedure, 247
    isotope fractionation with water clusters, 248–251f
    isotopomer structures of dideuterated water cluster, 248,250f
    water–ammonia cluster ion reactions, 256–258f
    water cluster formation, 247–248
    water cluster structures, 248,250f
Isotope fractionation
  ammonia clusters, 249–255
  methanol clusters, 255–256
  water clusters, 247–251
Isotope ratio of vapor, definition, 376
Isotope-specific photodissociation
  commercial applications, 12
  isotope effects on branching ratios, 10–11
  reagent excitation effect, 11
Isotope study of ozone formation mechanism
  calibration procedure, 172,174,177–179
  energy transfer mechanism, 176
  exchange process, temperature dependence, 176
  exchange reaction, 168–169
  experimental procedure, 176–177
  isotopomer ratio(s), 168
    dependence on experimental conditions, 169–171f
    simulation of time development, 171–173f
  isotopomers of ozone, abundances and enhancements, 174t,175
  mechanism determination, 171–173f
  molecular dioxygens, simulation of time development, 170,171f
  possible pathways, 167,168f
  radical complex mechanism, 176
  temperature dependence, 175–176
Isotope substitution, photodissociation process effect, 10–12
Isotopomer(s), strategies for laser-stimulated selective reactions and synthesis, 310–331
Isotopomer-selective photodissociations, 322,324f,325

Isotopomer-selective unimolecular processes, 312–323

J

Jeans escape, D–H fractionation mechanism in solar system, 375
Jeans formula, equation, 375

K

Kinetic energy, effects on reactions involving isotopes, 226–243
Kinetic isotope effect(s)
  *Cl$^-$ + CD$_3$Cl → CD$_3$*Cl + Cl$^-$, 27,29–30t
  *Cl(D$_2$O)$^-$ + CH$_3$Cl → CH$_3$*Cl + Cl(D$_2$O)$^-$, 30–33
  conventional transition-theory state, 17–18
  factorization, 22
  gas-phase muonium reactions, 111–133
  generalized transition-state theory, 17
  H + CD$_3$H → H$_2$ + CD$_3$, 22–28
  high-temperature studies, flash photolysis shock tube technique, 80–91
  reaction mechanism, 226
Kinetic isotope effect measurement for atmospheric methane budget studies
  accuracy, 397
  carbon kinetic isotope effect in principal sink process, 399–401
  chemical system, 393–394
  conditions for successful experiment, 396
  error vs. fractional conversions, 394,395f
  impact of kinetic isotope effect value on methane budget modeling, 401–403t
  k_{12}/k_{13} from ab initio perturbation molecular orbital theory, 399
  k_{12}/k_{13} from field measurements, 399
  precision, 397,399
  primary kinetic isotope effect in CH$_4$ + OH reaction, 402,404t,405
  reaction system design, 394–399
  secondary kinetic isotope effect vs. temperature, 397,398f
Kleinmann Low Nebula, sources for polyatomic molecules, 365

INDEX

L

Laboratory ozone isotope measurements
  development, 159–160
  enhancement
    pressure dependence, 160,161f
    temperature dependence, 160,162f
Large-curvature ground-state
  approximation, development, 21
Laser-assisted separation and preparation
  of molecular isotopomers, 310–311
Laser photolysis shock tube technique, *See*
  Flash photolysis shock tube technique
Laser-stimulated selective reactions and
  synthesis of isotopomers
  isotopomer-selective photodissociations,
    322,324–330
  isotopomer-selective unimolecular
    processes, 312–323
  organometallic compounds, 311–312
  other techniques, 331
  selective isomerizations, Cp(CO)$_2$FePHD
    vs. Cp(CO)$_2$FePH$_2$, 317,319–323
  selective vibrational high-overtone
    excitation, OD vs. OH, 312–318
  strategies, 311
Least-action ground-state approximation,
  development, 21
Lunar materials, oxygen isotopic
  composition, 139,140f

M

$\mu + C_2H_4$
  Arrhenius behavior, 131,132f
  kinetic isotope effects, 131
  observed reaction rate, 130–131
  reaction, 130
  tunneling, 131
$\mu + H_2$
  H$_2$ vs. D$_2$ experimental results, 120–122f
  reaction conditions, 120–121
  reaction rates and Arrhenius parameters
    for μ and H, 121,122t
$\mu + HX$
  abstraction reactions, 127–128
  halogen effect on kinetic isotope data,
    128–130
  kinetic data for H and μ, 128,129t
$\mu + NO$, kinetic isotope effects, 133

$\mu + X_2$
  Arrhenius behavior comparisons among
    halogens, 125,127
  experimental vs. calculated reaction
    rates, 125,126f
  rate constants, activation energies, and
    kinetic isotope effects, 123,124t
  reactions, 123
  tunneling, 123,125,126f
Mass, dependency of isotope effects, 139
Mass-independent isotopic fractionations
  applications, 151–153
  non-ozone systems, 147–150
  O + CO, 147–149
  occurrence, 141,151–153
  ozone formation, 141–143
  state of knowledge, 150–151
  sulfur isotopes, 149–150
  symmetry effect, 143
  temperature effect, 143
  theoretical state-of-the-art, 143–147
Mechanism, isotopic study of ozone
  formation, 167–179
Metal carbonyls, spectroscopic probes of
  relaxation dynamics, 338–339
Meteoritic materials, oxygen isotopic
  composition, 139,140f
Methane
  concentrations in atmosphere, 391t
  importance, 390
  reaction with OH radicals, 390
  residence time in atmosphere, 391t
  sources and carbon signatures, 400t
  weighted averaging of methane sources
    and sinks using $^{13}C/^{12}C$ ratios, 391,392t
Methyl radical
  ab initio calculations on Rydberg 3 s
    state, 297–298
  photochemistry studies, 297
Methyl radical Rydberg 3 s predissociation
  dynamics
  CH$_3$ [1000]–[0000] level structure and
    dynamics, 307,308f
  CH$_2$D and CD$_2$H [0000] level structure and
    dynamics, 307–308
  dynamics of [0100] level of CD$_3$, 305–307
  experimental procedure, 299–300
  isotope effects on tunneling rates, 303–305
  Raman excitation profile of CD$_3$
    and CH$_3$, 300,302f

Methyl radical Rydberg 3 s predissociation dynamics—*Continued*
  resonance Raman S branch rotational structure of CD_3 and CH_3, 300,301f
  rotational quantum number specific predissociation rates for CH_3 and CD_3, 300,303t
  rotationally specific excited state lifetime determination, 298–299
Mixed behavior, reactions of atomic ions with H_2, D_2, and HD, 207,208f
Mode specificity, description, 264
Molecular modeling techniques
  chemical dynamics of gas-phase reactions, 17
  quantitative treatment of kinetic isotope effects, 16
Multiphoton ionization dynamics within $(CH_3OH)_n Cr(CO)_6$ van der Waals clusters
  dynamical scheme, 350–352
  intracluster energy transfer, 352
  model heterocluster systems, rationale for selection, 337–340
  photoion yields
    ionization at 248 nm, 341–345f
    ionization at 350 nm, 344,346f–348f
    mass-resolved resonance-enhanced multiphoton ionization at 346–377 nm, 347,349f,350
  time-of-flight MS of $Cr(CO)_6$-containing heteroclusters following multiphoton ionization, 340–341
Muon interactions, validity of Born–Oppenheimer approximation, 112
Muon spin rotation, 114–117
Muonium
  comparison of properties with those of protium, deuterium, and tritium, 112,113t
  form of hydrogen, 112
  reaction kinetics studies, 112–113
Muonium kinetic isotope effects
  decay of muonium, 114–115
  formation of muonium, 114
  importance of studies, 114
  $\mu + C_2H_4$, 130–132f
  $\mu + H_2(D_2)$, 120–122t,f
  $\mu + HX$, 127–130
  $\mu + NO$, 133
  $\mu + X_2$, 123–127

Muonium kinetic isotope effects—*Continued*
  muon spin rotation technique, 114–117
  relative energies, 117–118
  time constraints, 114
  transition-state theory, 118–120
  tunneling, 118
  velocity of reactive particle, 118

N

Nearly Maxwellian, description, 227
Ni–C_2D_4 and Ni–C_2H_4, vibrationally mediated dissociations, 325–328
NO_3 reactions with atmospheric reduced sulfur compounds, mechanism studies using deuterium substitution, 97–99
Nonequilibrium isotope effects
  complex-forming bimolecular and termolecular reactions, 7–9
  occurrence, 2
  simple bimolecular reactions, 5–7
Non-mass-dependent isotope effects in O_4^+ formation
  bimolecular energy transfer reactions, 214–215
  experimental procedure, 212
  heavy ozone production, 221–222
  ionization scheme effects, 221
  isotopically substituted O_4^+ ion signals vs. ionizing electron energy, intensity ratios, 213,214f
  O_2^+ and O_4^+ ion signal intensity vs. ionizing electron energy, 212,213f
  permutation–inversion group analysis, 215–217
  permutation–inversion symmetry analysis
    basic functions, 217–218
    electronic state correlation, 219–221
    Pauli-allowed wave functions, 218,219t
    states of separated diatoms, correlation with electronic states of collisions complex, 215f
Non-ozone systems, mass-independent isotopic fractionations, 147–150

O

$O^- + CH_2D_2$
  branching fraction vs. average kinetic energy, 240,241f

INDEX

O⁻ + CH$_2$D$_2$—*Continued*
  minimum-energy reaction pathway, 240,242*f*
  rate constant vs. energy, 240
  thermochemistry of O⁻ reactions with isotopically substituted CH$_4$, 240,243*t*
O$_4$⁺ production
  appearance potential, 210–211
  termolecular association rate determination, 211–212
  two-step mechanism, 210
O + C$_2$D$_2$ and O + C$_2$H$_2$ products, isotope effects on rate constants, 83,85*t*,86
O + CO, mass-independent isotopic fractionations, 147–149
O + H$_2$ → OH + H, isotope effects on rate constants, 85*t*,87–90*f*
O⁻ + H$_2$ (and D$_2$ and HD)
  branching products vs. average kinetic energy, 236,238*f*
  mechanism, 236,239
  rate constants vs. temperature, 236,237*f*
  reaction, 235–236
  rotational energy vs. temperature, 239
O⁺ + HD
  branching fraction of OH⁺ produced vs. average kinetic energy, 233,234*f*
  kinetic energy vs. temperature, 234
  rate constant vs. average kinetic energy, 233
  rate constant vs. temperature, 233
  reaction, 232–233
  reactivity vs. potential, 233–234
  rotational energy vs. temperature, 234
O⁻ + N$_2$O
  NO⁻ production, contribution by individual NO⁻ isotopes vs. temperature, 230–232
  oxygen exchange vs. temperature, 229–230
  rate constant vs. temperature, 229
  rotational energy vs. temperature, 232
Ocean measurements, sulfur-containing compounds, 418–420*f*
OD and OH
  selective vibrational high-overtone excitation, 312–318
  vibrationally mediated dissociation, 322,324*f*,325
OH and OD reactions with CH$_3$CN and CD$_3$CN, 101–104*f*
OH reactions with atmospheric reduced sulfur compounds, 95–99

Optical measurements, procedure, 159
Orbital angular momentum conservation, reactions of atomic ions with H$_2$, D$_2$, and HD, 196,198
Organometallic isotopomers, vibrationally mediated dissociations, 322,324–330
Oxygen isotopic composition of lunar, meteoritic, and terrestrial materials, 139,140*f*
Ozone
  dissociation energy, 156
  electronic states, 163–165*t*
  formation mechanism, 167–179
  formation reactions, 155
  heavy isotope enrichment, mechanism, 156–165
  isotope ratios, 156
  loss processes, 156
  oxygen isotopic composition, 145–147
  symmetry-selective isotope effects, 160,162*f*
Ozone formation, mass-independent fractionation, 139,141–143

P

Permutation–inversion symmetry group analysis, O$_4$⁺ formation, 215–221
Phase change, D–H fractionation mechanism in solar system, 375
Photochemical reactions, isotope enrichment mechanisms, 264–277
Photochemistry, D–H fractionation mechanism in solar system, 377
Photochemistry of molecule, excited electronic state effect, 297
Photodissociation
  hydrogen isotope effects, 74–78
  isotope substitution effect, 10–12
  small molecules, isotope effects, 3
Photoionization lasers, multiphoton ionization, 341
Photophysics, transition metal carbonyls, 337–338
Picosecond IR laser pulses with analytical shapes, induction of isotopomer-selective unimolecular processes, 312–323
Polyatomic molecules, photofragmentation maps, 269–272
Potential surface, HN$_2$, 38,39*t*,41*f*

Predissociation dynamics of methyl radical, Rydberg 3 s, isotopic independence, 297–308
Primary isotope effect, definition, 119–120
Protostar region, description, 359

Q

Quantum mechanical tunneling, transmission coefficient effects, 20–21

R

Rampsberger–Kassel–Marcus theory, 2
Rayleigh distillation, D–H fractionation mechanism in solar system, 376
Reactions involving isotopes
  average kinetic energy calculation, 227–228
  branching fraction measurement procedure, 227
  experimental materials and procedure, 226–229
  importance of studies, 226
  $O^- + CH_2D_2$, 239–243
  $O^- + H_2$ (and D_2 and HD), 235–239
  $O^- + N_2O$, 229–232
  $O^+ + HD$, 232–235
  selected ion flow tube, 226–227
Reagent excitation, isotope-specific photodissociation effect, 11
Real-time isotopic distribution, measurements for sulfur-containing compounds using GC–MS–ILS, 421
Recrossing, handling by variational transition-state theory, 20
Reduced sulfur compounds
  reactions with Cl, 99–101
  reactions with NO_3, 97–99
  reactions with OH, 95–99
Relative energies, muonium kinetic isotope effects, 117–118
Resonance Raman excitation profiles, determination of rotationally specific exited-state lifetimes, 298–299
Resonance Raman spectroscopy, structure and dynamics probe of excited electronic states, 298
Rotational energy, reactions of atomic ions H_2, D_2, and HD, 198
Rotational line, intensity, 285
Rotational temperature, reactions involving isotopes, 226–243
Rydberg atoms, 181–182
Rydberg electron transfer
  free electron model, 182–183
  properties of Rydberg atoms, 183t
  rate constants for free electron attachment, 182–183
Rydberg 3 s predissociation dynamics of methyl radical, isotopic independence, 297–308

S

Scattering calculations, HN_2 and DN_2, 39–40
Secondary deuterium kinetic isotope effects, variational transition-state theory in reactions involving methane and chloromethane, 22–33
Secondary isotope effect, definition, 120
Selected ion flow tube
  experimental procedure, 226–227
  reactions involving isotopes, 226
Selective isomerizations, $Cp(CO)_2FePHD$ vs. $Cp(CO)_2FePH_2$, 317,319–323
Selective vibrational high-overtone excitation, OD vs. OH, 312–318
Simple bimolecular reactions, nonequilibrium isotope effects, 5–7
Small solar system bodies, observations of deuterium, 373–374
Solar system
  deuterium, 369–387
  formation pathways, 378,379f
Spectroscopic probes of relaxation dynamics, metal carbonyls, 338–339
Stabilization calculations, HN_2 and DN_2, 40
Stable isotope ratio measurements
  knowledge requirements, 138
  use as diagnostic probe, 138
Star(s)
  deuterium fractionation, 365
  formation, 365
  fractionation, 366
  surface chemistry, 365
Star formation region, description, 359
State-selected dissociation of $trans$-HONO(Ã) and -DONO(Ã)
  Ã-state potential energy surface, 280–282
  advantages of system for study, 279

State-selected dissociation of
  trans-HONO(Ã) and -DONO(Ã)—
  Continued
  alignment of OD, 285–288f
  alignment of OH, 285,289f
  alignment of OH (OD) fragment's
    half-filled 2pπ orbital, 290,292f
  experimental procedure and setup, 283,286f
  fragmentation dynamics, 293,295
  future research, 295–296
  photofragment rotational alignment,
    283,285,286f
  photofragmentation recoil velocity–
    rotational angular momentum
    correlation, 290–294
  polarization effects, 285,287f
  polarization vs. fragment Doppler profiles,
    291,293,294f
  product polarization vs. recoil direction,
    291,292f
  qualitative difference between Ã-state
    potential energy surfaces of DONO
    and HONO, 282,284f
  rotational energy distribution, 285,290
  vector analysis parameters, 293,294t
  vectorial parameters, 283,284f
  vibrational modes of HONO, 280t
Statistical behavior, reactions of atomic
  ions with H_2, D_2, and HD, 199–203
Stratospheric ozone isotope measurements
  in situ MS measurements, 157,158f
  optical measurements, 157,159t
  ozone sample collection, 157,159t
Stratospheric ozone layer, research, 155
Sulfur-containing compounds, determination
  in atmosphere and ocean using
  GC–MS–ILS, 409–421
Sulfur isotopes, mass-independent isotopic
  fractionations, 149–150
Surface catalysis, D–H fractionation
  mechanism in solar system, 377
Surface chemistry, stars, 365
Symmetry, isotope effects on CS_2
  formation, 189–191
Symmetry-related isotope effects
  discovery, 9
  formation of O_3 from O and O_2, 9
  mass independence, 9
  occurrence, 2–3
  origin, 2

Symmetry-related isotope effects—*Continued*
  ozone, pathways, 160,162f
  reactions, 9–10

T

Taurus molecular cloud 1, abundance
  ratios, 364t
Temperature, effects on reactions
  involving isotopes, 226–243
Termolecular association rate for O_4^+
  production, determination, 211–212
Termolecular reactions, nonequilibrium
  isotope effects, 7–9
Terrestrial materials, oxygen isotopic
  composition, 139,140f
Terrestrial planets
  chemical evolution, 380,385
  hydrogen contents, 382t
  observations of deuterium, 370,372
Thermochemical equilibrium, D–H
  fractionation mechanism in solar
  system, 376
Time-of-flight mass spectrometry of
  $Cr(CO)_6$-containing heteroclusters
  following multiphoton ionization
  experimental apparatus, 340–341
  photoionization lasers, 341
Transition metal carbonyls, photophysics,
  337–338
Transition-state theory
  description, 2
  muonium kinetic isotope effects,
    118–120
Translational energy, reactions of atomic
  ions with H_2, D_2, and HD, 196,197f
Tunneling, muonium kinetic isotope
  effects, 118

V

van der Waals clusters containing $M(CO)_6$,
  multiphoton ionization, 339–340
van der Waals cluster research
  development of models of macroscopic
    molecule dynamics, 336
  energy disposal processes, 336–337
  relaxation processes, 337

Variational transition-state theory for kinetic isotope effects
*Cl⁻ + CD$_3$Cl → CD$_3$*Cl + Cl⁻, 27,29–30t
*Cl(D$_2$O)⁻ + CH$_3$Cl → CH$_3$*Cl + Cl(D$_2$O)⁻, 30–33
comparison to quantum dynamics, 20
factorization of kinetic isotope effect, 22
H + CD$_3$H → H$_2$ + CD$_3$, 22–28
practical implementation, 19
quantum mechanical tunneling effects on transmission coefficient, 20–21
rate expression, 19
recrossing, 20
Velocity of reactive particle, muonium kinetic isotope effects, 118
Vibrational adiabaticity, description, 21

Vibrationally mediated dissociations
HCo(CO)$_4$ vs. DCo(CO)$_4$, 328–330
Ni–C$_2$H$_4$ vs. Ni–C$_2$D$_4$, 325–328
OH vs. OD, 322,324f,325
organometallic isotopomers, 322,324–330
Vibrational states, reactivity effects, 279–280

W

Water–acetonitrile cluster ions, isotope exchange reactions, 259f
Water–ammonia cluster ions, isotope exchange reactions, 256–258f
Water clusters, isotope fractionation, 247–251

Z

Zero point energy effects, reactions of atomic ions with H$_2$, D$_2$, and HD, 198–200f

Production: Betsy Kulamer
Indexing: Deborah H. Steiner
Acquisition: Barbara C. Tansill
Cover design: Amy Meyer Phifer

Printed and bound by Maple Press, York, PA